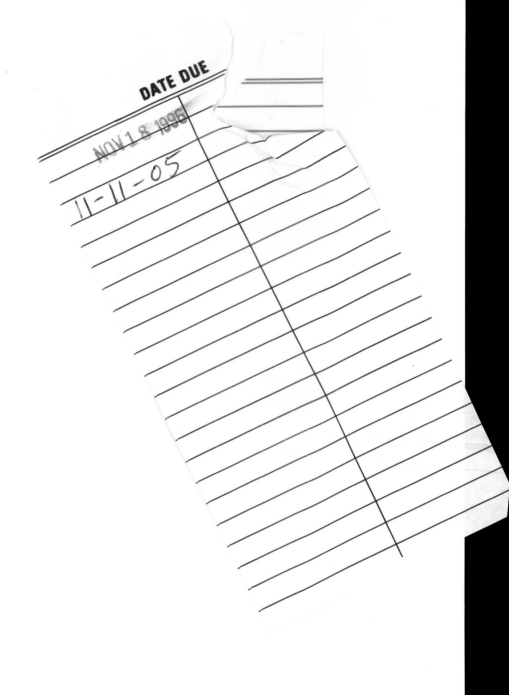

TECHNIQUES OF CHEMISTRY

VOLUME XXIII

LASER TECHNIQUES IN CHEMISTRY

Edited by

ANNE B. MYERS

University of Rochester
Rochester, New York

and

THOMAS R. RIZZO

École Polytechnique Federale de Lausanne
Lausanne, Switzerland

A WILEY-INTERSCIENCE PUBLICATION
JOHN WILEY & SONS, INC.

New York · Chichester · Brisbane · Toronto · Singapore

This text is printed on acid-free paper.

Library of Congress Cataloging in Publication Data:
ISBN 0-471-59769-4
ISSN 0082-2531

Printed in the United States of America

10 9 8 7 6 5 4 3 2 1

CONTRIBUTORS

G. A. BETHARDY, *Department of Chemistry, University of Akron, Akron, Ohio*

MICHAEL D. FAYER, *Department of Chemistry, Stanford University, Stanford, California*

PETER M. FELKER, *Department of Chemistry, University of California, Los Angeles, California*

JOHN HEPBURN, *Department of Chemistry, University of Waterloo, Waterloo, Ontario, Canada*

ROBIN M. HOCHSTRASSER, *Department of Chemistry, University of Pennsylvania, Philadelphia, Pennsylvania*

BRYAN E. KOHLER, *Department of Chemistry, University of California, Riverside, California*

W. A. MAJEWSKI, *Department of Chemistry, University of Pittsburgh, Pittsburgh, Pennsylvania*

RICHARD A. MATHIES, *Department of Chemistry, University of California, Berkeley, California*

ROGER E. MILLER, *Department of Chemistry, University of North Carolina, Chapel Hill, North Carolina*

ANNE B. MYERS, *Department of Chemistry, University of Rochester, Rochester, New York*

DAVID S. PERRY, *Department of Chemistry, University of Akron, Akron, Ohio*

ROMAN I. PERSONOV, *Institute of Spectroscopy, Academy of Sciences of Russia, Moscow, Russia*

J. F. PFANSTIEL, *Department of Chemistry, University of Pittsburgh, Pittsburgh, Pennsylvania*

D. F. PLUSQUELLIC, *Department of Chemistry, University of Pittsburgh, Pittsburgh, Pennsylvania*

DAVID W. PRATT, *Department of Chemistry, University of Pittsburgh, Pittsburgh, Pennsylvania*

ABHIJIT SENGUPTA, *Department of Chemistry, Stanford University, Stanford, California*

PEGGY A. THOMPSON, *Department of Chemistry, University of California, Berkeley, California*

GILBERT C. WALKER, *Department of Chemistry, University of Pittsburgh, Pittsburgh, Pennsylvania*

JÖRG C. WOEHL, *Department of Chemistry, University of California, Riverside, California*

INTRODUCTION TO THE SERIES

Techniques of Chemistry is the successor to Technique of Organic Chemistry and its companion, Technique of Inorganic Chemistry. The newer series reflects the fact that many modern techniques are applicable over a wide area of chemical science. All of these were originated by Arnold Weissberger and edited by him for many years.

Following in Dr. Weissberger's footsteps is no easy task, but every effort will be made to uphold the high standards he set. The aim remains the same: the comprehensive presentation of important techniques. At the same time, authors will be encouraged to illustrate what can be done with a technique rather than cataloging all known applications. It is hoped in this way to keep individual volumes to a reasonable size. Readers can help with advice and comments. Suggestions of topics for new volumes will be particularly welcome.

WILLIAM H. SAUNDERS, JR.

Department of Chemistry
University of Rochester
Rochester, New York

PREFACE

Thirty years ago, lasers existed mainly as *subjects* of research in the laboratories of a few physicists and engineers. Today they have become indispensible *tools* for research in chemistry, physics, biology, and a variety of other fields. In chemistry, lasers have had their greatest impact in spectroscopic applications, where the unique properties of laser radiation have both reinvigorated old spectroscopies and stimulated the invention of many new ones. In many cases, lasers have opened areas of chemical research that were previously unthinkable.

In this volume, 10 different laser spectroscopic techniques are described by one or more of their leading practitioners. We have made no attempt to cover "laser techniques in chemistry" comprehensively; this topic is far too broad for a single volume. Rather, we have chosen to focus on those techniques that are sufficiently established to be beyond the proof-of-principle stage, but have not yet become routine. Each chapter describes how the particular spectroscopy is carried out experimentally and/or interpretively, the types of systems to which it is applicable, and the type of information that can be learned from its application. Also included are some speculations about possible further developments and extensions of the techniques that one might expect to see in the near future.

The volume is targeted toward graduate students, postdoctorals, and senior scientists who are familiar with one or more types of optical spectroscopy and wish to learn about others. It is written in a sufficiently pedagogical style so as to allow someone uninitiated in a particular approach to get a start at applying it in the laboratory. We hope that our readers will come away with an increased understanding of the power and variety of laser spectroscopic techniques, an eagerness to apply some of these techniques to new problems, and, perhaps, ideas for even more novel and powerful techniques that might be developed in the future.

ANNE B. MYERS
THOMAS R. RIZZO

Rochester, New York
Lausanne, Switzerland
May, 1995

CONTENTS

TECHNIQUES OF CHEMISTRY

ARNOLD WEISSBERGER, *Founding Editor*
WILLIAM H. SAUNDERS, JR., *Editor*

VOLUME I
PHYSICAL METHODS OF
CHEMISTRY, in Five Parts
(INCORPORATING FOURTH COMPLETELY
REVISED AND AUGMENTED EDITIONS OF
PHYSICAL METHODS OF ORGANIC
CHEMISTRY)
*Edited by Arnold Weissberger and
Bryant W. Rossiter*

VOLUME II
ORGANIC SOLVENTS, Third Edition
*John A. Riddick, William B. Bunger,
and Theodore K. Sakano*

VOLUME III
PHOTOCHROMISM
Edited by Glenn H. Brown

VOLUME IV
ELUCIDATION OF ORGANIC
STRUCTURES BY PHYSICAL AND
CHEMICAL METHODS, Second
Edition, in Three Parts
Edited by K. W. Bentley and G. W. Kirby

VOLUME V
TECHNIQUE OF
ELECTROORGANIC SYNTHESIS,
in Three Parts
Parts I and II: *Edited by Norman L.
Weinberg*
Part III: *Edited by Norman L. Weinberg
and B. V. Tilak*

VOLUME VI
INVESTIGATIONS OF RATES AND
MECHANISMS OF REACTIONS,
Fourth Edition, in Two Parts
Edited by Claude F. Bernasconi

VOLUME VII
MEMBRANES IN SEPARATIONS
*Sun-Tok Hwang
and Karl Kammermeyer*

VOLUME VIII
SOLUTIONS AND SOLUBILITIES,
in Two Parts
Edited by Michael R. Duck

VOLUME IX
CHEMICAL EXPERIMENTATION
UNDER EXTREME CONDITIONS
Edited by Bryant W. Rossiter

VOLUME X
APPLICATIONS OF BIOCHEMICAL
SYSTEMS IN ORGANIC
CHEMISTRY, in Two Parts
*Edited by J. Bryan Jones, Charles J. Sih,
and D. Perlman*

VOLUME XI
CONTEMPORARY LIQUID
CHROMATOGRAPHY
R. P. W. Scott

VOLUME XII
SEPARATION AND
PURIFICATION,
Third Edition
*Edited by Edmond S. Perry and Arnold
Weissberger*

VOLUME XIII
LABORATORY ENGINEERING
AND MANIPULATIONS,
Third Edition
*Edited by Edmond S. Perry and Arnold
Weissberger*

VOLUME XIV
THIN-LAYER
CHROMATOGRAPHY,
Second Edition
Justus G. Kirchner

VOLUME XV
THEORY AND APPLICATIONS OF
ELECTRON SPIN RESONANCE
Walter Gordy

VOLUME XVI
SEPARATIONS BY CENTRIFUGAL
PHENOMENA
Hsien-Wen Hsu

VOLUME XVII
APPLICATIONS OF LASERS TO
CHEMICAL PROBLEMS
Edited by Ted R. Evans

TECHNIQUES OF CHEMISTRY

WILLIAM H. SAUNDERS, JR., *Series Editor*
ARNOLD WEISSBERGER, *Founding Editor*

VOLUME XXIII

LASER TECHNIQUES IN CHEMISTRY

Chapter **I**

FOURIER-TRANSFORM NONLINEAR SPECTROSCOPIES

Peter M. Felker
Department of Chemistry and Biochemistry
University of California, Los Angeles, California

Laser Techniques In Chemistry, Edited by Anne B. Myers and Thomas R. Rizzo.
Techniques of Chemistry Series, Vol. XXIII.
ISBN 0-471-59769-4   © 1995 John Wiley & Sons, Inc.

1.1.   INTRODUCTION

Fourier-transform (FT) spectroscopies (1) have proved to be very powerful methods in the study of chemical systems. The most prominent example, of course, is FT–infrared (IR) spectroscopy (2), which has developed into an indispensible tool for chemical analysis. Other techniques are also becoming more and more important. These include FT versions of optical absorption and emission (3), Raman spectroscopy (4), as well as methodological advances that have greatly facilitated the study of pulse-generated transient species (e.g., photolysis products, radicals, and ions) by FT techniques (5). These methods all have one thing in common— they are FT implementations of *linear* spectroscopies. An interferometer is used as the spectrum analyzer of photons involved in single-photon transitions.

In recent years, the development of FT versions of *nonlinear* spectroscopies has also been an area of active research (6–11). In these schemes an interferometer is used as the spectrum analyzer of the resonant *difference frequencies* characterizing two-photon resonant processes. The two-photon processes involved in these schemes are of two types. There are those that rely on stimulated Raman transitions [Fig. 1.1(a)] and those whose two-photon processes are the consequence of two single-photon resonant transitions in sequence [Fig. 1.1(b) and (c)]. In either case, the ω_1 and ω_2 excitation fields involved in the process pass through a Michelson interferometer prior to interacting with the sample (Fig. 1.2). The output of the interferometer then impinges on the sample, and an observable dependent on the pertinent two-photon process is measured as a function of interferometer delay. The

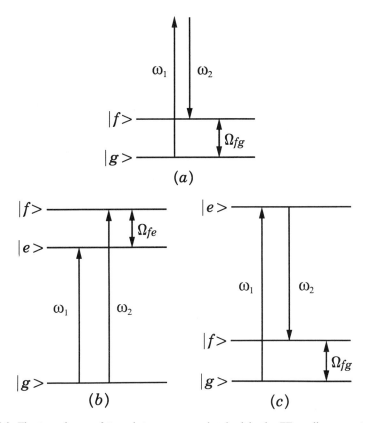

Figure 1.1. The two classes of two-photon processes involved in the FT nonlinear spectroscopies discussed in this chapter. (*a*) Stimulated Raman scattering. A two-color pulse (ω_1, ω_2) drives a rovibra-tional transition from $|g\rangle$ to $|f\rangle$, characterized by angular frequency Ω_{fg}. (*b*) and (*c*) Sequential one-photon resonant processes. The level diagram in (*b*) corresponds to ground-state hole burning. The ω_1 field depletes the population of $|g\rangle$ and the ω_2 field samples that depletion. The two-photon resonance frequency Ω_{fe} corresponds to the splitting between the excited states involved in the process. The level diagram in (*c*) corresponds to stimulated emission spectroscopy. The ω_1 field puts population into $|e\rangle$. The ω_2 field depletes that population by stimulated emission into ground-state $|f\rangle$. The two-photon resonance is at frequency Ω_{fg}.

"interferogram" that results from such an experiment is modulated by terms whose frequencies are resonances of the sample. In the case of the stimulated Raman schemes, the modulation frequencies are those ground-state rovibrational intervals associated with the energy differences between the initial and final states involved in the Raman transitions [Ω_{fg} in Fig. 1.1(*a*)]. Fourier transformation of such an interferogram yields a portion of the stimulated Raman spectrum of the sample. In the second kind of process the modulation frequencies in the interferogram correspond to the energy differences between those pairs of excited-state rovibronic levels connected to a single ground-state level [Ω_{fe} in Fig. 1.1(*b*)] or those pairs of ground-state levels connected to a single excited-state level [Ω_{fg} in Fig. 1.1(*c*)].

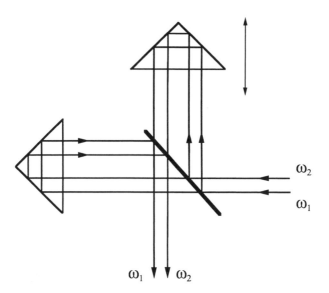

Figure 1.2. The interferometer arrangement employed for all of the FT nonlinear spectroscopies implemented in this laboratory. The ω_1 and ω_2 fields propagate through the interferometer parallel to one another (or colinear), such that they both experience the same delay. At the output the ω_1 fields from the two arms are combined colinearly, as are the ω_2 fields.

Fourier transformation of this type of interferogram produces a spectrum of these excited- or ground-state rovibronic frequencies.

Fourier transform nonlinear spectroscopies have several features that make them useful as spectroscopic tools. First, the spectral resolution available with the methods is independent of the properties of the light source employed. Instead, resolution depends on the range over which the interferometer is scanned in an experiment. This feature permits one to use convenient, high-power, short-pulsed laser sources to drive the relevant nonlinear processes without requiring any compromise on spectral resolution. Second, the frequency scale in interferometric FT spectroscopies can be made to be very accurate without a great deal of effort (1). This is because that scale is determined by the interferometer scan range, a distance that can be measured easily and accurately. Thus, when one is interested in absolute transition frequencies and/or small shifts in such frequencies from one species to another, a FT experiment is very well suited for obtaining the desired information. Third, the information content of the FT nonlinear spectroscopies that rely on sequential one-photon resonant transitions is different from that of their frequency-domain counterparts (11). When implemented in the frequency domain the spectroscopies measure single-photon vibronic spectra of labeled species. In contrast, the FT versions of these spectroscopies give the two-photon resonances directly. The upshot is that characteristics like rotational structure and Doppler broadening are not the same in the FT and frequency-domain versions. Depending on the experiment at hand and the information desired, these differences can render the

FT version the better alternative. Finally, in some circumstances there is information that is more readily obtained directly from an interferogram than from a spectrum. In particular, the spectroscopic manifestation of free rotation in a very large species is a large number of resonances in the frequency domain, whereas in the Fourier-conjugate domain (the interferogram) rotation is manifest as a small number of time localized, equally spaced transients (12) whose positions depend directly on the species' rotational constants. In these cases the measurement of such transients and their interpretation can be considerably easier than obtaining frequency-domain results of suitable quality and interpreting such results.

Our laboratory has developed a number of different methods of FT nonlinear spectroscopies. These are FT methods based on coherent Raman scattering (CRS) (6,10), ionization-detected stimulated Raman spectroscopy (IDSRS) (8,10), stimulated emission spectroscopy (SES), and hole-burning spectroscopy (HBS) (7,11). We have also demonstrated a variant of such schemes (down-shifting), which shifts high-frequency modulations in interferograms down to much lower frequency and thus facilitates their measurement (9). This chapter intends to provide a review of these techniques and the results obtained from them. The following section starts with a brief historical account of work relevant to this area. Then, Section 1.3 presents a pedagogical treatment to show how the two-photon resonances produce modulations in the interferograms whose frequencies are just the resonant difference frequencies. Section 1.4 outlines each of the methods, their information content, and their implementation in our laboratory. Following this, Section 1.5 presents representative results from these methods and discusses some of their implications. Finally, Section 1.6 closes with a brief discussion of the future in regard to the application of FT nonlinear spectroscopies.

1.2. HISTORICAL PERSPECTIVE

The first experiments involving the use of long pulses of incoherent light to achieve subpicosecond time resolution in dynamical studies were reported in the mid-1980s by several groups (13–17). Each of these experiments was an interferometric implementation of a resonant four-wave mixing scheme. The experimental arrangements are represented schematically by the diagram in Fig. 1.3. The output of a broadband laser source was split into two by a beamsplitter. One of the two parts was optically delayed with respect to the other. The two parallel beams were then focused at an angle into the sample. The intensity of a coherent beam generated by resonant degenerate, or nearly degenerate four-wave mixing in the sample was then detected as a function of the optical delay. The detected beam was the one propagating with wavevector $2\mathbf{k}_2 - \mathbf{k}_1$, where $\mathbf{k}_1$ represents the wavevector of the undelayed beam and $\mathbf{k}_2$ that of the delayed beam.

The results of these experiments displayed two notable features. First, transients evolving on timescales much faster than the pulse width of the light source used were observed (13,14,16). Second, when more than a single transition was spanned by the bandwidth of the excitation source, modulations having a frequency match-

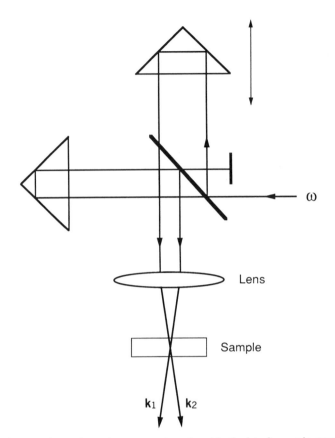

Figure 1.3. The general experimental arrangement employed in the interferometric spectroscopies of Refs. 13–17. The broadband field (ω) used to drive the resonant, degenerate (or near-degenerate) four-wave mixing process is split into two parts by a beam splitter. One part is optically delayed with respect to the other. The two, noncolinear beams are focused into the sample, where they intersect with different wavevectors ($\mathbf{k}_1$ and $\mathbf{k}_2$) and generate the four-wave mixing signal.

ing the frequency difference between the transitions appeared in the observable, despite the fact that the modulation period was much shorter than the excitation source pulse width (15,17). These results were explained (14,15) by noting that the pulses produced by an incoherent, or partially incoherent light source are actually comprised of a randomly phased series of very short, subpulses. Splitting the output of such a light source in two produces two pulses having the same pulse substructure—that is, two correlated series of ultrafast noise spikes. When a four-wave mixing process is driven by these two pulses in sequence, the pairs of correlated noise spikes between the pulses build up an accumulated photon echo, which is just the four-wave mixing signal in the experiment. The accumulated echo amplitude depends on the evolution of the sample during the time between correlated spikes. Thus, the four-wave mixing signal versus pulse delay can reflect sample dynamics on a timescale that is limited only by the duration of the noise spikes, not the overall pulse width of the light source.

In 1986 DeBeer et al. (18) reported an important extension to the experiments described above. In a near-degenerate four-wave mixing experiment on the D lines in Na, they used two, narrow-band dye-laser sources (5-GHz bandwidth, 7-ns pulses) to excite the two D-line transitions, respectively. The two laser beams were combined colinearly. Then the resulting two-color pulse was split in two. Next, the experiment proceeded in the same way as above: The undelayed ($\mathbf{k}_1$) and delayed ($\mathbf{k}_2$) pulses were focused at an angle into the sample and the beam generated at $2\mathbf{k}_2 - \mathbf{k}_1$ was detected as a function of delay. The interesting result was the observation of well-modulated beats having a period of 1.9-ps, beats corresponding to the energy splitting between the two upper states associated with the D-line transitions. Notably, these beats were not only much faster than the laser pulses, they were also much faster than the evolution of the noise substructure within those pulses. These experiments showed that very fast modulations reflecting energy differences between the states of a species (Bohr-frequency beats) could be observed in a nonlinear interferometric experiment employing two *narrow-band* light sources. Neither short nor very broadband pulses were required. This "ultrafast modulation spectroscopy" (UMS) (18) suggested a new, powerful approach for characterizing otherwise difficult-to-characterize energy splittings.

The work of DeBeer et al. (18) stimulated us to try to extend the interferometric UMS approach to four-wave mixing processes other than fully resonant ones. We subsequently demonstrated an interferometric version of coherent Raman scattering. This Fourier transform coherent Raman scattering (FTCRS) (6) has one procedural difference from UMS, aside from the fact that different four-wave mixing processes are involved in each. In FTCRS, just as in UMS, the outputs of two lasers are both split in two, and one split pair is delayed with respect to the other split pair. However, in FTCRS the two pulses of a given color from the two arms of the interferometer are recombined colinearly (see Fig. 1.2), unlike in UMS.

The analysis of FTCRS showed us that a fruitful way of viewing interferometric versions of nonlinear spectroscopies was as FT techniques analogous to linear methods (1) like FT–IR spectroscopy. This analysis also suggested that *any nonlinear spectroscopy involving a resonance condition of the form* $\omega_1 - \omega_2 \simeq \Omega_{ij}$ should be susceptible to implementation interferometrically as a FT spectroscopy. The general recipe for such implementation is to direct the ω_1 and ω_2 light beams parallel (or colinearly) through a Michelson interferometer, colinearly recombine the two ω_1 beams from the two interferometer arms and the two ω_2 beams from the two arms, direct the output of the interferometer to the sample, along with any other probe beams required, and measure the pertinent signal as a function of interferometer delay. In doing such an experiment one will obtain a signal versus delay trace (interferogram) that is modulated by the resonant difference frequencies Ω_{ij}. Fourier transformation produces the desired spectrum. Realizing this, we then conceived of and demonstrated the other FT nonlinear spectroscopies that are the subject of this chapter.

Finally, in 1988 a different class of interferometric four-wave mixing methods was reported by Dugan et al. (19,20). These methods give rise to interesting types of detuning oscillations, phenomena different from the Bohr frequency oscillations mentioned above. In the experiments the first (or second) and third wave (ω_2 and ω_2') in a coherent Raman process (see Fig. 1.4) are derived, respectively, from the two arms of the interferometer. The source for these two waves is a broadband dye oscillator. The source for the second (or first) wave (ω_1) is a narrow-band dye laser. The coherent signal beam is spectrally analyzed, and the observed interferograms depend on the detected frequency (ω_s or ω_{as}), the bandwidth of detection, and the bandwidth of the ω_1 wave. When only a single Raman band at frequency Ω_{ij} is involved, the interferograms so obtained are modulated at $\Delta_{ij} = 2\Omega_{ij} + \omega_s - \omega_1$ (for the Stokes scheme). When more than one band is involved, a sum of such

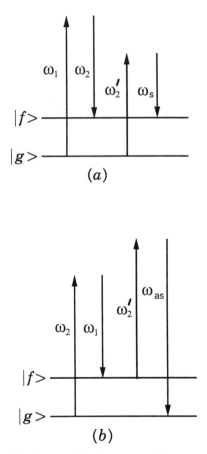

Figure 1.4. Level diagrams depicting two of the four-wave mixing processes involved in the interferometric schemes of Refs. 19 and 20. In both diagrams ω_1 represents a narrow-band field, and ω_2 and ω_2' are derived from the same broadband dye oscillator, with ω_2' optically delayed relative to ω_2. (a) Coherent Stokes Raman scattering. The signal field (ω_s) is subject to spectral filtering. (b) Coherent anti-Stokes Raman scattering. Again, the signal field (ω_{as}) is subject to spectral filtering.

modulation terms contributes to the interferogram. Data from such experiments have been used to measure vibrational frequency shifts and dephasing times in condensed-phase samples precisely. Dugan et al. (20) also theoretically characterized and reported the appearance of submaterial line width features in the spectrum of ω_s, when that wave is generated from ω_1, ω_2, and ω_2' at a fixed interferometer delay. A variety of other interesting phenomena has been predicted for this type of nonlinear interferometric scheme. For the remainder of this chapter, however, we focus only on those methods that have been developed in this laboratory.

1.3. GENERAL PHYSICAL PICTURE

To understand the basics of how FT nonlinear spectroscopies work, it is useful to analyze and compare simplified model systems representing the schemes. We start with a simple treatment of FT linear spectroscopies because the linear case provides a good point of departure for understanding the nonlinear case. We then consider the nonlinear schemes represented by Fig. 1.1(a) and by Fig. 1.1(b and c), respectively. We emphasize that the approach we take is one adopted for pedagogical purposes (21). Outlines of more rigorous treatments (10,11) are given in Section 1.4.

1.3.1. FT Linear Spectroscopy

Consider a one-photon absorption experiment on a sample of a species that has a single ground-state level $|g\rangle$ and a single excited-state level $|e\rangle$, and an infinitely sharp absorption resonance corresponding to the $|e\rangle \leftarrow |g\rangle$ transition (Fig. 1.5). Assume further that there is a way of detecting the rate of absorption processes occurring (e.g., detection of total fluorescence from $|e\rangle$) and that the detection process produces a signal I that is proportional to this rate. In a frequency-domain experiment designed to characterize the absorption spectrum of the sample one

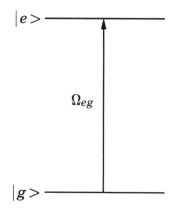

Figure 1.5. Level diagram depicting a simple linear spectroscopic process—absorption of light at angular frequency Ω_{eg} inducing a transition from $|g\rangle$ to $|e\rangle$.

scans the frequency of a narrow-band light source and measures I as a function of the frequency. The result is the absorption spectrum measured at a resolution given by the bandwidth of the light source employed. In a FT version of the same experiment the light source is first directed through a Michelson interferometer prior to the sample. The output of the interferometer then impinges on the sample, and I is measured as a function of the delay (τ) between the two arms of the interferometer to yield an interferogram − $I(\tau)$.

The observable $I(\tau)$ can be readily determined by noting two points. First, the probability of absorption, and thus the magnitude of I, is proportional to an integral over frequency of the spectral density of the light [$S(\omega, \tau)$] times the lineshape function for the absorption transition. If one takes the latter to be a delta function centered at Ω_{eg}, one has

$$I(\tau) \sim \int_{-\infty}^{\infty} S(\omega, \tau)\delta(\omega - \Omega_{eg})d\omega = S(\Omega_{eg}, \tau) \tag{1.1}$$

Second, as our notation suggests, the spectral density of the light source after the interferometer depends on τ as well as on ω. Assuming a balanced interferometer

$$S(\omega, \tau) \sim s(\omega)(1 + \cos[\omega\tau]) \tag{1.2}$$

where $s(\omega)$ is the spectral density of the light source prior to the interferometer (22). Combining Eqs. (1.1) and (1.2) one sees that

$$I(\tau) \sim s(\Omega_{eg})(1 + \cos[\Omega_{eg}\tau]) \tag{1.3}$$

The interferogram in this linear FT experiment is modulated at the absorption resonance frequency of the sample. Upon Fourier transformation of $I(\tau)$ one obtains a spectrum that is given by

$$I(\omega) = \int_{-\infty}^{\infty} I(\tau)e^{-i\omega\tau}d\tau \sim s(\Omega_{eg})\left[\delta(\omega) + \left(\frac{1}{2}\right)\delta(|\omega| - \Omega_{eg})\right] \tag{1.4}$$

where we have assumed that the range of τ covered in the experiment is infinite. One sees that this spectrum, aside from the zero-frequency peak, is just the absorption spectrum of the sample multiplied by an apparatus function—namely, the spectral density of the light source at frequency Ω_{eg}. For more realistic ranges of τ (i.e., finite ranges) the resolution in the Fourier spectrum is reduced to $\Delta\nu = 1/\tau$.

By an analysis like that above it is straightforward to show that in cases where the absorption spectrum of the sample $A(\omega)$ has more than a single absorption resonance

$$A(\omega) = \sum_{i} \alpha_i \delta(\omega - \Omega_i) \tag{1.5}$$

$I(\tau)$ is given by a sum of modulation terms whose frequencies are the absorption frequencies of the sample.

$$I(\tau) \sim \sum_i s(\Omega_i)\alpha_i[1 + \cos(\Omega_i)\tau] \tag{1.6}$$

Thus, in this more general case the FT of $I(\tau)$, apart from a zero-frequency peak, also gives the absorption spectrum of the sample, multiplied by an apparatus function that reflects the spectral profile of the light source.

$$I(\omega) \sim \sum_i s(\Omega_i)\alpha_i\left[\delta(\omega) + \frac{1}{2}\delta(|\omega| - \Omega_i)\right] \tag{1.7}$$

In summary, the frequency-dependent modulation of the light-source spectral density by the interferometer in a linear FT experiment allows one to measure the spectrum associated with the pertinent linear spectroscopic process. The spectrum measured covers a range determined by the bandwidth of the light source employed. The resolution of the spectrum is independent of the light source bandwidth and is determined, instead, by the range of τ scanned in the experiment.

1.3.2.   FT Nonlinear Raman Spectroscopy

Now consider an experiment designed to characterize a nonlinear (two-photon) resonance of a simple species. In particular, suppose that the species has two eigenstates, and one wishes to measure the stimulated Raman spectrum associated with transitions between these two states [Fig. 1.1(*a*)]. Assume that the Raman spectrum is characterized by a delta function at $\omega_1 - \omega_2 = \Omega_{fg}$. Assume also that there is a means by which to probe the rate of Raman transitions and that this probe produces a signal I that is proportional to that rate. In a frequency-domain approach to characterizing the stimulated Raman spectrum one would employ two light sources and measure I as a function of the frequency difference between these light sources. In the FT approach to this nonlinear spectroscopic experiment one sends the outputs of two light sources colinearly through a Michelson interferometer, directs the output of the interferometer to the sample, and then measures I as a function of interferometer delay. In this case the interferogram that results is given by

$$I(\tau) \sim \int_{-\infty}^{\infty}\int_{-\infty}^{\infty} S_1(\omega_1, \tau)S_2(\omega_2, \tau)\delta(\omega_1 - \omega_2 - \Omega_{fg})d\omega_1 d\omega_2$$

$$= \int_{-\infty}^{\infty} S_1(\omega_1, \tau)S_2([\omega_1 - \Omega_{fg}], \tau)d\omega_1 \tag{1.8}$$

where $S_1(\omega, \tau)$ is the spectral density of light source No. 1 after the interferometer and $S_2(\omega, \tau)$ is the same quantity associated with light source No. 2. Note that Eq. (1.8) is just a straightforward extension of Eq. (1.1). The key points of difference are that the nonlinear signal goes as the product of two spectral densities rather than being proportional to just one and that the resonance condition involves the *difference* between two photon frequencies rather than the frequencies themselves.

If one now applies Eq. (1.2) for the spectral densities of light sources 1 and 2 after the interferometer (22) and uses these expressions in Eq. (1.8), one obtains

$$I(\tau) \sim \int_{-\infty}^{\infty} (1 + \cos \omega_1\tau)(1 + \cos(\omega_1 - \Omega_{fg})\tau)s_1(\omega_1)s_2(\omega_1 - \Omega_{fg})d\omega_1 \quad (1.9)$$

which gives

$$I(\tau) \sim \left(1 + \frac{1}{2}\cos\Omega_{fg}\tau\right)\int_{-\infty}^{\infty} s_1(\omega_1)s_2(\omega_1 - \Omega_{fg})d\omega_1$$

$$+ \int_{-\infty}^{\infty}\left[\cos\omega_1\tau + \cos(\omega_1 - \Omega_{fg})\tau + \frac{1}{2}\cos(2\omega_1 - \Omega_{fg})\tau\right] \quad (1.10)$$

$$\times s_1(\omega_1)s_2(\omega_1 - \Omega_{fg})d\omega_1$$

where we have used $\cos a \cos b = \frac{1}{2}[\cos(a - b) + \cos(a + b)]$. There are three types of terms in Eq. (1.10). The first is independent of τ (the first term in the equation) and is uninteresting. Terms of the second type are τ dependent and are modulated at frequencies on the order of the frequencies of the light sources [the second integral in Eq. (1.10)]. By and large, these terms are also uninteresting and, moreover, are shifted well outside the frequency range of interest. The third type of term, proportional to $\cos\Omega_{fg}\tau$, is the important one. It modulates the interferogram at the Raman resonance frequency of the sample. Fourier transformation of Eq. (1.10), with neglect of the zero- and high-frequency terms, gives

$$I(\omega) \sim \delta(|\omega| - \Omega_{fg})\left[\int_{-\infty}^{\infty} s_1(\omega_1)s_2(\omega_1 - \Omega_{fg})d\omega_1\right] \quad (1.11)$$

Equation (1.11) shows that the stimulated Raman spectrum of the sample is obtained as the FT of the nonlinear interferogram. The spectral range covered in the Fourier spectrum is determined by an apparatus function that depends on the characteristics (bandwidths) of the light sources used. Furthermore, the resolution of the spectrum is independent of the light-source bandwidths. It depends, instead, on the range of τ scanned in the experiment.

Extension of the above analysis to a sample whose Raman spectrum $L(\omega_1, \omega_2)$ has more than one infinitely narrow Raman resonance

$$L(\omega_1, \omega_2) \sim \sum_i \alpha_i\delta(\omega_1 - \omega_2 - \Omega_i) \quad (1.12)$$

is straightforward. In this case one finds that the interferogram is modulated at all those Ω_i that are overlapped by the finite bandwidths of the light sources (as well as by high-frequency terms of the type discussed before). The FT of the interferogram gives the Raman spectrum of the sample over a spectral window determined by the bandwidths of the ω_1 and ω_2 light sources and at a resolution independent of those bandwidths

$$I(\omega) \sim \sum_i \left[\alpha_i \delta(|\omega| - \Omega_i) \times \int_{-\infty}^{\infty} s_1(\omega_1) s_2(\omega_1 - \Omega_i) d\omega_1 \right] \qquad (1.13)$$

where we have neglected the zero- and high-frequency components.

1.3.3. FT Versions of Sequential, One-Photon Resonant Spectroscopies

Figure 1.1(b) and (c) depict two-photon processes that rely on sequential single-photon resonant transitions. Figure 1.1(b) depicts hole-burning spectroscopy (23), while Fig. 1.1(c) depicts stimulated emission spectroscopy (24). Suppose for each of these cases the single-photon resonances are given by delta functions [e.g., $\delta(\omega_1 - \Omega_{eg})$ and $\delta(\omega_2 - \Omega_{fg})$ for the hole-burning case]. Suppose further that one has a means by which to detect whether the two-photon process has occurred and that the resulting signal I is proportional to the number of such processes. For hole burning I is given by

$$I(\tau) \sim \int_{-\infty}^{\infty} S_1(\omega_1, \tau) \delta(\omega_1 - \Omega_{eg}) d\omega_1 \times \int_{-\infty}^{\infty} S_2(\omega_2, \tau) \delta(\omega_2 - \Omega_{fg}) d\omega_2 \qquad (1.14)$$

where S_1 and S_2 are the spectral densities of the two light sources after the interferometer. By using Eq. (1.2) to substitute for these spectral densities, one can perform the integrations in Eq. (1.14) to obtain

$$I(\tau) \sim s_1(\Omega_{eg}) s_2(\Omega_{fg})(1 + \cos \Omega_{eg}\tau)(1 + \cos \Omega_{fg}\tau) \qquad (1.15)$$

This can be readily reexpressed as a sum of cosine terms

$$I(\tau) \sim s_1(\Omega_{eg}) s_2(\Omega_{fg}) \left[1 + \cos \Omega_{eg}\tau + \cos \Omega_{fg}\tau \right.$$
$$\left. + \frac{1}{2} \cos[(\Omega_{eg} + \Omega_{fg})\tau] + \frac{1}{2} \cos(\Omega_{fe}\tau) \right] \qquad (1.16)$$

where we have used $\Omega_{fg} - \Omega_{eg} = \Omega_{fe}$. This result is analogous to that of Eq. (1.10) for the stimulated Raman case. That is, there are zero- and high-frequency terms in $I(\tau)$, but there is also a modulation term at the frequency Ω_{fe}. The latter corresponds to the energy difference between the two excited-state levels connected in the two-photon process. In this case the FT of $I(\tau)$ gives back a spectrum exhibiting that two-photon resonance. The range of the spectrum is determined by an apparatus function that depends on the bandwidths of the ω_1 and ω_2 light sources. The resolution of the spectrum is independent of the light source characteristics but does depend on the delay range scanned in the experiment.

A similar analysis for the stimulated emission case [Fig. 1.1(c)] yields a very similar result. In this case, however, the resonant two-photon modulation term has a frequency equal to Ω_{fg}, which corresponds to the splitting between the two ground-state levels connected by the two-photon process. The FT of $I(\tau)$ in this case gives a spectrum exhibiting this resonance.

Generalization of the above to cases where multiple two-photon resonances are involved is straightforward. Suppose there is a whole manifold of $|g\rangle$ states, and similarly for $|e\rangle$ and $|f\rangle$. Then, for the hole-burning case $I(\tau)$ is given by

$$I(\tau) \sim P_g \alpha_{eg} \alpha_{fg} \sum_{g,e,f} \int_{-\infty}^{\infty} S_1(\omega_1, \tau)\delta(\omega_1 - \Omega_{eg})d\omega_1$$

$$\int_{-\infty}^{\infty} S_2(\omega_2, \tau)\delta(\omega_2 - \Omega_{fg})d\omega_2 \quad (1.17)$$

where P_g is the population of state $|g\rangle$ and α_{eg} and α_{fg} are measures of the strengths of the $|e\rangle \leftarrow |g\rangle$ and $|f\rangle \leftarrow |g\rangle$ transitions, respectively. Making use of Eq. (1.2), performing the integrations, and doing some trivial algebra gives

$$I(\tau) \sim \sum_{g,e,f} P_g \alpha_{eg} \alpha_{fg} s_1(\Omega_{eg}) s_2(\Omega_{fg}) \left\{ 1 + \cos \Omega_{eg}\tau + \cos \Omega_{fg}\tau \right.$$

$$\left. + \frac{1}{2} \cos[(\Omega_{eg} + \Omega_{fg})\tau] + \frac{1}{2} \cos(\Omega_{fe}\tau) \right\} \quad (1.18)$$

It is easy to see that the FT of $I(\tau)$ (again neglecting high-frequency and zero-frequence terms) yields a spectrum corresponding to the energy splittings between states in the $|e\rangle$ and $|f\rangle$ manifolds that are connected in the two-photon hole-burning process. The spectral range covered is determined by an apparatus function that depends on the spectral densities of the light sources. The resolution depends on the interferometer range scanned in the experiment. The analogous result obtains for the stimulated emission case when multiple resonances are involved. In this case the resonant modulations in the interferogram and the peaks in the Fourier spectrum correspond to splittings between states in the $|g\rangle$ and $|f\rangle$ manifolds that are connected to a common excited-state level by the two-photon stimulated emission process.

1.4. SPECIFIC METHODS

All the methods of concern to us here conform to the cases considered in Sections 1.3.2 and 1.3.3. Specific methods are distinguished by the way in which the pertinent two-photon event is detected. In this section we consider these various detection schemes. For each we describe the schemes, outline the perturbation theory analysis of them and the results of those analyses, and give a description of how they have been implemented in this laboratory.

1.4.1. FT Ionization-Detected Stimulated Raman Spectroscopies

1.4.1A. Description. Ionization-detected stimulated Raman spectroscopies rely on using resonantly enhanced multiphoton ionization (REMPI) to detect the population changes that result from stimulated Raman transitions. Two variants of IDSRS are pertinent here (see Fig. 1.6). In ionization-loss stimulated Raman spectroscopy (ILSRS) (25) the REMPI probe field (ω_3) is tuned to a vibronic resonance originating in the initial state of the stimulated Raman transition [$|g\rangle$ of Fig. 6(a)]. A finite photoion signal is always present. That signal decreases, however, when the stimulated Raman fields drive a Raman transition of the species. This decrease

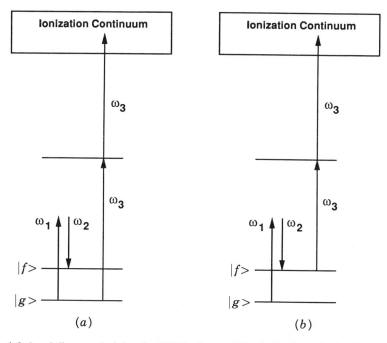

Figure 1.6. Level diagrams depicting the IDSRS schemes (a) Ionization-loss stimulated Raman spectroscopy (ILSRS) and (b) Ionization-gain stimulated Raman spectroscopy (IGSRS). In each case ω_1 and ω_2 excite a stimulated Raman transition and ω_3 probes the corresponding population change by resonant two-photon ionization. In mass-selective IDSRS the photoions are subjected to mass analysis.

reflects the loss of population of $|g\rangle$ due to the $|f\rangle \leftarrow |g\rangle$ Raman transitions. In ionization-gain stimulated Raman spectroscopy (IGSRS) (26) the REMPI probe is tuned to a transition originating in the final state of the Raman transition $[|f\rangle$ in Fig. 1.6(b)]. In this case, there is no photoion signal (ideally) unless the Raman fields drive a Raman transition. When Raman transitions do occur the increased population of $|f\rangle$ is reflected in a finite photoion signal. Both ILSRS and IGSRS are readily implemented with mass analysis of the photoions (27,28). Such analysis greatly increases the utility of the methods.

The FT implementations of ILSRS and IGSRS (8,10) involve (1) sending the ω_1 and ω_2 fields colinearly through a Michelson interferometer, (2) combining the output of the interferometer colinearly with the ω_3 field (which is delayed with respect to the Raman field), (3) focusing the resulting three-color beam into the sample, and (4) measuring the photoion signal at a given mass as a function of the delay between the arms of the interferometer. The resulting interferogram is modulated at the frequencies of those Raman resonances that (1) are effectively driven by the ω_1, ω_2 fields and (2) belong to species that contribute to the photoion signal at the detected mass.

1.4.1B.   Information Content.

The FT–IDSRS schemes have been analyzed by a perturbation theory/density-matrix approach employing double Feynman diagrams to represent the important density matrix terms (10). In this analysis it was assumed that the signal in ILSRS is given by the population of species in state $|g\rangle$ subsequent to interaction with the ω_1 and ω_2 fields. (That is, there was no attempt to model the photoionization process in detail.) Similarly, the signal in IGSRS was taken to be proportional to the population in $|f\rangle$ after interaction with the ω_1 and ω_2 fields. Under these assumptions the signals in both schemes are given by diagonal density matrix elements arising in fourth order in perturbation theory. In particular, the signal in ILSRS is given by $\rho_{gg}^{(4)}$. The signal in IGSRS is given by $\rho_{ff}^{(4)}$. Diagrams pertinent to these density matrix elements have been given in Ref. 10. Transcription of such diagrams yields general expressions for the signals in the two schemes.

Going beyond the general expressions for IDSRS signals referred to above requires specification of the properties of the ω_1 and ω_2 light sources. Two cases have been considered—one in which the light sources were taken to be effectively continuous and stochastic, and one in which they were taken to be transform-limited Gaussian pulses. The upshot of the analysis is a quantitative verification of the results from the simplistic treatment of Section 1.3.2. The most important results are as follows. The FT of a FT–IDSRS interferogram yields the Raman spectrum (spectra) of those species that contribute to the detected ion signal. The spectrum spans a frequency range determined by the bandwidths of the ω_1 and ω_2 light sources. The resolution of the spectrum is determined by the range over which the interferometer delay is scanned in the experiment. The pulse widths of the excitation sources do not limit the range over which the interferometer can be usefully scanned. (In other words, spectral resolution is not limited by short excitation pulses.) The line widths of the Raman resonances appearing in the FT spectrum are just what they would be in a frequency-domain IDSRS experiment. Finally, the Doppler broadening in the FT spectrum goes as $(\Omega_{fg}/c)\sqrt{k_B T/M}$, where c is the speed of light and M is the mass of the species.

1.4.1C. Implementation.

In this laboratory FT–IDSRS has been implemented (28) with a laser system composed of a single Nd:YAG laser (Spectra-Physics DCR-2A) pumping two dye lasers (Spectra-Physics PDL-2). Part of the frequency-doubled output of the Nd:YAG was used as the ω_1 field. The output of one of the dye lasers was employed as the ω_2 field. The bandwidth of this field could be varied from 0.3 to 3 to about 100 cm^{-1} by changing the order of the laser's diffraction grating. The REMPI probe field (ω_3) was generated by frequency doubling the output of the second dye laser. This field was delayed optically by about 10 ns to insure that it impinged on the sample after the Raman fields were gone. The ω_1 and ω_2 beams were combined colinearly on the 50% beam splitter of a Michelson interferometer. After retroreflection by solid corner cubes the beams in each arm of the interferometer were recombined colinearly on the beam splitter. The output of the interferometer was combined colinearly with the ω_3 beam on a dichroic mirror, and the resulting three-color beam was focused by a 25-cm focal length lens into the molecular beam sample. (Typical pulse energies at the sample were 10–25 mJ for the ω_1 and ω_2 fields and 10 μJ for the ω_3 field.) The interferometer delay was scanned by moving the stepper motor driven translation stage on which one of the retroreflectors was mounted. In performing a scan the stage was set running at a constant speed. The interference fringes from a frequency-stabilized He–Ne laser, which followed the same path in the interferometer as the ω_1 and ω_2 beams, were monitored by a photodiode. The output of the photodiode was then filtered, and zero crossings of the filtered signal were detected and counted electronically. After every n counts (n being preset) an electrical pulse was generated and sent to trigger the firing of the Nd:YAG laser. In this way the firing of the laser system was synchronized with the interferometer scan, and data points were collected at constant intervals of the interferometer delay.

The molecular beam was formed by expansion through an orifice controlled by a pulsed valve, the firing of which was synchronized with the laser firing. The expansion was skimmed several centimeters downstream of the valve. After being skimmed it entered the ionization region of a Wiley–McClaren time-of-flight (TOF) mass spectrometer (29), where it intersected the excitation fields at a right angle. Photoions were accelerated at right angles with respect to the laser and molecular beam propagation directions and detected by a dual microchannel plate assembly. The amplified signal from the microchannel plates was monitored by a boxcar integrator, whose gate was set to the ion mass peak of interest. The output of the boxcar was dumped to a computer each time the laser system was fixed. Throughout any given experiment ω_1, ω_2, and ω_3 were set to fixed values.

1.4.2. FT Coherent Raman Scattering

1.4.2A. Description.

Coherent Raman scattering (30) amounts to the detection of an ω_1, ω_2-induced stimulated Raman transition by using a third field (ω_3) to interact with the material coherence (off-diagonal density matrix element) that is created by that transition. This interaction produces a macroscopic polarization that generates a coherent beam of light. This light, the signal in the experiment, has an intensity that depends on how effectively the stimulated Raman process is driven and a frequency that equals $\omega_3 \pm \Omega_{fg}$, where Ω_{fg} is the Raman transition frequency. Figure 1.7 shows level diagrams depicting processes involved in coherent Stokes Raman scattering (CSRS) and coherent anti-Stokes Raman scattering (CARS).

In FT–CRS the temporally coincident ω_1 and ω_2 fields propagate colinearly or

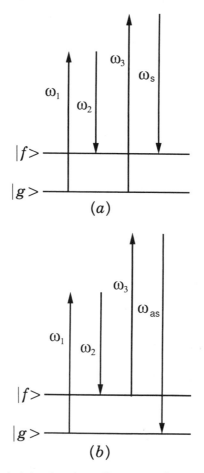

Figure 1.7. Level diagrams depicting the coherent Raman scattering processes involved in FT–CRS. (*a*) Coherent Stokes Raman scattering. (*b*) Coherent anti-Stokes Raman scattering. In both (*a*) and (*b*) the ω_1 and ω_2 fields drive a stimulated Raman transition. The ω_3 field interacts with the material coherence created by that process to produce the signal wave. In the FT versions of these processes the ω_1 and ω_2 fields traverse the interferometer. The ω_3 field need not do so. Also there is no requirement that the signal field be spectrally analyzed except in so far as is necessary to separate it from the excitation fields.

parallel through a Michelson interferometer. The output of the interferometer is combined with the ω_3 field on a beam splitter. The three beams are focused into the sample. The coherent signal beam is separated from the excitation fields by spatial, spectral, and/or polarization filtering, and its intensity is monitored as a function of the delay between the arms of the interferometer to yield an interferogram. The interferogram is Fourier transformed to give the desired spectrum.

1.4.2B. Information Content. FT–CRS has been analyzed (10) by perturbation theory with the aid of double Feynman diagrams to help identify the important matter–field interactions and density matrix elements. The pertinent diagrams for CRS correspond to third-order contributions to off-diagonal density matrix elements. These third-order coherences give rise to the nonlinear polarization in the sample, which in turn produces the coherent scattering signal. Transcription of the relevant diagrams, followed by the use of well-known recipes (30) for relating the third-order coherences to the intensity of the signal beam yielded a general expression for the signal intensity (10). The information content of FT–CRS was characterized by applying this expression to specific cases for the spectral densities of the light sources (the same ones assumed in the analysis of FT–IDSRS) and by accounting for the effect of the interferometer on the ω_1 and ω_2 fields. The important results (10) are similar to those obtained for FT–IDSRS. The FT of a FT–CRS interferogram yields the coherent Raman scattering spectrum of the sample (resonances at Ω_{fg}). The range of the spectrum is determined by the bandwidths of the ω_1 and ω_2 fields in the form of an apparatus function that multiplies the coherent Raman spectrum. The resolution of the spectrum is determined by the interferometer range scanned in the experiment. It is independent of the bandwidths and pulse widths of the ω_1 and ω_2 fields. (In regard to the pulse width issue it is important to point out that FT–CRS interferograms are modulated by Raman resonance frequencies up to and beyond the delay at which the two-color pulses from the two arms of the interferometer cease to overlap temporally.) The homogeneous line widths of the resonances in the spectrum depend on the damping rates associated with coherences between the $|g\rangle$ and $|f\rangle$ states, the ground states involved in the Raman processes. Finally, Doppler broadening is proportional to the Raman frequencies Ω_{fg}.

1.4.2C. Implementation. FT–CRS has been implemented in a number of different ways in this laboratory (6,9,10). In all cases the laser system was the same as that described above for FT–IDSRS. The interferometer was similar also, except the delay was not calibrated interferometrically. Instead, it was operated in a rudimentary step-scan mode. The stepper motor driving the movable arm was stepped a fixed number of steps, the signal was averaged over several laser shots, the averaged signal was dumped to a computer, and the stepper motor was stepped again to repeat the cycle. This mode of operation is suitable only for studying fairly low Raman frequencies (<200 cm^{-1}) because of vibration-induced interferometer-delay instabilities and variations in the size of the delay steps.

FT–CRS experiments can be implemented with two or three colors. In the former case, either one of the ω_1 or ω_2 fields serves double duty as the ω_3 field. In either case, only the ω_1 and ω_2 fields need be sent through the interferometer. Both the two- and three-color schemes have been used by us.

FT–CRS can also involve colinear or noncolinear excitation and signal fields. The former arrangement is easiest to align but can cause difficulty in separating the signal from the excitation fields, particularly for small Ω_{fg}. The latter produces a signal beam that propagates in a direction different than any of the excitation beams. The "folded BOXCARS" arrangement (31) is particularly useful in effecting the spatial separation of the signal beam. We have employed both colinear and noncolinear phase-matching geometries for FT–CRS. In the latter, the ω_1 and ω_2 beams propagate parallel, but not colinear to one another in and after the interferometer, until being focused with ω_3 into the sample. No matter what the geometry of the experiment, a monochromator was always employed to help extract the signal from the excitation fields. Detection of the signal beam involved a photomultiplier–boxcar integrator combination. The output of the boxcar was stored in a computer as a function of interferometer delay to yield an interferogram.

Both room temperature gases and supersonic molecular beams have been studied by FT–CRS. In the latter case, the apparatus employed was the same as that used for FT–IDSRS, except that the beam was not skimmed and interaction with the excitation fields occurred several millimeters from the expansion orifice.

1.4.3. FT–Stimulated Emission Spectroscopy

1.4.3A. Description. FT–stimulated emission spectroscopy (FT–SES) is an interferometric version of the stimulated emission pumping method (24). The relevant processes are depicted in Fig. 1.8(b). The ω_1 and ω_2 fields drive a *vibronically resonant* stimulated Raman process. The ω_1 field is resonant with the $|e\rangle \leftarrow |g\rangle$ transition, and the delayed ω_2 field is resonant with the $|e\rangle \rightarrow |f\rangle$ transition. The resonance Raman transitions from $|g\rangle$ to $|f\rangle$ are detected by monitoring the total spontaneous emission from the excited intermediate state $|e\rangle$. That emission intensity is depleted whenever ω_2 drives downward transitions. The FT–SES method entails tuning broadband ω_1 and ω_2 sources to overlap appropriate vibronic resonances, sending these same sources colinearly through a Michelson interferometer, directing the output of the interferometer to the sample, and detecting total emission as a function of interferometer delay to obtain an interferogram.

1.4.3B. Information Content   A perturbation theory analysis of FT–SES similar to that of the FT–IDSRS schemes has been performed (11). The key feature of this analysis is the recognition that the signal in SES by emission depletion basically is proportional to the population left in the excited intermediate state $|e\rangle$ after the ω_1 and ω_2 fields have encountered the sample. The relevant density-matrix elements are fourth-order diagonal corresponding to $|e\rangle - \rho_{ee}^{(4)}$. These matrix elements were calculated by double Feynman diagram methods to yield general expressions for the depletion of spontaneous emission due to the SES process. Characterization of FT–SES was then accomplished by applying these general

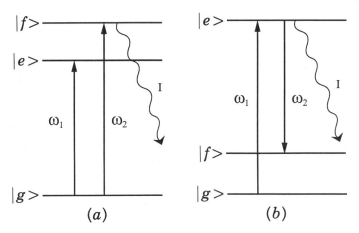

Figure 1.8. Level diagrams depicting fluorescence-detected (*a*) hole-burning spectroscopy (HBS) and (*b*) stimulated emission spectroscopy (SES). In (*a*) the total-emission signal (*I*) reflects the depletion of $|g\rangle$ by the ω_1 field. In (*b*) *I* reflects the depletion of $|e\rangle$ due to stimulated emission induced by the ω_2 field. The FT version of HBS produces resonant modulations corresponding to the splitting between $|e\rangle$ and $|f\rangle$. The FT version of SES produces such modulations corresponding to the splitting between $|f\rangle$ and $|g\rangle$.

expressions to particular cases for the spectral densities of the ω_1 and ω_2 sources (viz, continuous monochromatic, delta-function pulsed, and transform-limited Gaussian pulsed sources).

The main results (11) of the perturbation theory analysis are as follows. First, Fourier transformation of a FT–SES interferogram gives the spectrum of energy differences (Ω_{fg}) between those $|g\rangle$ and $|f\rangle$ states connected by resonant stimulated Raman transitions. Second, the effective range of the spectrum is determined by an apparatus function that depends on the spectral densities of the ω_1 and ω_2 light sources—the broader the bandwidths of these light sources, the larger the spectral range covered. Third, the resolution in the Fourier spectrum is dependent on the range over which the interferometer is scanned in the experiment. The light-source bandwidths and pulse widths play no role in this regard. Particularly significant are the facts that resolution is not limited either by (a) short pulse widths or (b) the magnitude of the delay between the ω_1 and ω_2 pulses. This is significant because it means that FT–SES can be implemented at high resolution with the short-pulse light sources that tend to be the most effective at driving resonant stimulated Raman processes. Fourth, the homogeneous line widths in the FT–SES spectrum are determined by the damping rates of the $|g\rangle$ and $|f\rangle$ states. They are not affected by fast decay of the excited intermediate state $|e\rangle$. Finally, Doppler broadening in the FT–SES spectrum depends on the ground-state frequencies Ω_{fg} and not on the vibronic resonance frequencies Ω_{eg} and Ω_{ef}.

1.4.3C. Implementation.   The apparatus used in this laboratory to implement FT–SES (11) is different from the one employed in FT–IDSRS and FT–CRS experiments. The laser system consists of a Q-switched, mode-locked Nd:YAG laser that synchronously pumps two etalon-tuned, cavity-dumped dye lasers. The frequency-doubled outputs of the two dye lasers provide the ω_1 and ω_2 fields. The excitation fields consist of pulses of $25-30$-ps duration, $2-3$-cm^{-1} bandwidth, and about 1-μJ energy at a repetition rate of 1 kHz. The ω_2 pulse train is typically delayed with respect to the ω_1 pulse train by $0.1-1$ ns. The two-pulse trains are combined colinearly on the 50% beam splitter of a Michelson interferometer. In addition to the beam splitter, the interferometer consists of two hollow, corner-cube retroreflectors, one of which is mounted on a stepper motor driven translation stage. The latter provides the moving arm of the interferometer. The output of the interferometer is mildly focused into the molecular beam sample. The continuous molecular beam is formed by passing a carrier gas (usually He) over the compound of interest in a heated Pyrex tube and allowing the mixture to expand into vacuum through a pinhole in the end of the tube. The molecular beam intersects the excitation fields several millimeters downstream of the expansion orifice. Total fluorescence is then detected with a photomultiplier, whose output is averaged by a boxcar integrator. The output of the integrator is dumped to a computer as a function of the delay between the interferometer arms. The interferometer is scanned in step–scan mode without any interferometric stabilization. As such, it is only useful in studies of low-frequency intervals (<200 cm^{-1}). Typically, signal from 100 to 300 laser pulses is averaged for each position of the interferometer.

1.4.4. FT Fluorescence-Detected Hole-Burning Spectroscopy

1.4.4A. Description.   Fluorescence-detected hole-burning spectroscopy (23) is depicted in the level diagram of Fig. 1.8(a). In the scheme the ground-state population hole burned by the ω_1 pulse via the $|e\rangle \leftarrow |g\rangle$ transition is probed by the delayed ω_2 pulse, which is resonant with the $|f\rangle \leftarrow |g\rangle$ transition. Total emission from $|f\rangle$ serves as the observable. The population transferred out of $|g\rangle$ by the ω_1 pulse gives rise to a depletion in the $|f\rangle$ emission. This emission depletion depends on two single-photon processes and, therefore, conforms to the scheme of Section 1.3.3. In the FT version of fluorescence-detected hole-burning spectroscopy (FT–HBS for simplicity) the ω_1 and ω_2 fields traverse a Michelson interferometer. The output of the interferometer impinges on the sample. Total emission is detected as a function of interferometer delay to yield an interferogram, which is then Fourier transformed.

1.4.4B. Information Content.   The perturbation theory analysis of FT–HBS (11) is very similar to that of FT–SES. The main point is that the fluorescence depletion is proportional to that part of the population of $|f\rangle$ that is affected by both pump and probe pulses. This is given by the diagonal density matrix element corresponding to $|f\rangle$ evaluated at fourth order in perturbation theory $- \rho_{ff}^{(4)}$. Transcription of the double Feynman diagrams that contribute to this quantity yields general expressions for the fluorescence depletion. Assuming specific cases for the properties of the light-source spectral densities allows one to analyze the information content of a FT–HBS spectrum. One finds results completely parallel to those for FT–SES. The FT of a FT–HBS interferogram yields the spectrum of energy differences between the pairs of excited states $|e\rangle$ and $|f\rangle$ connected to a given $|g\rangle$ state in the two-photon hole-burning process. The spectral range covered is determined by an apparatus function that reflects the finite bandwidths of the ω_1 and ω_2 light sources. The spectral resolution depends only on the interferometer delay range scanned in the experiment and not on light-source properties, such as bandwidth or pulse width. The homogeneous broadening of the Ω_{ef} resonances is given by the damping rates of the $|e\rangle$ and $|f\rangle$ states and does not depend on the decay of $|g\rangle$. Finally, the Doppler broadening in a FT–HBS spectrum is proportional to the excited-state difference frequencies Ω_{fe} and is not dependent on the frequencies of the vibronic transitions involved.

1.4.4C. Implementation. We implement FT–HBS with the identical apparatus employed for FT–SES: The only difference between the schemes is that ω_2 is tuned to an absorption transition in FT–HBS rather than the stimulated-emission transition to which it is tuned in FT–SES.

1.4.5. Down-Shifting of High-Frequency Modulations in FT Nonlinear Schemes

1.4.5A. Description. Fourier transform nonlinear spectroscopies must be implemented with pulsed light sources because of the intensity requirements for non-linear processes. At the same time, there is a maximal increment (giving rise to minimal sampling) that the interferometer can be stepped in a FT spectroscopy if one wants to avoid aliasing in the Fourier spectrum (1). The maximal interferometer increment is given by $1/(2\nu_{max})$, where ν_{max} is that resonance in the spectrum with the largest frequency. Clearly, the larger the maximal frequency, the larger the number of steps one must take to achieve a given spectral resolution (which is determined by the total delay range scanned). When low-repetition-rate light sources are used (e.g., 30 Hz), the data collection time required to characterize high-frequency resonances (e.g., C–H stretches) at high resolution (e.g., 0.01 cm^{-1}) can become prohibitive. This situation led us to think about trying to develop methods by which to convert high-frequency modulations into low-frequency ones to permit marked increases in maximal interferometer step sizes while still achieving better-than-minimal sampling. The scheme that we developed is called "down-shifting." We have since learned that high-quality interferometers can produce excellent FT spectra even with less-than-minimal sampling in the data collection (i.e., when interferometer steps are larger than $1/(2\nu_{max})$, if the spectral window covered in the experiment is small enough. Thus, in practical applications the down-shifting method will generally be obviated by other approaches. Nevertheless, the method has some intellectually appealing features, and for that reason we describe it here.

The basic idea (9) of down-shifting is that the τ dependence of the spectral density of a light field emerging from a Michelson interferometer should remain unchanged when the interferometer output is frequency shifted by stimulated Raman scattering in a high-pressure gas (30,32). That is, suppose the spectral density of the ω_2 field after the interferometer is given by $S_2(\omega_2, \tau) \sim s_2(\omega_2)(1 + \cos[\omega_2\tau])$, and suppose the frequency shift due to the stimulated Raman process is ω_s, then the spectral density of the Raman-shifted light should be

$$S_s(\omega_2 - \omega_s, \tau) \sim s_2(\omega_2)(1 + \cos[\omega_2\tau]) \tag{1.19}$$

If such is the case, then the use of the frequency-shifted light and the usual ω_1 field from the same interferometer to drive an appropriate nonlinear process (i.e., CRS, SES, HBS, or IDSRS) will produce resonant modulations in the interferogram at frequencies $\Omega_{ij} - \omega_s$ rather than at the Ω_{ij} resonance frequencies themselves. Clearly, when ω_s is close to Ω_{ij} this scheme can produce low-frequency modulations even when Ω_{ij} is very large, which is precisely one's aim.

1.4.5B. Information Content. Analysis of the down-shifting scheme (9) entails demonstrating that Eq. (1.19) does indeed characterize the effect of Raman shifting on the spectral density of an interferometer-modulated light source. If that equation is true, then an analysis like that of Section 1.3 clearly shows that modulations due to the resonances having frequencies Ω_{ij} will have frequencies *in the interferogram* equal to $\Omega_{ij} - \omega_s$. Equation (1.19) can be obtained by starting with the gain equations for the electric field amplitudes of the different frequency components of the Stokes-shifted light (32):

$$\partial|E_{sq}(z)|/\partial z = K|E_{2q}| \sum_n |E_{2n}||E_{sn}(z)| \tag{1.20}$$

where K is a constant, $|E_{2n}|$ is the electric-field magnitude of the nth longitudinal mode of the ω_2 field (frequency ω_{2n}), $|E_{sn}(z)|$ is the electric-field magnitude of the nth Stokes-shifted longitudinal mode of the field generated by stimulated Raman scattering (frequency $\omega_{2n} - \omega_s$), and z is a coordinate along the fields' propagation direction in the Raman cell (z varies from 0 to L, where L is the length of the cell). The coupled equations represented by Eq. (1.20) can be solved by assuming negligible depletion of the ω_2 field amplitudes due to the Raman shifting process. For the output from the cell one finds

$$|E_{sq}| = |E_{2q}|\beta[\exp(KI_2L) - 1] + |E_{sq}(0)| \tag{1.21}$$

where $I_2 \equiv \Sigma_n |E_{2n}|^2$ and $\beta \equiv I_2^{-1} \Sigma_n |E_{2n}||E_{sn}(0)|$. The spectral density of the Raman-shifted field can now be obtained by squaring Eq. (1.21). The result has a complicated dependence on τ because the I_2 and β factors involve sums containing the $|E_{2n}|$, which are proportional to $(1 + \cos \omega_{2n}\tau)$. However, because of destructive interference the sums over these cosine terms effectively average to zero after a delay of the order of $1/\Delta\omega_2$, where $\Delta\omega_2$ is the bandwidth of the ω_2 light source. Bearing this in mind, for long enough delays it is straightforward to show that the spectral density of the Raman-shifted light is, indeed, given by Eq. (1.19). Therefore, resonances at Ω_{ij} will give rise to interferogram modulations at $\Omega_{ij} - \omega_s$ when the ω_1 and $\omega_2 - \omega_s$ fields are used to drive the pertinent two-photon processes.

In the above, we have made the assumption that the Raman-shifting process is not characterized by a spectral width, that is, all the ω_{2n} are shifted in frequency exactly by an amount equal to ω_s. Thus, we have not considered the degree to which the finite spectral width of the Raman-shifting band influences the ultimate spectral resolution available with the down-shifting scheme. One notes, however, that due to the highly nonlinear nature of the Raman-shifting gain, there is good reason to believe that degradation of resolution arising from this source should be considerably less than the line width of that band.

1.4.5C. Implementation.

The down-shifting method was demonstrated (9) in a three-color FT–CRS experiment (see Fig. 1.9). The ω_1 and ω_2 fields were provided by two Nd:YAG-pumped dye lasers (both near 560 nm). The ω_3 field was a portion of the 532-nm output of the same Nd:YAG laser. The ω_1 and ω_2 fields were directed in parallel through a Michelson interferometer such that they experienced the same delay. The ω_2 output of the interferometer was focused into a 1-m-long gas cell containing 20 atm of methane ($\omega_s/2\pi \simeq 2916.7$ cm^{-1}) (33). The collimated first Stokes-shifted output of the Raman cell was separated by a Pellin–Broca prism from the ω_2 light and other light fields generated in the cell. The Raman-shifted light was focused together with the ω_1 and ω_3 fields into the sample cell, which contained methane at several hundred torr. The three beams conformed to the "folded-BOXCARS" phase-matched geometry (31). Coherent anti-Stokes Raman scattering generated in the sample cell (at angular frequency $\omega_1 + \omega_3 - (\omega_2 - \omega_s)$) was isolated by spatial and spectral filtering and detected with a photomultiplier. The photomultiplier output was averaged with a boxcar integrator and dumped to a computer as a function of the stepper motor driven interferometer delay to yield a down-shifted interferogram.

1.5. EXPERIMENTAL RESULTS

1.5.1. FT–IDSRS Results

Mass-selective IDSRS methods have some very significant advantages in application to the vibrational spectroscopy of species in molecular beams. These advantages, including high sensitivity, species- and size-selectivity, wide spectral range, and complementarity to other spectroscopies, have been discussed in detail elsewhere (28,34). The FT versions of mass-selective IDSRS provide the additional capabilities of (a) spectral resolution unlimited by laser bandwidth and (b) routinely accurate line-position measurements with essentially no additional effort.

1.5.1A. Bare Molecules and One-To-One Complexes.

We have studied a number of jet-cooled bare molecules and 1:1 complexes with FT–IDSRS (8,28,35). Some representative results are shown in Figs. 1.10 and 1.11. One notes the sharp ^qQ-branch features, which dominate the rotational structure in polarized Raman

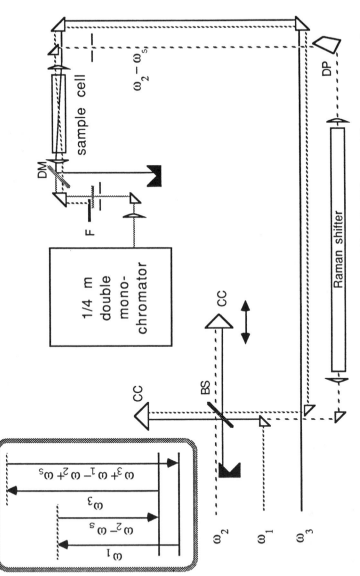

Figure 1.9. Schematic diagram of the apparatus used to obtain down-shifted FT–CRS spectra. The parameters ω_1 and ω_2 represent the outputs of two Nd:YAG-pumped dye lasers at about 560 nm, ω_3 represents light at 532 nm obtained by frequency doubling the fundamental of the Nd:YAG laser, and $\omega_2 - \omega_s$ represents light obtained by Raman shifting ω_2. Also, BS = 50% beam splitter, CC = corner-cube retroreflector, DP = dispersing prism, DM = dichroic mirror, and F = interference filter. The inset in the figure is a level diagram that depicts the role that each field plays in driving the coherent Raman process in the sample: ω_1, $\omega_2 - \omega_s$, and ω_3 are the excitation fields, and $\omega_3 + \omega_1 - \omega_2 + \omega_s$ is the anti-Stokes field generated in the process. [From Ref. 9 with permission. Copyright © 1988, Elsevier Science Publishers B.V.]

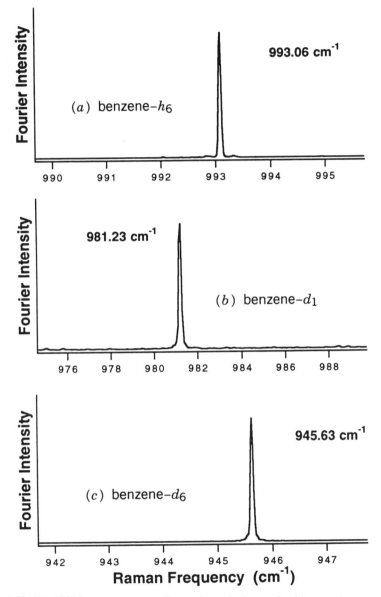

Figure 1.10. FT–IGSRS spectra corresponding to the ν_1 fundamentals of benzene isotopomers. The spectra were obtained by probing through the $S_1 \leftarrow S_0$ $1^0_1 6^1_0$ hot band. The frequencies of the resonances are given in the figure. The resolution in each spectrum is 0.05 cm^{-1}. [From Ref. 38 with permission. Copyright © 1992, American Institute of Physics.]

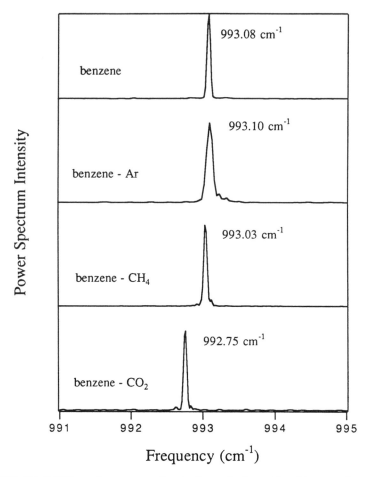

Figure 1.11. FT–IGSRS spectra corresponding to the ν_1 fundamentals of benzene, benzene–Ar, benzene–CH_4, and benzene–CO_2. The resolution in each spectrum is 0.05 cm^{-1}, except for the case of benzene–Ar, where it is 0.08 cm^{-1}. [From Ref. 28 with permission. Copyright © 1990, Optical Society of America.]

bands when measured with the ω_1 and ω_2 fields polarized parallel to one another (36). This aspect of the Raman spectra, together with the capabilities of the FT methods renders FT–IDSRS particularly powerful at characterizing vibrational transition frequencies, small frequency shifts (e.g., due to complexation), and splittings in vibrational bands. One sees clear examples of the first two capabilities in Figs. 1.10 and 1.11. The third is best exemplified by results on the benzene dimer as discussed in Section 1.5.1B.

1.5.1B. FT–IDSRS Results on Benzene Dimer and Trimer. Perhaps the
most important results obtained from application of a FT nonlinear spectroscopy

to a chemical system are those from FT–IDSRS experiments on isotopomers of benzene dimer (37,38). In this study the abilities of the FT methods to measure spectra at a resolution not limited by the laser bandwidths and to obtain the positions of vibrational resonances accurately were very significant factors for success.

The principal aim of the study was to learn about the geometry of the dominant dimer conformer. Several dimer structures had been proposed based on other spectroscopic results (39), but nothing definitive had been proved. Mass-selective IDSRS spectroscopy was applied to the ν_1 (totally symmetric ring-breathing mode) and ν_2 (totally symmetric C–H stretch) fundamentals of various isotopomers composed of perprotonated (d_0), perdeuterated (d_6), and monodeuterated (d_1) benzene monomer moieties. All of the FT results pertain to the ν_1 fundamentals.

Figures 1.12–1.14 show some pertinent FT–IDSRS results. Those in Fig. 1.12 relate to the two homodimers d_0–d_0 and d_6–d_6. (For comparison this figure also shows the highest resolution frequency-domain ILSRS spectra that could be obtained with our apparatus.) One notes two bands split by 0.5 cm^{-1} for each homodimer. The two FT–IGSRS spectra in Fig. 1.13(a) pertain to the d_0-localized ν_1 fundamental of the heterodimer d_0–d_6. These two spectra differ with respect to the position of the REMPI probe frequency ω_3 in the experiment [see Fig. 1.13(b)]. The important result is that two d_0-localized ν_1 bands, separated by about 0.25 cm^{-1}, are observed for this species. Shown in Fig. 1.13(b) are two vibronic spectra that correspond, respectively, to the excited vibrational levels reached via the Raman resonances appearing in Fig. 1.13(a). Analogous FT–IDSRS results, an example of which is shown in Fig. 1.14 for the d_0–d_1 isotopomer, were obtained for the other heterodimers.

The FT–IDSRS results on the benzene dimer are informative in two respects. First, their qualitative aspects point unequivocally to a dimer geometry characterized by two *inequivalent* benzene sites. This follows from the presence of two ν_1 bands for the two homodimers and two ν_1 bands per monomer moiety for the heterodimers, together with the fact that the ν_1 mode in the benzene monomer is nondegenerate. Second, the quantitative aspects of the data—splittings and absolute line positions relative to monomer frequencies—fit well to a perturbation theory model of the ν_1 level structure in dimers having inequivalent benzenes. This fit yields values for the site-shift and excitation-exchange matrix elements associated with the ν_1-excited states in the dimers. Such a quantitative match between theory and experiment would not have been possible had we been forced to rely on our frequency-domain capabilities to measure the IDSRS spectra. All in all, FT–IDSRS results represent an important part of a larger body of mass-selective IDSRS results on benzene dimer isotopomers. Taken together, these results not only prove that the dimer conformer studied has inequivalent benzene sites, they also provide strong evidence for an internally rotating "T"-shaped structure (38). Subsequent rotational spectroscopy on the species (40) has provided even better evidence that such a geometry does, indeed, characterize the dimer.

Compared to the benzene dimer study, the FT–IDSRS results on the benzene trimer are more limited (41). They comprise mass-selective FT–IGSRS spectra in the ν_1 fundamental region of the fully protonated species measured as a function of the REMPI probe frequency ω_3. Figure 1.15 shows a typical FT–IGSRS spec-

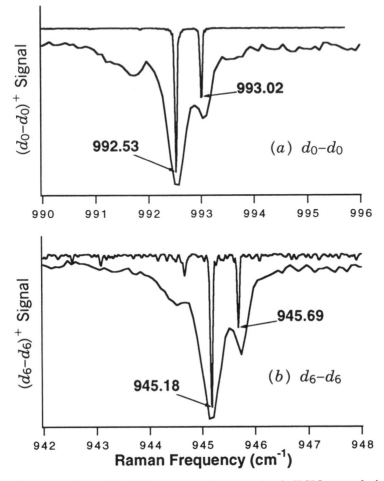

Figure 1.12. A comparison of FT–IDSRS spectra and frequency-domain IDSRS spectra for homodimers of benzene. (*a*) The ν_1 fundamental region of the fully protonated dimer. (*b*) The analogous region for the fully deuterated dimer. In both (*a*) and (*b*) the upper spectrum is from the FT method, and the lower one is from frequency-domain ILSRS implemented at the highest available resolution. Line positions are given in the FT spectra, which are plotted in negative-going fashion to facilitate comparison with the frequency-domain results. [From Ref. 38 with permission. Copyright © 1992, American Institute of Physics.]

trum. There are several points of note about the results. The FT–IGSRS spectra display only single peaks at 0.05-cm^{-1} resolution. The peaks do not shift around for different values of ω_3. And, the bands occur at a frequency of 991.63 ± 0.05 cm^{-1}, a shift of −1.43 cm^{-1} from the monomer ν_1.

The presence of only a single ν_1 fundamental in a species with three chromophores capable of producing ν_1 bands suggests that symmetry reduces the number of ν_1 transitions with Raman intensity from three to one. This can happen if all of

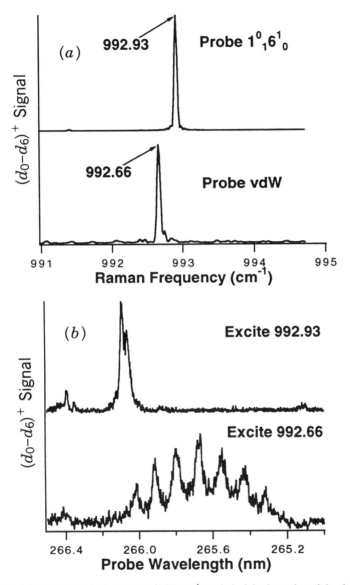

Figure 1.13. (*a*) Top: FT–IGSRS spectrum (0.05-cm^{-1} resolution) in the region of the d_0-localized ν_1 fundamental of the isotopically mixed d_0–d_6 benzene dimer. The spectrum was obtained with ω_3 set to the d_0-localized $1_0^0 6_0^1$ band of the species. Bottom: Same as the top except ω_3 was set to the second van der Waals band built off of the $1_1^0 6_0^1$ transition. (*b*) Top: Frequency-domain vibronic spectrum obtained by setting $\omega_1 - \omega_2$ to the Raman resonance of the d_0–d_6 species at 992.93 cm^{-1} and scanning ω_3. Bottom: Same but with $\omega_1 - \omega_2$ set to the Raman resonance at 992.66 cm^{-1}. [From Ref. 38 with permission. Copyright © 1992, American Institute of Physics.]

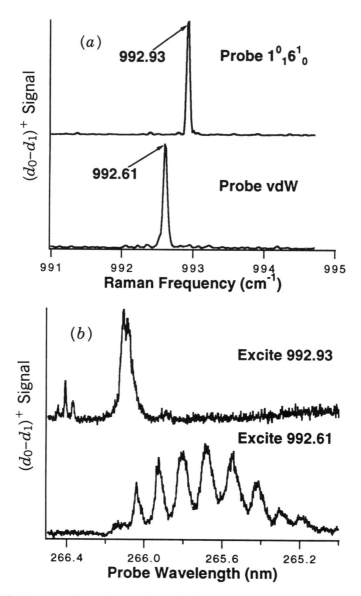

Figure 1.14. The same as Fig. 1.13 except the species is the d_0-d_1 isotopomer of benzene dimer. [From Ref. 38 with permission. Copyright © 1992, American Institute of Physics.]

the benzene moieties in the trimer are symmetrically equivalent, or nearly so. Thus, the FT–IGSRS data on the homotrimer provide evidence for a symmetrical species. Lower resolution, frequency-domain IDSRS data on various trimer isotopomers in the ν_1 and ν_2 regions support this conclusion, as do semiempirical calculations of the minimum-energy structure of the trimer (42). Beyond this, the shift of the

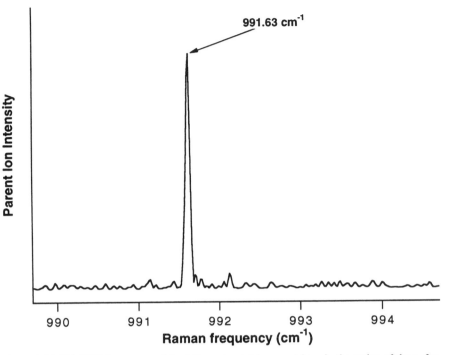

Figure 1.15. FT–IGSRS spectrum of the fully protonated benzene trimer in the region of the ν_1 fundamental. The spectral resolution is 0.05 cm^{-1}. The spectral range is about 5 cm^{-1}. The ω_3 field was tuned to the $S_1 \leftarrow S_0$ $1_1^0 6_0^1$ vibronic band of the species. [From Ref. 41 with permission. Copyright © 1993, American Institute of Physics.]

perprotonated trimer ν_1 band from the benzene monomer band, compared with the analogous shifts for the perprotonated dimer (-1.43 cm^{-1} relative to -0.04 and -0.53 cm^{-1}) suggests that there is a qualitative change in the perturbation of the ν_1 mode in going from the dimer to the trimer. Such a change might reasonably be attributed to a cyclic trimer structure in which each moiety interacts with the other two. In fact, such a structure must characterize a trimer with symmetrically equivalent sites. Thus, the qualitative and quantitative aspects of the FT–IGSRS data point to a trimer of this type (41).

1.5.2. FT–SES and FT–HBS Results

Experimental results from the application of FT–SES and FT–HBS (11) reveal features of these spectroscopies that are different from their frequency-domain counterparts (23,24), aside from the generic differences between the FT and frequency-domain approaches. The first pertains to rotational structure. Figure 1.16 shows FT–SES and FT–HBS spectra relating to ground- and excited-state torsional intervals, respectively, in *trans*-stilbene. One notes the same ^{q}Q-branch dominance in the rotational band that characterizes the FT–IDSRS results. This occurs because

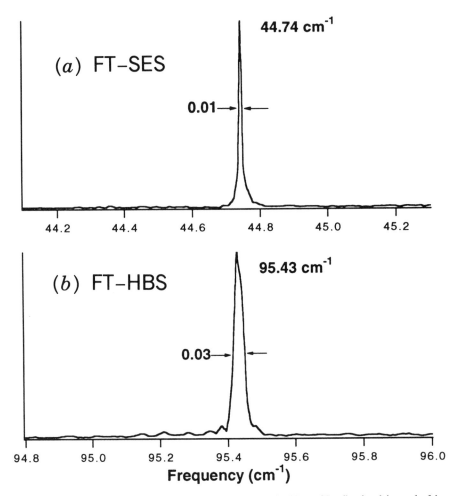

Figure 1.16. (*a*) FT–SES power spectrum corresponding to the $37_4 - 37_0$ vibrational interval of jet-cooled *trans*-stilbene. The excited intermediate state was the S_1 0^0 level. (*b*) FT–HBS power spectrum corresponding to the $37^2 - 37^4$ vibrational interval of *trans*-stilbene. The pertinent ground-state level in the HBS process was the 0_0 level. In both spectra the observed bands have widths limited by the resolution of the experiment. [From Ref. 11 with permission. Copyright © 1992, American Institute of Physics.]

the operator for the overall *two-photon* process in SES and HBS is very similar to the polarizability tensor operator associated with stimulated Raman transitions. Moreover, this tensor operator in SES (and in HBS) has an appreciable isotropic (rank-zero) component. This isotropic component only contributes intensity to the ${}^qQ$ branch. Finally, given the typical small differences in rotational constants between two vibrational levels of the same electronic state, all the lines that contribute to the ${}^qQ$ branch in a FT–SES or FT–HBS spectrum fall very close to, or on top of one another, while lines in the other branches are, of course, much more spread out. The frequency-domain SES (24) and HBS (23) methods, in contrast, give rise to *one-photon rovibronic* spectra of pump-labeled species. These spectra reflect electric-dipole rotational selection rules rather than Raman-like rules. Furthermore, they depend on the rotational constants of two vibronic states rather than on those of two vibrational states in the same electronic manifold. The upshot is that P-, Q-, and R-branch features all appear with comparable intensity in the frequency-domain schemes, and the ^qQ branch is more disperse in such spectra than in the FT spectra. In short, the rotational structure in the former spectra will tend to be much more complicated than in the latter. This is particularly true for large species, because the spectral simplification arising from rotational selectivity in the pump process becomes more and more difficult to achieve for larger and larger species.

The sparse rotational structure of FT–SES and FT–HBS spectra like those in Fig. 1.16, together with the general characteristics of FT nonlinear schemes, render FT–SES and FT–HBS particularly good at measuring accurate ground- and excited-state vibrational intervals, respectively, and at characterizing the shifts and splittings of vibrational bands. The drawback of this spectral simplicity is that the important information that rotational structure can provide is not available from the Fourier spectra. (However, it should be noted that rotational information is available, in principle, from the measured FT–SES or FT–HBS *interferogram*, if not from the Fourier spectrum. This is because in the interferogram the many rotational resonances that are spread out in the frequency domain can interfere to give rise to prominent recurrences. The positions of these recurrences, which are analogous to those that appear in rotational coherence spectroscopy (12), provide direct information on the rotational constants of the species. See Section 1.5.3.)

A second important feature of FT–SES in relation to its frequency-domain counterpart is that it can readily be applied to species with short-lived excited intermediate states. This is because FT–SES can be implemented with short-pulse sources, those capable of producing significant fluorescence depletions even when a short-lived excited state is involved, with no cost in resolution. Indeed, the spectrum of Fig. 1.16(*a*) pertains to a species whose S_1 state is short-lived enough (2.6 ns) to present very significant problems to those wanting to perform moderate-resolution frequency-domain SES on it. Yet, one sees from the figure that high quality FT–SES spectra can be obtained for it when picosecond lasers are used.

Besides *trans*-stilbene, we performed FT–SES and FT–HBS experiments relating to torsional transitions in tolane, and van der Waals transitions in carbazole –water and fluorene–benzene. All of these experiments have pertained to vibrational intervals less than 200 cm^{-1} because of the rudimentary nature of the interferometer employed. The results on these species are similar in form to those in Fig. 1.16. That is, they are dominated by sharp ^qQ-branch features. These results will be reported elsewhere.

1.5.3. FT–CRS Results

Our application of FT–CRS has not gone beyond demonstrating various aspects of the method. The reason for this is that we have been more interested in applying

FT nonlinear methods to the study of minority species in seeded, supersonic molecular beams. Coherent Raman scattering, because the signal scales with the square of the number density of the species of interest (30), does not have the sensitivity of IDSRS, SES, or HBS in application to such sparse samples. Nevertheless, FT–CRS data on both gas-phase and molecular-beam samples of bare molecules prove several important points. First, we have verified that the resolution in FT–CRS (as in all the other FT nonlinear spectroscopies) is not limited by short-pulse sources (10). This was demonstrated on N_2 in a bulb at 1–2 torr. With laser pulses having widths of about 2.5 ns, modulated interferograms pertaining to a pure rotational Raman transition in the species were observed at interferometer delays of about 7 ns. The point is that even when interferometer delays are so large that the fields from the two arms of the interferometer no longer overlap (the echo regime (18)) resonance modulations persist, and hence useful spectroscopic information can still be gathered.

A second significant aspect of FT–CRS was demonstrated in pure-rotational Raman FT–CRS experiments on O_2 and N_2 (43). In particular, it was shown that high spectral resolution over a broad Raman-spectral range in a single scan could be achieved with FT–CRS implemented in such a way. A representative Fourier spectrum demonstrating this point is shown in Fig. 1.17. The spectrum is of $O_2(g)$.

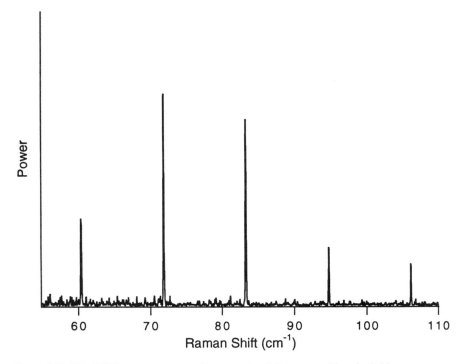

Figure 1.17. FT–CARS power spectrum of pure rotational Raman transitions in $O_2(g)$ at room temperature. See the text for further details.

It was obtained in a three-color experiment by using a narrow-band dye laser for ω_1, a very broadband dye oscillator for ω_2 (the dye laser grating was set to zeroth order), and 532-nm light from the Nd:YAG for ω_3. The CARS signal was detected through a monochromator operated with a broad detection bandwidth. Resolution in the spectrum is about 0.04 cm^{-1}.

One last feature of FT–CRS (and all FT spectroscopies) was demonstrated in experiments on jet-cooled benzene (43). Figure 1.18 shows FT–CRS interfero-

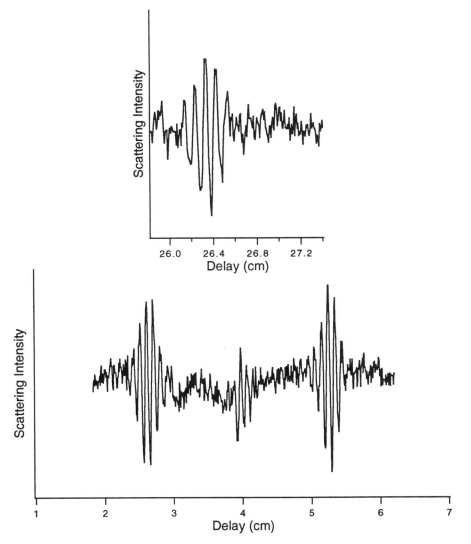

Figure 1.18. Portions of a FT–CARS *interferogram* obtained in pure rotational Raman experiments on jet-cooled benzene. At bottom are the first two rotational recurrences centered at delays of 1/(2B) and 1/B, respectively. At the top is the tenth recurrence centered at 5/B. For further details see the text.

grams corresponding to pure rotational CARS for this sample. One readily notes the presence of recurrences in these traces. The major recurrences are spaced by $1/(2B)$, where $B = 0.1897$ cm^{-1} is the nonunique rotational constant of benzene, an oblate symmetric top. They arise because the rotational Raman resonances in benzene are spaced by $2B$. Fourier transforming a spectrum containing equally spaced resonances produces constructive interferences in the resulting time domain trace, the spacing of these interferences being the inverse of the spacing of lines in the spectrum. [This fact is at the basis of the time-domain method of rotational coherence spectroscopy (12).] The point is that the manifestations of rotational structure can be very prominent in the interferograms obtained by FT spectroscopies. Indeed, these manifestations may be much more prominent and more readily interpreted than the manifestations of rotational structure in the Fourier spectrum. This is particularly true for large molecules because for these species the intensity is spread out among many features in the frequency domain but is localized in just a few widely separated recurrences in the interferogram. This aspect of the FT nonlinear spectroscopies could prove to be valuable in future structural studies of large species.

1.5.4. Down-Shifting Results

The down-shifting scheme described in Section 1.4.5 was demonstrated in FT–CARS experiments on CH$_4$(g) (9). The transitions studied were rotational members of the ν_3 C—H stretch fundamental of the molecule near 3000 cm^{-1}. Figure 1.19 shows an example of a down-shifted spectrum. The Raman resonance to which this spectrum corresponds is the $P^0(4)$ member of the ν_3 fundamental, which occurs at 2976.61 $\pm$ 0.05 cm^{-1} (44). The prominent peak appearing in Fig. 1.19 is at 60.15 $\pm$ 0.20 cm^{-1}. The latter frequency added to $\omega_s/2\pi$ should equal the Raman resonance frequency. Using the value for $\omega_s/2\pi$ quoted in Section 1.4.5, one obtains a value for the resonance frequency of 2976.87 cm^{-1} from the measured spectrum, which is within experimental error of the literature value. Also significant is the fact that the line width of the observed resonance is considerably less than the bandwidths of the ω_1 and ω_2 light sources (3 cm^{-1}). Obviously, this means that, like the ''normal'' FT nonlinear schemes, the down-shifting method allows for spectral resolution that is not limited by light source bandwidths. In the case of the spectrum in Fig. 1.19, we are confident, based on measurements made as a function of sample pressure, that the observed line width is limited by pressure broadening. As mentioned in Section 1.4.5, there is some question as to the effect of the Raman-shifting process on the resolution obtainable by the down-shifting method, but that question has not yet been fully addressed.

1.6. FUTURE OUTLOOK

In closing this chapter it is pertinent to speculate about future developments in and applications of FT nonlinear spectroscopies. As is evident from the preceding sections, the main focus in regard to these methods has been on their demonstration and characterization. To advance the techniques to the stage of routine application it is critical that they be implemented with high-quality, easy-to-use interferometers,

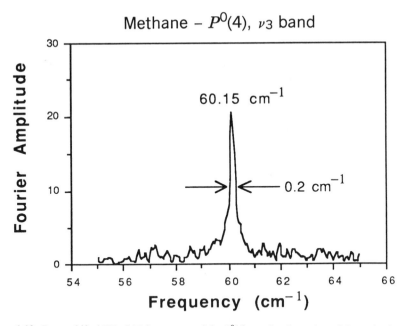

Figure 1.19. Down-shifted FT–CARS spectrum of the $P^0(4)$ rotational member of the ν_3 fundamental of methane. The 60.15-cm^{-1} value refers to the frequency of this rovibrational resonance as it appears in the down-shifted spectrum (after shifting by about 2916.7 cm^{-1}). [From Ref. 9 with permission. Copyright © 1988, Elsevier Science Publishers B.V.]

something that has not yet been done. Fortunately, advances in step–scan methodology (e.g., Ref. 5b) have now made the requisite instruments commercially available. With high-quality interferometers one gains at least three very important advantages. First, one obtains the ability to step the interferometer at intervals larger than allowed for minimal sampling and still obtain good Fourier spectra. This is a particularly important capability for the FT nonlinear methods because the data collection rate is often limited to the low repetition rate of a pulsed laser source. Second, one gains the ability to study the higher frequency intervals that are inaccessible when low-quality interferometers are employed. Third, information directly about Fourier amplitudes, rather than just about the squares of those amplitudes, becomes available. One can then use the various data processing methods that rely on knowledge of the phases of Fourier components. With a high-quality interferometer employed, the FT–SES, FT–HBS, and FT–IDSRS methods could likely become very prolific contributors in spectroscopic studies of gas-phase species and species in cold molecular beams.

A second area for potential development involves extending the spectral range of the FT nonlinear methods. For example, suppose that FT–HBS could be implemented in the vacuum ultraviolet (UV). One would then have the ability to measure excited-state splittings at a resolution unlimited by light-source bandwidths, with Doppler widths proportional to the splittings themselves rather than to the vacuum UV photon frequencies. This would represent a very significant increase in the capability to study states lying in this important spectral region. Or, suppose that FT–SES could be readily implemented in the far- and vacuum UV. One would then have the ability to measure the equivalent of resonance-Raman spectra at a resolution determined by the interferometer, not by the light source nor by a monochromator. Of course, the difficulty in extending these methods far into the UV involves finding appropriate optical materials for the interferometer. On the other hand, one notes that such an interferometer need not be designed for stability on the scale of the UV wavelengths to be used, as an interferometer for FT linear spectroscopy in the UV must be. Instead, the much-less-stringent criterion of stability on the scale of the wavelengths characterizing the excited-state (or ground-state) splittings (i.e., IR wavelengths) is the one that applies.

A third area for development involves demonstrating FT versions of other non-linear spectroscopies as well as inventing new classes of nonlinear interferometric schemes. In regard to the latter, one notes the work of Dugan et al. (19,20) as evidence that the numerous experimental parameters associated with a nonlinear spectroscopic process provide the opportunity to devise new variations of nonlinear interferometry that have important new features and advantages. Regarding the former, there are many nonlinear spectroscopies not touched on above that might be usefully implemented in a FT mode. These include, for example, nonlinear Raman methods, such as stimulated Raman gain and loss and Raman-induced Kerr effect spectroscopy (30), as well as SES and HBS methods utilizing photoionization (45), or degenerate four-wave mixing (46) as the probe process.

A full exploration of the use of the interferogram itself (i.e., without Fourier transforming it) to obtain information is a fourth promising area for future work. We have mentioned above that such an approach could prove a powerful one for measuring the rotational constants of large species because rotational information is "concentrated" into localized recurrences in the interferogram. The small effective spectral ranges that generally characterize the results from FT nonlinear spectroscopies (usually laser bandwidth-limited to several wavenumbers) work to one's advantage in this regard, because one can legitimately filter out interferogram noise at frequencies outside the pertinent spectral range. Such filtering techniques might be expected to enhance the observability of rotational recurrences in interferograms considerably. Decay times represent information of a second type that might best be derived directly from the interferogram. As we mentioned in Section 1.2, this was the impetus for the early development of nonlinear interferometric methods (13–16) and has been investigated for other schemes (19,20). Little has been done in this regard, however, in conjunction with the FT spectroscopies that we have demonstrated in our laboratory.

Finally, while the usefulness of the methods discussed herein has been clearly demonstrated in gas-phase and molecular-beam studies, their applicability and potential advantages in studies of condensed-phase samples has yet to be addressed experimentally. One can imagine that the methods might be particularly useful in regard to very accurate measurements of solvent-dependent vibrational frequency shifts and in the characterization of vibrational dephasing times. In addition, the double-resonance FT–SES and FT–HBS methods applied to condensed-phase systems would combine the chromophore sensitivity of resonance Raman spectroscopy with all the resolving power that one would ever need in such studies. Hence, one can foresee the possibility that these methods might allow vibrational studies of complicated systems (e.g., proteins) at an unprecedented level of detail and accuracy. We hope to explore these possibilities in the years ahead.

ACKNOWLEDGMENTS

It is a pleasure to acknowledge the work of the members of my research group who have been involved with the development and application of Fourier transform nonlinear spectroscopies. Principal among these are Greg Hartland and Bryan Henson. Leslie Connell, Timothy Corcoran, Robert Hertz, Paul Joireman, Shane Ohline, and Vincent Venturo have also contributed. This work was supported by the U.S. Department of Energy, Office of Basic Energy Sciences through grant No. DE-FG03-89-ER 14066.

REFERENCES

1. For example, R. J. Bell, *Introductory Fourier Transform Spectroscopy*, Academic, New York, **1972**.

2. P. B. Felgett, *J. Phys. Radium*, **1958**, *19*, 187. P. Jacquinot, *Rep. Prog. Phys.*, **1960**, *23*, 267.

3. For example, P. Luc and S. Gerstenkorn, *Appl. Opt.*, **1978**, *9*, 1327.

4. For example, T. Hirschfeld and B. Chase, *Appl. Spectrosc.*, **1986**, *40*, 133. V. M. Hallmark, C. G. Zimba, J. D. Swalen, and J. F. Rabolt, *Spectroscopy*, **1987**, *2*, 40.

5. For example, (a) D. J. Donaldson and S. R. Leone, *Chem. Phys. Lett.*, **1986**, *132*, 240. (b) G. V. Hartland, D. Qin, and H.-L. Dai, *J. Chem. Phys.*, **1993**, *98*, 2469.

6. P. M. Felker and G. V. Hartland, *Chem. Phys. Lett.*, **1987**, *134*, 503. G. V. Hartland and P. M. Felker, *J. Phys. Chem.*, **1987**, *91*, 5527.

7. P. M. Felker, B. F. Henson, T. C. Corcoran, L. L. Connell, and G. V. Hartland, *Chem. Phys. Lett.*, **1987**, *142*, 439.

8. G. V. Hartland, B. F. Henson, L. L. Connell, T. C. Corcoran, and P. M. Felker, *J. Phys. Chem.*, **1988**, *92*, 6877.

9. T. C. Corcoran, L. L. Connell, G. V. Hartland, B. F. Henson, and P. M. Felker, *Chem. Phys. Lett.*, **1988**, *147*, 517.

10. G. V. Hartland, B. F. Henson, and P. M. Felker, *J. Chem. Phys.*, **1989**, *91*, 1478.

11. G. V. Hartland, P. W. Joireman, L. L. Connell, and P. M. Felker, *J. Chem. Phys.*, **1992**, *96*, 179.

12. For example, see P. M. Felker, *J. Phys. Chem.*, **1992**, *96*, 7844.

13. S. Asaka, H. Nakatsuka, M. Fujiwara, and M. Matsuoka, *Phys. Rev. A*, **1984**, *29*, 2286.

14. N. Morita and T. Yajima, *Phys. Rev. A*, **1984**, *30*, 2525.

15. R. Beach, D. DeBeer, and S. R. Hartmann, *Phys. Rev. A*, **1985**, *32*, 3467.

16. T. Hattori, A. Terasaki, and T. Kobayashi, *Phys. Rev. A*, **1987**, *35*, 715.

17. J. E. Golub and T. W. Mossberg, in *Ultrafast Phenomena V*, Fleming, G. R. and Siegman, A. E. (Eds.), Springer, Berlin, **1986**, p. 164.

18. D. DeBeer, L. G. Van Wagenen, R. Beach, and S. R. Hartmann, *Phys. Rev. Lett.*, **1986**, *56*, 1128.

19. M. A. Dugan, J. S. Melinger, and A. C. Albrecht, *Chem. Phys. Lett.*, **1988**, *147*, 411.

20. M. A. Dugan and A. C. Albrecht, *Phys. Rev. A*, **1991**, *43*, 3877 and 3922.

21. The treatment is less than rigorous because it does not account for realistic lineshapes, and it does not begin at the level of quantum mechanical and excitation-field amplitudes.

22. Depending on the way in which the light beams are directed in and out of the interferometer the spectral densities can go as $(1 + \cos \omega\tau)$ and/or $(1 - \cos \omega\tau)$. For our purposes here the appropriate sign is not important. For simplicity, we shall always assume the $+$ case to obtain.

23. See, for example, S. A. Wittmeyer and M. R. Topp, *Chem. Phys. Lett.*, **1989**, *163*, 261.

24. C. Kittrell, J. L. Kinsey, S. A. McDonald, D. E. Reisner, R. W. Field, and D. H. Katayama, *J. Chem. Phys.*, **1981**, *75*, 2056. C. E. Hamilton, J. L. Kinsey, and R. W. Field, *Annu. Rev. Phys. Chem.*, **1986**, *37*, 493.

25. W. Bronner, P. Oesterlin, and M. Schellhorn, *Appl. Phys. B*, **1984**, *34*, 11.

26. P. Esherick and A. Owyoung, *Chem. Phys. Lett.*, **1983**, *103*, 235. P. Esherick, A. Owyoung, and J. Pliva, *J. Chem. Phys.*, **1985**, *83*, 3311.

27. B. F. Henson, G. V. Hartland, V. A. Venturo, and P. M. Felker, *J. Chem. Phys.*, **1989**, *91*, 2751.

28. G. V. Hartland, B. F. Henson, V. A. Venturo, R. A. Hertz, and P. M. Felker, *J. Opt. Soc. Am. B*, **1990**, *7*, 1950.

29. W. C. Wiley and I. H. McClaren, *Rev. Sci. Instrum.*, **1955**, *26*, 1150.

30. See, for example, Y. R. Shen, *The Principles of Nonlinear Optics*, Wiley, New York, **1984**.

31. J. A. Shirley, R. J. Hall, and A. C. Eckbreth, *Opt. Lett.*, **1980**, *5*, 380.

32. W. R. Trutna, Y. K. Park, and R. L. Byer, *IEEE J. Quantum Electron.*, **1979**, *QE-15*, 648.

33. A. Owyoung, in *Chemical Applications of Nonlinear Raman Spectroscopy*, Harvey, A. B. (Ed.), Academic, New York, **1981**, pp. 281–320.

34. P. M. Felker, in *Molecular Dynamics and Spectroscopy by Stimulated Emission Pumping*, Dai, H.-L. and Field, R. W. (Eds.), World Scientific (in press).

35. G. V. Hartland, B. F. Henson, V. A. Venturo, and P. M. Felker, *J. Phys. Chem.*, **1992**, *96*, 1164.

36. For example, D. A. Long, *Raman Spectroscopy*, McGraw-Hill, New York, **1977**.

37. B. F. Henson, G. V. Hartland, V. A. Venturo, R. A. Hertz, and P. M. Felker, *Chem. Phys. Lett.*, **1991**, *176*, 91.

38. B. F. Henson, G. V. Hartland, V. A. Venturo, and P. M. Felker, *J. Chem. Phys.*, **1992**, *97*, 2189.

39. K. C. Janda, J. C. Hemminger, J. S. Winn, S. E. Novick, S. J. Harris, and W. Klemperer, *J. Chem. Phys.*, **1975**, *63*, 1419. J. B. Hopkins, D. E. Powers, and R. E. Smalley, *J. Phys. Chem.*, **1981**, *85*, 3739. P. R. R. Landridge-Smith, D. V. Brumbaugh, C. A. Haynam, and D. H. Levy, *J. Phys. Chem.*, **1981**, *85*, 3742. K. O. Börnsen, H. L. Selzle, and E. W. Schlag, *J. Chem. Phys.*, **1986**, *85*, 1726. K. Law, M. Schauer, and E. R. Bernstein, *J. Chem. Phys.*, **1984**, *81*, 4871. M. Schauer and E. R. Bernstein, *J. Chem. Phys.*, **1985**, *82*, 3722.

40. E. Arunan and H. S. Gutowsky, *J. Chem. Phys.*, **1993**, *98*, 4294.

41. B. F. Henson, V. A. Venturo, G. V. Hartland, and P. M. Felker, *J. Chem. Phys.*, **1993**, *98*, 8361.

42. B. W. van de Waal, *Chem. Phys. Lett.*, **1986**, *123*, 69.

43. B. F. Henson, G. V. Hartland, and P. M. Felker, (unpublished results).

44. J. Herranz and J. P. Stoicheff, *J. Mol. Spectrosc.*, **1963**, *10*, 448.

45. D. E. Cooper, C. M. Klimcak, and J. E. Wessel, *Phys. Rev. Lett.*, **1981**, *46*, 324.

46. Q. Zhang, S. Kandel, T. Wasserman, and P. H. Vaccaro, *J. Chem. Phys.*, **1992**, *96*, 1640.

Chapter **II**

NEAR-INFRARED LASER-OPTOTHERMAL TECHNIQUES

Roger E. Miller
Department of Chemistry,
University of North Carolina, Chapel Hill, North Carolina

Laser Techniques In Chemistry, Edited by Anne B. Myers and Thomas R. Rizzo.
Techniques of Chemistry Series, Vol. XXIII.
ISBN 0-471-59769-4   © 1995 John Wiley & Sons, Inc.

2.1. INTRODUCTION

Infrared (IR) spectroscopy is well known for its generality, which comes from the fact that most molecules possess at least one IR allowed vibrational mode. For this reason, IR spectroscopy is potentially of great importance as a state selective probe in the study of chemical dynamics. Despite this potential, however, it is only quite recently that it has been used for this purpose in conjunction with molecular beam methods. The main difficulty has been one of sensitivity. Due to the low energy associated with IR photons, the detector sensitivities in this spectral region are much smaller than those characteristic of photomultipliers typically used in the visible and ultraviolet (UV) regions of the spectrum. Despite this and other difficulties, considerable progress has recently been made and a number of approaches are now being used to great advantage in the study of vibrational dynamics. Although a disadvantage in some respects, the long fluorescence lifetimes associated with the excited vibrational states of most molecules ensures that the spectral resolution is very high, particularly when continuous wave (CW) lasers are used. High spectral resolution is an important advantage from both the point of view of spectroscopic characterization and state selective probing of molecules.

The conventional approach to obtaining the IR spectra of molecules in the gas phase involves the measurement of the attenuation of a light source due to molecular absorption. Due to their high spectral brightness and purity, CW IR lasers are particularly well suited to this type of measurement. The primary sensitivity limitation in such an experiment arises from the fact that the laser output is detected directly so that any amplitude noise associated with the laser appears as noise in the spectrum. A number of approaches have been used to minimize this problem (1–3), based upon the idea of shifting the signal to high frequencies where the lasers are relatively quiet. These methods can provide near shot noise limited detection corresponding to an absorbance of approximately 10^{-6} (1). The primary application of this method has been to the spectroscopic study of molecules cooled in a free-jet expansion. In order to obtain a high density of molecular absorbers, the expansion is usually generated by a pulsed molecular beam source (4). A complete discussion of these direct absorption methods is given in Chapter III in this volume.

It is now standard practice to use various indirect detection methods to overcome some of the noise problems resulting from laser amplitude fluctuations. Such ''action'' spectra involve the measurement of some property of the system, which changes as a result of laser excitation. The associated gain in sensitivity can be quite high. In fact, the vast majority of laser-based techniques that have been used to study chemical dynamics are based on some form of indirect detection. Perhaps the best known of these is laser induced fluorescence (LIF) (5), which is used extensively in the visible and UV regions of the spectrum where the excited-state lifetimes are rather short and fluorescence occurs before the molecules have time to move out of the observation volume. The fluorescence that results from laser excitation is collected by an optical system and focused onto a detector, normally a photomultiplier. Chapter IV in this volume discusses this method in detail.

The long fluorescence lifetimes associated with vibrationally excited states (typically in the millisecond regime) give the molecules time to move out of the optical collection volume so that the emission becomes spread out in space, causing a decrease in the collection efficiency of the light. When compounded with the low sensitivity of the detectors in this spectral region, the overall sensitivity of LIF in the IR is greatly reduced relative to the visible–UV regions of the spectrum. Nevertheless, if the molecular density is high enough, an IR fluorescence spectrum can be recorded, as recently demonstrated by Klemperer and co-worker (6). It remains to be seen if this method will have the sensitivity needed for use as a state selective probe in a chemical dynamics experiment.

In addition to causing a molecule to fluoresce, vibrational excitation increases the total energy of a molecule. For high-frequency vibrations, such as O—H and C—H stretches, this vibrational energy is large ($3000-4000$ cm^{-1}) compared with kT (200 cm^{-1}) and we can expect that a thermal detector used to monitor the energy of a beam of molecules would easily see the laser induced energy change. The method obviously takes advantage of the fact that the excited-state lifetime is long with respect to the flight time of the molecules from the laser excitation volume to the thermal detector (bolometer). This form of indirect detection is the subject of this chapter and is referred to as the optothermal detection technique (7,8). When a vibrationally hot molecule impinges on the detector, the associated vibrational energy is relaxed, causing an increase in the temperature of the detector. Since the bolometer detector does not ''see'' the laser output directly, the method does not suffer from the laser noise problems discussed above. For the same reason the method is also background free. As we will see in the following sections, the result is a much lower minimum detectable absorption, making the method applicable to the very low density environments associated with highly collimated molecular beams. The method is therefore applicable to not only spectroscopic studies, but also a wide range of dynamical experiments in single or crossed molecular beams. A number of these methods are discussed in some detail below. Before doing so, however, it is necessary to quantify the characteristics of the thermal detectors currently available and to discuss their applicability for the present purposes.

2.2. THE BOLOMETER

2.2.1. Performance Specifications

For the present purposes there are two main specifications that define the performance of a thermal detector, namely, its sensitivity and response time. The sensitivity is normally characterized in terms of the responsivity (S), which is a measure of the detector output (in volts per watt of power in), and the detector noise (N) measured in volts per (hertz)$^{1/2}$. These two can be combined to give the noise equivalent power (NEP = N/S) in watts per (hertz)$^{1/2}$. As illustrated below, the sensitivity of these thermal detectors can be quite high in comparison with the energy absorbed by the molecules in a beam from a CW laser. Liquid helium cooled bolometers operating near their theoretical limit can have an NEP as low as 1×10^{-14} W Hz$^{-1/2}$.

For a number of reasons, the operating temperature of the detector can have a large influence on its noise equivalent power. First, as the temperature of the device is lowered the Johnson noise also decreases. Second, the heat capacities of the detector materials generally decrease rapidly with decreasing temperature, such that the same power input results in a larger temperature rise. Finally, for bolometers that operate at very low temperatures (<10 K) the electrical response is very strongly dependent on temperature. As a result, the highest sensitivity bolometers are those that operate near absolute zero. Nevertheless, thermal detectors operating at room temperature can have NEP values that are sufficient for some molecular beam applications. These are summarized below.

2.2.2. Room Temperature Thermal Detectors

Although there are clear sensitivity advantages to using liquid helium cooled detectors in optothermal spectroscopy, there are also the complexities associated with operating the necessary cryogenics, which may not be necessary for all applications. It is therefore important to consider what can be done using room temperature detectors. Although a number of different types of thermal detectors exist that operate at or near room temperature, the pyroelectric detectors (9) seem best suited for the present purposes. They have relatively high sensitivity (NEP = 3 $\times$ 10^{-10} W Hz$^{-1/2}$ and are quite small (often the size of a transistor can). In many cases these detectors come with a built in preamplifier to help reduce the noise. Although these detectors are three to four orders of magnitude less sensitive than cryogenically cooled bolometers, they have been used to record spectra in collimated beams (10). Some of the loss in sensitivity can be compensated for by moving the detector closer to the molecular beam source, thus increasing the beam flux. Since the detector is less sensitive than the liquid helium cooled bolometers, it saturates less easily and can tolerate the higher beam fluxes. Perhaps the best application for these detectors is in monitoring situations, to ensure that the laser induced vibrational excitation is optimized, such as in a scattering experiment involving vibrationally excited species. In spectroscopic applications it is limited to species that can be generated in high concentrations and that have large IR transition moments. As an example, the spectrum of HF has been recorded in this way using an F-center laser, yielding a signal-to-noise (S/N) ratio of approximately 100–1000:1 with the detector placed roughly 10 cm from the molecular beam source. There is at least one experiment reported in the literature (11) where such a detector has been used to monitor the laser induced vibrational excitation in a crossed molecular beam experiment. The initial costs associated with a pyroelectric detection system are approximately an order of magnitude less than those of a cryogenic bolometer and the operating costs are essentially zero compared with the substantial operating costs associated with liquid helium cooled bolometers.

2.2.3. Semiconductor Detectors

By far the most common bolometers used in optothermal experiments are those based on the use of doped semiconductor materials, usually silicon or germanium (12,13). The impurity concentrations can be varied over a limited range in order to define the operating temperature and resistance of the bolometer. The best overall performance is achieved when the liquid helium is pumped below the lambda point, using a small backing pump to lower the vapor pressure over the liquid. This not only lowers the temperature, thus reducing noise and heat capacities, but also ensures that the thermal bath to which the bolometer is attached is very stable in temperature, because the helium is superfluid at these temperatures. Composite bolometers formed by attaching a small active element (typically 0.5×0.5 mm) to a collecting diamond or sapphire plate (typically 2×6 mm) provide additional advantages since the latter materials have lower heat capacities at these temperatures than the doped silicon or germanium.

The operating principle of the bolometer is simple. The bolometer is biased in the manner shown in Fig. 2.1(a), where the voltage source is a stable mercury battery. By choosing the load resister to be large with respect to the resistance of

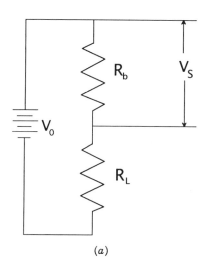

(a)

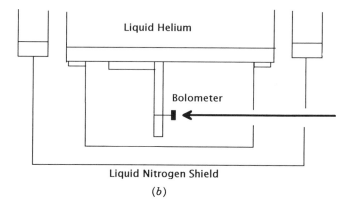

(b)

Figure 2.1. (a) Bolometer electrical biasing and (b) the physical mounting geometry.

the bolometer, the current is kept nearly constant. In this way, resistance changes associated with temperature are transformed into voltage changes that can be easily measured. Since the bolometer is a rather high impedance device ($\approx 1-10$ MΩ), first stage amplification is normally done on the cold surface of the cryostat using a cold FET based amplifier. As shown in Fig. 2.1(b), the bolometer must be shielded from background IR radiation in order to obtain the best performance. This is accomplished using two copper shields, one thermally grounded to the helium cryostat and the second to the liquid nitrogen cryostat. The entrance holes through which the molecular beam passes are kept as small as possible to avoid unnecessary thermal loading of the bolometer.

The commercially available bolometers (14) have operating characteristics that are very near the theoretical limit. Therefore, when designing a system one is forced to tradeoff sensitivity and response time to meet the particular application. A bolometer having a time constant of approximately 5 ms will have an NEP of approximately 1×10^{-13} W Hz$^{-1/2}$ under the conditions needed for molecular beam applications. For comparison, the F-center lasers typically used in optothermal experiments have a single mode output power of approximately 100 mW at 3700 cm^{-1}. From this we can see that the minimum detectable absorption is approximately 10^{-12}, or six orders of magnitude lower than what has been achieved by direct absorption. This additional sensitivity is what enables the method to be applied to highly collimated molecular beam experiments where the molecular densities are much lower than in the free-jet expansion itself. The somewhat lower NEP quoted in an earlier section of this chapter corresponds to the bolometer being totally in the dark. Since the bolometer is necessarily exposed to some room temperature background radiation in a molecular beam application, the NEP is always somewhat degraded from the theoretical limit. Loading of the bolometer due to exposure to a high molecular beam flux will further degrade its sensitivity. For this reason, several approaches have been used to avoid exposing the bolometer to the full molecular beam, as discussed below.

2.2.4. Superconducting Detectors

Metal film superconducting bolometers take advantage of the fact that at the superconducting transition the resistance of the film decreases rapidly with temperature (15). Due to the very low heat capacity of the thin-metal film and its good thermal contact with the substrate, these detectors are noted for their fast response time. A typical response time for a Sn superconducting bolometer is approximately 1 μs with an NEP of approximately 10^{-12} W Hz$^{-1/2}$ (13). To maintain the bolometer at the superconducting transition, the temperature of the thermal bath must be carefully controlled. This is simply done by wrapping a heating wire around the copper block to which the bolometer is mounted. The copper block is weakly thermally coupled to the liquid helium bath, which is usually maintained below the lambda point to provide the best thermal stability possible. These devices are fast enough that they can be used in conjunction with pulse laser systems (16). It is important to note, however, that if the bolometer is exposed to the full molecular beam flux, of the bolometer, the response time soon becomes limited by the frost built up on the surface rather than by the heat capacity of the film itself. At that point the bolometer must be warmed up in order to drive off the frost.

With the recent developments that have occurred in the field of high-temperature superconductors, there has been some progress in using such materials to build bolometers that operate at liquid nitrogen temperatures (17). The NEP of these bolometers [2.4×10^{-11} W Hz$^{-1/2}$ for a 1×1-mm element (17)] can be higher than that of the room temperature detectors discussed above. Improvements in this performance are likely forthcoming, particularly for molecular beam applications where the blackening agents normally used to ensure that the IR light is absorbed by the detector are not required, thus reducing the heat capacity of the detector. Although these detectors are still not commercially available, they may eventually provide a low-cost alternative to those operating at liquid helium temperatures. Work is also underway to develop microbolometer arrays from these materials (18), which may have interesting applications in the study of photofragment spectroscopy. These microbolometers (6×13 μm) have somewhat better NEP values (4.5×10^{-12} W Hz$^{-1/2}$) due simply to their smaller size.

2.3. THE OPTOTHERMAL METHOD: TECHNICAL DETAILS

The fact that the optothermal method has sufficient sensitivity to obtain IR spectra of highly collimated molecular beams is important given that both pressure and Doppler broadening can be largely eliminated in the beam and the geometry of the laser–molecule interaction region can be tailored to the particular application. When combined with the low rotational temperatures that can be achieved using a free-jet expansion source, the spectra obtained in this way normally provide complete rotational resolution. Figure 2.2 shows a schematic diagram of what has be-

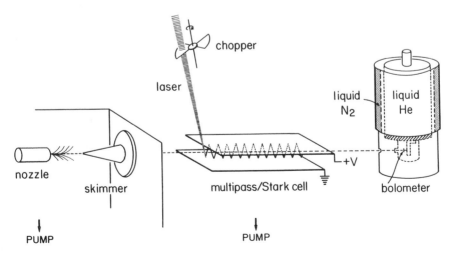

Figure 2.2. Optothermal spectrometer with a parallel plate multipass cell.

come a rather standard configuration for an optothermal spectrometer. The vacuum system consists of a double differentially pumped set of chambers. The source chamber is typically pumped by a diffusion pump with a pumping speed of approximately 5000 L s^{-1}, while the pump on the detector chamber can be somewhat smaller (2000 L s^{-1} or less). A conical skimmer, with an aperture of approximately 0.5 mm, separates the two chambers and samples the central portion of the free-jet expansion emanating from a pinhole nozzle of approximately 50-μm diameter. The resulting molecular beam is collision free in the second chamber. In cases where a semiconductor bolometer is used, it is placed approximately 50–100 cm from the skimmer. In the apparatus shown in Fig. 2.2 the laser crosses the molecular beam approximately 50 times as a result of multiple reflections between two parallel mirrors. By chopping the laser and using phase sensitive detection of the bolometer signal, the dc signal associated with the kinetic energy of the molecular beam and the adsorption energy of the molecules to the detector can be eliminated. In this way, a zero baseline spectrum can be recorded by scanning the laser in frequency. Nonzero signals are only seen when the laser is absorbed by molecules in the beam, thus modifying their energy relative to the dc background.

It is important to emphasize that although the optothermal method is an indirect detection technique, given that the bolometer does not "see" the laser, it does detect the full intensity of the molecular beam. This detection can present some difficulties, particularly when using cryogenically cooled bolometers. At the operating temperatures of these bolometers the vapor pressure of most species is essentially zero, which means that when the molecules strike the bolometer they freeze to its surface. Over a period of time this leads to a build up of frost on the surface, resulting in an increase in the heat capacity of the detector, and thus a slower response time and decreased sensitivity. In addition, since the molecules tend to freeze onto the surface to form an amorphous solid, there is a tendency over time for crystallization of this layer to occur. In view of the high sensitivity of the bolometer, this can be a major problem since the associated energy release can be quite large. We have found from experience that if the molecular beam is composed entirely of condensable gases, then crystallization of the frost layer occurs in random and sudden events that lead to noise "spikes." These noise spikes resemble sharp transitions in a spectrum. To avoid this difficulty, the species of interest is normally diluted to approximately 1–10% in helium. This process not only decreases the deposition rate of the frost on the bolometer, but the helium carrier gas also seems to prevent the crystallization from occurring, even at the same overall condensable species exposure.

Another difficulty that can arise when the full molecular beam flux is directed onto the bolometer is microphonic noise. Due to the presence of the skimmer and the various collimators mounted on the cryostat shields, vibrations can lead to a modulation of the molecular beam flux, thus causing noise. For this reason, it is important to carefully align the instrument so that small motion of these collimators will not change the flux reaching the bolometer. This can be done by ensuring that the collimators are somewhat larger or smaller than the bolometer. In the first case the bolometer remains fully illuminated by the molecular beam even when the collimators move, while in the latter the motion of the collimators simply changes where the beam hits the bolometer. It is also important to point out that efforts need to be made to ensure that the bolometer does not see any of the scattered laser light, as this will cause both an offset of the baseline and noise.

The problem of frost buildup on the surface of the bolometer can be overcome in several ways. For example, a set of quadrupole fields can be used to state specifically focus the off axis component of the molecular beam onto the bolometer, while a beam stop is used to shield the bolometer from the beam itself, as shown in Fig. 2.3. The result is a considerable reduction in the flux of condensable species onto the bolometer, as well as the elimination of the carrier gas. This electric resonance optothermal spectroscopy (EROS) can have S/N advantages, but is of course limited to polar molecules that can be focused. Transition intensities observed in this mode of operation must be viewed with caution, however, since they depend not only on the state populations and transition moments, but also on the focusing characteristics of the individual states (19).

For molecules that dissociate upon IR excitation (typical of weakly bound van der Waals complexes) the IR spectrum can be obtained by positioning the bolometer off to one side of the molecular beam where only the recoiling photofragments can be detected. This has the advantages that the large flux of molecules from the primary beam is not detected and there is no restriction to polar systems. As discussed in section 2.6, this approach has a number of interesting additional advantages in the study of the dissociation dynamics.

The majority of work that has been carried out to date using the optothermal method has been done with an F-center laser (FCL) operating in the region from

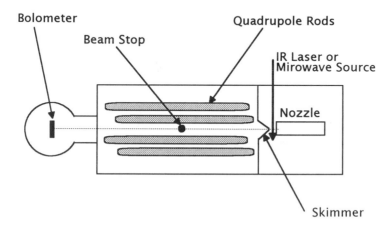

EROS Apparatus

Figure 2.3. Schematic diagram of an electric resonance optothermal spectrometer.

2900 to 4500 cm^{-1}. Although commercially available, these lasers are not supplied with the tuning capabilities needed for high-resolution spectroscopy experiments. Nevertheless, individual groups have addressed this problem in a number of different ways (20–22) giving systems that provide continuous coverage over the stated spectral region with a resolution of 1–5 MHz. Power outputs vary from 1 to 100 mW over this range. More recently, a CW YAG pumped FCL has become commercially available in the 5800–6780-cm^{-1} region and has been used in optothermal experiments (23). This region is favorable for studying vibrational overtone transitions in molecules, and the related area of the study of intramolecular vibrational energy redistribution (IVR) in molecules (24,25). The lower transition moments associated with these overtone transitions are compensated in part by the higher power (~500 mW) associated with the lasers in this region. For high overtone transitions, CW dye or Ti:Sapphire lasers have also been used in an optothermal spectrometer (26,27).

Diode lasers have also been used in optothermal experiments (7,28). In fact, the original demonstration of the method was done using a diode laser pumping the fundamental in CO at 2200 cm^{-1} (7). Nevertheless, the low output power and imperfect spectral coverage characteristic of these lasers makes them somewhat less suitable for this application, at least at present. The real advantage of these devices is that they are available for a very wide range of near-IR wavelengths. Given the rapid advances that are occurring in diode laser technology and the possibility of using very high Q resonant build up cavities to enhance the circulating power, these devices will undoubtedly have an important place in future optothermal experiments.

2.4. SPECTROSCOPIC APPLICATIONS OF THE OPTOTHERMAL METHOD

As an illustration of the capabilities of the optothermal technique, Fig. 2.4 shows a spectrum of the ν_1 band of the HCN dimer recorded using an F-center laser (29). This spectrum was obtained by expanding a 1% HCN in helium mixture from a pinhole nozzle of 35-μm diameter. It is clear from the spectrum that the HCN dimer is linear, given that the strong transitions in the spectrum form simple P and R branches. The C—H stretching vibration excited in this case corresponds to the ''free'' C—H bond on the proton acceptor molecule. The weaker lines have been assigned (29) to hot band transitions associated with the low-frequency symmetric and asymmetric bending modes of the hydrogen bond. What is particularly interesting about this spectrum is the fact that the two HCN monomer transitions appearing in this region have the opposite sign to those of the dimer. Since we know that the vibrationally excited state of the HCN monomer is long lived, excitation of the monomer must increase the energy delivered by the molecules to the bolometer. This implies that vibrational excitation of the dimer results in a decrease in the energy seen by the bolometer. This can be understood if one considers that the binding energy of these complexes is considerably less than the photon energy in the near-IR. Consequently, the dimer can dissociate into two monomers that subsequently recoil out of the molecular beam, thus missing the bolometer altogether. The result is a laser induced decrease in the molecular beam flux, and hence a reduction in the energy delivered to the bolometer.

The resolution in the spectrum shown in Fig. 2.4, although clearly very high, is limited by Doppler broadening resulting from the nonorthogonality of the laser molecular beam crossings, as shown in Fig. 2.2. Since the crossings are not orthogonal, the distribution of velocities in the molecular beam gives rise to a spread in Doppler shifts. For the parallel plate multipass cell used in this case, the width of the transitions are approximately 10 MHz full width at half-maximum (fwhm). As Fig. 2.5 demonstrates, the dimer transitions associated with the ν_1 band of the dimer have the same line width as that of the monomer, which indicates that the dissociation lifetime of the dimer is quite long [>140 ns (29)], such that lifetime broadening is small with respect to the observed Doppler width.

The situation is quite different for the ν_2 band (corresponding to excitation of the hydrogen bonded C—H stretching vibration), also shown in Fig. 2.5. In this case, the transitions associated with the dimer are much broader than those of the monomer and the lineshape, instead of being Gaussian, is Lorentzian. More precisely, the lineshape is a Voigt profile, since the contribution to the width from Doppler and lifetime broadening are comparable. The Lorentzian component of the broadening can be accounted for by a dissociation lifetime of the complex of 6.1 ns (29). It is clear from measurements such as these that the dissociation dynamics of these complexes is highly dependent on the particular mode excited. As discussed in somewhat more detail in section 2.6, the study of such mode specific effects can reveal the detailed dynamical pathways. The high resolution of the optothermal method is clearly useful not only for obtaining fully rotationally resolved spectra, but also for carrying out careful lineshape measurements that can be used to determine dissociation lifetimes of these complexes. A large number of such studies have been carried out using the optothermal method (8,30–32).

It is important to point out that even though the near-IR spectra being discussed here primarily probe the intramolecular degrees of freedom of the complex, the intermolecular vibrations can also be studied by observing both weak hot band transitions of the type seen in Fig. 2.4 and combination bands involving the intermolecular modes (33,34). In cases where both are observed, it is not only possible to determine the intermolecular frequencies corresponding to the monomer units

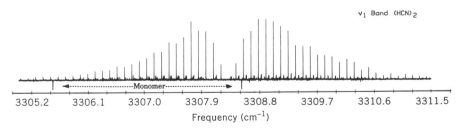

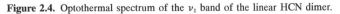

Figure 2.4. Optothermal spectrum of the ν_1 band of the linear HCN dimer.

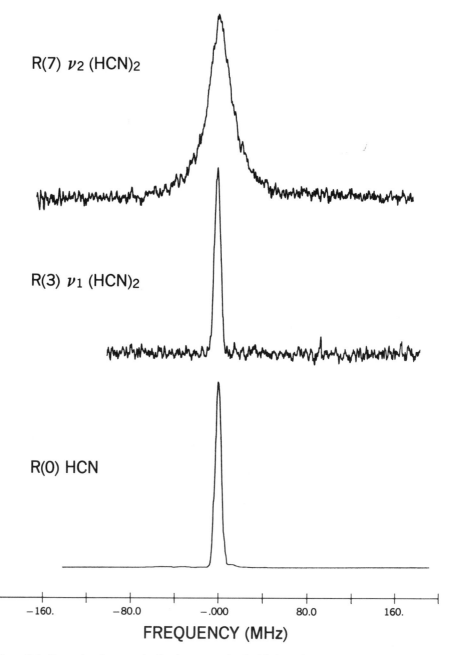

Figure 2.5. Comparison between the lineshapes associated with the HCN monomer and the ν_1 and ν_2 bands of the dimer.

being in the ground state, as obtained by far-IR spectroscopy (35), but also their dependence on the intramolecular vibrational state. This latter information is important in understanding the nature of the coupling between the low- and high-frequency vibrations in the complex, which are ultimately responsible for their dissociation.

2.5. DOUBLE-RESONANCE METHODS

As the HCN dimer example illustrates, even the high resolution available from the optothermal method is sometimes insufficient to observe the homogeneous broadening associated with the dissociation of the excited state. In addition, hyperfine structure at the level below 1 MHz is normally lost in the Doppler broadening. One way of improving the resolution is to combine IR and microwave spectroscopies, so that the states of interest can be accessed using the IR, while at the same time having the resolution characteristic of the microwave. Recent optothermal IR–microwave double-resonance experiments (36) have been reported on the v_1 mode of the HCN dimer using an apparatus similar to the one shown in Fig. 2.2, with the addition of a modified microwave E-bend to launch the microwave radiation between the plates of the multipass cell. The resolution available in the microwave transitions associated with the upper vibrational state was sufficient to determine the lifetime of the complex to be 1.7(5) μs (36), well in excess of what can be measured from the IR line widths. The gain in resolution comes primarily from the fact that the Doppler broadening at microwave frequencies is small.

Infrared–infrared double-resonance spectroscopy is also of interest in that it provides access to higher lying vibrational states in a stepwise manner. For this to be viable, the intermediate vibrational levels must have a long enough lifetime so that dissociation does not occur before a second photon can be absorbed. An example of such a system is Ar–HF (37). Optothermal spectra of this complex have been obtained for the fundamental H–F stretch, and the results show that the vibrationally excited complex reaches the bolometer without dissociating (37). This corresponds to an excited-state lifetime in excess of 0.3 ms. By directing a second F-center laser into the multipass cell shown in Fig. 2.2, but in the opposite direction, it is possible to excite the $v = 1$ molecules produced by the first laser into $v = 2$. Since the power per unit bandwidth available from the F-center laser is sufficient to saturate these transitions, the double-resonance signal can be seen with high S/N. A double-resonance spectrum resulting from such a pumping scheme is shown in Fig. 2.6 for the HF monomer. Since in this case both $v = 1$ and $v = 2$ are long-lived states, the bolometer signals have the same sign for both transitions. What is surprising about the Ar–HF double-resonance signals is that the sign of the signal from the first laser is opposite that of the second laser. The implication is that, while the complex is long lived in $v = 1$, it dissociates in a time that is short with respect to the molecular flight time in $v = 2$. What is even more interesting is that if the first (pump) laser is tuned to a Σ or Π bending combination band (38), which effectively removes the hydrogen atom from the Ar–HF bond, the associated

HF Monomer Double Resonance

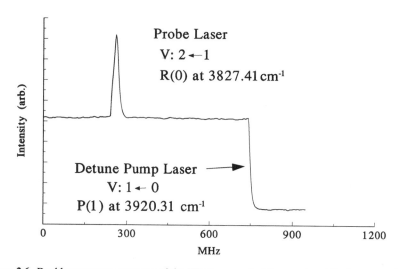

Figure 2.6. Double-resonance spectrum of the HF monomer. In this experiment the laser pumping the $v = 1 \leftarrow 0$ transition is modulated and the signals are observed with a phase-sensitive detector. The large dc offset is therefore a result of tuning the first laser in resonance with the $R(0)$ transition in the $v = 1 \leftarrow 0$ band. With this laser locked to the fundamental transition, the second laser is scanned through the $R(1)$ transition associated with $v = 2 \leftarrow 1$.

double-resonance signals into the corresponding $v = 2$ states again have the same sign as those associated with the pump laser. Clearly by decoupling the H-atom motion from the bond that must be broken, the dissociation rate is greatly reduced. There are a number of possible explanations why the rate increases in going from $v = 1$ to $v = 2$, including changes in the anharmonic coupling and energy gaps. A detailed account of this work is given elsewhere (39).

The spectroscopy of high-lying vibrational states is of fundamental interest because this is the energy regime where chemical bond rupture occurs. For this reason, there is a continued interest in understanding how vibrational states couple in this high energy, high density of states regime. High-resolution double-resonance spectroscopy can obviously help address such questions. The fact that the molecules are also cooled in a molecular beam expansion is a plus, so far as reducing the rotational congestion that might otherwise obscure the more interesting vibrational structure at high energies. Double-resonance optothermal experiments designed to access such states in molecules of medium complexity have recently been reported (40). The experimental setup used in these experiments is essentially identical to the one shown in Fig. 2.2, with the two lasers directed into the multipass cell as discussed above. Experiments have been carried out on propyne for $v = 1$ (using a single FCL), $v = 2$ (using two 3-μm FCLs), and $v = 3$ (using one 3-μm FCL and one 1.4-μm FCL) (40). Thus the effects of vibrational mode coupling are being explored as a function of the density of vibrational states within the same molecule. The prospects are very good for going to even higher states in double resonance, using CW dye and Ti:Sapphire lasers.

A spectroscopic application of the optothermal method that has yet to be explored to any great extent is that of recording electronic spectra of molecules that are dark, meaning that they do not fluoresce strongly, either due to predissociation of the excited state or simply low oscillator strengths. High laser powers would obviously be necessary to excite such long-lived electronically excited states in isolated molecules. The advantage of the optothermal method in such studies is that the signals increase in proportion to the number of molecules excited. Some preliminary work of this type has been reported (41) and it is likely that more developments will be forthcoming. Of course, when the molecules have high fluorescence yields, then the LIF methods discussed in Chapter IV in this volume will usually be more sensitive.

2.6. APPLICATION TO THE STUDY OF PHOTODISSOCIATION DYNAMICS

As indicated above, the dissociation lifetimes of these complexes can often be obtained directly from spectroscopy, assuming of course that the resolution is sufficient to observe the homogeneous broadening. This assumes, as is often the case (31,32), that the coupling is directly from the initially excited state to the dissociative continuum. Although this information is extremely valuable and has been used to obtain important new insights into the nature of the energy-transfer processes involved in the rupture of the weak bond (8,30–32), there remain many important dynamical questions that cannot be answered from the dissociation rates alone. Some of these can be addressed using modifications of the above methods to study the detailed state-to-state dynamics of these systems.

2.6.1. Angular and State Resolved Experiments

From the large body of spectroscopic work that has been carried out on these weakly bound complexes, it is now clear that the associated vibrational dynamics is highly nonstatistical and varies widely from system to system. Unfortunately, our understanding of these processes is still insufficient to predict with confidence which will be most important for any given system. To achieve this level of understanding we must clearly look more deeply into the photodissociation process. In particular, methods are needed for determining the final state condition of the fragments produced by dissociation of the parent molecule. A number of such methods have recently been developed (42–46) that are based upon various detection schemes. The emphasis in our laboratory has been to try to extend the IR technology to experiments of this type. Once again the advantage in using IR spectroscopy is its generality. As an illustration of the first of these methods we consider the dissociation of the HF dimer upon excitation of one of the H—F stretches. This system is an ideal starting point because the density of photofragment channels is very low. Since these fragment states are widely separated in energy, the kinetic energy released to the fragments depends strongly on the particular rotational channel. As a result, the angle to which the fragments recoil in the laboratory frame depends on the internal states of the two HF fragments. For this reason, the rotational state distribution can be determined by measuring the angular distribution of the photofragments using a rotatable apparatus like the one shown in Fig. 2.7. In this case, the bolometer detects the fragment intensity as a function of the scattering angle. A multipass cell is used in these experiments to give many laser–molecular beam crossings in a small photolysis region. This is done by positioning two spherical mirrors at their radius of curvature (a schematic diagram is given somewhat later in the chapter, Figure 2.10). A small hole is drilled in the center of one of the mirrors through which the laser enters the cell. By misaligning the other mirror slightly from the axial configuration, so that the first laser reflection does not simply leave the cell through the same hole, a large number (typically 60) of focused laser–molecular beam crossings are obtained midway between the two mirrors. The overall spot size is approximately 1 mm in radius.

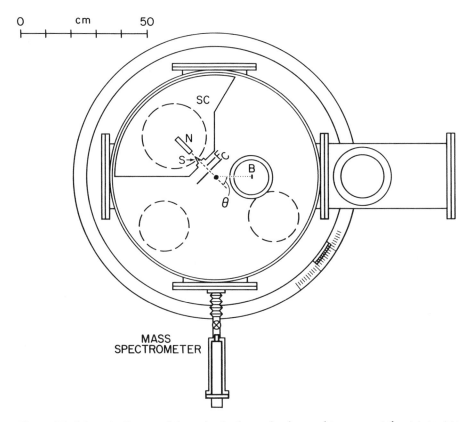

Figure 2.7. Schematic diagram of the molecular beam chamber used to carry out the state-to-state photodissociation experiments.

To make the connection between the scattering angle and the recoil energy of the photofragments, it is necessary to determine the velocity of the parent dimer. This is done by measuring a Doppler spectrum resulting from two laser–molecular beam crossings, an orthogonal crossing to define the zero position, and a non-orthogonal crossing to give the Doppler shift. Figure 2.8 shows an example of such a spectrum, from which the Doppler shift between the two peaks can be used to determine the velocity and the lineshape of the transition associated with the non-orthogonal crossing determines the velocity distribution.

Figure 2.9 shows an example of an angular distribution for HF dimer, with the observed structure being assigned to the individual rotational channels, labeled by the rotational quantum numbers associated with the two HF fragments (j_1 and j_2). The solid line through the data points is a fit, using a Monte Carlo technique to average over the experimental geometry (47). As expected, the states corresponding to high rotational energy appear at small angles in the laboratory frame, while those of lower rotational energy scatter to larger angles. Since these experiments allow us to determine both the internal and kinetic energies of the fragments, we can use conservation of energy to determine the dissociation energy of the complex (47,48). We have published a series of papers on the state-to-state dissociation of the HF dimer that include information on the initial-state dependence of the final-state distributions (44,47,48), including excitation of the inter-molecular bending and stretching vibrations in combination with the H—F stretches (49).

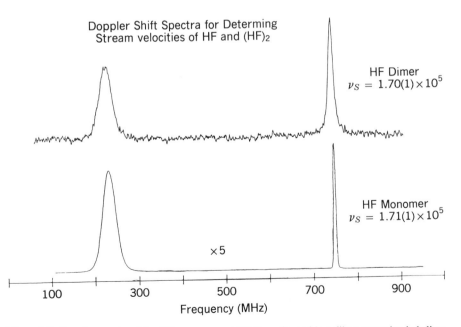

Figure 2.8. Doppler spectra of the HF monomer and HF dimer formed in a dilute expansion in helium. There is clearly very little velocity slip under these conditions.

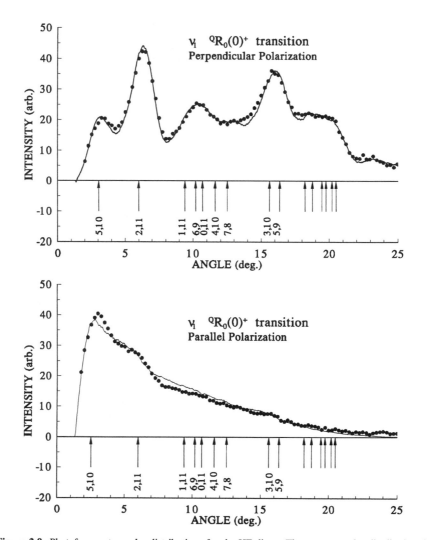

Figure 2.9. Photofragment angular distributions for the HF dimer. The upper angular distribution that shows the pronounced peaks was recorded with a laser polarization direction chosen so that the fragments recoil perpendicular to the molecular beam. The lower panel shows an angular distribution for the same transition but with the laser polarization favoring dissociation along the molecular beam direction. This dramatic polarization effect results from the strong correlation between the transition moment of the parent complex and the direction of recoil of the fragments.

It is important to note that each fragment channel is determined by a correlated pair of rotational quantum numbers, the correlation being established by the need to conserve energy. Scalar correlations of this type are extremely helpful in trying to understand the nature of the dissociation process. For example, the preference for channels with one fragment in a high j state and the other having little or no rotational excitation has led us to explain the dissociation of HF dimer in terms of an impulsive process occurring suddenly from a structure not unlike that of the ground-state dimer (47).

Although experiments of the type discussed above provide much more detail concerning the dissociation dynamics than spectroscopy alone, there is still more to be learned. Since the transition moment of the parent molecule and the velocity of recoil and the rotational angular momenta of the fragments are all vector quantities, vector correlations (50) may exist that can shed further light on the problem. Since the IR spectrum of the HF dimer is completely resolved in the above experiments, individual eigenstates are dissociated. As a result, even though the excited-states lifetimes are rather long (tens of nanoseconds) in comparison with the rotational period of a molecule, overall rotation of the complex does not destroy any alignment imposed by the interaction of a polarized laser with the transition moment of the parent molecule. Thus the vector correlation between the transition moment of the parent molecule and the recoil velocity vector of the fragments can be determined (44) even for these long-lived states. The angular momentum algebra needed to interpret the polarization dependence of the angular distribution is given elsewhere (44). The essence of the experiment can be understood by noting that if the polarization of the laser is aligned perpendicular to the molecular beam and the transition is parallel, the excited molecules will be preferentially aligned with their transition moments perpendicular to the molecular beam, given that the probability of excitation goes as $\mu \cdot E$. In the case of HF dimer, this corresponds to the F—F axis being preferentially aligned perpendicular to the molecular beam. As a result, if the molecule dissociates along this axis, the fragments will recoil into the maximum possible laboratory scattering angle allowed by the particular recoil energy. This gives rise to the peaks in the angular distribution shown in Fig. 2.9. By rotating the polarization of the laser to be parallel to the molecular beam, dissociation occurs preferentially along the molecular beam and the resulting angular distribution peaks near zero degrees. Experiments carried out in both parallel and perpendicular polarization can therefore be used to determine the anisotropy parameter (β) (44) characteristic of the dissociation process. This additional information on the center-of-mass angular distribution for the photodissociation process can be extremely helpful in understanding the dynamics. For example, in the case of HF dimer this vector correlation reveals that the fragments are generated preferentially along the A axis of the complex, consistent with the impulsive dissociation mechanism proposed above (44). Work is underway in the author's laboratory to measure other vector correlations involving the rotational angular momentum vectors of the two fragments. For this it will be necessary to use a spectroscopic method to probe the fragments.

2.6.2. Optothermal Pump–Probe Experiments

The experimental methods discussed above are clearly very powerful and provide important new information on the predissociation dynamics of complexes that have low densities of states. Systems we have studied to date include $(HF)_2$ (44,47,48), HF–DF, DF–HF (51), Ar–CO_2 (42,45), and H_2,D_2–HF (52). For complexes formed from heavier monomer units the above method quickly fails due to insufficient angular resolution to observe the individual photofragment internal channels. This problem has recently been overcome by using a second F-center laser, downstream of the photolysis region, to state selectively probe the fragments by optothermal spectroscopy. Such a two laser IR pump–IR probe scheme can provide the generality needed to study a wide range of complexes with widely varying dynamics.

The experimental apparatus developed for this purpose is shown schematically in Fig. 2.10. Two spherical multipass cells are used to optimize the pump and probe efficiency for the corresponding transitions. The first laser is tuned to a rovibrational transition in the complex of interest, which leads to the production of fragments scattered into various laboratory angles. As discussed above, these fragments contain considerable translational and internal energy, and therefore can be detected by the liquid helium cooled bolometer. The second F-center laser can then be tuned to various monomer fragment transitions in order to probe their internal states. Excitation of the fragments results in an increase in their internal energy and thus a bolometer signal that is proportional to the population in the probed level.

Results from such a pump–probe experiment are shown in Fig. 2.11 for N_2–HF. The large dc offset is the result of tuning the pump laser into resonance with the complex, such that the bolometer detects the photofragments. At a laboratory angle of 6°, probe laser signals are only observed for transitions arising from $j = 7$, while at larger angles the only transitions observed are those associated with HF produced in $j = 12$. Conservation of energy then suggests that these two states

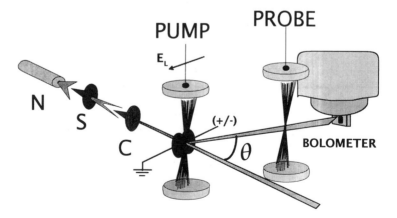

Figure 2.10. Schematic diagram of the pump–probe geometry used to spectroscopically determine the internal state distributions for the fragments.

are produced in coincidence with the $v = 1$ and $v = 0$ states of the N_2 fragment, respectively. These two channels therefore arise from two very different dissociation mechanisms, the first involving the intermolecular V—V transfer of energy and the second the conversion of excess vibrational energy into primarily rotation of the HF fragment. These two mechanisms sample very different portions of the potential energy surface and led one to ask why it is that they have comparable probabilities. The answer will come with the aid of theoretical calculations on both the potential energy surface and the dynamics. Clearly the ability of this experimental method to pull apart the angular distribution and to determine the internal states of the fragments spectroscopically will make it extremely powerful in the study of these processes in many different species of varying complexities, including trimer and tetramer systems. A more complete discussion of N_2—HF complex is given elsewhere (53).

2.6.3. Orientation and Alignment

In recent years a number of groups (54–56) have made use of the fact that rotationally cold molecules produced in a free-jet expansion can be oriented in a

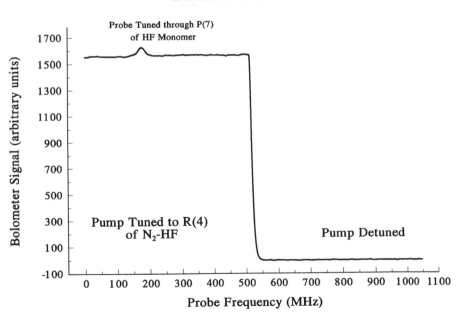

Figure 2.11. Spectrum of the HF photofragment resulting from vibrational predissociation of the N_2—HF complex. The large dc offset is the total photofragment signal at this angle. When the probe laser vibrationally excites the HF fragment the energy it delivers to the bolometer is increased, giving rise to the observed signal.

large uniform electric field, simply because the energy of interaction with the field, namely, $\mu \cdot E$, can be made comparable to, or even greater than, the rotational energy. For optimum orientation one requires a large dipole moment and a small rotational constant. The linear HCN trimer (57) is ideally suited in this regard, and we have carried out a series of experiments designed to explore the use of large electric fields in achieving orientation and modifying molecular spectra (58). The corresponding theory for the trimer has also been published (59). For this molecule it is possible to make the energy of interaction with the field over 300 times larger than the rotational constant, resulting in complete quenching of the free rotation of the molecule.

In a strong electric field the only good rotational quantum number is m. The hybridization of the rotational energy levels to produce states that undergo pendular type motion about the field direction (58,59) results in dramatic changes in the observed rovibrational spectrum. For the case of the HCN dimer with the laser and static electric fields parallel, the normal linear molecule spectrum collapses into what appears to be a Q branch at the origin, as shown in Fig. 2.12. The transition in this series that appears at the lowest frequency results from $m = 0$, the most strongly oriented state. Figure 2.13 shows the $m = 0$ wave function as a function of the electric field, corresponding to a molecule with a dipole moment of 1 D and

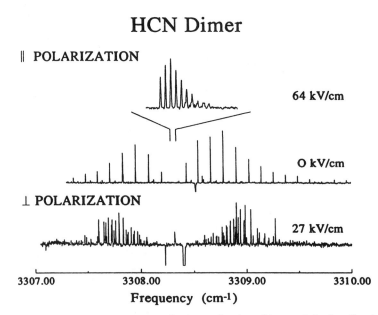

Figure 2.12. Pendular spectra of the HCN dimer as a function of laser polarization direction. In the upper spectrum the laser polarization direction is parallel to the static electric field such that the selection rule is $\Delta m = 0$. This spectrum, which appears at the origin of the C—H vibrational band of the dimer, gains intensity as the electric field is increased. The transition in this series appearing at the lowest frequency results from $m = 0$, the most strongly oriented state.

a rotational constant of 0.1 cm^{-1}. A field of approximately 25 kV cm^{-1} is sufficient to achieve essentially complete right–left orientation of this state and for many systems the degree of orientation can be much greater.

In Section 2.6.2 we discussed how laser polarization could be used to *align* molecules and to determine vector correlations. The question we can now ask is "What new dynamical information can be obtained if we can *orient* the complexes prior to their dissociation?" By placing an electrode on either side of the photolysis region shown in Fig. 2.10, a linear polar complex can be oriented so that it points sideways in the molecular beam. For the HCN dimer, the result is that the proton-donor molecules are all positioned to one side, while the proton acceptors are on the other. If the molecule now dissociates along its axis, the proton donor will recoil in one direction and the proton acceptor in the other. Thus, the right–left asymmetry that is imposed on the system, affords us some very interesting possibilities.

For example, let us reconsider the N_2–HF complex. Although the results discussed above were sufficient to assign the rotational state of the HF and the vibrational state of the N_2 fragment, the rotational state of the N_2 fragment could not be uniquely determined, owing to the congestion in the angular distribution resulting from the fact that the bolometer detects both the N_2 and the HF fragments. From laser polarization studies we already know that this complex dissociates essentially along its axis, as was the case for the HF dimer discussed above. There-

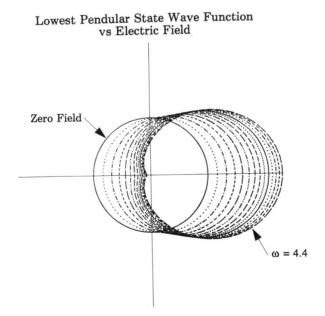

Lowest Pendular State Wave Function
vs Electric Field

Zero Field

$\omega = 4.4$

Figure 2.13. The angular dependence of the wave function for the lowest pendular state as a function of electric field. These calculations correspond to a molecule with a dipole moment of 1 D and a rotational constant of 0.1 cm^{-1}. $\omega = \mu(D) \times E(\text{kV cm}^{-1}) \times 0.168/\text{B(cm}^{-1})$.

fore, if a pendular field is applied to the photolysis region so that the two fragments recoil in opposite directions it should be possible to separate the angular distributions associated with two fragments. The reduced congestion that results from this type of experiment has allowed us to assign all of the channels and obtain a complete set of correlated state distributions for this system.

Similar experiments carried out on the HF dimer have allowed us to independently detect the proton donor and acceptor. In light of our model for the dissociation of this complex, which rationalizes the high j–low j correlation in terms of the large torque experienced by the proton donor and small torque on the acceptor, the implication is that the internal energy of the two fragments is very different. By orienting the complex we have confirmed this by noting that the bolometer signal is much larger when detecting the donor than the acceptor. There are clearly many possible applications of this method, not only to weakly bound complexes, but also to the photochemistry of chemically bound species. Taken together, these various forms of the optothermal technique provide us with a rather complete arsenal of methods for studying the photodissociation of these weakly bound complexes.

2.7. APPLICATION TO THE STUDY OF COLLISIONAL ENERGY TRANSFER

2.7.1. Gas–Gas Collisions

As noted above, one of the important advantages of the optothermal method over some of the other IR detection methods is that it provides sufficient sensitivity so that state-specific detection is possible in highly collimated molecular beams, where the molecular fluxes are very low. For this reason it is ideally suited to crossed beam scattering experiments. The first such application involved the measurement of resonant rotational energy transfer between two HF molecules (60). The primary beam was formed by strongly expanding a dilute mixture of HF in He so that the majority of the molecules are cooled to $j = 0$. Higher rotational states of this beam were then monitored downstream of the scattering region using the optothermal method. The secondary beam was formed by expanding pure HF from a much lower source pressure so that higher rotational states were thermally populated. Resonant transfer of rotational energy is then possible between the molecules in the two beams $((0, j) \rightarrow (j, 0))$ resulting from the very long range dipole–dipole interactions associated with this system. Normally cross sections of this type cannot be measured experimentally since the two HF molecules are indistinguishable. However, in a molecular beam experiment the two molecules are distinguishable because the cross sections are highly peaked in the forward direction and the velocities of the two molecules are in different directions. The experiment was carried out by monitoring the population of a given rotational state in the primary beam as a function of the secondary molecular beam intensity, and hence the magnitude of the primary beam attenuation. Since the vast majority of the primary beam HF molecules are initially in $j = 0$, this state is simply attenuated by the secondary beam. For $j =$

1 and $j = 2$, however, the population increases initially as the secondary beam intensity increases due to the resonant transfer of molecules out of $j = 0$ into these two states. This study has prompted a number of theoretical studies (61–63), which give results that are in good agreement with the experimentally determined first-and second-order dipole–dipole cross sections.

More recently, the optothermal method has been applied to the study of state-to-state differential scattering of HF by Ar (64,65). In these experiments an HF chemical laser was used to probe the internal states of the scattered HF molecules as a function of the laboratory scattering angle. In this way the differential scattering cross sections for the elastic and the rotationally inelastic channels could be measured independently. Differential cross sections have been reported for $j = 0, 1, 2, 3,$ and 4. Data of this type can provide exacting tests of the intermolecular potential surface in regions that are complimentary to those probed by the spectroscopy of the van der Waals molecule.

2.7.2. Gas–Surface Collisions

To date there has been very limited application of the optothermal method to the study of gas–surface scattering. The one study that has appeared in the literature involved the scattering of HF from LiF (66), where an HF chemical laser was used to probe the scattered molecules. Rotational accommodation was found to be incomplete for this system. In view of the generality of the optothermal method, this is an area of application that will certainly become more important in the future. Indeed, work underway in our laboratory is designed to explore this application for a variety of systems. The primary advantage will come when either complex molecules or systems that do not possess easily accessed electronically excited states that can be probed by LIF are studied. For example, the scattering of methane from a wide variety of surfaces, including catalytic surfaces that can induce bond cleavage, is ideally suited to the optothermal method.

2.8. SUMMARY

This chapter has provided a summary of the various applications of the optothermal method that have recently been developed. It is obvious from the widely varied subjects covered that this method has general applicability to a large number of problems. Its high sensitivity and resolution provide us with new capabilities that are, so far, unique to the optothermal method. Despite all the progress, there is every indication that the method is still in its infancy and that many new applications will be forthcoming. Some of these applications have been alluded to in this chapter, such as the study of "dark" electronically excited states and the study of inelastic and reactive surface scattering. When considered together with the more mature fields of IR spectroscopy and state-to-state photodissociation dynamics, it is clear that this method will continue to have a prominent place in the study of chemical reaction dynamics.

ACKNOWLEDGMENTS

I am grateful to the students and postdoctorals with whom I have worked and who made this work possible. Support for this research is gratefully acknowledged from the National Science Foundation, Grant No. CHE-89-00307 and the Donors of the Petroleum Research Fund (administered by the ACS).

REFERENCES

1. C. M. Lovejoy and D. J. Nesbitt, *J. Chem. Phys.*, **86**, 3151 (1987).
2. S. W. Sharpe, D. Reifschneider, C. Wittig, and R. A. Beaudet, *J. Chem. Phys.*, **94**, 233 (1991).
3. J. S. Go, T. J. Cronin, and D. S. Perry, *Chem. Phys.*, **175**, 127 (1993).
4. C. M. Lovejoy and D. J. Nesbitt, *Rev. Sci. Instrum.*, **58**, 807 (1987).
5. D. H. Levy, *Sci. Am.*, **250**, 96 (1984).
6. H. C. Chang and W. Klemperer, *J. Chem. Phys.*, **98**, 9266 (1993).
7. T. E. Gough, R. E. Miller, and G. Scoles, *Appl. Phys. Lett.*, **30**, 338 (1977).
8. R. E. Miller, *Science*, **240**, 447 (1988).
9. E. H. Putley, *Optical and infrared detectors*, Springer-Verlag, New York, 1977, p. 71.
10. R. E. Miller, *Rev. Sci. Instrum.*, **53**, 1719 (1982).
11. C. V. Boughton, R. E. Miller, and R. O. Watts, *Aust. J. Phys.*, **35**, 611 (1982).
12. M. A. Kinch, *J. Appl. Phys.*, **42**, 5961 (1971).
13. M. Zen, *Accommodation, Accumulation and other Detection methods, Atomic and molecular beam methods*, G. Scoles (Ed.), Oxford University Press, New York, 1988, pp. 254–275.
14. Infrared Laboratories Inc., Tucson, Arizona.
15. R. O. Smith, B. Serin, and E. Abahams, *Phys. Lett. A*, **28**, 224 (1968).
16. D. Bassi, A. Boschetti, G. Scoles, M. Scotoni, and M. Zen, *Chem. Phys.*, **71**, 239 (1982).
17. S. Verghese, P. L. Richards, K. Char, and S. A. Sachtjen, *IEEE Trans. Magn.*, **27**, 3077 (1991).
18. S. Verghese, P. L. Richards, K. Char., D. K. Fork, and T. H. Geballe, *J. Appl. Phys.*, **71**, 2491 (1992).
19. G. T. Fraser, A. S. Pine, W. A. Kreiner, and R. D. Suenram, *Chem. Phys.*, **156**, 523 (1991).
20. J. V. V. Kasper, C. R. Pollock, R. F. Curl, Jr., and F. K. Tittel, *Appl. Opt.*, **21**, 236 (1982).
21. Z. S. Huang, K. W. Jucks, and R. E. Miller, *J. Chem. Phys.*, **85**, 3338 (1986).
22. D. D. Nelson Jr., A. Schiffman, K. R. Lykke, and D. J. Nesbitt, *Chem. Phys. Lett.*, **153**, 105 (1988).
23. H. Meyer, E. R. T. Kerstel, D. Zhuang, and G. Scoles, *J. Chem. Phys.*, **90**, 4623 (1989).
24. K. K. Lehmann, B. H. Pate, and G. Scoles, *J. Chem. Soc. Faraday Trans.*, **86**, 2071 (1990).
25. B. H. Pate, K. K. Lehmann, and G. Scoles, *J. Chem. Phys.*, **95**, 3891 (1991).
26. C. Douketis, D. Anex, G. Ewing, and J. P. Reilly, *J. Phys. Chem.*, **89**, 4173 (1985).
27. C. Douketis and J. P. Reilly, *J. Chem. Phys.*, **96**, 3431 (1992).
28. T. E. Gough, R. E. Miller, and G. Scoles, *J. Chem. Phys.*, **69**, 1588 (1978).
29. K. W. Jucks and R. E. Miller, *J. Chem. Phys.*, **88**, 6059 (1988).
30. R. E. Miller, *AIP Conf. Proc., Volume Date 1987*, **172**, 365 (1988).
31. R. E. Miller, *NATO ASI Ser., Ser. C*, **212**, 131 (1987).
32. R. E. Miller, *Acc. Chem. Res.*, **23**, 10 (1989).
33. K. W. Jucks and R. E. Miller, *Chem. Phys. Lett.*, **147**, 137 (1988).
34. D. J. Nesbitt and C. M. Lovejoy, *Faraday Discuss. Chem. Soc.*, **86**, 13 (1988).
35. R. J. Saykally, *Acc. Chem. Res.*, **22**, 295 (1989).

36. E. R. T. Kerstel, K. K. Lehmann, J. E. Gambogi, X. Yang, and G. Scoles, *J. Chem. Phys.*, **99**, 8559 (1993).

37. Z. S. Huang, K. W. Jucks, and R. E. Miller, *J. Chem. Phys.*, **85**, 6905 (1986).

38. C. M. Lovejoy and D. J. Nesbitt, *J. Chem. Phys.*, **91**, 2790 (1989).

39. P. A. Block and R. E. Miller, *Chem. Phys. Lett.*, **226**, 317 (1994).

40. G. Scoles (private communication).

41. U. Buck, J. Kesper, R. E. Miller, A. Rudolph, and J. Vigue, *Chem. Phys. Lett.*, **125**, 257 (1986).

42. E. J. Bohac, M. D. Marshall, and R. E. Miller, *J. Chem. Phys.*, **97**, 4890 (1992).

43. M. P. Casassa, J. C. Stephenson, and D. S. King, *NATO ASI Ser., Ser. B*, **171**, 367 (1988).

44. M. D. Marshall, E. J. Bohac, and R. E. Miller, *J. Chem. Phys.*, **97**, 3307 (1992).

45. E. J. Bohac, M. D. Marshall, and R. E. Miller, *J. Chem. Phys.*, **97**, 4901 (1992).

46. J. R. Hetzler, M. P. Casassa, and D. S. King, *J. Phys. Chem.*, **95**, 8086 (1991).

47. E. J. Bohac, M. D. Marshall, and R. E. Miller, *J. Chem. Phys.*, **96**, 6681 (1992).

48. D. C. Dayton, K. W. Jucks, and R. E. Miller, *J. Chem. Phys.*, **90**, 2631 (1989).

49. E. J. Bohac and R. E. Miller, *J. Chem. Phys.*, **99**, 1537 (1993).

50. Paul L. Houston, *J. Phys. Chem.*, **91**, 5388 (1987).

51. M. Wu, R. Bemish, E. J. Bohac, and R. E. Miller (in preparation).

52. E. J. Bohac and R. E. Miller, *J. Chem. Phys.*, **98**, 2604 (1993).

53. E. J. Bohac and R. E. Miller, *Phys. Rev. Lett.*, **71**, 54 (1993).

54. H. J. Loesch and A. Remscheid, *J. Chem. Phys.*, **93**, 4779 (1990).

55. B. Friedrich and D. R. Herschbach, *Z. Phys. D-Atoms Mol. Clusters.*, **18**, 153 (1991).

56. D. P. Pullman, B. Friedrich, and D. R. Herschbach, *J. Chem. Phys.*, **93**, 3224 (1990).

57. K. W. Jucks and R. E. Miller, *J. Chem. Phys.*, **88**, 2196 (1988).

58. P. A. Block, E. J. Bohac, and R. E. Miller, *Phys. Rev. Lett.*, **68**, 1303 (1992).

59. J. M. Rost, J. C. Griffin, B. Friedrich, and D. R. Herschbach, *Phys. Rev. Lett.*, **68**, 1299 (1992).

60. P. F. Vohralik and R. E. Miller, *J. Chem. Phys.*, **83**, 1609 (1985).

61. K. Takayanagi and T. Wada, *J. Phys. Soc. Jpn.*, **54**, 2122 (1985).

62. G. D. Billing, *Chem. Phys.*, **112**, 95 (1987).

63. P. F. Vohralik, R. O. Watts, and M. H. Alexander, *J. Chem. Phys.*, **91**, 7563 (1989).

64. L. J. Rawluk, Y. B. Fan, Y. Apelblat, and M. Keil, *J. Chem. Phys.*, **94**, 4205 (1991).

65. J. A. Barnes, M. Keil, R. E. Kutina, and J. C. Polanyi, *J. Chem. Phys.*, **76**, 913 (1982).

66. D. Ettinger, K. Honma, M. Keil, and J. C. Polanyi, *Chem. Phys. Lett.*, **87**, 413 (1982).

Chapter **III**

DIRECT ABSORPTION IN SUPERSONIC FREE-JETS

David S. Perry and G. A. Bethardy
Department of Chemistry,
University of Akron, Akron, Ohio

Laser Techniques In Chemistry, Edited by Anne B. Myers and Thomas R. Rizzo.
Techniques of Chemistry Series, Vol. XXIII.
ISBN 0-471-59769-4 © 1995 John Wiley & Sons, Inc.

3.1. INTRODUCTION

The tremendous spectral simplification achievable for supercooled molecules has made the use of nozzle expansions one of the most powerful and widely used techniques in modern spectroscopy. The number of rotational lines per vibrational band in a nonlinear molecule scales as $T^{3/2}$. Therefore, at a typical temperature in a jet, $T \approx 3$ K, the number of rotational lines is reduced a thousandfold relative to room temperature. Furthermore, even for relatively small molecules, only a small fraction of the sample is in the vibrational ground state at room temperature. For example, in ethanol, which has only three heavy atoms, only one-seventh of the molecules are in the vibrational ground state at room temperature. Accordingly, many spectra that show only a smooth contour at room temperature become well resolved under jet-cooled conditions, and spectra that are extremely complicated become easy to assign.

This chapter reviews direct absorption methods for recording spectra of jet-cooled molecules. Other chapters in this volume review a number of the other methods applicable to ultracold molecules. By direct absorption, we mean any technique that detects the attenuation of light that passes through the jet. Since most polyatomic molecules have absorption bands distributed throughout the optical spectrum, from the far-infrared (IR) to the vacuum ultraviolet (UV), direct absorption is a very general approach for recording their spectra. Section 3.2 is a survey of the light sources that have been used for direct absorption spectroscopy of jets in the various regions of the spectrum.

The challenge of direct absorption methods stems from the rarefied nature of a jet and from the limitations of laboratory scale vacuum equipment. For molecules with strong absorptions or with simple spectra, large fractional absorptions can be obtained. However, for many interesting problems, the extinction coefficient is small, which implies weak absorption signals in the jet.

The sensitivity of the direct absorption method can be enhanced in a number of ways. Nozzle geometries (Section 3.3) and optical multipass schemes (Section 3.4) can be chosen to maximize the optical path length over which the absorption occurs. The principal source of noise in direct absorption experiments is almost always intensity fluctuations of the light source. Baseline subtraction and a variety of modulation schemes (Section 3.5) allow a substantial noise reduction—sometimes almost to the detector noise limit. The principal components of a generic direct absorption experiment are illustrated schematically in Fig. 3.1. Sections 3.2–3.5 can be used as a sort of menu of options for designing a single-resonance direct absorption experiment.

Direct absorption can also be used as the detection mechanism for double-resonance experiments in a free-jet (Section 3.6). Double-resonance methods can provide an additional measure of spectral simplification, unambiguous spectral assignments, and allow new states to be reached. For these advantages, one must pay a price in terms of the complexity of the experimental apparatus, but good sensitivity can still be obtained.

Finally, in Section 3.7 three applications are described to illustrate the range of

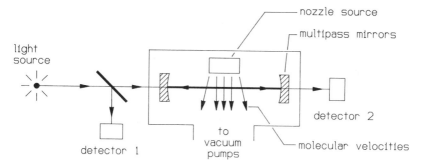

Figure 3.1. A schematic setup for a free-jet absorption experiment.

applicability of the direct absorption method and the quality of the insights that have been obtained. In keeping with the didactic nature of this volume, this chapter is intended primarily as a tutorial for scientists interested in the direct absorption method. We will not attempt to reference all work that has been done by direct absorption.

It should be noted that there are many cases where other methods of obtaining spectra of jet-cooled molecules offer substantial advantages over the direct absorption approach. For example, for electronic transitions with a high-fluorescence quantum yield, laser-induced fluorescence is much more sensitive. The optothermal method is advantageous where the laser sources are sufficiently intense and it offers the possibility of applying molecular beam techniques, such as quadrupole focusing and Stark spectroscopy.

3.2. LIGHT SOURCES

Light sources for direct absorption experiments include high-resolution continuous wave (CW) lasers, pulsed lasers, and incoherent sources.

The bandwidth of a single mode CW laser is typically two orders of magnitude or more narrower than a typical gas-phase Doppler width at the laser frequency. Therefore, the resolution of spectra recorded with such lasers is limited only by the distribution of velocities in the jet. Furthermore, CW lasers may be effectively combined with pulsed or modulated jets for effective noise suppression.

In the far-IR, CW lasers will operate only at discrete molecular transition frequencies. Tunability is obtained by mixing these fixed frequencies with microwaves (1). This approach can also be used in the mid-IR with CO_2 lasers (2) to obtain very convenient and reliable high-resolution sources. The main problem with the microwave mixing approach is that there are a limited number of fixed-laser frequencies, which leaves inconvenient gaps in the spectral coverage. The commercially available lead–salt diode lasers can cover the whole fingerprint region in the mid-IR. However, many diodes are needed for this coverage and individual diodes have gaps in their tuning curves. The tunable power available from microwave mixing or from lead–salt diodes is typically in the microwatt range, sometimes up to a milliwatt. Frequency difference lasers based on $LiNbO_3$ are very broadly tunable in the near-IR (3). These lasers produce only a few microwatts of power but there are no gaps in the spectral coverage. Difference frequency generation in $AgGaS_2$ is now being developed to extend this technology further into the mid-IR (4). Within the ranges where they operate (~3μ and ~1.5 μ) color center lasers offer higher power (10 or 100 mW), complete spectral coverage, and ease of operation. In the visible and very near IR dye and Ti:sapphire lasers offer continuous coverage and enough power to allow doubling into the near UV. If it is not already, this brief summary will soon be out of date. The point is that very high resolution sources suitable for direct absorption spectroscopy are available, with a few gaps, in every spectral region except the deep UV.

Pulsed lasers have also been used (5) and are an essential aspect of the ringdown technique described in Section 3.5.5.

Broad-band incoherent sources, such as a glow bar or deuterium lamp, are used in combination with Fourier transform (FT) spectroscopy or in the vacuum UV where CW laser sources are not available. Pulsed xenon lamps have been effective in combined direct absorption–fluorescence experiments that measure absolute fluorescence quantum yields (6).

3.3. NOZZLE GEOMETRICS

3.3.1. Effusive Microchannel Arrays

In 1975 Chu and Oka (7) published the first report of direct absorption spectroscopy in an expanding gas. An effusive gas source was used consisting of a bundle of tubes 1.8 mm long each with a 5-μm radius. By definition, in an effusive expansion the source pressure is low enough (typically a torr or less) so that the mean-free-path λ is much larger than the radius of the tube (8). Unfortunately, the gas density in the expansion is correspondingly low (10^{13} molecules cm^{-3}), which results in weak absorption signals. Collisions are not important in this molecular flow regime and there is no cooling of the expanding gas. The Boltzmann distribution of internal energies is determined by the temperature of the gas reservoir.

A microchannel array source offers sub-Doppler resolution because of the geometry of the capillary tubes. Those molecules inside the source with velocities parallel to the capillaries pass cleanly through into the vacuum; molecules with velocities in other directions strike the inside walls of the tubes, and therefore do not pass through in great numbers. Observed residual Doppler widths of 3 (7) and 7 MHz (9) have been obtained. These line widths are about 3 and 5% of the room temperature gas-phase Doppler width.

The effective absorption path length and, hence, the sensitivity can be enhanced by using a number of microchannel arrays. Pine and Nill (9) achieved an effective path length of 16 cm by multipassing a diode laser beam directly in front of four such capillary arrays.

Despite the excellent resolution that is achievable, microcapillary arrays have not been widely used because of the low density of the expansion and the lack of cooling of the internal degrees of freedom.

3.3.2. Circular Nozzles

The tremendous utility of gas expansions for absorption spectroscopy became evident when researchers began using supersonic free-jet expansions. The immediate advantages are gas densities that are orders of magnitude greater than those from an effusive source and concentration of the absorption intensity into a drastically reduced number of lines.

Beginning with the studies of Jensen et al. (10) in 1976 and Travis et al. (11) in 1977, there have been many reports of direct absorption spectroscopy employing axisymmetric free-jet expansions formed from a circular (12–14) nozzle. Workers in the field have used nozzles of their own design, as well as commercial varieties including modified automobile injector valves (15,16) and nozzles manufactured specifically for usage in supersonic expansions (13). These designs vary in their complexity from a simple pinhole in a thin metal sheet (16), to more complicated mechanisms that allow for pulsed modes of operation (14). In order to increase the effective optical path length, Hojer et al. (17) passed their diode laser through the free-jets formed from a row of 10 circular nozzles.

There are a number of review articles addressing the theory of supersonic expansions (18–20). The cooling that occurs in a supersonic jet is the result of collisions in the expanding gas.

To understand the cooling process, it is convenient to categorize the molecular degrees of freedom as vibrational motion v, rotational motion r, and translational motion parallel and perpendicular to the flow lines. The average total energy per molecule in a small volume element may be written as

$$\epsilon = \epsilon_v + \epsilon_r + \epsilon_\perp + \epsilon_\parallel + \epsilon_s \tag{3.1}$$

with

$$\epsilon_\perp = \tfrac{1}{2} m(\langle v_x^2 \rangle + \langle v_y^2 \rangle) \tag{3.2}$$

$$\epsilon_\parallel = \tfrac{1}{2} m\langle (v_z - v_s)^2 \rangle \tag{3.3}$$

$$\epsilon_s = \tfrac{1}{2} m\, v_s^2 \tag{3.4}$$

Here the flow velocity v_s, is the average velocity of molecules in a particular volume element. The z direction is taken to be parallel to the flow direction. The translational energies $\epsilon_\perp$ and $\epsilon_\parallel$ represent only motions relative to the flow velocity.

Consider first an effusive expansion in which there are no collisions in the throat of the nozzle or at any point during the expansion. Velocities perpendicular to the flow arise only from the solid angle subtended by the nozzle opening and this angle decreases with the distance from the nozzle. This purely geometric effect causes $\epsilon_\perp$ to decrease with distance from the nozzle. However, without collisions, the cooling of $\epsilon_\perp$ does not spread to $\epsilon_\parallel$, ϵ_r, and ϵ_v. The average energies in these degrees of freedom remain at the values characteristic of the nozzle temperature.

In a supersonic expansion, collisions cause $\epsilon_\perp$ to begin to equilibrate first with $\epsilon_\parallel$ and then with ϵ_r and ϵ_v. Accordingly, $\epsilon_\parallel$, ϵ_r, and ϵ_v decrease with distance from the nozzle. The collision rate becomes less and less with increasing distance from the nozzle. Eventually the energy-transfer processes cease. Thus the energy distribution becomes "frozen," that is, it no longer changes with distance from the nozzle. Generally, vibrational relaxation is the slowest so ϵ_v freezes first followed by ϵ_r and $\epsilon_\parallel$. The parameter $\epsilon_\perp$ does not freeze but goes on decreasing until the Mach disk or a wall of the vacuum chamber is reached.

At many locations in the expansion, the degrees of freedom are not in equilibrium, but often each may be approximately characterized by a temperature. One typically (21–23) finds $T_v > T_r \approx T_\parallel > T_\perp$. Non-Boltzmann rotational distributions have been reported for OCS (24), HF (25), and CH_3F (26). Nonetheless, rotational temperature is still a useful concept because the distribution among the more populated levels usually remains close to Boltzmann and a good first approximation to the intensity distribution in molecular spectra is readily obtained. While there has been some evidence for mode-selective vibrational cooling in a supersonic expansion (27), the vibrational temperature is often low enough that virtually all of the population is in the vibrational ground state.

Large molecules with many vibrational and rotational degrees of freedom bring a substantial heat content into the expansion, all of which must ultimately be relaxed into $\epsilon_\perp$. Therefore, the lowest temperatures are obtained with a dilute mixture of the seed gas in an inert carrier gas. The heavier carrier gases are more effective at removing rotational and vibrational energy from large molecules (28) and are also more effective at conformational relaxation (29). One can obtain $T_r \approx 5-7$ K and $T_v < 50$ K in Ar expansions with a stagnation pressure-nozzle diameter product, $p_0 d_0 \approx 3$ torr-cm (28). However, clusters form more easily in Ar expansions than with the lighter rare gases (30), such as He, because of the stronger intermolecular attraction. The binding energy of the clusters is released into the jet as translational energy, thereby reheating the expansion. Therefore, for the lowest possible temperatures, extremely dilute mixtures in high-pressure He are used. For example, $T_r = 410$ mK has been obtained for an expansion of 2 ppm s-tetrazine in 70 atm of He through a 25-μm nozzle (31). Temperatures less than 1 mK have been obtained in pure He expansions (32).

The lowest rotational temperature is not necessarily the best for absorption spectroscopy. The most precise spectroscopic constants and the most insightful dynamical information may be obtained when a large number of rotational levels are available for analysis. Therefore it is advisable to choose the highest temperature that will yield a fully analyzable spectrum. For a complicated spectrum, initial assignments can be easiest with a preliminary spectrum at the lowest achievable temperature (33).

The expansion conditions that will enhance cluster formation are evident from Knuth's sudden freeze model (34):

$$x \approx 0.5 \left[n_0 \sigma^3 \left(\frac{\epsilon}{kT_0} \right)^{7/5} \left(\frac{d_0}{\sigma} \right)^{2/5} \right]^{5/3}$$ (3.5)

Here, x is the terminal mole fraction of dimers, n_0 is the number density in the source, T_0 is the source temperature, d_0 is the nozzle diameter, ϵ is the well depth, and σ is the collision radius. Knuth (34) found Eq. (3.5) to be reliable for the inert gases when $x < 0.01$; Vasile and Stevie (35) found it useful but not always completely accurate for a molecular seed gas. Trimers (34) and larger clusters (36) will become important when x is larger. Nonetheless, the scaling (37)

$$x \propto T_0^{-4} \, p_0^{5/3} \, d_0^{2/3}$$ (3.6)

is a useful guide, especially when combined with the scaling of the terminal temperature.

$$T_1 \propto T_0^{\beta} \, (p_0 \, d_0)^{-\alpha}$$ (3.7)

and the rate of gas flow through the nozzle.

$$\frac{dN}{dt} \propto T_0^{-1/2} \, p_0 \, d_0^2$$ (3.8)

The exponents in Eq. (3.7) depend on the heat capacity ratio of the gas. Experimentally, α is found to be 1.0 for the inert gases (He, Ne, or Ar) and 0.6 for neat polyatomics (SF_6 or C_2H_4); β is 1.3 and 0.8, respectively (38). The values of α and β for the inert gases are in good agreement with theory (39).

Since the gas flow is limited by the throughput of the pumping system, Eqs. (3.7) and (3.8) imply that the lowest temperatures will be achieved with a relatively small nozzle diameter and a very high stagnation pressure. Equation (3.6) indicates that substantial cluster formation is expected under these conditions. If it is desired to minimize the fraction of clusters, a larger d_0 can be used at a more moderate p_0, and the nozzle can be heated.

The spectroscopic resolution is limited by the Doppler width arising from the angular distribution of molecular velocities in the jet. For a $\cos^4\theta$ flux distribution (40) the Doppler width (fwhm) is given by $\Delta\nu = 1.2 \, \nu v_s / c$, where ν is the frequency and c is the speed of light. This width is typically a little greater than the thermal Doppler width for the seed gas at the nozzle temperature. For dilute mixtures in an inert carrier gas, the flow velocity is $v_s \approx (5RT/M)^{1/2}$, where M is the average molecular weight; therefore, the heavier carrier gases offer an advantage in resolution. The resolution disadvantage of a light carrier gas is only partly mitigated by aerodynamic focusing (41) of a heavy seed gas toward the axis of the jet. Such aerodynamic effects as well as enhanced cluster formation in the denser regions of the jet can give rise to flat-topped absorption lines and even double-peaked absorption features (14,16,42,43).

3.3.3. The Sliced-Jet

One approach to narrowing the Doppler width of direct absorption lineshapes is to collimate the jet into a molecular beam. While this has been done (44), the absorption intensities are weak because the insertion of a skimmer between the nozzle and the laser beam causes the nozzle to be placed much further away from the laser beam. The density of absorbers at the laser beam is also reduced as a result of the interaction of the skimmer with the jet.

As an alternative, the conceptual inverse of skimming, that is, removing the molecules from the center of the gas expansion rather than the edges, has been demonstrated to give sub-Doppler resolution. This technique, called sliced-jet spectroscopy (45), employs a small blade placed between the circular nozzle and the laser crossing to obstruct molecules that would otherwise absorb at the center of the Doppler profile [see Fig. 3.2(d)]. The appearance of the lineshapes in sliced-jet spectroscopy is similar to those in the familiar Lamb dip spectroscopy. The difference is that in sliced-jet spectroscopy the ''hole'' produced in the velocity distribution is formed by physically blocking a portion of the molecules in the free-jet rather than by the optical hole burning as in the case of the Lamb dip (46). The high-resolution spectrum appears as a series of dips in the irregular contour of the free-jet absorption spectrum.

The advantage of the sliced jet over a skimmed beam comes from the very small size of the blade so that absorption measurements may be made close to the nozzle in a jet that is only slightly perturbed.

The sliced-jet technique is based on the correlation between the position in the jet and the direction of the absorber velocities. Because of collisions in the region near the nozzle opening, the correlation is not perfect and accordingly, the depth of the hole in the absorption contour is not 100%. For acetylene spectra in the 3-μ region, 50% holes 20 MHz wide have been recorded and the narrowest holes are 12 MHz wide (45).

The slit nozzle, described below, offers much stronger absorption intensity as well as sub-Doppler resolution, and therefore is generally the preferred method. Since the blade causes only a light perturbation in the jet, the sliced jet does provide access to the extremely low temperatures obtainable with a circular nozzle and it is easily implemented. A modulated version of the sliced jet has been devised (47).

3.3.4. Slit Nozzles

Use of a long narrow slit-shaped nozzle fundamentally changes the dimensionality of the expansion. Three slit nozzle designs are shown in Fig. 3.2. In all cases, the laser beam propagates at least approximately parallel to the length of the slit opening. In contrast to a free-jet formed from a circular nozzle, where the gas expands in two spatial dimensions, the expansion from a slit occurs only in the

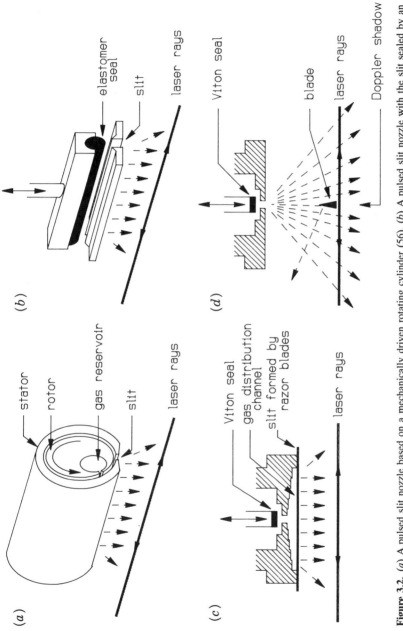

Figure 3.2. (a) A pulsed slit nozzle based on a mechanically driven rotating cylinder (56). (b) A pulsed slit nozzle based on a mechanically driven rotating cylinder (57). (c) A pulsed slit nozzle based on a two-stage expansion. [Figure adapted with permission from Ref. 33.] (d) Schematic for a sliced jet experiment (45).

dimension that is perpendicular to the slit. For the ideal case of an infinitely long slit, the expansion of the gas in the direction of the slit is forbidden by symmetry. In practice, the small residual dispersion in the molecular velocities parallel to the slit axis gives rise to a Doppler width of tens of megahertz in the IR. This residual Doppler width is typically a factor of 5–10 less than for an expansion of the same gas mixture through a circular nozzle. Line narrowing in a planar jet was first reported by Veeken and Reuss (48) in 1985.

An expansion from a slit nozzle also enhances the peak absorption signal, and hence the sensitivity (49). First, the concentration of the absorbers into a narrower Doppler width enhances the absorption intensity by the same factor. Second, when the length of the slit is greater than the distance of the laser beam from the nozzle, the path length of the light through the absorbers is greater than for a circular nozzle.

Because of the reduced dimensionality of the expansion, the gas density only decreases as R^{-1} (as opposed to R^{-2} for a circular nozzle) and the adiabatic cooling is correspondingly slower, (50,51). While extremely low temperatures are best obtained with a circular nozzle (52), temperatures as low as 4 K are still possible with a slit nozzle (49,53). Sulkes et al. (52) compared the cooling of internal degrees of freedom in planar jets with that in axisymmetric jets. They found that a slit nozzle with a modest aspect ratio (5:1) could produce lower temperatures (2 K) than an ideal (large aspect ratio) planar jet while retaining some of its advantages.

The slower planar expansion gives rise to more three-body collisions, which enhances the formation of weakly bound complexes (54). The ratio of monomers to complexes in the jet can be a very sensitive function of the expansion conditions as the authors found to be the case in ethanol (53). Veeken and Reiss (48) studied the kinetics of the formation of CO_2 clusters. Hagena (55) studied the nucleation and growth of larger clusters. Qualitatively, the same principles govern the choice of the expansion conditions as for circular nozzles, although the details of the scaling relationships (Eqs. 3.6–3.8) will be different because the expansion is slower.

The high throughput of slit nozzles has led most workers to adopt pulsed operation. The gas flow is reduced by a factor of the duty cycle thus reducing the pumping capacity required. Three methods of rapidly switching the gas flow through a slit-shaped orifice are illustrated in Fig. 3.2.

The first slit-jet source used in conjunction with direct absorption spectroscopy was employed by Amirav et al. in 1981 to record the jet-cooled spectra of aniline and 9,10-dichloroanthracene in the region of their $S_0 \rightarrow S_1$ band origins (56). Their source was constructed of two concentric cylinders each of which had a slit cut into it as illustrated in Fig. 3.2(a). The inner cylinder, which was 90 mm long with a 50-mm diameter and a 0.5-mm wall thickness, had an outside diameter that matched the inner diameter of the outer cylinder to within a tolerance of 0.02 mm. The inside cylinder was rotated at 5–13 Hz to produce gas pulses that were a couple of hundred microseconds long with the slits in the two cylinders overlapped. The 90-mm × 0.3-mm slit nozzle could be heated to 220° C or cooled to obtain the desired concentration of seeded molecules.

Lovejoy et al. (57) designed a pulsed slit nozzle to record the IR spectra of jet-cooled monomers and clusters. Interchangeable nozzles 12.5 mm long by about 0.125 mm wide are attached to a nozzle holder that fits inside of the valve body. In a later version (58), the slit length was increased to 40 mm. The slit orifice [Fig. 3.2(*b*)] is sealed by an elastomer which is held against the sharp lip of the nozzle by a leaf spring. The valve is driven by an electromagnetic solenoid actuator. Current pulses through the solenoid produce a magnetic field gradient that accelerates the ferromagnetic plunger away from the nozzle thereby opening the valve. The valve can be operated at repetition rates up to 60 Hz producing pulse durations in the range 150–600 μs. Heating and cooling of the nozzle is accomplished by circulating liquid through a chamber that surrounds the solenoid.

Veeken and Reuss (48) first realized that it is not necessary to seal the slit itself in order to obtain pulsed operation. Carrick et al. (59) used a tapered single-stage orifice with a square cross section near the seal and a slit-shaped cross section (25:1 aspect ratio) at the vacuum interface. A simple variant of this concept designed in our laboratory (33) is shown in Fig. 3.2(*c*) A two-stage expansion is employed to gate the gas flow through the slit quickly and repetitively. The first expansion occurs at a 1.5-mm diameter hole which is opened and closed by a piezoelectrically activated plunger. From this aperture the gas flows into a 2-mm × 2-mm × 20-mm gas distribution channel and then through the slit opening which is formed by a pair of razor blades. The adjustable separation between the blades is set with an automotive feeler gauge to form a slit that is typically 20 mm × 0.1 mm (60) (200:1 aspect ratio). The design concept in Fig. 3.2(*c*) is very general and may be implemented by fitting a single machined part to any standard valve. With this simple design, we were able to match the signal-to-noise (S/N) and resolution of previously reported (58) 1-butyne spectra.

Busarow et al. (61) used a continuous planar expansion in their far-IR direct-absorption spectrometer at the expense of having to resort to large vacuum pumps. Their studies on van der Waals clusters employed a 38-mm × 0.025-mm slit (1500:1 aspect ratio) which could be cooled to 77 K.

Because it simultaneously offers advantages in resolution and sensitivity, the slit nozzle has become the workhorse for recording direct absorption spectra of jet-cooled species, both monomers and clusters.

3.4. OPTICAL MULTIPASS ARRANGEMENTS

3.4.1. The White Cell

A dramatic increase in the path length of the light through the absorbers can be obtained by reflecting the light many times through the jet. The White cell (62) has been used for many years for this purpose in the context of gas-phase absorption spectroscopy. The cell consists of one front mirror and two back mirrors [Fig. 3.3(*a*)] and offers the following advantages:

1. A large number of passes are possible (> 100).

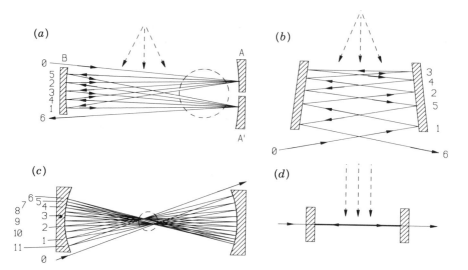

Figure 3.3. Multipass configurations for free-jet direct absorption experiments: (*a*) the White cell, (*b*) two plane mirrors, (*c*) two spherical mirrors, and (*d*) a high-finesse etalon. The path of the light rays is shown by solid lines with arrows. The numbers indicate the sequence of reflections on one mirror. Where the molecular velocities lie in the plane of the page, they are shown as dashed lines with arrows; where they lie approximately normal to the page, they are indicated by dashed circles. In (*a*) two alternative configurations are illustrated.

  2. The cell has a large angular acceptance aperture that is determined by the area of the back mirrors. Therefore, it is usable with incoherent light sources.

  3. The alignment of the cell is stable because of the confocal mirror spacing.

As seen from Fig. 3.3(*a*), the White cell is most effectively used with a gas-phase sample that fills the entire volume between the mirrors. The two different geometries shown in Fig. 3.3(*a*) have been used in order to combine the White cell with a free jet (59,63). In cases where it is permissible to sample a relatively large volume within the jet, the White cell may be the preferred multipass cell.

3.4.2. Two Plane Mirrors

  The first cell (13,64) used with direct absorption in a free-jet consisted of two plane mirrors [Fig. 3.3)*b*)]. A small angle between the two mirrors causes the progression of reflections to turn around so that the exit beam emerges from the same side of the mirrors as the entrance beam. If the plane mirrors were parallel, a cell with the same number of passes would sample a volume four times larger. Because the mirrors cannot focus the light, the acceptance angle is rather small and diffraction of the laser beam will limit the number of passes. Although a classical ray analysis (65) suggests that a larger number of passes are possible, practical applications (13,64) used about 25 passes. This is a simple, yet effective, multipass cell for circumstances where a large volume of absorbers may be sampled.

3.4.3. Two Spherical Mirrors

A multipass cell using two spherical mirrors (66) in a near concentric alignment [Fig. 34.3(c)] offers the following advantages:

1. All rays are forced through a small collective waist in the center of the cell which enables effective coupling to a small volume of absorbers.
2. A large number of passes (40–50) can be obtained and this number is easily adjustable.
3. The alignment of all rays in a plane makes the cell useful for sub-Doppler spectroscopy. Line widths as narrow as 12 (66) and 5 MHz (67) have been reported.

The cell is a limiting form of the Herriott cell (68). In the full concentric limit, the cell is formally similar to the plane mirror cell shown in Fig. 3.3(b). The pattern of reflections on each mirror, the small acceptance angle, and the beam divergence are all similar. However, a small deviation from concentricity increases the maximum number of passes by strengthening the confinement of laser beam while still maintaining the small waist of rays in the center of the cell.

The maximum number of passes N_{max} is determined by the need to couple the beam into and out of the cell cleanly, which limits the number of reflections that can be crowded onto each mirror. A Gaussian beam analysis (66) shows that

$$N_{max} = 2 \left(\frac{h^2 \pi^4}{18 L \lambda} \right)^{1/3} + 1 \tag{3.9}$$

When the mirror diameter is $h = 50$ mm, the mirror spacing is $L = 200$ mm and the wavelength is $\lambda = 3$ μm, N_{max} is predicted to be 57 passes. Stable alignments have been obtained in the range of 40–50 passes. When combined with baseline subtraction, we found a maximum S/N near 25–30 passes (33). Far-IR experiments have been done with 8–16 passes (69).

3.4.4. High-Finesse Etalon

The effectiveness of a resonant high-finesse etalon in enhancing direct absorption signals has been demonstrated by Coe et al. (70) in their high-resolution spectra of fast ion beams. When used with short-pulse lasers, the high-finesse etalon need not be resonant with the incoming light. This latter technique is the ringdown approach, which will be discussed separately in Section 3.5.

In this section, we will assume that the etalon is tuned to the frequency of a narrow CW laser and can be scanned with the laser to record an absorption spectrum. A possible geometry for combination with a free-jet is shown in Fig. 3.3(d). Although use of a tuned etalon to enhance direct absorption in a jet has not been

reported, the principles are undertood from related work (67,70,71) and the difficulties and potential advantages will be outlined here.

To estimate the enhancement of the absorption signal with a resonant etalon, we
consider the reflectivity $\mathcal{R}$, the transmissivity $\mathcal{T}$, and the absorption $\mathcal{A}$, of each
mirror in Fig. 3.3(d):

$$\mathcal{R} + \mathcal{T} + \mathcal{A} = 1 \qquad (3.10)$$

The transmissivity of the jet is $\mathcal{T}^\dagger = 1 - \alpha$, where α is the fractional absorption
of light in one pass through the jet. For the geometry of Fig. 3.3(d), α is the same
for light traveling in either direction. By summing the amplitudes from each pass,
the intensity $I^{(t)}(\alpha)$ transmitted through the etalon and jet is found to be (72)

$$I^{(t)}(\alpha) = \frac{\mathcal{T}^2 \mathcal{T}^\dagger I^{(i)}}{(1 - \mathcal{R}\mathcal{T}^\dagger)^2} \left[\frac{1}{1 + \dfrac{4\mathcal{R}\mathcal{T}^\dagger}{(1 - \mathcal{R}\mathcal{T}^\dagger)^2} \sin^2 \dfrac{\delta}{2}} \right] \qquad (3.11)$$

where δ is the phase difference between successive passes and $I^{(i)}$ is the incident
intensity. On resonance, the factor in square brackets is unity. Equation (3.11)
applies for flat or curved mirrors but not to a confocal geometry.

If it is assumed that $\alpha \ll 1 - \mathcal{R}$, then all terms in α^2 and higher powers of α
can be neglected and the peak transmission on resonance is

$$I^{(t)}(\alpha) = \frac{\mathcal{T}^2}{(1 - \mathcal{R})^2} \left[1 - \frac{\alpha(1 + \mathcal{R})}{(1 - \mathcal{R})} \right] I^{(i)} \qquad (3.12)$$

the apparent absorption coefficient is then

$$\alpha_{\text{app}} \equiv \frac{I^{(t)}(0) - I^{(t)}(\alpha)}{I^{(t)}(0)} \qquad (3.13)$$

$$= \alpha \left(\frac{1 + \mathcal{R}}{1 - \mathcal{R}} \right)$$

Therefore a tuned etalon enhances the absorption signal by a factor of
$(1 + \mathcal{R})/(1 - \mathcal{R})$. Since $1 - \mathcal{R}$ may be as small as 10^{-4}, the potential enhancement
of weak absorptions is very large.

There are a number of conditions necessary to realize the enhancement predicted
by Eq. (3.13). First, the circulating power inside the etalon should not saturate the
transition. If it does, the enhancement will be limited by the equalization of upper
and lower state populations. If some transitions are saturated and others are not,
then the relative intensities will be difficult to interpret.

Second, the frequency jitter of the laser must be small compared to the bandpass
of the etalon so that the resonance condition in Eq. (3.11) may be satisfied through-

out the experiment. The bandpass of the etalon is

$$\Delta\nu_{\text{bandpass}} = \frac{c}{2nLF} \tag{3.14}$$

where c is the speed of light and L is the distance between the mirrors. For this application, the index of refraction n between the mirrors is $n = 1$. The finesse F is given by

$$\frac{1}{F^2} = \sum_i \frac{1}{F_i^2} \tag{3.15}$$

where summation combines the various contributions to the finesse. In deriving Eq (3.13), it was implicitly assumed that the reflectivity finesse

$$F_{\mathcal{R}} = \frac{\pi\mathcal{R}^{1/2}}{1 - \mathcal{R}} \tag{3.16}$$

is a good approximation to the overall finesse.

Third, any frequency jitter on the incoming laser beam is seen as intensity noise at a detector placed after the etalon. In practical cases, this may be the factor that limits the S/N enhancement that can be obtained. Clearly, use of a very narrow bandwidth laser will help, and high-frequency modulation schemes can be effective in discriminating against noise resulting from laser frequency jitter.

Finally, the incoming laser beam must be mode-matched to the TEM_{00} mode of the etalon. For a Gaussian beam, this means controlling its waist position and its divergence angle. Unfortunately mode-matching means that any light back-reflected from the etalon is also mode-matched right back into the source laser. Therefore optical isolation between the laser and the etalon may be required for stable operation of the laser.

3.5. MODULATION SCHEMES AND SIGNAL PROCESSING

Free-jet direct absorption spectrometers often employ filtering, signal averaging, and modulation schemes to suppress laser intensity noise and thereby to improve sensitivity. These schemes involve a lock-in amplifier for phase sensitive detection of a modulated signal, a gated integrator–boxcar averager, or both. Modulation schemes based on amplitude and frequency modulation techniques have been used. In practice, the best results are obtained by modulating at the highest frequency possible because of the inverse dependence of the laser noise on the modulation frequency.

3.5.1. Baseline Subtraction

One approach to eliminating laser intensity noise is to measure it explicitly with one detector (detector 1 in Fig. 3.1) and then subtract it from the signal on the detector used to monitor absorption by the jet (detector 2). Care must be taken to scale the signals to the same level so that in the absence of absorption by the jet, the difference is zero. The two detectors and their preamplifiers should be matched for frequency response near the modulation frequency of the experiment. The subtraction itself is done conveniently with a differential amplifier with high common mode rejection (73). It is possible to combine baseline subtraction with a suitable optical multipass arrangement. However, the best S/N ratio may occur at less than the maximum number of passes because any noise that results from the alignment stability of the multipass cell occurs only on the signal channel (detector 2) and so may not be subtracted out.

With a relatively low-power difference frequency laser, Lovejoy and Nesbitt (73) were able to use baseline subtraction to approach the shot noise limit. With a color center laser, we obtained a five-fold noise reduction from baseline subtraction combined with 25 passes in a multipass cell for a total gain of about 125 times (33).

3.5.2. Amplitude Modulation

Amplitude modulation of the absorption signal has been accomplished by chopping the laser source, by using a pulsed-jet expansion, or in some cases by using both techniques simultaneously.

Continuous wave lasers can be amplitude modulated mechanically by using a rotating chopper or tuning fork. They can also be conveniently modulated at high frequencies using acoustooptical or electrooptical modulators. Hojer et al. (17) amplitude modulated their diode laser beam at 1 kHz with a rotating chopper and employed phase sensitive detection.

Although the signal can be modulated by chopping the laser beam, weak absorptions can be detected more effectively by modulating the absorption of the beam. In 1980, Vaida and McClelland (12) chopped a continuous free-jet expansion from a circular nozzle with a tuning fork; however, now most laboratories rely on a pulsed-jet apparatus to modulate the density of absorbers. The various mechanisms for driving a pulsed valve include an ac motor (56,109), an air driven piston (48), electromagnetic (solenoidal) (57), and piezoelectric (74). A piezoelectric valve offers relatively short uniform pulses and can be operated at high repetition rates. The other varieties offer higher peak flow rates. New pulsed-valve designs allow high-temperature source conditions to be attained (75,76).

Operation of the valve at a low duty cycle allows the use of gated detection to reject noise that occurs between the gas pulses. When the flow of absorbers can be concentrated into short pulses, the theoretical gain in sensitivity is the reciprocal of the square root of the duty cycle. In practice, the desired jet temperature and concentration of clusters as well as the limitations of the particular pulsed valve will determine the optimum duty cycle. The effective modulation frequency in a pulsed experiment is determined by the FT of the pulse shape. Usually two gated intervals are sampled. The first gate is timed to coincide with the maximum absorption; the second is set just before or after the pulse of absorbers. The difference between these two signals is recorded to discriminate against low-frequency fluctuations in the laser intensity. Since laser intensity noise usually decreases with frequency, it is desirable to use the shortest pulses possible consistent with the required total gas flow. Pulse lengths of 200–500 μs are typical (13,33,57), which implies effective modulation frequencies in the range of 1–2 kHz.

3.5.3. Frequency Modulation

Frequency modulation schemes may be based on either Stark modulation of the molecular transitions or on frequency modulation of the laser source.

The first method to be used in conjunction with free-jet direct absorption experiments was Stark modulation. The effusive microchannel array experiments of Chu and Oka (7) employed two Stark plates separated by 0.85 cm. Square-wave fields of several hundred volts per centimeter were used to frequency modulate the absorption transitions of NH_3 and $^{13}CH_3F$. Phase sensitive detection at the modulation frequency results in a derivative lineshape. Mizugai et al. (16) realized the importance of modulating at high frequency to avoid low-frequency noise associated with the laser source. They used a Stark modulation at 100 kHz in symmetric (16) and spherical tops (77), to improve their sensitivity by two orders of magnitude to make single-pass absorptions as small as 10^{-6} detectable. Stark tuning is first order in the electric field in symmetric and spherical tops, but it is second order —and therefore much weaker—in asymmetric top molecules. Since there is a strong dependence of the Stark coefficients on the rotational quantum numbers involved and on the possible presence of perturbations, the relative intensities in Stark-modulated spectra can be hard to interpret. For these reasons, Stark modulation is reserved for a few specific applications where it is most appropriate.

The effusive microchannel array experiments of Pine and Nill (9) were based on the frequency modulation of their single mode diode laser. Since diode lasers are current tuned, frequency modulation could be accomplished by adding a small ac signal at 8 kHz to the ramp current. Currently, Busarow et al. (61) used the frequency modulation technique to record direct absorption spectra in the far-infrared. The laser source consists of a line tunable far-IR laser that is tuned by mixing its output with tunable microwave radiation in a GaAs Shottky barrier diode. This produces sidebands at the sum and difference of the two frequencies. Frequency modulation of the microwaves at 50 kHz results in a corresponding frequency modulation of the sidebands. Phase sensitive detection referenced at twice the modulation frequency produces a second derivative lineshape, as shown in Fig. 3.4.

The modulated sliced jet (47,78) is a third way of producing a frequency modulation. The blade in the sliced jet [Fig. 3.2(d)] is replaced by a 0.2-mm diameter vibrating string. The string vibrates, left to right in Fig. 3.2(d) to frequency modulate the hole in the Doppler profile. The metal string can be driven at up to 10 kHz with the aid of a high-field magnet and an alternating current in the string.

3.5.4. Fourier Transform Spectroscopy

Although the earliest FT spectra of jet-cooled molecules were recorded at relatively low resolution (79–82), more recent work has employed high-resolution equipment to obtain line widths on the order of the gas-phase Doppler limit (83,84). The best commercially available equipment has retardations of 2.5 and 5.0 m, which corresponds to apodized line widths of 0.004 and 0.002 cm^{-1} respectively. Fourier transform spectroscopy (FTS) offers its well-known advantages:

1. With only a few changes of sources, detectors, and beam splitters, commercial instruments will cover the whole range from about 10 cm^{-1} in the far-IR to 62,000 cm^{-1} in the UV. Free-jet FTS experiments have been concentrated in the mid- to near-IR (750–4000 cm^{-1}), but the technique has also been extended into the UV (85). In the far-IR Doppler limited spectra have not been obtained with FTS because FTS resolution for a particular retardation is independent of frequency but the Doppler width is proportional to frequency. Therefore, in the far-IR, the instrumental line width will be much larger than the Doppler width in the jet and the apparent absorption coefficient would be correspondingly reduced.

2. Since all frequencies in a spectrum are observed simultaneously, the multiplex advantage provides a sensitivity enhancement over scanned operation of about the square root of the number of spectral elements. The total light intensity at the detector is monitored as one of the mirrors in the Michelson interferometer is moved. All of the laser-based techniques rely on scanning the laser frequency through the spectrum to record the absorption at one frequency at a time. The multiplex advantage applies in the IR where the detectors are background limited but is not effective in the visible and UV, where the noise derives from the statistics of photon arrivals from the source. Nonetheless, FTS still offers high throughput and has been effectively applied

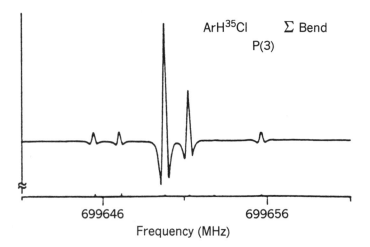

Figure 3.4. A portion of the frequency modulated ($2f$) spectrum of the Ar–HCl complex at 23 cm^{-1} in the far-IR. [Adapted with permission from Ref. 61.]

to the electronic spectra of jet-cooled species in the visible and near UV regions of the spectrum (85).

3. A broad section of spectrun ($\sim$ 300 cm^{-1}) can be recorded in a single experiment at high resolution (84).

Limitations of free-jet FTS spectroscopy derive from the requirement of a continuously operating light source with a smooth spectrum covering the entire frequency range of the experiment. These requirements limit the technique to incoherent light sources that have a relatively low-spectral brightness. Up to now, optical multipass arrangements have not been combined with free-jet FTS spectroscopy. Only the White cell is usable with an incoherent source and this cell would sample a relatively large volume within the jet. With the distance from the nozzle in the range of 1.5–11 mm (42) only a small volume of absorbers are available. The waist size of the FTS beam (typically ~2 mm) is many times the diffraction limit, which makes it particularly difficult to crowd a large number of reflections into a small volume.

The majority of free-jet FTS experiments have used pinhole nozzles with a circular aperture but a slit jet has been used to improve the resolution by about a factor of 2 (to 230 MHz) in the C–H fundamentals of CH_4 and CF_3H near 3030 cm^{-1} (42). In the mid-IR, the resolution of the best FTS instruments is well matched to the Doppler width in an axially symmetric jet; therefore, there is little advantage to concentrating the absorbers into a narrower lineshape with a slit jet. Even for the highest frequency fundamentals, the twofold improvement in resolution with the slit jets is much less than the order of magnitude improvement that a slit jet would allow.

Most applications have been to neat expansions of small molecules with relatively high vapor pressure including NH_3 (79), CH_3Cl (80), CO, NO, acetylene, propyne (42), N_2O, $CBrF_3$, CF_3I (86), and $CHCl_2F$ (87). For benzene (81) and pyridine (82), concentrated mixtues in Ar carrier gas were used. Recently, spectra of dimers [$(NO_2)_2$] (88) and clusters [$(HF)_5$] (89) have also been reported.

3.5.5. Cavity Ringdown

Cavity ringdown laser absorption spectroscopy (CRLAS) (90) involves the use of a short-pulse laser in combination with a very high-finesse etalon containing the absorbing species [Fig. 3.3.(d)]. When the pulse length is shorter than the roundtrip time in the etalon cavity, no interference occurs, so the light need not be resonant with one of the cavity modes. Rather, the photon wave packet may be viewed in a classical ballistic sense as bouncing back and forth between the mirrors. Of the light incident on the cavity, a fraction $\mathcal{T}$ enters the cavity. Each time the packet strikes a mirror, a fraction $\mathcal{T}$ is transmitted and a fraction $\mathcal{R}$ is reflected. With very high-reflectivity mirrors, a detector placed after the cavity sees a series of say 10,000 pulses exponentially decreasing in intensity. When an absorbing medium, for example, a free-jet (91), is placed between the mirrors, the absorption is detected as a decrease in the decay time of the pulses.

Since absorption by the jet is detected by attenuation of the transmitted light, CRLAS is a direct absorption technique and therefore falls within the scope of this chapter. However, the practical aspects of CRLAS are quite different from traditional absorption spectroscopy, and the application of CRLAS to free-jets is described in detail elsewhere (92). Therefore only a brief summary is included here to enable the reader to put it into the context of the other direct absorption methods described in this chapter.

Many of the advantages of CRLAS derive from the effective use of pulse lasers:

1. Sensitivity as good as a fractional absorption of $\alpha = 2 \times 20^{-7}$ per pass per laser shot has been reported (92). This amazing sensitivity relies on the use of a "supercavity" with a finesse of $F \approx 10,000$.

2. Each shot is a separate measurement. Therefore pulse-to-pulse intensity fluctuations, which have been the limiting factor in applying pulsed lasers to direct absorption experiments, are not relevant to the sensitivity.

3. Since the timescale for the ringdown is typically a few tens of microseconds, the technique may be effectively combined with pulsed methods of absorber production. For example, pulsed lasers can be used to produce radicals in high concentration, or to vaporize atoms, molecules, or clusters from a surface (92).

4. Tunable pulsed coherent light sources are available in every region of the spectrum from the IR through into the vacuum UV.

Up to now the ringdown method has been limited to the visible and near IR by the availability of extremely high reflectivity coatings. However, this problem can be addressed by improvements in coating technology. Even in spectral regions where the best available finesse is somewhat less than 10,000, the ringdown method may still be very sensitive.

When the pulse length is longer than the rountrip time in the cavity, as is typically the case with ordinary Q-switched lasers, interference may occur between successive roundtrips in the cavity and this limits the time resolution with which the decay can be measured. The sensitivity could be compromised but the effects may not be noticeable if the coherence length of the pulse is shorter than the cavity dimensions. Nonetheless, an input beam with good transverse mode quality is required and it must be mode-matched to the TEM_{00} cavity mode.

The FT principle dictates that there is a tradeoff between the coherence length of the light and its bandwidth. Therefore the highest resolution experiments will require long cavity lengths and suitably matched laser pulses. Scherer et al. (92), used 1.2-GHz bandwidth pulses with a 0.5 m cavity length to record electronic spectra of Cu_2, Cu_3, and AlAr.

3.6. DOUBLE-RESONANCE METHODS

As outlined in the introduction, the reason for using supersonic jets in spectroscopy is the tremendous spectral simplification that results at very low temperatures. With the sub-Doppler resolution and sensitivity possible with a slit nozzle, discrete absorption spectra of a great many molecules may be resolved and assigned. However, as the energy and the molecular size and flexibility increases, the limit is soon reached where the spectra become unresolvable or unassignable.

Spectra in this range often have additional complexities. Some spectra may be heavily perturbed with each zero-order bright state fractionated into many molecular eigenstates. Thus each eigenstate acquires some oscillator strength which adds a great deal of spectral congestion. The irregular distribution of line spacings and intensities makes assignment by traditional means a very grueling process. In other systems, such as hydrogen-bonded and van der Waals complexes, large amplitude motion complicates the spectrum. For radicals, the presence of an unpaired electron multiplies the number of allowed transitions. Indeed, complexities such as these represent the frontier areas of molecular spectroscopy.

Considerable spectral simplification can be obtained by double-resonance spectroscopy. Ideally one laser beam, the "pump," is used to label a single quantum state among those that are populated in the jet. A second laser frequency, the "probe," can then be scanned to record a spectrum only of those transitions that share a common level with the pump transition. This is a *three-level* double-resonance experiment that assumes there is no collisional energy transfer between the pump and probe events.

Often knowledge of the pump region from previous work makes assignment of the probe region trivial. When the pump beam interacts with more than one quantum state or when the pump region is unassigned, the situation is more complicated, but the number of assignment possibilities is still greatly reduced relative to a single-resonance experiment.

The use of two photons also allows a wider variety of final states to be reached than is possible with a single photon.

When collisions occur between the pump and probe events, a double-resonance signal may be obtained for some probe transitions that do not share a common level with the pump transition. Such *four-level* double-resonance experiments can probe state-to-state collisional energy-transfer rates.

The earliest direct-absorption-detected IR double-resonance studies in jets or beams employed two high-power line-tunable gas lasers to study the collisional relaxation of ammonia (93). Both N_2O lasers were Stark-tuned to IR transitions of ammonia. The first laser was chopped to label the pump transition, and the probe transition was Stark modulated at 112 kHz for noise suppression. At first an effusive source was used, but later (94) a skimmed nozzle source was introduced. Cross sections for collision-induced rotational transitions of ammonia with several polyatomic partners were measured.

The Nijmegen group also uses a high-power line-tunable laser (CO_2) as the pump but a continuously tunable F-center laser as the probe (95). In recent work, they have also used a far-IR probe (96). Their double-resonance method has been used to characterize the jet (97) and to study collisional relaxation (98), hot band spectroscopy (99), and coherence effects (100).

A double-resonance technique (Fig. 3.5.) has been developed in our laboratory

(67,101), which uses two continuously tunable lasers. Because a tuned etalon is used with the pump beam, the power requirements are low and there are continuously tunable lasers in many regions of the spectrum that would be suitable. The states involved in the pump transitions are labeled by high-frequency (400 kHz) amplitude modulation of the pump beam. Modulation transfer to the probe transition is detected with a lock-in amplifier. Modulation at 400 kHz is very effective at suppressing the intensity noise on the probe beam that is normally the dominant noise source in direct absorption experiments. The multipass cell on the probe beam is the two-spherical-mirror type described above, and the build-up cavity on the pump beam is a tuned etalon of finesse 100. The laser crossings are located 1 cm from a 0.15-mm diameter nozzle.

A propyne spectrum taken with two F-center lasers operating in the 3-μ region is shown in Fig. 3.6. In this case the $K = 1$ and 2 pump transitions are only about 5 MHz apart and so are unresolved by the pump. Since the pump transitions are not exactly superimposed, it was possible to distinguish $K = 1$ and 2 probe features by first pumping one side of the unresolved pump feature, then the other. The structure of the spectrum reveals the mechanism of intramolecular vibrational redistribution (IVR) in the propyne $2\nu_1$ band near 6500 cm^{-1}. The best resolution obtained with this technique is 5 MHz, but it is more typically in the 15–25 MHz range.

The cost of a double-resonance method is in the extra equipment, the extra experimental complexity, and the extra scanning time. In this technique, the vacuum system is very simple. The complexity is all in the lasers and the optics that are amenable to technological advances. The payoff for us has been spectra that were fully assigned at the moment they were recorded.

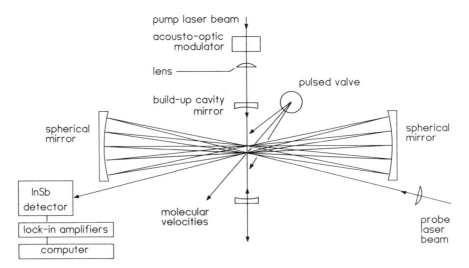

Figure 3.5. Schematic for a direct-absorption-detected IR double-resonance experiment. [Adapted with permission from Ref. 101.]

3.7. APPLICATIONS

Applications of the direct absorption method are diverse both in terms of the spectral region involved and in terms of the physical content of the problems that are addressed. A comprehensive review of each applications area is best left to reviews that focus on the conceptual problems rather than on the technique used to address them. Therefore, this section will highlight only three specific examples that will illustrate the scope of the technique and its level of development.

3.7.1. Electronic Spectroscopy

Most electronic spectroscopy of jet-cooled species is done with more sensitive techniques, such as laser-induced-fluorescence. When the fluorescence quantum yield is unity or at least a constant in the spectral interval studied, the fluorescence excitation spectrum is equivalent to a direct absorption spectrum. However, radiationless transitions often cause a substantial dependence of the quantum yield on the vibrational and rotational quantum numbers in the upper state. In this case, the two types of spectra may be very different in appearance and the ratio of the intensities provides a measure of the relative fluorescence quantum yield. When radiationless processes are so fast that the quantum yield is essentially zero, the direct absorption method provides access to otherwise inaccessible spectra. This is often the case for the higher electronic states, particularly in the vacuum UV. In the vacuum UV very fast radiationless processes frequently result in broad spectral features that can be effectively studied with low-resolution incoherent light sources (102,103).

The direct absorption method has been applied to the electronic spectra of both

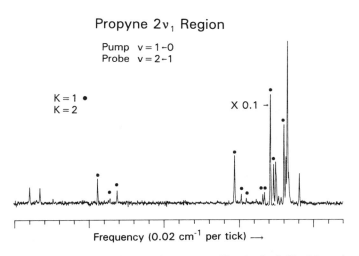

Propyne $2\nu_1$ Region

Pump $v = 1 \leftarrow 0$
Probe $v = 2 \leftarrow 1$

$K = 1$ •
$K = 2$

X 0.1 →

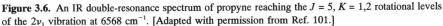

Frequency (0.02 cm^{-1} per tick) →

Figure 3.6. An IR double-resonance spectrum of propyne reaching the $J = 5$, $K = 1,2$ rotational levels of the $2\nu_1$ vibration at 6568 cm^{-1}. [Adapted with permission from Ref. 101.]

organic (104–106) and inorganic (107) systems. The story of pyrazine (108) illustrates the kinds of results and insights that can be obtained:

1. The fluorescence quantum yield from the S_1 manifold is low at the band origin and decreases exponentially with vibrational energy (109). This result was interpreted in terms of the nonradiative decay of S_1–triplet eigenstates.

2. The quantum yield decreases with the rotational quantum number J, showing a $(2J + 1)^{-1}$ dependence (56,109). With the aid of a vector coupling model, the rotational temperature and the Coriolis coupling parameter could be extracted (110).

3. The quantum yield was found to be highest at the center of rotationally resolved features, which led to the conclusion that the resolved features were superimposed on an unresolved "grass" with a low quantum yield (111).

3.7.2. Infrared Spectroscopy

In the IR, fluorescence from vibrationally excited states is weak and often diluted by intramolecular interactions (112). Therefore, the optothermal and direct absorption methods are more sensitive than fluorescence detection. The direct absorption method has been used to obtain accurate vibrational and rotational constants for a variety of organic (58,59,113) and inorganic (114,115) compounds. Infrared spectra of radicals have also been reported (116).

When IVR is active, individual rotational levels of vibrationally excited states are fragmented into a clump of molecular eigenstates. The number of discrete transitions in the spectrum becomes many times larger than would be expected for an unperturbed spectrum. The frequencies and intensities of these transitions carry information about the rate and mechanism of intramolecular vibrational and rotational relaxation processes.

Spectra of the C–H region of ethanol near 2990 cm^{-1} demonstrate the type of results that can be obtained (53):

1. The number of observed molecular eigenstates per upper state rotational level varies from about 2 to about 40 between $J = 0$ and $J = 4$, $K_a = 2$ indicating that extensive intramolecular dynamics would follow a hypothetical coherent excitation of a clump of molecular eigenstates.

2. The lifetime of a coherently prepared C–H stretch depends on the rotational quantum numbers and has the average value of 59 ps.

3. Several mechanisms are competing in the same band including anharmonic, a-axis Coriolis, and b,c-axis Coriolis coupling. Random matrix calculations (117) were used to estimate the root-mean-square average coupling matrix elements for each mechanism.

4. Insight into the ultimate fate of the initial excitation comes from the density of eigenstates in each clump. The observed level density is greater than the vibrational level density and approaches the total rotation–vibration level density. This result would indicate that the dynamics explore all of the

vibrational phase space, including both gauche and trans conformations, and much of the rotational phase space consistent with conservation of total angular momentum.

3.7.3. Clusters

Direct absorption work on clusters spans the range of both IR (69,77,118–121) and electronic spectra (122,123). Specialized systems such as exciplexes (124) and carbon clusters (125) have also been explored.

A principal goal of such studies is to accurately determine the intermolecular potential. In Ar–HCl there are two intermolecular coordinates, the distance between the centers of mass of the constituents and an angle specifying the orientation of the HCl. The Ar–H$_2$O complex is a bench-mark system for complexes with three intermolecular coordinates. The coordinates are now two angles specifying the orientation of the water molecule as well as the center-of-mass distance.

Since hydrogen-bonded and van der Waals complexes are weakly bound, excitations of the intermolecular coordinates occur in the far-IR part of the spectrum. Accordingly, vibration–rotation–tunneling (VRT) spectra in the far-IR are the key to mapping out the multidimensional intermolecular potentials.

A portion of the far-IR spectrum of Ar–HCl is shown in Fig. 3.4. The complexes were formed in a planar expansion. The tunable far-IR light was produced by mixing microwaves with the output of line-tunable far-IR gas lasers and then multipassed through the jet with the two spherical mirror cell described above.

For Ar–H$_2$O, 9 VRT bands have been recorded with this apparatus (126). When combined with combination differences from direct absorption spectra in the OH region (127) and microwave spectra (128), 12 VRT bands have been analyzed. The data set on Ar–D$_2$O includes 4 VRT bands (128,129).

As is the case for many weakly bound systems, the three intermolecular coordinates of Ar–H$_2$O are not separable, and accurate calculations of the VRT energy levels requires an explicit simultaneous treatment of all three dimensions. Cohen and Saykally (130) employed the collocation method to calculate the eigenvalues. This calculation was nested within a least-squares fit to obtain the best potential energy surface. The current best surface, labeled AW2, has substantial angular–radial coupling and is by far the most accurate three-dimensional intermolecular surface obtained to date.

ACKNOWLEDGMENTS

The authors are grateful to those who provided reprints of their work and to A. V. Chirokolava and T. J. Cronin for helpful comments on the manuscript. Support for this work was provided by the Division of Chemical Sciences, Office of Basic Energy Sciences, Office of Energy Research, United States Department of Energy under grant DE-FG02-90ER14151. This support does not constitute endorsement by DOE of the views expressed in this chapter.

REFERENCES

1. G. A. Blake, K. B. Laughlin, R. C. Cohen, K. L. Busarow , D.-H. Gwo, C. A. Schmuttenmaer, D. W. Steyert, and R. J. Saykally, *Rev. Sci. Instrum.*, **62**, 1693 (1991); 1701 (1991).

2. P. K. Cheo, *IEEE J. Quantum Electron*, **QE-20**, 700 (1984).

3. A. S. Pine, *J. Opt. Soc. Am.*, **64**, 1683 (1974).

4. R. F. Curl, paper WA01 givenat the OSU International Symposium on Molecular Spectroscopy, Columbus, Ohio, June 14–18, 1993.

5. E. Villa, A. Amirav, and E. C. Lim, *J. Phys. Chem.* **92**, 5393 (1988).

6. M. Sonnenschein, A. Amirav, and J. Jortner, *J. Phys. Chem.*, **88**, 4214 (1984).

7. F. Y. Chu and T. Oka, *J. Appl. Phys.*, **46**, 1204 (1975).

8. G. Scoles (Ed.), *Atomic and Molecular Beam Methods Vol. 1*, Oxford University Press, New York, 1988, p. 84.

9. A. S. Pine and K. W. Nill, *J. Mol. Spectrosc.*, **74**, 43 (1979).

10. R. J. Jensen, J. G. Marinuzzi, C. P. Robinson, and S. D. Rockwood, *Laser Focus*, May 1976, p. 51.

11. D. N. Travis, J. C. McGurk, D. McKeown, and R. G. Denning, *Chem. Phys. Lett.*, **45**, 287 (1977).

12. V. Vaida and G. M. McClelland, *Chem. Phys. Lett.*, **71**, 436 (1980).

13. G. D. Hayman, J. Hodge, B. J. Howard, J. S. Muenter, and T. R. Dyke, *Chem. Phys. Lett.*, **118**, 12 (1985).

14. A. M. de Souza, D. Kaur, and D. S. Perry, *J. Chem. Phys.*, **88**, 4469 (1988).

15. D. G. Leopold, V. Vaida, and M. F. Granville, *J. Chem. Phys.*, **81**, 4210 (1984).

16. Y. Mizugai, H. Kuze, H. Jones, and M. Takami, *Appl. Phys. B*, **32**, 43 (1983).

17. S. Hojer, H. Ahlberg, S. Lundqvist, J. Davidsson, and L. Holmlid, *Infrared Phys*, **27**, 261 (1987).

18. J. B. Anderson, in *Molecular Beams and Low Density Gasdynamics*, P. P. Wegener (Ed.), Marcel Dekker, New York, 1974.

19. D. H. Levy, *Annu. Rev. Phys. Chem.*, **31**, 197 (1980).

20. Hakuro Oguchi, *Rarefied Gas Dynamics*, Vols. I and II, University of Tokyo Press, Tokyo, 1984.

21. U. Borkenhagen, H. Malthan, and J. P. Toennies, *J. Chem. Phys.*, **63**, 3173 (1975).

22. G. M. McClelland, K. L. Saenger, J. J. Valentini, and D. R. Herschbach, *J. Phys. Chem.*, **83**, 947 (1979).

23. C. E. Klots, *J. Chem. Phys.*, **72**, 192 (1980).

24. S. G. Kukolich, D. E. Oates, and J. H. S. Wang, *J. Chem. Phys.*, **61**, 4686 (1974).

25. T. E. Gough and R. E. Miller, *J. Chem. Phys.*, **78**, 4486 (1983).

26. C. Douketis, T. E. Gough, G. Scoles, and H. Wang, *J. Phys. Chem.*, **88**, 4484 (1984).

27. A. K. Jameson, H. Saigusa, and E. C. Lim, *J. Phys. Chem.*, **87**, 3007 (1983).

28. A. Amirav, U. Even, and J. Jortner, *Chem. Phys.*, **51**, 31 (1980).

29. R. S. Ruoff, T. D. Klots, T. Emilsson, and H. S. Gutowsky, *J. Chem. Phys.*, **93**, 3142 (1990).

30. T. Emilsson, T. C. Germann, and H. S. Gutowsky, *J. Chem. Phys.*, **96**, 8830 (1992).

31. R. E. Smalley, L. Wharton, D. H. Levy, and D. W. Chandler, *J. Chem. Phys.*, **68**, 2487 (1978).

32. J. Wang, V. A. Shamamian, B. R. Thomas, J. M. Wilkinson, J. Riley, C. F. Giese, and W. R. Gentry, *Phys. Rev. Lett.*, **60**, 696 (1988).

33. G. A. Bethardy and D. S. Perry, *J. Chem. Phys.*, **98**, 6651 (1993).

34. E. L. Knuth, *J. Chem. Phys.*, **66**, 3515 (1977).

35. M. J. Vasile and F. A. Stevie, *J. Chem. Phys.*, **75**, 2399 (1981).

36. O. F. Hagena, in *Molecular Beams and Low Density Gasdynamics*, P. P. Wegener, (Ed.), Marcel-Dekker, New York, 1974.

37. Obtained from Eq. (3.5) using the ideal gas law $n_0 = p_0/kT_0$.

38. Johannes M. P. Geraedts, Ph. D. Thesis, Katholieke Universiteit te Nijmegen (1983), p. 25.

39. J. P. Toennies and K. Winkelmann, *J. Chem. Phys.*, **66**, 3965 (1977).

40. G. M. Stewart and J. D. McDonald, *J. Chem. Phys.*, **78**, 3907 (1983).

41. J. F. de la Mora and J. Rosell-Llompart, *J. Chem. Phys.*, **91**, 2603 (1989).

42. A. Amrein, M. Quack, and U. Schmitt, *J. Phys. Chem.*, **92**, 5455 (1988).

43. G. Baldacchini, S. Marchetti, and V. Montelatici, *Lett. Nuovo Cimento*, **41**, 439 (1984).

44. D. Kaur, A. M. de Souza, J. Wanna, S. A. Hammad, L. Mercorelli, and D. S. Perry, *Appl. Opt.*, **29**, 119 (1990).

45. L. R. Mercorelli, S. A. Hammad, and D. S. Perry, *Chem. Phys. Lett.*, **162**, 277 (1989).

46. M. D. Levenson and S. S. Kano, *Introduction to Nonlinear Laser Spectroscopy*, Academic, New York, 1988, p. 73.

47. G. A. Bethardy, Ph. D. Thesis, University of Akron, Akron, Ohio (1993).

48. K. Veeken and J. Reuss, *Appl. Phys. B*, **38**, 117 (1985).

49. A. McIlroy and D. J. Nesbitt, *J. Chem. Phys.*, **91**, 104 (1989).

50. D. R. Miller and R. P. Andres, *J. Chem. Phys.*, **46**, 3418 (1967).

51. A. E. Beylich, *Z. Flugwiss. Weltraumforsch.*, **3**, 48 (1979); *Prog. Astronaut Aeronaut.*, **74**, 710 (1981).

52. M. Sulkes, C. Jouvet, and S. A. Rice, *Chem. Phys. Lett.*, **87**, 515 (1982).

53. G. A. Bethardy and D. S. Perry, *J. Chem. Phys.*, **99**, 9400 (1993).

54. C. M. Lovejoy and D. J. Nesbitt, *J. Chem. Phys.*, **86**, 3151 (1987).

55. O. F. Hagena, *Surf. Sci.*, **106**, 101 (1981).

56. A. Amirav, U. Even, and J. Jortner, *Chem. Phys. Lett.*, **83**, 1 (1981); A. Amirav, C. Horowitz, and J. Jortner, *J. Chem. Phys.*, **88**, 3092 (1988).

57. C. M. Lovejoy, M. D. Schuder, and D. J. Nesbitt, *Chem. Phys. Lett.*, **127**, 374 (1986); C. M. Lovejoy and D. J. Nesbitt, *Rev. Sci. Instrum.*, **58**, 807 (1987).

58. A. McIlroy and D. J. Nesbitt, *J. Chem. Phys.*, **92**, 2229 (1990).

59. P. Carrick, R. F. Curl, M. Dawes, E. Koester, K. K. Murray, M. Petri, and M. L. Richnow, *J. Mol. Struct.*, **223**, 171 (1990).

60. The slit width was incorrectly reported as 10 μm in Ref. 33.

61. K. L. Busarow, G. A. Blake, K. B. Laughlin, R. C. Cohen, Y. T. Lee, and R. J. Saykally, *J. Chem. Phys.*, **89**, 1268 (1988).

62. J. U. White, *J. Opt. Soc. Am.*, **32**, 285 (1942).

63. M. Hepp, I. Pak, K. M. T. Yamada, E. Herbst, G. Winnewisser, *J. Mol. Spectrosc.*, **166**, 66 (1994).

64. D. Prichard, J. S. Muenter, and B. J. Howard, *Chem. Phys. Lett.*, **135**, 9 (1987).

65. P. G. Lethbridge and A. J. Stace, *Rev. Sci. Instrum.*, **58**, 2238 (1987).

66. D. Kaur, A. M. de Souza, J. Wanna, S. A. Hammad, L. Mercorelli, and D. S. Perry, *Appl. Opt.*, **29**, 119 (1990).

67. J. Go, T. J. Cronin, and D. S. Perry, *Chem. Phys.*, **175**, 127 (1993).

68. D. Herriott, H. Kogelnik, and R. Kompfner, *Appl. Opt.*, **3**, 523 (1964); D. R. Herriott, and H. J. Schulte, *Appl. Opt.*, **4**, 883 (1965).

69. N. Pugliano and R. J. Saykally, *J. Chem. Phys.*, **96**, 1832 (1992).

70. J. V. Coe, J. C. Owrutsky, E. R. Keim, N. V. Agman, D. C. Hovde, and R. J. Saykally, *J. Chem. Phys.*, **90**, 3893 (1989).

71. C. Douketis, D. Anex, G. Ewing, and J. P. Reilly, *J. Phys. Chem.*, **89**, 4173 (1985).

72. The derivation of this equation is a generalization of the development on pp. 323–325 of M. Born and E. Wolf, *Principles of Optics*, Permagon Press, New York (1975).

73. C. M. Lovejoy and D. J. Nesbitt, *J. Chem. Phys.*, **86**, 3151 (1987).

74. J. B. Cross and J. J. Valentini, *Rev. Sci. Instrum.*, **53**, 38 (1982).

75. D. W. Kohn, H. Clauberg, and P. Chen, *Rev. Sci. Instrum.*, **63**, 4003 (1992).

76. M. Fink, J. Hager, D. Glatzer, and H. Walther, *Rev. Sci. Instrum.*, **64**, 3020 (1993).

77. S. Yamamoto, M. Takami, and K. Kuchitsu, *J. Chem. Phys.*, **81**, 3800 (1984); M. Takami, Y. Ohshima, S. Yamamoto, and Y. Matsumoto, *Faraday Disc. Chem. Soc.*, **86**, 1 (1988).

78. G. A. Bethardy and D. S. Perry, Rev. Sci. Instrum., in press.

79. D. L. Snavely, S. D. Colson, and K. B. Wiberg, *J. Chem. Phys.*, **74**, 6975 (1981).

80. D. L. Snavely, K. B. Wiberg, and S. D. Colson, *Chem. Phys. Lett.*, **96**, 319 (1983).

81. D. L. Snavely, V. A. Walters, S. D. Colson, and K. B. Wiberg, *Chem. Phys. Lett.*, **103**, 423 (1984)

82. V. A. Walters, D. L. Snavely, S. D. Colson, K. B. Wiberg, and K. N. Wong, *J. Phys. Chem.*, **90**, 592 (1986).

83. A. Amrein, M. Quack, and U. Schmitt, *Z. Phys. Chem. Neue Folge*, **154**, 59 (1987).

84. M. Quack, *Annu. Rev. Phys. Chem.*, **41**, 839 (1990).

85. E. C. Richard, C. T. Wickham-Jones, and V. Vaida, *J. Phys. Chem.*, **93**, 6346 (1989); E. C. Richard and V. Vaida, *J. Chem. Phys.*, **94**, 153 (1991).

86. A. Amrein, H. Hollenstein, M. Quack, and U. Schmitt, *Infrared Phys.*, **29**, 561 (1989).

87. M. Snels and M. Quack, *J. Chem. Phys.*, **95**, 6355 (1991).

88. D. Luckhaus and M. Quack, *Chem. Phys. Lett.*, **199**, 293 (1992); *J. Mol. Struct.*, **293**, 213 (1993).

89. M. Quack, U. Schmitt, and M. A. Suhm, *Chem. Phys. Lett.*, **208**, 446 (1993).

90. A. O'Keefe and D. A. G. Deacon, *Rev. Sci. Instrum.*, **59**, 2544 (1988).

91. A. O'Keefe, J. J. Scherer, A. L. Cooksy, R. Sheeks, J. Heath, and R. J. Saykally, *Chem. Phys. Lett.*, **172**, 214 (1990).

92. J. J. Scherer, J. B. Paul, A. O'Keefe, and R. J. Saykally, in *Advances in metal and semiconductor clusters*, Vol. III, M. A. Duncan, (Ed.), in press.

93. F. Matsushima, N. Morita, S. Kano, and T. Shimizu, *J. Chem. Phys.*, **70**, 4225 (1979); F. Matsushima, N. Morita, Y. Honguh, and T. Shimizu, *Appl. Phys.*, **24**, 219 (1981).

94. Y. Honguh, F. Matsushima, R. Katayama, and T. Shimizu, *J. Chem. Phys.*, **83**, 5052 (1985).

95. K. Veeken and J. Reuss, *Appl. Phys.*, **B34**, 149 (1984).

96. M. Havenith, H. Linnartz, E. Zwart, A. Kips, J. J. ter Meulen, and W. L. Meerts, *Chem. Phys. Lett.*, **193**, 261 (1992).

97. N. Dam and J. Reuss, *Appl. Phys.*, **B49**, 39 (1989).

98. N. Dam, S. Stolte, and J. Reuss, *Chem. Phys.*, **135**, 437 (1989).

99. N. Dam, R. Engeln, J. Reuss, A. Fayt, and A. S. Pine, *J. Mol. Spectrosc.*, **139**, 215 (1990).

100. N. Dam, L. Oudejans, and J. Reuss, *Chem. Phys.*, **140**, 217 (1990).

101. J. Go and D. S. Perry, *J. Chem. Phys.*, **97**, 6994 (1992).

102. A. Amirav and J. Jortner, *J. Chem. Phys.*, **82**, 4378 (1985).

103. M. I. McCarthy and V. Vaida, *J. Phys. Chem.*, **90**, 6759 (1986); *ibid.* **92**, 5875 (1988).

104. T. S. Zwier, E. Carrasquillo M., and D. H. Levy, *J. Chem. Phys.*, **78**, 5493 (1983).

105. V. Vaida, *Acc. Chem. Res.*, **19**, 114 (1986).

106. A. Amirav and J. Jortner, *Chem. Phys. Lett.*, **95**, 295 (1983); O. Sneh, A. Amirav, and O. Cheshnovsky, *J. Chem. Phys.*, **91**, 3532 (1989).

107. A. Amirav, A. Penner, and R. Bersohn, *J. Chem. Phys.*, **90**, 5232 (1989); A. Penner, A. Amirav, S. Tasaki, and R. Bersohn, *J. Chem. Phys.*, **99**, 176 (1993).

108. A. Amirav, *J. Phys. Chem.*, **92**, 3725 (1988).

109. A. Amirav and J. Jortner, *J. Chem. Phys.*, **84**, 1500 (1986).

110. A. Amirav and Y. Oreg, *Chem. Phys.*, **126**, 343 (1988).

111. A. Amirav, *Chem. Phys.*, **126**, 327 (1988).

112. G. M. Stewart and J. D. McDonald, *J. Chem. Phys.*, **78**, 3907 (1983).

113. P. R. Brown, P. B. Davies, G. M. Hansford, and N. A. Martin, *J. Mol. Spectrosc.*, **158**, 468 (1993).

114. P. B. Davies, N. A. Martin, M. D. Nunes, D. A. Pape, and D. K. Russell, *J. Chem. Phys.*, **93**, 1576 (1990).

115. M. Takami and Y. Matsumoto, *Mol. Phys.*, **64**, 645 (1988).

116. R. F. Curl, K. K. Murray, M. Petri, M. L. Richnow, and F. K. Tittel, *Chem. Phys. Lett.*, **161**, 98 (1989).

117. J. Go, Ph. D. Thesis, University of Akron, Akron, OH (1993).

118. D. J. Nesbitt, *Chem. Rev.*, **88**, 843 (1988).

119. T. A. Hu, D. G. Prichard, L. H. Sun, J. S. Muenter, and B. J. Howard, *J. Mol. Spectrosc.*, **153**, 486 (1992).

120. A. C. Legon and A. P. Suckley, *J. Chem. Phys.*, **91**, 4440 (1989).

121. H. Linnartz, A. Kips, W. L. Meerts, and M. Havenith, *J. Chem. Phys.*, **99**, 2449 (1993).

122. D. J. Donaldson, V. Vaida, and R. Naaman, *J. Phys. Chem.*, **92**, 1204 (1988).

123. A. Amirav and A. Penner, *J. Phys. Chem.*, **94**, 7739 (1990).

124. A. Amirav, M. Castella, F. Piuzzi, and A. Tramer, *J. Phys. Chem.*, **92**, 5500 (1988).

125. A. Van Orden, H. J. Hwang, E. W. Kuo, and R. J. Saykally, *J. Chem. Phys.*, **98**, 6678 (1993).

126. R. C. Cohen, K. L. Busarow, K. B. Laughlin, G. A. Blake, M. Havenith. Y. T. Lee, and R. J. Saykally, *J. Chem. Phys.*, **89**, 4494 (1988); R. C. Cohen, K. L. Busarow, Y. T. Lee, and R. J. Saykally, *J. Chem. Phys.*, **92**, 169 (1990); R. C. Cohen and R. J. Saykally, *J. Chem. Phys.*, **95**, 7891 (1991).

127. R. Lascola and D. J. Nesbitt, *J. Chem. Phys.*, **95**, 7917 (1991).

128. G. T. Fraser, F. J. Lovas, R. D. Suenram, and K. Matsumura, *J. Mol. Spectrosc.*, **144**, 97 (1990).

129. S. Suzuki, R. E. Bumgarner, P. A. Stockman, P. G. Green, and G. A. Blake, *J. Chem. Phys.*, **94**, 824 (1991).

130. R. C. Cohen and R. J. Saykally, *J. Chem. Phys.*, **98**, 6007 (1993).

Chapter **IV**

HIGH RESOLUTION OPTICAL SPECTROSCOPY IN THE ULTRAVIOLET

W. A. Majewski, J. F. Pfanstiel, D. F. Plusquellic, and D. W. Pratt
Department of Chemistry,
University of Pittsburgh, Pittsburgh, Pennsylvania

Laser Techniques In Chemistry, Edited by Anne B. Myers and Thomas R. Rizzo.
Techniques of Chemistry Series, Vol. XXIII.
ISBN 0-471-59769-4   © 1995 John Wiley & Sons, Inc.

4.1. INTRODUCTION

Tunable, single-mode, continuous-wave (CW) dye lasers with intracavity frequency doubling are among the most versatile light sources known (1). When
combined with a molecular beam, a sensitive, spatially selective fluorescence detector, and a high-speed data acquisition system, lasers of this type can be used to
obtain optical spectra in the ultraviolet (UV) with a resolution approaching 1 part
in 10^9. At this resolution, even relatively large molecules exhibit *eigenstate-resolved*
spectra, which give information about *all* of the quantum states that are accessible
to the system. In this chapter, we describe the construction of a spectrometer that
merges all of these component technologies into a single apparatus. We also describe its application to a variety of chemically relevant problems in molecular
structure and dynamics.

Prior to the development of this apparatus, CW dye lasers operating in the visible
had been used for years in atomic beam studies of light-matter interactions at high
resolution (2). Later, Demtröder and co-workers (3) introduced external frequency
doubling and utilized the single-mode CW dye laser and a seeded molecular beam
to obtain sub-Doppler optical spectra of several small molecules like SO_2 in the UV.
But the practical realization of this approach (with intracavity frequency doubling)
for studies of the eigenstate-resolved spectra of large molecules in the UV was first
demonstrated by Dymanus, Majewski, Meerts, and co-workers (4–10) in their pioneering experiments on pyrazine, pyrimidine, naphthalene, and van der Waals complexes of fluorene and tetracene with rare gas atoms. It was the spectacular success
of these experiments that motivated the construction of our own instrument.

4.2. APPARATUS

The overall layout of the high resolution CW laser/molecular beam spectrometer
is shown in Fig. 4.1. The frequency doubled light beam from a single frequency
CW ring dye laser crosses a seeded supersonic molecular beam 15- or 110-cm
downstream from the nozzle, at right angles to the molecular beam. The laser-
induced fluorescence excitation spectrum (FES) of probe molecules seeded in this
beam is collected and detected at right angles to both beams. The FES signal as
well as others are acquired as a function of the laser frequency by a high-speed
data acquisition system. Excitation of the sample with a narrow bandwidth laser,
in combination with the preparation of the sample in a molecular beam and the
high spatial selectivity of the light collection system, give the spectrometer a resolving power of 2.5 parts in 10^9. At an excitation wavelength of about 300 nm,
the frequency bandwidth resolution of the spectrometer is 2.5 MHz.

The description of the apparatus is divided into its five primary components;
the single frequency CW laser (A), the molecular beam machine and light
collection–detection system (B), the frequency calibration sources (C), and two
computers (D and E) to record, process, and analyze the acquired signals. A detailed
description of these components follows a short discussion of the overall layout.

The CW dye laser and its peripheral devices rest on a vibrationally isolated laser

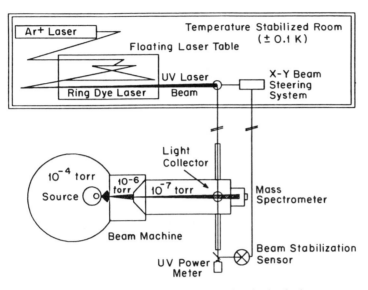

Figure 4.1. The overall layout of the high resolution CW laser/molecular beam spectrometer.

table (Newport 12 × 5 × 1.5 ft). Its three table legs are equipped with 11 vibration isolators (Berry Model SLM-3) to damp building vibrations. This floating config-uration necessarily requires a beam locking system to ensure the spatial stability of the laser beam at the laser–molecular beam crossing region. The beam lock is accomplished by using two orthogonal beam steering systems mounted on the laser table to direct the UV beam to the crossing region, and a four-quadrant photode-tector mounted on the beam machine to detect a small portion of the UV beam after it passes through the crossing region (Fig. 4.1).

Each beam steering system (11), operating independently, consists of a galva-nometer (General Scanning Model 325D) and PID servo electronics (General Scan-ning Model A6325K). Each galvanometer rotates a UV reflectively coated mirror in response to a differential error output signal taken from two opposite quadrants of the four-quadrant photodetector. The error signal, having both a sign and mag-nitude proportional to the direction and degree of the off-center beam displacement, adds at a biasing amplifier inside the PID controlling loop of the galvanometer circuit and thereby causes the drive section of the galvanometer to return the UV beam to the center of the photodetector. The two orthogonal systems ensure a position stability of better than 0.1 mm from the center of the beam crossing region. An *xy* translational stage, housing the four-quadrant photodetector, provides for fine tuning of the laser beam position with respect to the molecular beam to minimize scattered light and maximize fluorescence signal strength.

4.2.1.   The CW Ring Dye Laser

The CW laser is a modified version of the Spectra-Physics Model 380D ring dye laser (6). A schematic diagram of its principal components is shown in Fig. 4.2. The

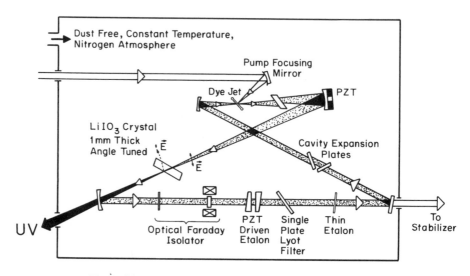

Figure 4.2. The principal components of the CW ring dye laser.

laser was modified in a number of ways. All hand-driven mirror adjusters were replaced with micrometers accessible from outside the laser cavity. A doubling crystal assembly was positioned at the auxiliary beam waist. A Plexiglas cover replaces the metal cover to allow visual alignment of the laser while maintaining a dust-free nitrogen atmosphere within the laser cavity.

The 514.5-nm line of an argon ion laser (Spectra-Physics Model 171) is focused on a dye jet containing R6G (595–630 nm), Kiton Red (620–640 nm), and/or DCM (630–700 nm) dyes dissolved in viscous solvents. The dye solutions provide the broadband lasing medium through which the CW laser is tuned. The principal components that enforce single frequency and traveling wave operation of the dye laser are also shown in Fig. 4.2. The figure-eight laser cavity consists of four mirrors and has two foci. The dye jet is positioned at one focus and the second harmonic generating (SHG) crystal made from $LiIO_3$ (Gsänger) resides at the other. Intracavity placement of the doubling crystal in the high intensity fundamental wave (~ 1 W) offsets the inherently low conversion efficiency (~ 10^{-3}) of the nonlinear SHG process. Two types of crystal cuts are typically used and require different intracavity orientations with respect to the incident beam. A Brewster angle cut is used to ensure small reflective losses within the cavity while a perpendicular incidence cut requires an antireflective coating for low reflective loss operation. A detailed discussion of the Brewster-cut $LiIO_3$ crystal has been given by Majewski (12). A detail of the intracavity configuration is shown in Fig. 4.3.

The other components in the cavity enforce single frequency operation and include a single-plate birefringent filter (Lyot filter), a galvanometer-scanned thin-plate etalon (TPE), and a piezo-scanned thick etalon (TE). These elements act as optical frequency bandpass filters that successively narrow the range of the lasing frequencies until only one 200 MHz cavity mode remains above threshold. The

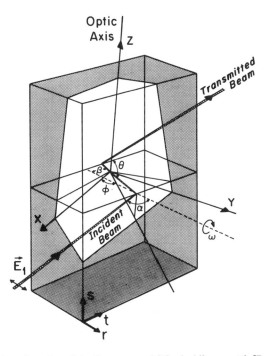

Figure 4.3. Intracavity orientation of the Brewster-cut LiIO₃ doubling crystal. [Reprinted with permission from Ref. 12.]

free spectral ranges and finesses of these components are shown in Fig. 4.4. Nearly all intracavity components are positioned at Brewster's angle with respect to the horizontal cavity polarization.

The cavity mode is scanned by changing the cavity length with two rotatable quartz plates mounted on galvanometers near Brewster incidence. A voltage ramp is applied to the galvanometers' controlling electronics causing both to rotate but in opposite senses to compensate for each plate's beam shift. Since typical frequency scans range from 90 to 120 GHz in the fundamental, both the TPE and TE transmission curves must follow the cavity mode so that it remains near the peak of the transmission curves. The single-plate Lyot filter's bandpass is sufficiently broad for these scan ranges, and near peak power is maintained without scanning this element. The transmission curve shown in Fig. 4.4 is that of the commercially available triple-plate filter. The single-plate curve is roughly a factor of 3 broader than the one shown.

The galvanometer-controlled TPE replaces the fixed-plate configuration in the 380D laser. This galvanometer scans in response to the same voltage ramp that scans the quartz plate galvanometers. Both an offset bias potentiometer, used to rotate the thin plate to lower or higher order with respect to the intracavity wave, and a gain potentiometer, used to reduce or enhance the voltage ramp gain applied

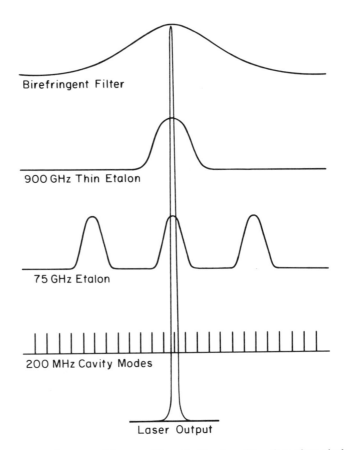

Birefringent Filter

900 GHz Thin Etalon

75 GHz Etalon

200 MHz Cavity Modes

Laser Output

Figure 4.4. Free spectral ranges and finesses of the optical bandpass filters that enforce single frequency operation of the dye laser (not to scale).

to the TPE PID controlling circuitry, provide the user with tuning control over five or six TE modes.

The unmodified TE is scanned with a high voltage ramp that is applied to a piezoelectric spacer. Superimposed on this ramp is a 2 kHz modulated voltage (1 V peak-to-peak). The modulated etalon induces amplitude modulation of the intra-cavity power when not on top (zero slope) of the etalon transmission curve. A phase sensitive detector, utilizing a photodiode to monitor the modulated laser intensity, senses the slope of the transmission curve and drives the high voltage ramp, causing the TE to track the cavity mode when the laser is scanned.

One additional optical filter is present to enforce traveling wave operation. In-tracavity counterpropagating beams produce undesirable standing wave patterns within the cavity, resulting in spatial hole burning within the dye jet. Hole burning significantly reduces the efficiency of the laser induced emission. A unidirectional device (or optical diode) (13) consisting of a Faraday rotator and an optically active quartz plate rotates (~ 4°) the polarization vector of the counterclockwise propagating electromagnetic wave and thereby induces anti-Brewster angle reflective losses. The clockwise propagating wave experiences no net phase retardation and retains above threshold amplification at the selected frequency.

The intracavity components permit single frequency scans over a range of 90–120 GHz. Phase matching the SHG crystal doubles the scan range (180–240 GHz, or 6–8 cm^{-1}). The frequency doubled UV light (300–340 nm) is coupled out of the laser cavity and reflected through the beam machine onto a power meter (Laser Precision Corp. Model RK-5200). The laser cavity is aligned to optimize the UV power.

Single frequency scan operation of the dye laser is a prerequisite for its narrowband stabilization. The narrowband operation is made possible by using stabilization electronics (Spectra-Physics Model 388) and the external reference station (Spectra-Physics Model 389), collectively referred to as Stabilok. The components that make Stabilok operational are an intracavity piezoelectric driven mirror (PDM) and an external reference station that includes both reference cavity (RC) and slave cavity (SC) interferometers. The RC and SC interferometers have finesses of 3 and free spectral ranges of 500 MHz and 10 GHz, respectively. The RC and SC outcoupled fringe signals detected with a photodiode are shown in Fig. 4.5. The function of the interferometers is to convert frequency changes to amplitude changes.

Stabilok provides single frequency scan operation of the laser even in the event of mode hops. At the moment Stabilok is activated, the system scans the frequency

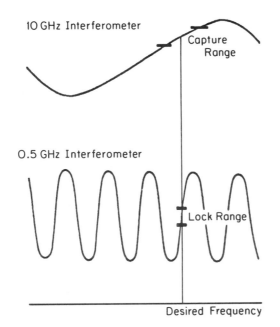

Figure 4.5. Outcoupled slave cavity (upper) and reference cavity (lower) interferometer fringe signals.

of the laser cavity (LC) by rotation of the quartz plates until the RC photodiode voltage (error signal) is on the steepest portion of a fringe within a narrow capture range about 0 V (see Fig. 4.5). The slave cavity length also is changed to bring its photodiode voltage to within its capture range. When these two conditions are met, Stabilok activates the intracavity PDM. The LC galvanometers stabilize frequency noise in the power band range from 0 to 200 Hz and the PDM provides stabilization up to 10 kHz. The cooperative effect of these two elements is to narrow the laser bandwidth to less than 750 kHz.

After lock is established, the laser is scanned by a voltage ramp that drives the RC quartz plate, changing its length and FSR. The laser cavity frequency is made to track the FSR changes of the RC by ensuring that the RC error signal remains within the capture range. The SC, acting passively, is made to follow its own error signal. In the event of a mode hop, the RC voltage jumps outside of the capture range and lock is broken. At that instant, the SC and RC lengths are temporarily frozen. The SC error signal causes the LC galvanometers to scan so as to return them to within their capture range. At this point, the RC regains control of the LC galvanometers to correct for its error signal. Lock is again established and the laser frequency returned to where it was before lock was broken.

The Stabilok system has undergone a number of significant modifications to improve its performance beyond that commercially provided. The major advances involved modifying the RC and SC interferometers to extend the scan range from 1 to 4 cm^{-1} in the fundamental. This was accomplished by replacing the RC galvanometer and its PID controlling electronics with a larger model (General Scanning Model 325D) to increase the quartz plate angular displacement from 25° to 50° and thereby to increase the scan range of the interferometer. The piezoelectric material driving the SC mirror also was replaced to enhance its scan range. The voltage ramp driving the RC galvanometer was improved with respect to linearity by replacing it with an 18-bit digital-to-analog ramp.

4.2.2. The Beam Machine and Detection System

The molecular beam machine is shown in Fig. 4.6. It provides for the preparation of sample molecules in a well-collimated supersonic molecular beam and for the detection of FES signals from a spatially selective region at the laser and molecular beam crossing point. The beam machine consists of three differentially pumped vacuum chambers referred to as the source, buffer, and research chambers. The three chambers are separated by two 1-mm skimmers through which the molecular beam travels. The skimmers partition the beam flux among the three chambers so as to maintain a differential pressure across them. In this arrangement, the gas pressure in each chamber is maintained so that the mean free path of the carrier gas is larger than the distance the beam travels in each chamber. The conical shape of the skimmers prevents recoil collisions from disturbing the collision-free environment in the molecular beam after the expansion. A quadrupole mass spectrometer located 120-cm downstream of the molecular beam source is used to align the molecular beam through the skimmers and to monitor the molecular beam intensity during a scan.

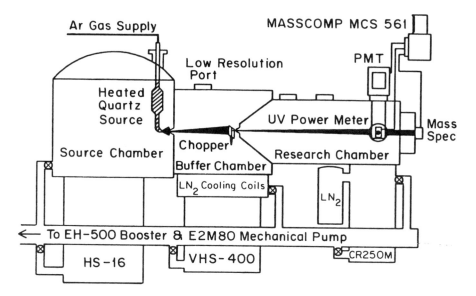

Figure 4.6. Diagram of molecular beam machine showing the three differentially pumped chambers and the light collection optics, 1.1-m downstream of the nozzle.

The source chamber houses the quartz molecular beam source. The source (see Fig. 4.7) consists of three sections, each of which is heated by 0.5-mm platinum coaxial wire allowing temperatures as high as 400°C to be obtained. The temperature of each section is monitored with a thermocouple (NiCr/CuNi). Provision has been made to allow for sample loading while maintaining vacuum conditions in the source chamber. The top section is used to hold the sample. The middle section provides a trap for particles to prevent the nozzle from becoming blocked. Two sources are available, one with a 240-µ diameter nozzle and another with a 90-µ diameter nozzle, to allow for wider variation of the backing gas pressures and rotational temperatures. An optimum argon backing gas pressure of about 500 torr is found for the 240-µ nozzle and 1–10 atm for the 90-µ nozzle. A high pressure gas chamber also is available for the preparation of gas samples or binary gas mixtures.

The buffer and research chambers each provide a region to cross the molecular beam with the laser beam. Figure 4.7 also shows a detail of the two configurations. A set of collection optics is positioned at each of the molecular and laser beam crossing regions (only one set is shown). The collection optics provide spatial filtering of the FES signal. The effective collection volume imaged on the photomultiplier tube (PMT) is a 1-mm diameter sphere.

The limiting Doppler resolution of the apparatus is obtained from

$$\delta\nu_{FWHM} = 2\sqrt{\ln 2}\ \nu_0\ \frac{d}{z}\frac{v}{c} \tag{4.1}$$

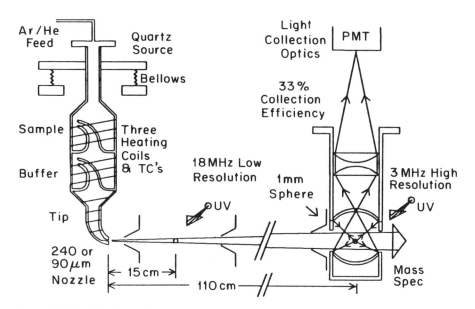

Figure 4.7. Detailed views of the molecular beam source and collection optics. An independent set of collection optics resides above the low resolution port.

where v_0 is the excitation frequency, d is the collection volume diameter, z is the distance downstream of the source, FWHM is the full-width at half-maximum, and v and c are the molecular beam and light velocities, respectively. Here, it is assumed the laser beam intensity and molecular beam density cross sections have Gaussian profiles. The measured beam velocity of argon is 500 m s^{-1} and, at an excitation frequency of 33,000 cm^{-1}, the calculated Doppler linewidth in the high resolution port ($z = 1100$ mm) is then

$$\delta v_{\text{FWHM}} = 2\sqrt{\ln 2}\ 9 \times 10^{14}\ \text{MHz} \left(\frac{1\ \text{mm}}{1100\ \text{mm}}\right) \left(\frac{500\ \text{m s}^{-1}}{3 \times 10^8\text{m s}^{-1}}\right) = 2.0\ \text{MHz} \quad (4.2)$$

The calculated linewidth in the low resolution port ($z = 150$ mm) is 18 MHz. The observed values (2.5 and 20 MHz) are only slightly larger than the calculated values indicating that only minor contributions to the linewidth exist due to the laser bandwidth or from transit time and/or power saturation effects.

Deciding which port to use depends on several factors; the natural linewidth of the molecular transition, the oscillator strength of the molecular transition, and the relative percentage of the molecule/complex in the molecular beam under the expansion conditions employed. When the natural linewidth of the transition is greater than about 30 MHz, excitation and detection in the low resolution port is employed. When linewidths are less than 30 MHz, the oscillator strength and beam density determine which port is utilized.

The light collection system, shown in greater detail in Fig. 4.8, has been de-

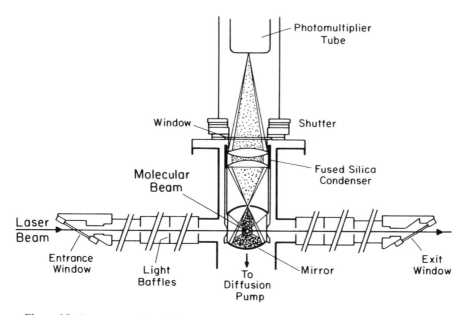

Figure 4.8. A transverse view of the excitation region. [Reprinted with permission from Ref. 6.]

scribed elsewhere (6). Two spherical mirrors with equal radii image the fluorescence through a small hole in the upper mirror. The upper spherical mirror has a focus at the beam crossing and reflects all fluorescence onto the lower mirror. All direct and reflected light on the lower mirror is reflected onto the PMT through a condensing assembly. In this configuration, the size of the hole in the upper mirror (~ 2 mm) determines the extent of spatial filtering at the beam crossing.

The PMT (EMI 9813QB) has a quantum efficiency of 20%. Thermionic emission of electrons along the dynode amplification chain and from the photocathode contribute to the background photon count (dark counts). Pulse amplitude discrimination of dark counts from the dynode chain and PMT cooling to reduce spontaneous photocathode emission are provided to minimize the dark count contribution to the total background count. Other sources of noise include light scattering from the entrance and exit windows and Rayleigh scattering from the molecular beam. Typically, total noise counts of less than 200 counts s^{-1} are observed.

4.2.3. Calibration Sources

Complete frequency calibration of a spectrum is performed using an interferometer, or etalon, to establish a relative frequency scale and the iodine absorption spectrum to give an absolute reference standard (14). Figure 4.9 illustrates the configuration of these elements.

The etalon spacer is a 0.5-m long and 5-cm diameter cylinder made of Zerodur ceramic with a 1.25-cm diameter hole bored through the center. On each end is mounted a f = 0.25-m concave mirror coated for 97% reflectance from 550 to 700

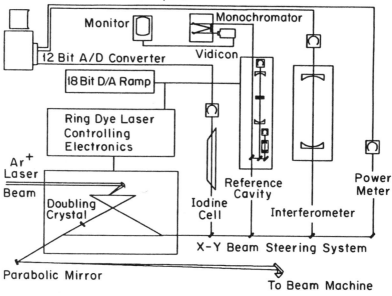

Figure 4.9. Dye laser calibration sources and controlling electronics.

nm. The mirrors also serve to completely seal the cavity from pressure changes and thus eliminate gas density changes in the cavity. The Fabry–Perot confocal etalon has a measured free spectral range of 149.8760 ± 0.0005 MHz (given by FSR = $c/4L$, where c is the speed of light and L is the length of the cavity) and a finesse of 100. The etalon is aligned with a small portion of the fundamental so that it behaves as a planar etalon. This arrangement is achieved when the optical axes of the spherical mirrors and the fundamental ray are collinear. In this arrangement, the FSR, given by $c/2L$, is 299.7520 MHz.

Thermally induced length changes of the etalon spacer give rise to a thermal drift of the marker frequency positions in time. The magnitude of this effect may be assessed based on the following considerations. Zerodur has a thermal expansion coefficient of 3×10^{-8} K^{-1}. Since about 3×10^6 λ's of 600-nm light are required for one complete roundtrip in the 0.5-m cavity, thermal changes induce length changes of about 0.1 λ K^{-1}. This value is equivalent to 15 MHz K^{-1} since shortening or lengthening the spacer by 1 λ is equivalent to 1 FSR (150 MHz). On the timescale of the experiment, the temperature stabilization of the spacer material needs to be in the milli-Kelvin range if a measurement of line positions to an accuracy of ±0.1 MHz is desired. This was accomplished passively by placing the etalon in a dewar and actively by thermally stabilizing the dewar shell to ± 0.05 K by regulating the ambient room temperature. The dewar thus serves to dampen the air temperature fluctuations.

Direct assessment of this error has been made by measuring the changes in the relative frequency positions of an etalon marker and an overlapping molecular transition as a function of time. Figure 4.10 shows two of these scans taken 45 min apart. The shift in the marker position (upper trace) relative to the molecular transition (lower trace) in this time interval is 3 MHz, the linewidth of the molecular transition. The results indicate that the thermal drift rate is on the order of 4 MHz h^{-1}. This error is systematic and limits the accuracy of the measured rotational constants to ±0.1 MHz. For any particular experiment, the magnitude of this error will depend on factors such as the spectral resolution, the density of molecular transitions, and the timescale of the experiment. An alternative stabilization technique has recently been reported which makes use of a cavity that is actively stabilized to a He–Ne laser (15).

The absolute frequency standard is the iodine absorption spectrum covering a range of 9000 cm^{-1} from 11,000 to 20,000 cm^{-1}. Most line positions (centers of mass) are given with an accuracy of ±30 to ±45 MHz (14), which limits the accuracy of the absolute frequency calibration to ±0.02 cm^{-1}.

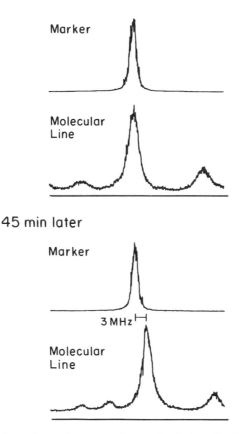

Figure 4.10. Two scans of an etalon marker compared to a molecular transition, illustrating the thermal drift of the etalon over a 45-min time interval.

Frequency matching the CW laser to the low resolution pulsed laser to locate the desired molecular signal is performed using a 0.5-m monochromator and vidicon to monitor the dispersed light. The vidicon signal is displayed on a TV monitor. The system also is used to provide a coarse frequency calibration of the CW laser and to ensure a single mode scan when being aligned.

4.2.4. Data Acquisition Software

A MASSCOMP computer (Model MCS561) is used to monitor the performance of the laser before and during the course of the scan as well as to digitize all signals necessary to obtain a normalized and calibrated spectrum. Figure 4.11 illustrates the four signals acquired during the scan.

The need for a specialized computer becomes apparent when the following issues are considered. A typical spectrum acquired with this apparatus spans about 5 cm^{-1} or 150×10^3 MHz. If our spectral resolution is 3 MHz and we desire 10 points per resolution interval (0.3 MHz digital accuracy), the spectrum will require up to 5×10^5 data points or 1 Mbyte of storage (2 bytes/data point). As illustrated in Fig. 4.11, three other data files of equal length are needed to normalize and

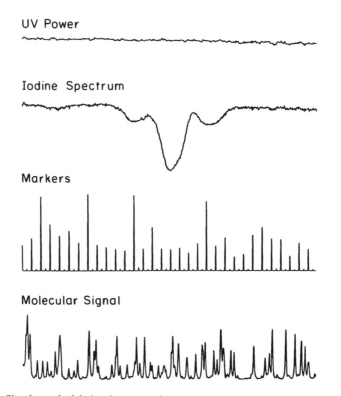

UV Power

Iodine Spectrum

Markers

Molecular Signal

Figure 4.11. Signals acquired during the course of a scan. The molecular signal is acquired by a parallel interface; the remaining signals are sampled by A/D converters.

calibrate the spectrum, requiring a total of 4 Mbytes per scan. Typical scan times range from 20 to 40 min. Thus, in a 30-min scan, the acquisition bandwidth must be at least 2.2 kbytes s^{-1}. The actual bandwidth used is roughly a factor of 8 times larger for signal averaging reasons.

In addition to meeting the acquisition bandwidth and storage requirements, the computer must also monitor the performance of the laser and simultaneously update a graphics display with all the acquired data during the course of a scan. The necessity in real-time applications of having simultaneous activity at the I/O interface, graphics display, and central processor has prompted MASSCOMP to provide dedicated peripheral processors to work asynchronously with the central processor in order to facilitate these tasks. The architectural configuration of their computer system has the characteristics of a tightly coupled multiprocessor. Communication and synchronization between these processors occurs through a bus-based shared memory system. The master–slave relationship of the central–peripheral processors allows this system to accomplish these concurrent activities without extensive need to buffer data on job queues. The processors forming this triangular configuration include a data acquisition and control processor (DACP), a graphics processor (GP), and a host processor (HP) running the UNIX operating system. A diagram of the main system components and their interconnects across the triple bus structure is shown in Fig. 4.12. Additionally, the UNIX-based operating system provides pipelining mechanisms for concurrent processing on the HP as well as a means for prioritizing interrupt handling to ensure reliable delivery and service of scheduled routines in a real-time environment.

Our data acquisition program, *jba*, written in the C programming language, is menu driven and display oriented to permit the user to optimize the scan conditions with respect to laser stability, scan rate, and digital accuracy. In order to implement these tasks, the program utilizes all of the computer's resources. A flow chart illustrating the essential features of this program is shown in Fig. 4.13.

The DACP services all interrupts from the parallel interface (PI) and analog-to-digital converter (ADC) devices. The buffering scheme utilized by the DACP is shown at the top of Fig. 4.13. Data collected by the DACP are written directly into main memory where access is possible from both the DACP and HP. Upon completion of each buffer, the DACP interrupts the HP at which point the HP services a buffer completion routine (BCR). The molecular signal from the photon counter (an external device) is acquired in digital form with the PI. All analog signals are sampled with an ADC device. (Sampling refers to a process by which a digital value is computed and collected from an analog signal at regularly timed intervals.) Since two devices are used, two BCR routines are serviced by the HP. Each device has its own computational requirements.

The photon counting device transmits the accumulated photon count over a preset time interval to the PI device in the form of a binary coded decimal (BCD) signal. Sixteen digital lines are used to transfer three 4-bit decimal digits (0000–1001$_2$), a 3-bit exponent, and a sign bit. When the sign bit is ignored, only 5000 values in the range from 0.00×10^2 to 9.99×10^7 counts can be transmitted to the PI by the photon counter. A constant two-dimensional array acts as a translation

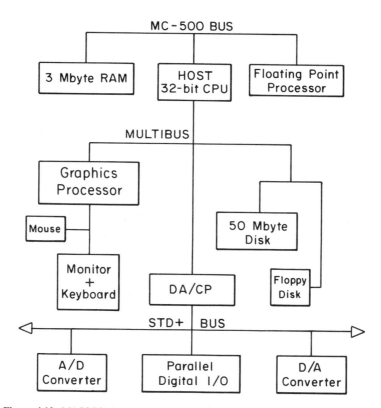

Figure 4.12. MASSCOMP peripheral devices and the bus interconnect architecture.

table to convert the transmitted BCD values into a 2's complement (binary) form. The array value is dereferenced with the use of the BCD value by applying the exponent-2 to the column and the 2's complement form of the 12-bit mantissa to the row. (Note that the row dimension is $1001\ 1001\ 1001_2$ or 2457_{10}.) Furthermore, in order to store these data in short-integer form (16-bit), the array values represent the square root of the accumulated count, first scaled by 32,767 before rounding. Thus, the largest possible value of 1,000,000 counts is dereferenced as 32,767, and so on.

All other acquired signals are analog and are digitized with six 12-bit ADCs. The six ADCs are sequentially digitized at a rate of 1 MHz and constitute a set. Signal averaging is accomplished by time multiplexing up to eight separate sets over the delay interval between two consecutive PI samples. Since the DACP routes the ADC values to sequential memory, the array is first sorted according to each individual ADC. The value saved for each ADC is the sum of eight samples, one from each set. Short integer storage also is possible for each ADC since the maximum possible value is 2^{12}bits $\times$ 8 sets = 32,767. The PI and the six ADC data sets are moved into separate arrays until 512 values are acquired at which time the data are written out to individual prenamed files on disk in block (512 byte) form.

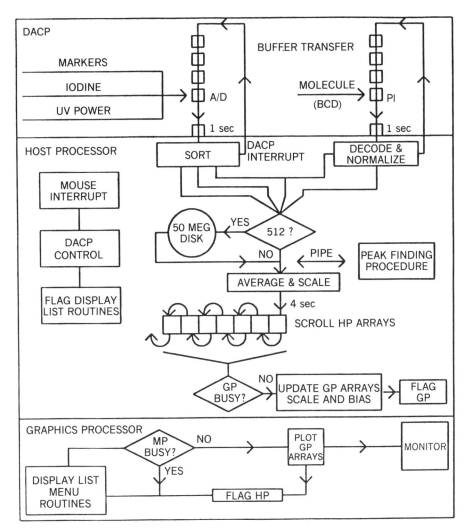

Figure 4.13. Main features of the data acquisition program, *jba*. The triple processor architecture provides for simultaneous data acquisition, manipulation, and presentation during the course of a scan.

The buffer completion routines also scale and offset the data so the GP can immediately process and display the data in graphic form. When the buffer size is less than the screen width, the displayed data are scrolled by shifting old data to the left before the new data are inserted. Once prepared, the HP flags the GP to update the display.

The laser scan performance is monitored by a procedure that performs short-range interpolation over the etalon marker spacings and notifies the user in the event that there is an inconsistency in these relative spacings. The process to perform this function is first forked and then overlayed and, as such, is run as a concurrent process on the HP. A two-way UNIX pipe supports the interprocess communication and data transfer between the concurrently running BCR and the peak-finding processes.

The GP is programmed using routines that resemble assembly language. At run time, subroutines called display lists are loaded into arrays (RAM). At this point, the array addresses are made known to the GP. The base addresses of shared data memory between the HP and GP are also established at this time. The RAM, utilized by the HP and/or DACP to store acquired data, thereby becomes addressable *via* direct memory access by the GP. In addition to plotting the acquired data, all menus are called using display lists. Mouse and/or keyboard events interrupt the HP to service routines that flag the GP to display parameter input and program control menus.

All of the files acquired for a single scan are stored in a prenamed directory to prevent accidental file deletion. Access to the spectral data is made by simply specifying the directory name. All additional files used for the calibration and analysis also reside in this directory.

4.2.5. Data Analysis Software

The data analysis program, *jb* is written in the C programming language. The primary goal of this software is to implement fitting strategies for the analysis of complex spectra. The strategy pursued is one of pattern recognition, quantum number assignment, and linear least-squares fitting. We currently use a MCS5600 computer system for these calculations.

A complete spectral analysis begins with the calibration and normalization of the molecular spectral data. The second step involves visual inspection of the observed rotational structure. The inspection process permits the evaluation of the band type and, therefore, the determination of the spectroscopic selection rules, a prerequisite to the detailed process of assigning lower and upper state quantum numbers to the observed transition frequencies. The third step requires generation of a simulated spectrum based on the rotational constants obtained from an *ab initio* structure and/or from the results for similar molecules. A visual comparison of the simulated spectrum with the experimental one on a graphics display permits the refinement of the rotational constants used to generate the simulation. This refinement is accomplished either by a simple stepwise adjustment of the parameters or by a least-squares fit after making quantum number assignments to experimental line frequencies. The collection of all such assignments constitutes an assignment file that is used in the fitting program to minimize the observed minus calculated standard deviation of the experimental and simulated line frequencies and intensities. Each of these steps is described in more detail below.

The calibration procedure begins with the determination of the relative frequency scale. A peak-finding procedure reads the marker file and locates the file positions at the HWHM of each marker found in this file. All marker positions as well as information about the other data files are saved in a special ASCII file known as the "working directory" file.

The frequency spacing (FSR) in the UV between consecutive marker positions has been very accurately determined to be 599.5040 MHz. This spacing represents the finest division of the frequency "ruler." However, the time-based acquisition of spectral data is a nonlinear function of frequency. The top portion of Fig. 4.14 illustrates the deviation from linearity given as the percentage change in marker position spacing B relative to adjacent spacings A and C as a function of marker number. The lower panel of Fig. 4.14 shows two different regions of the experimental spectrum (upper traces) with the corresponding portions of the best-fit simulated spectrum (lower traces). The simulated line positions provide additional divisions on the frequency ruler and illustrate the nonlinearity of the scan between markers. The spectral region shown on the right reveals a 1 MHz deviation in the experimental line frequency relative to the "true" simulated line position. In contrast, the spectral region shown on the left is located near a marker where the lines show no such deviation. Both positive and negative deviations appear randomly in other regions of the spectrum and give rise to a statistical error of 1 MHz in the

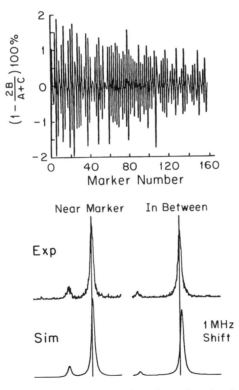

Figure 4.14. Scan nonlinearities. The upper portion shows the nonlinearity of a single scan, expressed as the percentage change in marker spacing B relative to adjacent marker spacings A and C. The lower portion shows the difference between the experimental (top) and true, or simulated (bottom) line frequencies in two different regions of the spectrum.

overall (observed minus calculated) standard deviation and a 5 kHz standard deviation in the rotational constants. The scan nonlinearity is a result of optical coupling of the reference cavity mode with etalon modes generated from beam splitters in the beam path. Root-mean-square (RMS) values of 0.3–3% (1–9 MHz) are commonly observed and are significant enough to warrant the use of interpolation routines to help offset these errors. These interpolation routines utilize linear, cubic spline, or polynomial functions and return the interpolated frequency given a file position and *vice versa*. The use of these routines circumvents the need to manipulate the raw spectral data in order to linearize the data in frequency.

The absolute frequency of the spectrum is established by comparing the recorded iodine spectrum with that documented in the iodine spectral atlas (14). Typically, the atlas frequency of the most symmetric line is entered at its center of mass. This information is also saved in the "working directory" file. Thus, the "working directory" file is used to keep a record of all information necessary for the complete calibration of the spectrum. Only it and the power normalized molecular signal file need to be saved in order to review and analyze the spectrum.

Assignment of a calibrated spectrum begins with graphical methods of inspection at successive levels of horizontal scale expansion. Three different levels of view are possible and "hot key" selected, expanding (or contracting) the data about the cursor location. The method of digital averaging used at each level depends on its compression factor. The first level typically uses a scale compression factor so that the entire band can be displayed on a single screen. The averaging method simply finds the maximum intensity in each predefined frequency interval since digital averaging at this resolution would otherwise "wash out" all spectral features. The compression factor for the second level is user adjustable and intermediate between level one and the raw spectral data. It is typically set to give a 10-fold increase in the digital resolution. The averaging method includes options to find either the maximum intensity or the cubic spline interpolated intensity across each frequency interval. In order to expedite the reviewing process, the averaging routines for these two levels of view are run only once after which the averaged data are written to memory and accessed directly. These data also are saved to file in order to circumvent the repetition of the averaging process each time the spectrum is reviewed.

The compression factor for the third level of view again is user adjustable but is usually set to display the raw data. Again, this is typically a 10-fold increase in digital resolution. Since the storage and manipulation of the raw data in main memory is precluded due to insufficient space, only the data that are to be displayed are accessed from file. When averaging is necessary at this level, compression factors are usually small since the processing time needed for the on-line averaging can significantly slow the reviewing process.

A three-button mouse is used to successively page through the data at each level of view and to position the cursor cross-hair for the expansion and compression of data when changing between the levels.

The implementation of the pattern recognition fitting strategy requires displaying a simulated spectrum on the same screen as the experimental spectrum. A comparison of the subbranch patterns revealed in both permits tentative assignments to be made and the refinement of the parameters used to generate the simulated spectrum. Figure 4.15 shows a portion of an experimental spectrum (upper trace) and the two deconvoluted simulated spectra (lower traces) that comprise it.

Each simulated spectrum consists of a line set. Each line has a simulated frequency and intensity, a set of six (or more) quantum numbers, and a band type. All simulated line intensities are coadded in the displayed array at the frequency interpolated position. Each line shown in Fig. 4.15 was first convoluted with a 3 MHz Lorentzian function. Additional lineshape options include Gaussian and Voigt profiles. Two possible line sets may be displayed simultaneously with options to frequency shift, scale, and offset one with respect to the other. An option to coadd both simulated line sets allows hybrid band ratio factors to be determined. Here, each simulation channel contains one of the two different band type simulations as illustrated in Fig. 4.15. The simultaneous display of two line sets also is very useful

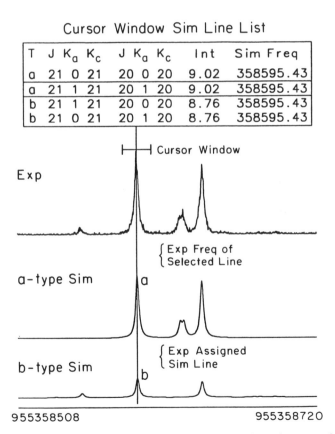

Figure 4.15. Using two simulated spectra (bottom) to assign an experimental spectrum (middle). Assignments of the *a*- and *b*-type (simulated) lines within the cursor window are shown at the top as they appear on the graphics terminal.

for the deconvolution of overlapping spectral bands. Additional options are provided for each simulation channel to display only P, Q, or R branch transitions and/or transitions having specific J'', K_a'', K_c'', J', K_a', or K_c' quantum numbers for easy identification of subbranch patterns and/or for creation of specific subbranch assignment files.

Experimentally assigned simulated lines are marked on the display with a letter according to band type. The band types, quantum numbers, simulated frequencies, and intensities of all lines within the cursor window are easily displayed and sequentially selected with the space bar. Experimentally assigned simulated lines that are selected reveal the assigned frequency on the display with a vertical line. The cursor position frequency is easily assigned/reassigned to the simulated line selected.

Assignment files consist of a line set, each line identified by the band type, the quantum numbers, and the assigned experimental frequency. A new assignment file consisting of all the assigned lines may be updated/created from either simulated spectrum. This file is then used to refine the Hamiltonian parameters in a least-squares analysis.

Simulations are generated using a C language computer program based on the options and parameters listed in a special menu. The menu displays both a primary and an auxiliary set of parameters. These parameter sets are used in different ways by the rotor program. The primary parameter values are used to generate the initial line set but are never automatically modified by the rotor program. On the other hand, the auxiliary parameter values are set to those returned by the rotor program after a least-squares fit of the assigned line set. Any number of parameters may be toggled on or off in the least-squares analysis and are indicated as such by variation indicators. The standard deviation (1 σ) of each parameter as well as the observed minus calculated standard deviation are returned and displayed.

In addition, direct access to the auxiliary parameter values is provided from the display of the spectrum. Changing any parameter value causes a new simulation to be generated and displayed every 5 s. The procedure generating the simulation makes use of a special file created by the rotor program that contains the energy derivatives of each simulated line with respect to all of the parameters. A new simulation is generated based on the derivative approximation (16). When used in conjunction with subbranch filter options to match patterns, this procedure permits rapid but approximate parameter values to be obtained. The line intensities are not modified by this procedure and therefore are only approximate. A best-fit simulated spectrum may then be generated by performing a least-squares fit of the parameter values using the assignment file.

The menu includes additional options to select the pure band types a, b, or c or the hybrid band types ab, ac or bc; to set the rotational temperature; and to specify the maximum J, K, and ΔK order to calculate. For hybrid band types, the polarization ratio is specified in terms of the transition moment orientation angle θ. Provision to rotate the Hamiltonian prior to diagonalization is provided for one or both states in terms of the Euler angles. This permits the fitting of perturbation terms in the rotational Hamiltonian not lying along a principal axis. Complete discussions of the rotor programs have been given elsewhere (17).

Often, it is desirable to make rotational assignments of bands that have severely perturbed upper states. When the ground state parameter values are known, assignments of rotational transition energies to perturbed excited state levels may be made based on the equal energy displacements of the common excited state level. Upon selecting a particular simulated line, the program searches the line set for all transitions that terminate in this level. All transitions are listed along with the observed minus calculated energy differences. Upon selection of any particular transition from the list, the appropriate portion of the experimental data containing this line is automatically displayed. Repositioning the cursor updates the observed minus calculated energy difference and permits rapid association of simulated lines with perturbed upper state transition energies. This procedure permits reliable assignments to be made, a necessary step for the formulation of a deperturbation model based on the dependencies of the perturbed energies on the rotational quantum numbers, J and K.

An additional feature allows two different experimental spectra to be viewed simultaneously on the same screen. Often this feature is useful when comparing spectra of the same molecule acquired under different experimental conditions. For example, the identification of subbranch patterns associated with particular low values of J (and K) may be made based on the line intensity changes due to the acquisition of bands at different rotational temperatures. This feature also permits the concatenation of experimental spectra that span more than one continuous scan of the laser (~ 8 cm^{-1}). The secondary set of experimental files displayed is relocatable with the use of the cursor and has an independent set of scaling factors that permit the match of line intensities in the overlapping region. An interpolation routine ensures a correct frequency match at the point of attachment. A concatenated spectrum composed of any number of files may be produced with successive applications of this procedure.

In addition, a Fourier transform data filter is available to permit digital smoothing of noisy data. A frequency bandwidth parameter is provided to control the degree of smoothness desired.

"X-windows" versions of these programs are in preparation (18).

4.3. APPLICATIONS

Figure 4.16 shows the first spectrum taken with this apparatus, the fully resolved fluorescence excitation spectrum of the 0_0^0 band in the $S_1 \leftarrow S_0$ electronic transition of 1-fluoronaphthalene (1FN) at about 310 nm (19). Illustrated there are the three different levels of view of the spectrum; the entire band, the Q branch, and a detailed view of the most congested portion of the Q branch. Factors of 20 were used in each scale expansion; the digital resolution in the bottom panel is 270 kHz. Also shown in the bottom panel are the spectral assignments, $^{\Delta K_a}\Delta J_{K_a K_c}(J)$. All 4000 + observed lines have been assigned. Each exhibits a width (FWHM) of 3 MHz which, when fit to a Voigt profile, yields a homogeneous linewidth of 1.4 MHz ($\tau_r = 110$ ns). The remaining width (~ 2 MHz) is Doppler in origin. A single rotational temperature of 9 ± 1 K accurately reproduces the relative intensities of all

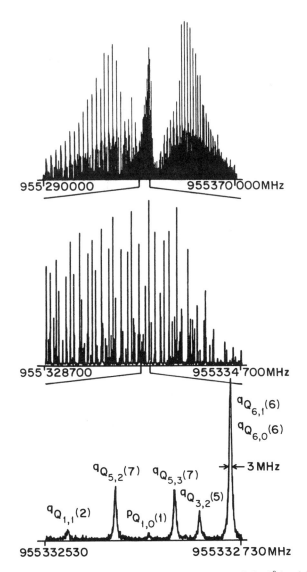

Figure 4.16. Rotationally resolved fluorescence excitation spectrum of the 0_0^0 band in the $S_1 \leftarrow S_0$ transition of 1-fluoronaphthalene, at about 314 nm. [Reprinted with permission from Ref. 19.]

lines. The entire band is an *ab* hybrid band with 75 ± 2% *a* character and 25 ± 2% *b* character, yielding $\theta = \pm 30 \pm 2°$ for the angle between the $S_1 \leftarrow S_0$ transition moment vector and the *a* inertial axis. The corresponding transition in naphthalene is *x* axis polarized; the *a* axis in 1FN makes an angle of 17° with respect to *x*. The significantly larger value of θ in 1FN is believed to be a consequence of S_1/S_2 state mixing.

Table 4.1 lists the derived values of the rotational constants of the zero-point vibrational levels of the S_0 and S_1 states of 1FN. These are accurate to ±0.1 MHz. Also listed in Table 4.1 are the rotational constants of naphthalene and several other substituted naphthalenes, also examined by high resolution methods (20–27). Without exception, all spectra could be fit to rigid-rotor, asymmetric top Hamiltonians for both electronic states, yielding rotational constants that are accurate to ±0.1 MHz. In 1FN, there are no signs of centrifugal distortion for values of J and K_a up to 25; similar results were obtained for the remaining molecules in both electronic states.

All molecules examined to date exhibit significantly different ground state rotational constants, even the two rotational isomers of the two naphthols 1HN and 2HN (cf., Table 4.1). This result demonstrates the power of this technique for purely analytical applications (28). Moreover, the changes in these rotational constants on electronic excitation are all different, demonstrating the power of this technique for studies of molecular structure and dynamics in electronically excited states. Typically, ΔA and so on, are all negative in naphthalene and its derivatives, reflecting the general increase in size of the molecule that occurs on $\pi \rightarrow \pi^*$ excitation. The only exception is 1-aminonaphthalene (1AN), which has $\Delta A > 0$, believed to be a consequence of the shortening of the R–NH$_2$ bond in the S_1 state (22).

Table 4.1. Rotational Constants (in MHz)[a] of the Zero-Point Vibrational Levels of the S_0 and S_1 States of Several Naphthalenes

	S_0			S_1[b]			
	A	B	C	ΔA	ΔB	ΔC	Reference
N-h_8	3105.1	1231.4	883.9	−77.5	−17.8	−15.2	6
N-d_8	2517.5	1094.8	772.2	−60.0	−16.5	−13.3	6
1FN	1920.6	1122.2	708.5	−29.1	−20.1	−11.9	19
2FN	2844.8	808.4	629.6	−81.1	−5.8	−7.5	20
1HN (cis)	1947.6	1124.3	713.1	−23.7	−18.5	−10.6	21
1HN (trans)	1942.1	1133.6	716.0	−20.5	−22.4	−11.7	21
2HN (cis)	2849.3	824.7	639.8	−80.2	−2.9	−5.8	21
2HN (trans)	2845.1	825.4	640.0	−73.0	−2.8	−5.4	21
1AN	1933.8	1127.6	713.1	45.6	−28.1	−5.9	22
1MN	1894.6	1126.3	709.7	−32.8	−16.9	−11.3	23
2MN	2805.3	822.7	638.7	−65.0	−8.1	−8.6	23
1,4DMN	1178.2	1111.9	576.5	−14.4	−17.0	−8.0	24
2,3DMN	2170.7	668.4	514.4	−45.5	−7.5	−6.9	25
1CNN	1478.7	956.8	581.0	−21.4	−13.3	−8.1	26
1NA	1361.7	703.3	465.4	−19.1	−5.9	−6.0	27

[a]Errors are ± 0.1 MHz (1σ).
[b]$\Delta A = A' - A''$, and so on.

4.3.1. Determining Atomic Positions in Large Molecules

A key finding in our early experiments was that the high resolution technique could be used to determine the center-of-mass (COM) coordinates of specific atoms in large molecules, based on the well-known sensitivities of inertial parameters to mass and distance. The first application of this technique was to the *cis–trans* isomerization equilibrium in 1HN and 2HN.

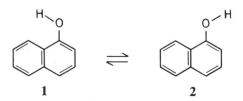

1   **2**

Now, the S_0 barrier to such isomerizations is known to be high, on the order of 1200 cm^{-1} (29). Thus, in the collision-free environment of a cold molecular beam, both isomers should be present, providing their zero-point energies are not too different. Indeed, we found (21) [as had others (30)] that there are two electronic origins in the low resolution spectra of 1HN and 2HN, separated by 274 and 317 cm^{-1}, respectively. Johnson *et al.* (21) then showed that the two bands in each molecule could be assigned to specific OH rotamers by recording the spectra of the 0_0^0 bands of 1HN/1DN and 2HN/2DN, and comparing their rotational constants to determine the COM coordinates of the hydroxy hydrogen atoms. (Here, 1DN and 2DN are the OH deuterated naphthols.) This comparison was made using (the planar) Kraitchman's equations (31)

$$|x|_{\text{COM}} = \left(\frac{(I_y' - I_y)(I_x' - I_y)}{\mu(I_x - I_y)} \right)^{1/2} \tag{4.3a}$$

$$|y|_{\text{COM}} = \left(\frac{(I_x' - I_x)(I_y' - I_x)}{\mu(I_y - I_x)} \right)^{1/2} \tag{4.3b}$$

(where I_x', I_y' are the in-plane moments of the deuterated molecules), well known for their application to similar problems in microwave spectroscopy (32). A clear distinction between the two isomers was made. For example, as deduced from Eqs. (4.3), the *cis* and *trans* rotamers of 2HN have the experimental hydroxy atom coordinates $(|x|, |y|)$ = (3.44, 1.56 Å) and (3.95, 0.14 Å), respectively, compared to the *ab initio* (3–21G basis) values (3.51, 1.55 Å) and (4.03, 0.02 Å).

An unexpected result of these experiments was the finding that the difference in the rotational constants of the two isomers is different from that expected on the basis of a simple inertial model. Thus, if one assumes that *only* the H atom moves in the isomerization; that is, the motion is a simple rotation about the C–O bond, then one expects, for example, that B″ (*cis*-1HN) > B″ (*trans*-1HN). Experimentally (*cf.* Table 4.1), the reverse order is found. Clearly, then, there are other motions occurring during the isomerization. Calculations suggest that the principal additional motion is a displacement of the heavier oxygen atom in the opposite direction, owing to the attraction of its lone pair(s) to the neighboring C—H hydrogen atom (21). Thus, extremely subtle properties of the potential energy surface are probed by the high resolution experiment.

Atomic COM coordinates have been determined in several other molecules using this method. These include 2-hydroxyquinoline (33), 2-pyridone (34), the 2-pyridone dimer (35), 2-methyl-1-naphthol (2M1HN) (36), hydroquinone (37), 2-pyridone-H_2O (38), 1-naphthoic acid (1NA) (27), several van der Waals complexes (39–41), and several methyl and ethyl esters of phenol and aniline (42). The two OH rotamers (C_{2h} and C_{2v}) of hydroquinone were also distinguished by the difference in their nuclear spin statistical weights (37).

4.3.2. Probing Motions Along Low Frequency Coordinates

Replacement of the fluorine atom in 1FN and 2FN with isoelectronic groups containing more than one atom, like hydroxy (OH), amino (NH_2), or methyl (CH_3), introduces other vibrational degrees of freedom into the problem. Examples include the OH torsion in 1/2HN, the NH_2 inversion in 1/2AN, and the CH_3 hindered internal rotation in 1/2MN. We next show how the high resolution technique may be used to explore motion along vibrational coordinates of this type.

In one class of problems, the mode in question possesses nascent angular momentum that can couple to overall rotation. This produces changes in the energy level structure from which one can extract information about the potential energy surface along the vibrational coordinate, in different electronic states. A nice example is provided by the data in Fig. 4.17, which shows a 600-MHz portion of

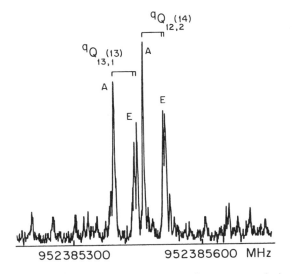

Figure 4.17. A 600 MHz portion of the rotationally resolved fluorescence excitation spectrum of the 0^0_0 band in the $S_1 \leftarrow S_0$ transition of 1-methylnaphthalene, showing the splittings of the lines by the tunneling motion of the methyl group and the first-order torsion–rotation interaction. [Reprinted with permission from Ref. 43.]

the rotationally resolved FES of the 0_0^0 band of the $S_1 \leftarrow S_0$ transition of 1MN (23,43). In this spectrum, each rovibronic line is split into two lines, A and E, and the E lines are further split into a doublet. The A lines are the rovibronic transitions connecting the lowest A torsional levels of the S_0 and S_1 states, and the E lines are the rovibronic transitions connecting the lowest E torsional levels of the S_0 and S_1 states (1MN belongs to the molecular symmetry group G_6). The A and E lines appear at different optical frequencies because the $A-E$ tunneling splitting in each state is different, being governed by threefold barriers that are different. The observed $A-E$ splitting, $\Delta\nu_1 = 50.3 \pm 0.2$ MHz, is the *difference* in these tunneling splittings. Additionally, as shown in Fig. 4.17, the E lines are further split by J- and K-dependent interactions between the torsional motion and the overall rotational motion of the molecule. In 1MN, this splitting obeys the empirical relation $\Delta\nu_2 = (0.493 \pm 0.02) K_a$ MHz. The observation of *two* splittings in the spectrum $\Delta\nu_1$ and $\Delta\nu_2$ makes possible the determination of the torsional barriers in *both* electronic states.

The essential physics in the problem can be captured by adding to the asymmetric rotor Hamiltonian the first-order perturbation terms D_a and D_b,

$$\hat{H}_{\text{eff}} = AJ_a^2 + BJ_b^2 + CJ_c^2 + D_aJ_a + D_bJ_b \qquad (4.4)$$

Here, $D_a = FW_E^{(1)} \rho_a$ and $D_b = FW_E^{(1)} \rho_b$, where F is the internal rotor constant, $W_E^{(1)}$ is a first-order torsion–rotation perturbation coefficient [tabulated by Hayashi and Pierce (44)], and the $\rho_{a,b}$ are direction cosines, describing the relative orientations of the inertial axes and the axis of internal rotation. The operators J_a, and so on, are the components of the total rotational angular momentum J which, in this case, includes contributions from internal rotation. [The rotational constants A, B, C in Eq. (4.4) also differ slightly from the usual rigid-rotor constants, owing to the contributions of higher order perturbations] (23). The A levels are not affected by the first-order perturbation terms. But the E levels are split by the torsion–rotation interaction. If initially degenerate, and if $\rho_a > \rho_b$, then the two levels are split by $2D|K_a|$, which is proportional to K_a, as observed. Thus, roughly speaking, the tunneling methyl group distinguishes the two possible directions of overall molecular rotation, since the magnitude of the perturbation depends on whether the two coupled angular momenta "add" or "subtract."

Clearly, the magnitudes of the perturbations also depend on the barrier heights, in *both* electronic states. This dependence has been exploited by us and by others to determine the torsional barriers of several restricted rotors in the two electronic states connected by the photon. In 1MN, we find $V_3(S_0) = 809$ cm^{-1} and $V_3(S_1) = 565$ cm^{-1} (23). Also studied to date are the methyl rotors in 2MN (23), 1,4DMN (24), 2,3DMN (25), *cis*- and *trans*-2M1HN (36), nonatetraene (45), decatetraene (45), and *p*-toluidine (46), and the NH$_3$ rotors in the hydrogen bonded complexes *cis*- and *trans*-2-naphthol-NH$_3$ (47) and 2-pyridone-NH$_3$ (48). The information gained from these experiments has given us a much deeper understanding of the torsion–rotation interaction, and of the subtle electronic interactions that are responsible for the hindered rotation of methyl groups. In addition, the work on *p*-

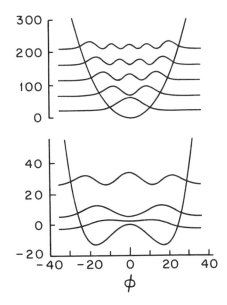

Figure 4.18. The R—COOH torsional potential energy surfaces of 1-naphthoic acid in its S_0 (bottom) and S_1 (top) electronic states. The vertical axis is in cm^{-1}. [Reprinted with permission from Ref. 27.]

toluidine [a G_{12} molecule with $V_6(S_0) = 5.6$ cm^{-1} and $V_6(S_1) = 43.9$ cm^{-1} (46)] has provided us with the first known example of a "precessing" methyl group; that is, a group that tilts away from the rotor axis in the course of its internal rotation. A review of these contributions (at both low and high resolution) has recently appeared (49).

In another class of problems, electronic excitation produces a significant change in the equilibrium geometry of the molecule, resulting in an extensive Franck–Condon progression in one (or more) low frequency coordinate(s). Rotationally resolved studies of each of the bands in the progression(s) then provides new information about the potential energy surfaces along this (these) coordinate(s). A nice example is provided by recent results on 1NA (Structure **3**) (27). Here,

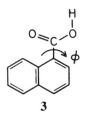

3

several vibronic bands that appear within 300 cm^{-1} of the electronic origin were examined at full rotational resolution, in both the protonated (—COOH) and deuterated (—COOD) molecules. The data show that all bands belong to the *s-cis* isomer of 1NA. They also show that all bands are torsional in nature; that is, they involve displacements along either the S_0 or the S_1 carboxyl torsional coordinate ϕ, or both. Unambiguous assignments of the bands follow from the observed inertial defects ($\Delta = I_c - I_a - I_b$, rigorously zero for a nonvibrating planar molecule), from which the torsional potential energy surfaces of both electronic states have been derived (Fig. 4.18). A similar approach is being used by us to study the NH_2 inversions in 1/2AN (22), the conformational properties of 1/2-ethylnaphthalene (50) and 1/2-vinylnaphthalene (45), and the low frequency modes of 4,4'-dimethylaminobenzonitrile (the TICT molecule) (51).

4.3.3. Inertial Axis Reorientation (Axis Switching)

To completely describe a polyatomic molecule rotating in space, one must use two reference frames; a space (or laboratory)- fixed coordinate system and a molecule (or body)- fixed coordinate system. Typically, the experimental conditions define both the origin and the direction of the axes of the space-fixed frame. And, also typically, the origin of the molecule-fixed frame is chosen to be at the COM of the molecule, and the axes of this frame are chosen to be coincident with the molecule's principal axes of inertia. In this way, by using the Euler angles ϕ, θ, and χ to specify the orientation in space of the moving, molecule-fixed frame, one can use the same angles to specify the orientation in space of the angular momentum J produced by the molecule's rotational motion. This convenient prescription is desirable for many purposes, including the determination of spectroscopic selection rules (52).

The orientation of J in the molecule-fixed frame depends on the geometry of the molecule (53). This geometry (*i.e.*, that which is defined by the equilibrium positions of the nuclei) usually changes when the molecule undergoes an electronic transition. If the symmetry of the molecule is low enough, a change in its geometry can produce a reorientation of its inertial axes, and of its molecule-fixed coordinate system with respect to the space-fixed frame. The convenient set of axes to use changes "abruptly" when the photon is absorbed. Two sets of moving coordinates (and two sets of Euler angles) will then be required to describe the rotational motion of the molecule, one for each electronic state. If this occurs, a modification of the spectroscopic selection rules also may be required.

Figure 4.19 shows a beautiful example of inertial axis reorientation in the fully resolved $S_1 \leftarrow S_0$ spectrum of 2-pyridone (2PY), the keto tautomer of 2-hydroxypyridine (34). The frequencies of the lines can be fit with reasonable rigid-rotor Hamiltonians for the ground and excited states, but the calculated intensities of the lines do not agree with experiment, for *any* unique orientation of the optical transition moment. The $S_1 \leftarrow S_0$ spectrum of the 0_0^0 band of 2PY is an *ab* hybrid band, with $\theta = \pm 51°$. In Held and Champagne's (HCs) (34) analysis of this spectrum, they found that no single value of $|\theta|$ gave the correct relative intensities of the a- and b-type lines that appear. Simulations of the spectrum with one value of θ gave good agreement with experiment in the P branch but not in the R branch, whereas another value of θ gave good agreement in the R branch but not in the P

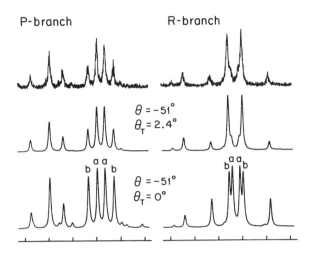

Figure 4.19. Interference effects in the $S_1 \leftarrow S_0$ spectrum of 2-pyridone produced by "axis switching." The spacing between the frequency markers is 100 MHz. [Reprinted with permission from Ref. 34.]

branch. However, a careful examination of all assigned rovibronic lines in the spectrum showed that the total line strength and intensity from a given initial state is conserved. From this observation, HC concluded that there were interference effects in the spectrum caused by inertial axis reorientation on $S_1 \leftarrow S_0$ excitation. A value of $\theta_T = 2.4°$, the angle by which the inertial axes are rotated in the *ab* plane, was deduced from a rigorous fit of the data (Fig. 4.19).

A first-principles treatment of this phenomenon, called "axis switching" by some, was first given by Hougen and Watson (54). The specific example considered by them was the linear → bent, $\tilde{A}\ ^1A_u \leftarrow \tilde{X}\ ^1\Sigma_g^+$ transition of acetylene, in which anomalous rovibronic line intensities also were observed. Axis switching also has been observed in other systems (8,55). Hougen and Watson's approach was first to diagonalize the asymmetric rotor Hamiltonian of each state in its respective principal axis system, then to rotate the coordinate system of the excited state into the coordinate system of the ground state, and finally to relate the two sets of eigenfunctions by a similarity transformation, using a "switching matrix." Held and Champagne's (34) approach was the converse. Instead of rotating the excited-state eigenfunctions into the coordinate system of the ground state, they first transformed the Hamiltonian of the excited state, yielding

$$\hat{H}'_{eff} = A'J_a^2 + B'J_b^2 + C'J_c^2 + D'(J_aJ_b + J_bJ_a) \tag{4.5}$$

Here, A and so on are the excited state rotational constants, J_a and so on are the components of the excited state rotational angular momentum operator *expressed in the inertial frame of the ground state*, and D' is a "perturbation," expressing the fact that the photon-induced change in the geometry of the molecule introduces additional off-diagonal terms in the excited state Hamiltonian [*cf.* Eq. (4.4)]. These terms mix the asymmetric rotor functions and produce anomalous intensities in the spectrum. The parameter D' is a measure of the degree of this state mixing. By fitting the spectrum, using the eigenfunctions determined by diagonalizing Eq. (4.5), HC determined D' and deduced the value of θ_T. Two values of θ_T are possible, $\pm\, 2.4°$, but one can be eliminated if the orientation of the transition moment vector is known. The possible light-induced changes in the structure of 2PY that are responsible for this axis reorientation have been discussed (34).

The informed reader will recognize inertial axis reorientation as being the rotational analog of the Duschinsky effect (56), in which the composition of the normal modes of vibration changes on electronic excitation. Both effects can have significant dynamical consequences, as also discussed elsewhere (34).

4.3.4. Hydrogen Bonded Complexes

2-Pyridone actually exhibits two origins in its $S_1 \leftarrow S_0$ spectrum, separated by about 100 cm^{-1}. These two bands are the electronic origins of two conformers of 2PY, differing in the degree of nonplanarity at the nitrogen atom. (Each of these bands is axis switched, owing to the aforementioned in-plane distortions.) But there is another strong band in the $S_1 \leftarrow S_0$ spectrum of 2PY, displaced 945 cm^{-1} to the blue of the first monomer origin. As Held (57) first showed, this band belongs to the 2PY dimer $(2PY)_2$ (Structure 4). The 2PY dimer is an extremely interesting

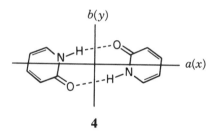

4

species as it exhibits two hydrogen bonds that link up the two monomer units in a ''*cis*-peptide'' configuration. Configurations like this play an important role in the intermolecular recognition processes that are crucial to many biological functions.

Proof that this higher frequency band is that of $(2PY)_2$ was provided by examining it at higher resolution, sufficiently high to expose the underlying rotational structure. Experimental values of the rotational constants were used to optimize two geometrical properties of the dimer, the separation of the COM's of the two monomer units, and the angle between this vector and the C=O bonds. Interpreting these in terms of the hydrogen bonding parameters R (N–H– – –O) and θ [C=O– – –(H)–N], Held (57) found that $R = 2.75 \pm 0.03$ Å and $\theta = 122 \pm 2°$. The corresponding solid state values, determined by X-ray crystallography (58), are $R = 2.77 \pm 0.03$ Å and $\theta = 136 \pm 2°$. This result shows that, at least in the case of $(2PY)_2$, intermolecular interactions in the condensed phase have little or no influence on the primary properties of the hydrogen bond. Electronic excitation of

2PY does, however, affect these properties. Analysis of the fully resolved spectrum of $(2PY)_2$ shows that R increases by 0.08 Å in the S_2 state, relative to the S_0 state. This finding shows that the intermolecular hydrogen bonds in $(2PY)_2$ decrease in strength on $S_2 \leftarrow S_0$ excitation, a result that is consistent with the large blue shift of the dimer band relative to those of the monomer.

Utilizing the changes in moments of inertia that occur when one atom is replaced by another of different mass, Held (35) also discovered that there are distortions of the hydrogen bond geometries in $(2PY)_2$ when the hydrogen atoms in the bonds are replaced by deuterium atoms, in d_1-$(2PY)_2$ and d_2-$(2PY)_2$. Values of $\Delta R = 0.008$ Å, $\Delta\theta = 0°$, and $\Delta\phi$ (the dihedral angle) $= 0.96°$ in S_0 d_2-$(2PY)_2$ and $\Delta R = 0.003$ Å, $\Delta\theta = 0°$, and $\Delta\phi = 0.86°$ in S_2 d_2-$(2PY)_2$ were deduced from the observed rotational constants. Though small, these changes in the geometrical parameters of the two hydrogen bonds lead to a failure of Kraitchman's equations, owing to the effects of vibrational averaging. In $(2PY)_2$, the hydrogen bonded hydrogen atoms participate in low frequency intermolecular vibrational modes, and the amplitudes of these change when hydrogen is replaced by deuterium, leading (in the case of anharmonic potentials) to a change in the vibrationally averaged position of the atom. This change is smaller in S_2 $(2PY)_2$ because the hydrogen bonds are weaker (59).

High resolution techniques also have been used to probe the properties of hydrogen bonds in several other systems, including an intramolecular hydrogen bond in 2-hydroxyquinoline (33) and the intermolecular hydrogen bonds in the NH_3 complexes of cis- and trans-2-naphthol (47), the H_2O complexes of 2PY (38), and the NH_3 complexes of 2PY (48). The data on 2PY-H_2O (Structure **5**) and 2PY-NH_3 (Structure **6**) show that the water and ammonia molecules each form two planar,

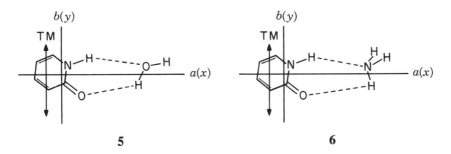

5   **6**

nonlinear hydrogen bonds with the amine hydrogen and the carbonyl oxygen of 2PY, in both states. Again, there are blue shifts on $S_1 \leftarrow S_0$ excitation but these shifts ($+633$ cm^{-1} in 2PY–H_2O and $+176$ cm^{-1} in 2PY–NH_3) are smaller than that in $(2PY)_2$. By comparing the rotational constants of 2PY–H_2O with those of 2PY, it was shown that the structure of the solute changes significantly on solvation by H_2O, in the direction of a zwitterionic structure. Similar results were found for 2PY–NH_3. Additionally, this hydrogen bonded complex exhibits significant torsion–rotation perturbations in its $S_1 \leftarrow S_0$ spectrum. The barriers to internal rotation of the attached NH_3 group were determined to be $V_3(S_0) = 424.3$ and

$V_3(S_1) = 274.4$ cm^{-1}, respectively. Since these values are much larger than those exhibited by the NH$_3$ molecules in cis- and trans-2HN—NH$_3$ (47) (where only steric, van der Waals interactions contribute, see below), it was concluded that in 2PY–NH$_3$ the single ammonia molecule acts as both a proton acceptor and donor, the high barrier being a consequence of a relatively weak (C=)O— — —H—N donor bond. The corresponding (C=)O— — —H—O bond in 2PY–H$_2$O is much stronger. Again, all intermolecular hydrogen bonds weaken on $S_1 \leftarrow S_0$ excitation.

Interestingly, Kraitchman's equations appear to be valid in 2PY–H$_2$O. Independent determinations of the amine hydrogen atom position, the two hydrogen atom positions in water, and the oxygen atom position in water using isotopic substitution methods all gave the same results. However, the test for 2PY–H$_2$O is less sensitive than the test for (2PY)$_2$ with respect to this issue since (a) 2PY–H$_2$O has a smaller moment arm and (b) the substituted positions lie closer to an inertial axis or plane in the reference structure.

4.3.5. van der Waals Complexes

In cases where it occurs, significant vibrational averaging of observed rotational constants can provide a unique view of large amplitude motions and the anisotropy of intermolecular forces. Nowhere is this more apparent than in the study of van der Waals complexes. The van der Waals ''bond'' is extremely weak; thus, the attached atom or molecule experiences large amplitude motion in all canonical directions. A rotationally resolved spectrum is an extremely sensitive probe of these motions, and how they depend on the electronic structure of the ''host'' molecule to which the ''guest'' atom or molecule is attached.

We have studied several such complexes, including the Ne and Ar complexes of 1/2FN (40), trans-stilbene (tS) (39), and all-trans-1,4-diphenyl-1,3-butadiene (DPB) (40). The results for tS are typical. We began by recording the rotationally resolved $S_1 \leftarrow S_0$ spectra of both the bare molecule (see below) and its van der Waals complex. In the case of Ar–tS, two prominent vibronic bands appear about 40 and 63 cm^{-1} to the red of the 0_0^0 band of the parent molecule; similar results were obtained for both bands. (The band at $0_0^0 - 63$ cm^{-1} is presumed to be the 0_0^0 band of the Ar–tS complex.) Then we fit the spectra, obtaining rotational constants for both tS and Ar–tS in both electronic states. Comparisons of these constants *and* the orientations of the $S_1 \leftarrow S_0$ transition moments (all spectra are predominantly a-axis polarized) then show that the complex contains a single rare gas atom and that a preferred binding site exists, above (or below) one of the benzene rings.

One further assumption, that the equilibrium geometry of the parent molecule is unaffected by complex formation, then makes possible a determination of the coordinates of the Ar atom in the tS inertial frame, using Kraitchman's equations (31). As we have seen, the conventional application of these equations utilizes the changes in the moments of inertia (and principal axes of the inertia tensor) resulting from isotopic substitution to determine an atom's coordinates. But a careful study of the classical mechanics underlying these equations reveals that any mass change which results in a movement of the principal frame can be used to determine the position of the affected mass, whether substituted *or* added. The key assumption is that the substitution, *or* addition, does not change the positional relationships of the mass centers that make up the original inertial frame. Though not tested, this assumption seems reasonable in the case of weakly bound van der Waals complexes.

A sketch of the geometry of the Ar–*t*S complex is shown below (Structure **7**).

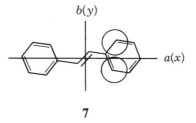

7

Owing to the r^2 dependence of moments of inertia, there is an inherent sign ambiguity in all of the coordinates; x, y, and z; resulting in eight possible Ar atom positions. However, there are only two unique ones, depending on $\pm y$, since $\pm x$ and $\pm z$ are symmetry-related coordinates in *t*S itself. These two positions are shown as open circles in Structure **7**. The Ar atom in Ar–*t*S is localized over one of the benzene rings, above (or below) the ring plane by 3.01 ± 0.02 Å, but is displaced away from the ring center by more than 1.0 Å in both transverse directions, a surprising result.

Model calculations of the Ar–*t*S potential have been performed and are described in detail elsewhere (39). Briefly, it was found that the usual pairwise-additive atom–atom potential leads to Ar geometrical coordinates, properly averaged over displacements in all three dimensions, that are in very poor agreement with experiment [e.g., $(|x|, |y|, |z|)_{calc} = (1.1, 0.7, 3.5$ Å$)$ *versus* $(|x|, |y|, |z|)_{obs} = (2.9, 1.2, 3.0$ Å$)$]. Only when the potential was modified by (a) making the aromatic carbon atoms more attractive and (b) making the potential significantly flatter along y was it possible to bring the calculated coordinates into better agreement with experiment [e.g., $(|x|, |y|, |z|)_{calc} = 2.8, 1.0, 3.3$ Å]. Thus, anisotropies in the potential are quite important.

Recently, we discovered a new effect in the study of the high resolution spectra of several substrates to which a molecule, rather than an atom, is attached by a weak van der Waals bond. A nice example is provided by the CH_4 complexes of 1/2FN (20). Like the corresponding Ar complexes, two bands are observed to the red of the 0^0_0 band of the bare molecule, with comparable shifts. But each of the CH_4–1/2FN complex bands exhibits, at high resolution, three closely spaced rovibronic bands, each of which may be fit to a rigid-rotor Hamiltonian. Our analyses of these bands show that the positions of the attached CH_4 are the same in each band, and are not very different from that of the attached Ar in the corresponding Ar–1/2FN complexes. But the three bands have different intensities at all backing pressures, approximately 5:9:10. Since this intensity ratio is that expected for the

$J = 0$, 1, and 2 states of a freely rotating CH_4, the results suggest that the binding energy of these three states is slightly different in the S_0 and S_1 states of the "surface" to which the CH_4 is attached. Further probes of this behavior should provide intriguing information about the potential of interaction that governs the motion of CH_4 (and other molecules) in the vicinity of such surfaces.

4.3.6. Acid–Base Chemistry

Motions of another type have been explored in other high resolution experiments. We have in mind experiments on "precursor geometry-limited" complexes (60), complexes in which two atoms or groups are linked by a weak van der Waals or hydrogen bond. If one excites, electronically, one of the two atoms or groups, one can initiate a bimolecular reaction with the second atom or group, constrained to interact along the weak bond linking the two moieties together. A nice example of this kind of experiment is provided by the work of Jouvet *et al.* (61) on the van der Waals complex Hg–H_2. In this case, excitation of the resonant Hg $6^3P_1 \leftarrow 6^1S_0$ transition leads to the formation of HgH. Thus, the absorption (or excitation) spectrum of the complex can be an exquisite probe of motion along the reaction coordinate.

Our first application of this technique was to the *cis* (Structure **8**) and *trans* (Structure **9**) hydrogen bonded complexes of ammonia and 2-naphthol (2HNA)

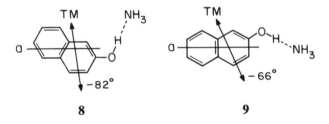

8   **9**

(47). Unlike the complexes of 2PY, these complexes exhibit red-shifted electronic origins, $0_0^0 - 586$ cm^{-1} (*c*-2HNA) and $0_0^0 - 626$ cm^{-1} (*t*-2HNA). This shows that the two (linear) hydrogen bonds strengthen on $S_1 \leftarrow S_0$ excitation. Indeed, it is known that the condensed phase pK_a of 2HN decreases significantly on S_1 excitation, from 9.8 to 2.8. It also is known that excited state proton transfer occurs in higher order clusters, both in the condensed phase (62) and in the gas phase (63). Thus, with only one NH_3 attached to 2HN, we do not expect the reaction to occur when 2HNA is irradiated with light. Indeed, no naphthoate ion emission has been observed. Nonetheless, we can study the high resolution spectrum of 2HNA in great detail, and learn a great deal about the *initial* motion of the acid and base along the reaction coordinate.

Figure 4.20 shows a 1-GHz segment of the 0_0^0 band in the $S_1 \leftarrow S_0$ spectrum of *t*-2HNA and its component parts. Surprisingly, there are two closely spaced origins in this band, separated by 1.51 cm^{-1}. (The corresponding separation in *c*-2HNA is 0.80 cm^{-1}.) These two origins are the *A*- and *E*-subband origins, split by the tunneling motion of the attached NH_3 group. Thus, fits of the spectra (see Fig.

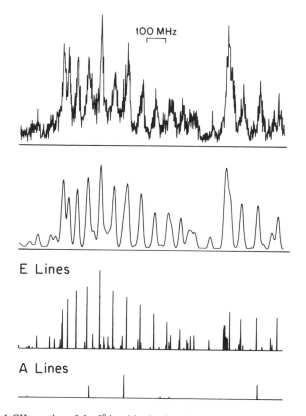

Figure 4.20. A 1-GHz portion of the 0_0^0 band in the $S_1 \leftarrow S_0$ spectrum of t-2HNA near the E-subband origin. The central portion is a progression in J with $\Delta J = \Delta K_a = \Delta K_c = 0$. The second panel shows the sum of the A- and E-line calculated spectra in the lower two panels, each of which has been convoluted with a 20 MHz FWHM Voigt lineshape function (18 MHz Gaussian component, 5 MHz Lorentzian component). [Reprinted with permission from Ref. 47.]

4.20) yield the rotational constants of both electronic states *and* the torsion–rotation perturbation constants D_a and D_b. From these, Plusquellic *et al.* (47) determined the vibrationally averaged values of $R(\text{O–H---N})$, $\theta[\text{C–O (H)---N}]$, the threefold barrier heights, and the orientation of the rotor axis in the inertial frame, in *both* complexes in *both* electronic states.

Two of these results are particularly interesting. One is that there is a significant *decrease* in the heavy atom separation on $S_1 \leftarrow S_0$ excitation, from $R = 2.77$ to $R = 2.62 \pm 0.02$ Å in c-2HNA and from $R = 2.79$ to $R = 2.57 \pm 0.02$ Å in t-2HNA. The second is that, concomitantly, there is a significant *increase* in the threefold barriers hindering the motion of the attached NH_3 molecules (from $V_3 = 41.1$ to 53.8 cm^{-1} in c-2HNA and from $V_3 = 34.2$ to 58.2 cm^{-1} in t-2HNA). Calculations show that these barriers are primarily steric in origin; hence, their increase with decreasing R is not surprising.

Perhaps more surprising is the information that the light-induced change in R provides about the mechanism of the proton transfer reaction. Shown in Fig. 4.21 are one-dimensional slices of model potential energy surfaces for t-2HNA in its S_0 and S_1 states; the corresponding surfaces for c-2HNA are qualitatively similar. These surfaces were modeled using the Lippincott–Schroeder potential (64), as described elsewhere (47). Each surface exhibits two minima when plotted as a function of r(O–H). The inner minimum corresponds to the covalent structure O–H– – –N. The outer minimum corresponds to the ionic, proton transferred structure O$^-$– – –H–N$^+$. Both structures contribute to the properties of the hydrogen bonds in c- and t-2HNA [and (2PY)$_2$, etc., see above] in both electronic states. In the ground state, the covalent structure dominates, but the ionic structure gains in importance on excitation of both c- and t-2HNA to their S_1 states. Thus, photoexcitation increases the strength of the hydrogen bond, decreasing R. But *little hydrogen atom motion occurs*. Instead, the NH$_3$ group moves towards the hydrogen atom, and the magnitude of the orbital overlap between the empty o^* orbital of the OH group and the lone pair electrons of the nitrogen atom increases, on the way towards breaking the OH bond. The energy difference between the canonical co-

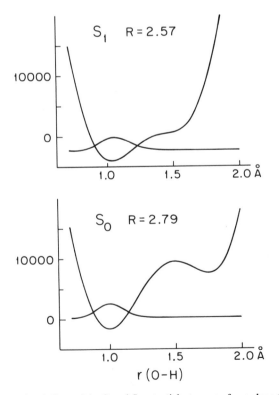

Figure 4.21. One-dimensional slices of the S_0 and S_1 potential energy surfaces along the proton transfer coordinate in t-2HNA. The vertical axis is in cm^{-1}. [Reprinted with permission from Ref. 47.]

valent and ionic structures decreases, the distance from the reactant to product along the reaction coordinate decreases, and the barrier to proton transfer decreases. As a result, the zero-point vibrational wave function of the S_1 state extends well into the proton transfer region, exploring the transition state for the excited state proton-transfer reaction.

We currently are extending studies of this type to other vibrational coordinates in the 1/2HNA system, to other acid–base pairs, to higher order clusters, and to other precursor geometry-limited complexes, to mimic the behavior of other fundamentally important, bimolecular reaction types. The results promise to shed considerable light on the potential energy surfaces that govern the dynamical behavior of isolated molecules, making at the same time a strong connection to chemistry in the condensed phase.

Remarkably, Plusquellic (65) also discovered another interesting quantum interference effect in his analysis of the fully resolved $S_1 \leftarrow S_0$ spectrum of t-2HNA. This interference effect is a consequence of quantum state mixing in *both* electronic states. The mixing is produced by the coupling between overall molecular rotation and the torsional motion of the NH_3 group. Consequently, the symmetries of the rovibronic Hamiltonians are reduced [*cf.* Eq. (4.5)], and the two distinguishable channels for $S_1 \leftarrow S_0$ excitation (the 0_0^0 band in t-2HNA is an *ab* hybrid band) are converted into indistinguishable ones (as in axis switching, see above). Plusquellic (65) showed that one could use this effect to determine the *absolute* orientation of the $S_1 \leftarrow S_0$ transition moment vector in the inertial frame of the complex. Comparison of this result with that for the $S_1 \leftarrow S_0$ transition of the bare molecule then shows that the orientation of its transition moment in the molecular frame is unaffected by complex formation.

4.3.7. Photoisomerization Reactions

Unimolecular reactions induced by light, such as photoisomerizations, also can be studied by rotationally resolved electronic spectroscopy. This is because the absorption of light changes the shape of the molecule. As we have seen, the derived values of the rotational constants are extremely sensitive to this shape. Now that we are learning how to interpret such changes in terms of vibrational motions along different coordinates, there is the prospect that careful studies of the spectrum of an isomerizing molecule will provide detailed information about which vibrational coordinates are involved, as a function of its energy content.

To date, our attention has been focused on two molecules, *trans*-stilbene (tS) and all-*trans*-1,4-diphenyl-1,3-butadiene (DPB). Both have played pivotal roles in the development of our current understanding of photoisomerization reactions (66). The accepted model for the isomerization process is that, following excitation and rapid equilibration to a planar minimum, the molecule passes over a barrier on the S_1 surface into a twisted, less-conjugated configuration, which then decays rapidly into *cis* and *trans* ground state structures. The barrier in tS is believed to lie at an excess energy above the S_1 origin of $\Delta E \sim 1200$ cm^{-1}, whereas that in DPB is at $\Delta E \sim 500$ cm^{-1}. Both the barrier crossing and the subsequent relaxation of the twisted form are strongly solvent dependent.

Despite its larger size, DPB (Structure **10**) proved more tractable than tS in

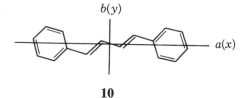

10

our first study of its high resolution $S_1 \leftarrow S_0$ spectrum in the gas phase (67). Principally, this is because of its longer radiative lifetime, $\tau_r = 63$ ns, giving rise to a homogeneous width of 2.4 MHz for individual rovibronic lines of the first one-photon-allowed band, 37.2 cm^{-1} to the blue of the two-photon origin of the S_1 state at 29,615.3 cm^{-1} (68,69). Pfanstiel et al. (67) found that this band (Band 1) exhibits a fully resolved spectrum, spanning about 1 cm^{-1} and containing more than 2000 lines (FWHM ~ 4 MHz), as shown in Fig. 4.22. Despite the apparent lack of a central Q branch, Band 1 is a parallel-type transition polarized entirely (> 95%) along the a axis of the near prolate top ($\kappa'' = -0.9936$). All lines were fit to a rigid-rotor Hamiltonian for both vibronic levels, yielding rotational constants for both electronic states (the smallest of these, C'', is 137.5 MHz). The results showed that no large conformational distortion occurs on excitation of Band 1 of DPB. The equilibrium geometries of both states are planar, with relatively large out-of-plane vibrational amplitudes ($\Delta I'' = -10.3$, $\Delta I' = -11.3$ amu Å^2). The principal structural changes that occur are (a) a slight expansion of the molecule in directions perpendicular to a ($\Delta A = -97.0$ MHz) and (b) a still smaller shortening of the molecule along a. These changes are consistent with a description of the S_1 state that is partially bond-order reversed, and has equal contributions from single and double excitations with just greater than one-half of the optical electron density being localized on the phenyl groups, with nodes along the bonds in the phenyl rings that are most perpendicular to a (70).

The $S_1 \leftarrow S_0$ spectrum of the electronic origin of tS (Fig. 4.23) (71) is very similar to that of the (one-photon) origin of DPB. Both spectra exhibit well-developed K_a subbands, each with its own P, Q, and R branch. Additionally, both spectra are strongly influenced by relatively large ΔA values, which shift entire K subbands to the red by $\Delta \nu \sim \Delta A K_a^2$. But these shifts are less apparent in tS because of its much larger homogeneous width. (The measured fluorescence lifetime of the zero-point vibrational level of S_1 tS is ~ 2.6 ns (72), corresponding to a FWHM linewidth of ~ 70 MHz.) Nonetheless, the shifts do appear; note in Fig. 4.23 how the Q-branch transitions in each successive K_a subband shift to the red in a K_a^2-dependent way. Less obvious in the tS spectrum is another consequence of the large ΔA, the mixing of J families. Not all K_a belonging to a particular J appear together. Instrumental in the analysis of these effects was the software described earlier, making it possible to work with portions of the experimental spectrum, the simulated spectrum with an adjustable linewidth, and a deconvoluted "stick" spectrum simultaneously on a graphics monitor. With this software, Champagne et al.

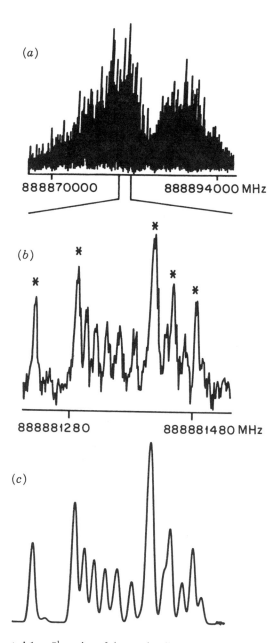

Figure 4.22. The central 1 cm^{-1} portion of the rotationally resolved fluorescence excitation spectrum of Band 1 in the "S_1" $\leftarrow S_0$ one-photon transition of all-*trans*-1,4-diphenyl-1,3-butadiene at about 337 nm. Shown in B is the most congested part of the Q branch, simulated in C using the derived inertial parameters. The dominant pattern in B and C is a Q-branch progression in J with $K_a = 6$, beginning with $J = 6$. Overlapping features are indicated with asterisks and include P-branch lines. In addition, each line in this portion of the spectrum is an unresolved K_a doublet. [Reprinted with permission from Ref. 67. Copyright 1989 by the AAAS.]

996338000   996368000 MHz

996350500   996355500 MHz

996352500   996353000 MHz

(71) were able to separate out specific K_a subbands, manipulate their relative frequencies, and coadd them to produce simulated spectra of the desired width and shape (*cf.* Fig. 4.23).

From these spectral fits, Champagne *et al.* (71) concluded that, like DPB, *t*S is a rigid, near-prolate ($\kappa'' = -0.9812$), asymmetric top in both electronic states. Again, ΔA is relatively large and negative ($\Delta A = -71.1$ MHz), suggesting that there is also an expansion of this molecule in directions perpendicular to *a*. Additionally, the inertial defects are relatively large, $\Delta I'' = -15.3$ and $\Delta I' = -12.4$ amu Å^2. From these results, it was concluded that *t*S is planar in both vibrational levels but exhibits relatively large amplitude, out-of-plane zero-point motions. These results are consistent with force-field (QCFF/PI) calculations (73).

Intriguing things happen at higher excess energies in the S_1 manifolds of both DPB and *t*S. As the vibrational energy of S_1 DPB is increased, perturbations appear in the fully resolved spectra (45). Figure 4.24, a portion of Band 4 at 0_0^0 (S_1) + 263 cm^{-1}, shows a representative example. Like Band 1, Band 4 is completely resolved and can be fit to rigid-rotor Hamiltonians for both states. The S_1 rotational constants derived from these fits are slightly different (*e.g.*, $\Delta A = -91.7$ MHz) from those of Band 1. But, perhaps more importantly, each observed K_a subband is displaced further to the red than predicted by the rigid-rotor "fit," as shown in Fig. 4.24. These perturbations become increasingly important with increasing K_a and with increasing ΔE. Nonetheless, the spectrum continues to exhibit sharp lines, though a broad background begins to contribute to the intensity at $\Delta E \sim 500$ cm^{-1}.

Behavior of a somewhat different type occurs in *t*S (40). Here, S_1 bands up to $\Delta E = 400$ cm^{-1} can be fit with rigid-rotor Hamiltonians. Different S_1 levels exhibit significantly different rotational constants. Above $\Delta E = 400$ cm^{-1}, the principal vibronic bands begin to show sidebands, which are also rigid-rotor-like. These sidebands become more pronounced above $\Delta E = 600$ cm^{-1} and rigid-rotor models no longer statisfactorily account for the line positions, in a K_a-dependent way (as in DPB). Above 1000 cm^{-1}, the individual line structure begins to disappear.

We believe that the behavior in the S_1 states of both DPB and *t*S can be explained by an interacting "dark-state" model. That is, increasingly as a function of excess energy, dark states begin to interact with the "bright states" carrying the oscillator

---

Figure 4.23. The rotationally resolved fluorescence excitation spectrum of the 0_0^0 band in the $S_1 \leftarrow S_0$ transition of *trans*-stilbene, at about 310 nm. The top panel shows the overall experimental band contour (top trace), which spans about 2 cm^{-1}, and the corresponding computer simulation (bottom trace). The middle panel shows the center portion of the experimental spectrum at higher resolution (top trace) and the corresponding computer simulation (bottom trace). This portion of the spectrum contains contributions from (left) P-, (center) Q-, and (right) R-branch transitions. The bottom panel shows an expanded portion of the Q branch, at full experimental resolution (top trace), and two corresponding computer simulations, one with an assumed line width of 70 MHz (middle trace) and one with an assumed line width of 1 MHz (bottom trace). The two strongest lines in the experimental trace are the K_a subband Q-branch transitions with $K_a' = 2$ (left) and $K_a' = 1$ (right). The temperature used in the computer simulations is 6 K. The weak progression of band structure that occurs on the red side of the P branch (top panel) is more prominent at higher temperatures and is caused by the formation of K_a subband heads at or near $J' = 50$. [Reprinted with permission from Ref. 71.]

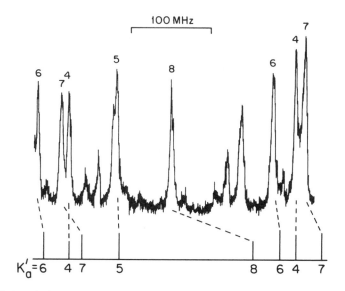

Figure 4.24. Perturbations in the fully resolved spectrum of Band 4 in all-*trans*-1,4-diphenyl-1,3-butadiene.

strength. The origin of these dark states remains unclear. They could either be other, Franck–Condon forbidden, S_1 levels of the *trans* structure, or S_1 levels of a *cis*-like structure, on the other side of the isomerization barrier. We currently are attempting to deperturb some of these spectra, and to fit the deconvoluted spectra with Watson's distortable rotor Hamiltonian (74) in order to answer some of these questions. We also are probing the photoisomerization behavior of other molecules, like all-*trans*-octatetraene.

4.3.8. Intramolecular Relaxation Dynamics

In some sense, we have come full circle. Mention of dark states brings to mind the case of pyrazine, with which our ''adventure'' in high resolution spectroscopy began. It will be recalled (75) that in pyrazine, the $S_1 \leftarrow S_0$ spectrum contains many more lines than expected for an ordinary allowed electronic transition. Figure 4.25 (76) shows a typical example. This spectrum, a fluorescence excitation spectrum of the P1 transition, exhibits at least 36 lines, rather than the expected *one* line. (The P1 transition terminates in $J' = 0$.) This additional structure originates from a coupling between a *single* rovibronic level of the S_1 state and many *quasi*-isoenergetic rovibronic levels of the lowest triplet (T_1) state. Similar behavior has been observed for several $J' \neq 0$ levels of the zero-point level of S_1 pyrazine (77). As is well known, the existence of such states, the so-called molecular eigenstates, provides the frequency-domain equivalent explanation of intersystem crossing (ISC) in the isolated molecule. Other time-dependent phenomena, such as intra-molecular vibrational relaxation, may be explained in the same way (78).

What makes the results on DPB and *t*S so intriguing is the possibility that such

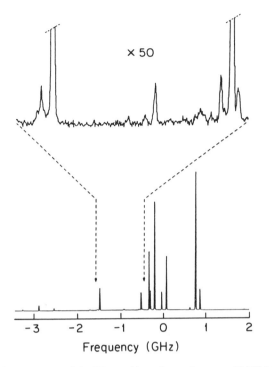

× 50

Frequency (GHz)

Figure 4.25. Excitation spectrum of the P1 transition of pyrazine, at −12,192 MHz relative to the origin of the $S_1 \leftarrow S_0$ transition. [Reprinted with permission from Ref. 76.]

mixed states are appearing in their high resolution spectra. In this case, the mixed states are those of a reactant and product molecule, coupled to each other along a *reaction coordinate*. The corresponding states in ISC are produced in much the same way. Thus, the mixed states of pyrazine are produced by the coupling of zero-order singlet and triplet states along *many* vibrational coordinates. But before we can make this connection between intramolecular reaction and relaxation dynamics, we need to learn how to interpret such state mixing in simpler systems, where perhaps only one or a few vibrational coordinates are involved. Our own high resolution work in the near future will be directed towards achieving this objective.

4.4. SUMMARY

On the molecular level, a chemical reaction occurs when atoms or groups of atoms rearrange themselves into a new configuration in response to changes in the local electronic environment. It is well known that light may be used to induce changes in the local electronic environment, by exciting a molecule to a different electronic state. In this chapter, we have described the design and construction of a new CW laser/molecular beam apparatus that can be used to perform such excitations, at chemically relevant energies in the UV, with unparalleled spectral resolution. We also have described in some detail the computer software that has been developed to record and analyze the resulting spectra. Finally, we have shown how the information obtained from these spectra relates to a variety of structural and dynamical properties of large, isolated molecules in the gas phase.

The specific examples discussed include naphthalene chromophores to which different substituents are attached, tautomeric systems exhibiting hydrogen bonds, van der Waals complexes, precursor geometry-limited acid–base reactions, and systems susceptible to photoisomerization and other intramolecular relaxation processes. At low excess energies, most spectra exhibit rigid-rotor behavior, from which information about the preferred structures and conformations of the molecules in both electronic states is obtained. The intermolecular forces responsible for weak van der Waals bonds and stronger hydrogen bonds have been probed using this technique, as well as motions of the molecule and/or complex along different vibrational coordinates. Potential energy surfaces along reaction coordinates are being explored in other systems. At higher excess energies, perhaps near a barrier, the observed spectra begin to exhibit deviations from rigid-rotor behavior, possibly owing to a coupling of the prepared state with other dark states in the vicinity. Current efforts are increasingly devoted to determine which interactions are responsible for these couplings, and how they depend on molecular structure and conformation.

ACKNOWLEDGMENTS

We thank our co-workers Blaise Champagne, Dennis Clouthier, Andy Held, Willy van Herpen, Sue Humphrey, Surya Jagannathan, Jeff Johnson, Richard Judge, Leo Meerts, Xue-Qing Tan, and Jeff Tomer for their many contributions to the work that is described herein. We also thank the National Science Foundation for its continuing financial support. This chapter is dedicated to Jan Kommandeur, fellow ''molecular eigenstater,'' on the occasion of his retirement from the University of Groningen.

REFERENCES

1. Duarte, F. J. and Hillman, L.W. (Eds.). *Dye Laser Principles*, Academic, San Diego, **1990.**

2. Jacquinot, P. in *High Resolution Laser Spectroscopy*, Shimoda, K. (Ed.), Topics in Applied Physics, Vol. 13, Springer-Verlag, Berlin, **1976**, p. 51.

3. Demtröder, W. *Laser Spectroscopy*, Springer Series in Chemical Physics 5, Springer-Verlag, Berlin, **1982**.

4. van der Meer, B. J., Jonkman, H. Th., Kommandeur, J., Meerts, W. L., and Majewski, W. A. *Chem. Phys. Lett.* **1992**, *92*, 565.

5. Meerts, W. L., and Majewski, W. A. in *Laser Spectroscopy 6*, Weber, H. P. and Lüthy, W. (Eds.), Springer-Verlag, Berlin, **1983**, p. 147.

6. Majewski, W. A., and Meerts, W. L. *J. Mol. Spectrosc.* **1984**, *104*, 271.

7. Meerts, W. L., Majewski, W. A., and van Herpen, W. M. *Can. J. Phys.* **1984**, *62*, 1293.

8. Meerts, W. L. and Majewski, W. A. *Laser Chem.* **1986**, *5*, 339.

9. van Herpen, W. M., Meerts, W. L. and Dymanus, A. *Laser Chem.* **1986**, *6*, 37.

10. van Herpen, W. M., Meerts, W. L., and Dymanus, A. *J. Chem. Phys.* **1987**, *87*, 182.

11. Majewski, W. A. and Plusquellic, D. F. U.S. Patent No. 4,994,661, Feb. 19, 1991.

12. Majewski, W. A. *Optics Commun.* **1983**, *45*, 201.

13. Johnston, Jr., T. F. and Profitt, W. *IEEE J. Quant. Electron.* **1980**, *QE16*, 483.

14. Gerstenkorn, S. and Luc, P. "Atlas du spectroscopie d'absorption de la molecule d'iode," CNRS, Paris, **1978** and **1982**.

15. Riedle, E., Ashworth, S. H., Farrell, T. T., and Nesbitt, D. J. *Rev. Sci. Inst.* **1994**, *65*, 42.

16. Birss, F. W. and Ramsay, D. A. *Comp. Phys. Comm.* **1984**, *38*, 83. See also Feynman, R. P. *Phys. Rev.* **1939**, *56*, 340.

17. Plusquellic, D. F. Ph.D. Thesis, University of Pittsburgh **1992**.

18. Plusquellic, D. F. et al. (in preparation).

19. Majewski, W. A., Plusquellic, D. F., and Pratt, D. W. *J. Chem. Phys.* **1989**, *90*, 1362.

20. Champagne, B. B. Pfanstiel, J. F., Pratt, D. W., and Ulsh, R. C. *J. Chem. Phys.*, in press **1995**.

21. Johnson, J. R., Jordan, K. D., Plusquellic, D. F., and Pratt, D. W. *J. Chem. Phys.* **1990**, *93*, 2258.

22. Plusquellic, D. F. and Pratt, D. W. (in preparation).

23. Tan, X.-Q., Majewski, W. A., Plusquellic, D. F., and Pratt, D. W. *J. Chem. Phys.* **1991**, *94*, 7721.

24. Uijt de Haag, P., Spooren, R., Ebben, M., Meerts, W. L., and Hougen, J. T. *Mol. Phys.* **1990**, *69*, 265.

25. Tan, X.-Q., Clouthier, D. J., Judge, R. H., Plusquellic, D. F., Tomer, J. L., and Pratt, D. W. *J. Chem. Phys.* **1991**, *95*, 7862.

26. Berden, G., Meerts, W. L., and Kriener, W. *Chem. Phys.* **1993**, *174*, 247.

27. Jagannathan, S. and Pratt, D. W. *J. Chem. Phys.* **1994**, *100*, 1874.

28. See, for example, Ebben, M., Spooren, R., ter Meulen, J. J., and Meerts, W. L. *J. Phys. D., Appl. Phys.* **1989**, *22*, 1549.

29. Kim, K.-S. and Jordan, K. D. *Chem. Phys. Lett.* **1994**, *218*, 261 and references cited therein.

30. Hollas, J. M. and bin Hussein, M. *Z. J. Mol. Spectrosc.* **1988**, *127*, 497 and references cited therein.

31. Kraitchman, J. *Am. J. Phys.* **1953**, *21*, 17.

32. Gordy, W. and Cook, R. L. *Microwave Molecular Spectra*, 3rd ed., Wiley-Interscience, New York, **1984**.

33. Held, A., Plusquellic, D. F., Tomer, J. L., and Pratt, D. W. *J. Phys. Chem.* **1991**, *95*, 2877.

34. Held, A., Champagne, B. B., and Pratt, D. W. *J. Chem. Phys.* **1991**, *95*, 8732.

35. Held, A. and Pratt, D. W. *J. Chem. Phys.* **1992**, *96*, 4869.

36. Tan, X.-Q. and Pratt, D. W. *Chem. Phys. Lett.* **1993**, *207*, 510.

37. Humphrey, S. J. and Pratt, D. W. *J. Chem. Phys.* **1993**, 99, 5078.

38. Held, A. and Pratt, D. W. *J. Am. Chem. Soc.* **1993**, *115*, 9708.

39. Champagne, B. B., Plusquellic, D. F., Pfanstiel, J. F., van Herpen, W. M., and Meerts, W. L. *Chem. Phys.* **1991**, *156*, 251.

40. Champagne, B. B. et al. (in preparation).

41. Walker, M. J. et al. (in preparation).

42. Hepworth, P. A. et al. (in preparation).

43. Tan, X.-Q., Majewski, W. A., Plusquellic, D. F., Pratt, D. W., and Meerts, W. L. *J. Chem. Phys.* **1989**, *90*, 2521.

44. Hayashi, M. and Pierce, L. *J. Chem. Phys.* **1961**, *35*, 1148.

45. Pfanstiel, J. F. et al. (in preparation).

46. Tan, X.-Q. and Pratt, D. W. *J. Chem. Phys.* **1994**, *100*, 7061.

47. Plusquellic, D. F., Tan, X.-Q., and Pratt, D. W. *J. Chem. Phys.* **1992**, *96*, 8026.

48. Held, A. and Pratt, D. W. *J. Am. Chem. Soc.* **1993**, *115*, 9718.

49. Spangler, L. H. and Pratt, D. W. in *Jet Spectroscopy and Molecular Dynamics*, Hollas, J. M. and Phillips, D. (Eds.), Chapman & Hall, London, **1995**.

50. Tan, X.-Q. et al. (in preparation).

51. Fujita, I. et al. (in preparation).

52. Zare, R. N. *Angular Momentum*, Wiley-Interscience, New York, **1988**.

53. The idea that molecules have well-defined shapes is surely one of the most pervasive concepts in science. For an interesting perspective on this idea, see Trindle, C. *Isr. J. Chem.* **1980**, *19*, 47.

54. Hougen, J. T. and Watson, J. K. G. *Can. J. Phys.* **1965**, *43*, 298. See also Huet, T. R., Godefroid, M., and Herman, M. *J. Mol. Spectrosc.* **1990**, *144*, 32.

55. Konings, J. A., Majewski, W. A., Matsumoto, Y., Pratt, D. W., and Meerts, W. L. *J. Chem. Phys.* **1988**, *89*, 1813.

56. Duschinsky, F. *Acta Physicochim. U.R.S.S.* **1937**, *7*, 551.

57. Held, A. and Pratt, D. W. *J. Am. Chem. Soc.* **1990**, *112*, 8629.

58. Penfold, B. R. *Acta Crystallogr.* **1953**, *6*, 591.

59. The expansion of strong intermolecular hydrogen bonds that occurs when deuterium is substituted for hydrogen has been studied extensively in X-ray crystallography. See, for example, Ubbelohde, A. R. and Gallagher, K. J. *Acta Crystallogr.* **1955**, *8*, 71.

60. For an early review, see Bernstein, R. B., Herschbach, D. R., and Levine, R. D. *J. Phys. Chem.* **1987**, *91*, 5365.

61. Jouvet, C., Boivineau, M., Duval, M. C., and Soep, B. *J. Phys. Chem.* **1987**, *91*, 5416.

62. Brucker, G. A. and Kelley, D. F. *J. Chem. Phys.* **1989**, *90*, 5243.

63. Droz, T., Knochenmuss, R., and Leutwyler, S. *J. Chem. Phys.* **1990**, *93*, 4520.

64. See Saitoh, T., Mori, K., and Itoh, R. *Chem. Phys.* **1981**, *60*, 161 and references cited therein.

65. Plusquellic, D. F. and Pratt, D. W. *J. Chem. Phys.* **1992**, *97*, 8970.

66. For a review, see Fleming, G. R. *Chemical Applications of Ultrafast Spectroscopy*, Oxford University Press, New York, **1986**. See also Waldeck, D. H. *Chem. Rev.* **1991**, *91*, 415.

67. Pfanstiel, J. F., Champagne, B. B., Majewski, W. A., Plusquellic, D. F., and Pratt, D. W. *Science* **1989**, *245*, 736.

68. Shepanski, J. F., Keelan, B. W., and Zewail, A. H. *Chem. Phys. Lett.* **1983**, *103*, 9.

69. Horwitz, J. S., Kohler, B. E., and Spiglanin, T. A. *J. Chem. Phys.* **1985**, *83*, 2186.

70. Bennett, J. A. and Birge, R. E. *J. Chem. Phys.* **1980**, *73*, 4234.

71. Champagne, B. B., Pfanstiel, J. F., Plusquellic, D. F., Pratt, D. W., van Herpen, W. M., and Meerts, W. L. *J. Phys. Chem.* **1990**, *94*, 6.

72. Syage, J. A., Felker, P. M., and Zewail, A. H. *J. Chem. Phys.* **1984**, *81*, 4706.

73. Negri, F., Orlandi, G., and Zerbetto, F. *J. Phys. Chem.* **1989**, *93*, 5124.

74. Watson, J. K. G. in *Vibrational Spectra and Structure*, Durig, J. R. (Ed.), Elsevier, Amsterdam, **1977**, Vol. 6, p. 1 and references cited therein.

75. For a review, see Kommandeur, J., Majewski, W. A., Meerts, W. L., and Pratt, D. W. *Annu. Rev. Phys. Chem.* **1987**, *38*, 433.

76. van Herpen, W. M., Meerts, W. L., Drabe, K. E., and Kommandeur, J. *J. Chem. Phys.* **1987**, *86*, 4396.

77. Siebrand, W., Meerts, W. L., and Pratt, D. W. *J. Chem. Phys.* **1989**, *90*, 1313.

78. See, for example, Perry, D. S. in *Laser Techniques in Chemistry*, Rizzo, T. and Myers, A. J. (Eds.), Wiley, New York, **1995**.

Chapter **V**

GENERATION OF COHERENT VACUUM ULTRAVIOLET RADIATION: APPLICATIONS TO HIGH-RESOLUTION PHOTOIONIZATION AND PHOTOELECTRON SPECTROSCOPY

John W. Hepburn
Department of Chemistry,
University of Waterloo, Waterloo, Ontario, Canada

5.1. INTRODUCTION

The use of frequency mixing techniques to generate short wavelength coherent light is very well established in optical physics, and has been for more than two decades. Although not a common experimental technique in chemical physics, the application of coherent short wavelength light to studies in chemical dynamics and spectroscopy is now fairly widespread. This chapter provides a practical guide to frequency mixing techniques, and discusses the use of the resulting coherent light in high-resolution photoionization and photoelectron spectroscopy experiments. Before beginning detailed discussions about generating and applying coherent short

Laser Techniques In Chemistry, Edited by Anne B. Myers and Thomas R. Rizzo.
Techniques of Chemistry Series, Vol. XXIII.
ISBN 0-471-59769-4 © 1995 John Wiley & Sons, Inc.

wavelength light, we survey the short wavelength domain, which is terra incognita for most chemists (1).

In this chapter, we shall be concerned with the region of the electromagnetic spectrum called the vacuum ultraviolet (VUV), wavelengths shorter than 2000 Å, but longer than about 200 Å (2). At wavelengths below 2000 Å, the oxygen in air is strongly absorbing, and as the wavelength gets shorter, all molecules start to absorb, making it impossible to propagate VUV through air, hence the name. Another practical dividing line is the limit of transmission of refractive optics, the LiF cutoff at about 1050 Å. Although not a universally accepted nomenclature, the spectral region at wavelengths shorter than this limit is often called the extreme ultraviolet, or XUV. The importance of the distinction between VUV and XUV is purely practical, since both correspond to excitation and ionization of valence electrons in molecules. For simplcity, we shall use VUV to refer to the entire spectral region from 2000 to 200 Å, and use XUV when referring specifically to the range of wavelengths shorter than 1050 Å. The absence of windows and refractive optics in the XUV means that experimental life is more complicated, as one must make extensive use of differential pumping and use reflective optics that are not very efficient at these short wavelengths at anything other than grazing incidence.

Historically, work at these wavelengths required the use of discharge lamps and vacuum monochromators or spectrographs. The introduction of synchrotron radiation light sources led to an expansion of work at short wavelengths, but did not revolutionize experimental capabilities in the VUV because until recently, the intensity of synchrotron radiation has been comparable to discharge lamps in the VUV range (3). To provide a point of comparison for our discussion of coherent VUV light sources, we briefly consider the properties of synchrotron radiation in this spectral range. At a second generation synchrotron light source, such as the Super-ACO positron storage ring in Orsay, one can expect to get on the order of 10^{10} photons s^{-1} in the ionization volume, focused to a spot on the order of 1 × 0.1 mm, at a spectral resolution of 0.2 Å (4). These figures are typical for a bending magnet beamline with a normal incidence monochromator (3-m focal length at Orsay). The photon energy range is about 5–40 eV for such a beamline, and the figures quoted are for a photon energy of about 20 eV. The projected figures for third generation synchrotron radiation sources are better than this, due to their increased brightness.

As an example of a third generation machine, the Advanced Light Source (ALS) at Berkeley can be considered (5). The performance of the bend magnet beamlines at the ALS will be similar to those quoted above, although the resolution and spot size will be better because of the increased brightness. Thus, on the order of 10^{10} photons s^{-1} at 20 eV with a resolution of 0.05 Å (13 cm^{-1}, 1.6 meV) will be available for experiments. The use of undulators for synchrotron radiation generation allows for even higher photon fluxes at improved resolution. As an example, if we consider the specifications for the U8 undulator beamline at the ALS, a flux of 10^{13} photons s^{-1} is expected at 20 eV, with a resolution of 0.05 Å. Higher resolutions will be possible, potentially 10 times better, with a concomitant decrease in photon flux. Thus, at the current state of the art for synchrotron radiation sources, the best performance in the spectral range of interest to this chapter will be on the order of 10^{12} photons s^{-1} at a resolution of 2 cm^{-1}. As shall be discussed in Section 5.2, current coherent VUV sources can achieve photon fluxes of 10^{12} photons s^{-1} with resolutions better than 0.5 cm^{-1} over the 6–19-eV photon energy range.

A coherent light source at any wavelength has many advantages over conventional sources, beyond the obvious benefits of higher intensity and directionality. The VUV generated by frequency mixing techniques has the coherence properties of the lasers used to provide the fundamental light, typically pulsed dye lasers. Since pulsed lasers can be made single mode, meaning a resolving power of about 4×10^6 at visible wavelengths for typical pulsewidths of 5 ns, it is possible to have the same photon energy resolution in the VUV. This is an improvement of about an order of magnitude over the best monochromators available (6), and more than two or three orders of magnitude over state-of-the-art synchrotron sources. The polarization of the coherent VUV is entirely determined by the easily controlled fundamental polarization. Thus, one can have coherent VUV with any polarization characteristics, and the VUV polarization can be changed easily. It is now relatively straightforward to produce very short pulses of tunable visible light (pulse widths < 10^{-14} s), meaning that coherent VUV can also be generated with these very narrow pulse widths for time-resolved experiments. The spectral purity of coherent VUV is also very high, as only discrete, well-defined frequencies are generated in the frequency mixing process. This eliminates problems with higher order radiation, which is a severe problem with monochromatized synchrotron radiation, particularly if the radiation comes from undulator sources. Since the short wavelength coherent light is generated from visible/UV lasers whose wavelength can be easily calibrated, it is straightforward to calibrate the VUV wavelength to an absolute accuracy of better than 1 part in 10^5, which is critical for experiments measuring ionization energies.

There are many possible comparisons between coherent VUV and synchrotron radiation, usually involving per time interval units. These time integrated specifications ignore the pulsed nature of the coherent light, which can be very useful for some of the experiments described in this chapter. In this spectral range, photon absorption often leads to ionization. A pulsed photoionization light source is ideal for time-of-flight (TOF) spectroscopy, either for mass spectrometry or photoelectron spectroscopy. The advantages of TOF spectroscopy are twofold: simplicity of design and construction, and sensitivity, because of the multiplexed detection. In addition, for the pulsed field ionization experiments described below, one needs a substantial delay between excitation and field ionization, and this is easy with a pulsed coherent source. Finally, because one can generate short pulses of VUV by frequency mixing short pulse lasers, time resolved measurements with femtosecond time resolution are possible with coherent VUV.

This chapter is intended neither as a comprehensive review of chemical physics research using short wavelength coherent light, nor as a detailed pedagogical work on the physics of frequency mixing and short wavelength spectroscopy, but rather as a practical guide to frequency mixing techniques, particularly for XUV wavelengths. The power of coherent VUV techniques will be demonstrated by considering its application to high-resolution photoionization and photoelectron spectroscopy. Particular attention will be paid to the application of coherent VUV to ultrahigh resolution photoelectron spectroscopy, using the newly developed technique of pulsed field ionization–zero kinetic energy electron spectroscopy, usually abbreviated as PFI–ZEKE spectroscopy. The marriage of coherent VUV with PFI–ZEKE spectroscopy is powerful, and it nicely illustrates the possibilities made available by coherent VUV.

The application of coherent VUV to photochemistry and photoionization experiments has been reviewed recently (7), as has its use for high-resolution spectroscopy (8). Readers are referred to these reviews for more information about generation and applications of coherent VUV.

5.2. GENERATION OF TUNABLE VUV BY FREQUENCY MIXING

The generation of broadly tunable coherent VUV is done by frequency mixing in gases, most commonly by a process known as four-wave frequency mixing. The theory of four-wave frequency mixing has been thoroughly treated in other works (9,10), and four-wave mixing has been reviewed in many places (7,11,12). Given this extensive background material the purpose of this section is to review the basic idea of frequency mixing and provide practical examples of some of the better ways of generating coherent light at various wavelengths in the VUV. Because technical details are not always obvious from published papers, this discussion is limited to cases where our group has direct experience, or examples where these details are known to us. An attempt will be made to provide a critical look at the positive and negative aspects of various frequency mixing methods, albeit from our own point of view.

Frequency mixing techniques at moderate laser intensities (13) result from the nonlinear response of all media to applied electric fields, a response that is normally expressed as a Taylor series expansion in terms of the applied electric field:

$$P(\omega) = \chi^{(1)} \cdot E(\omega) + \chi^{(2)} \cdot E(\omega) \cdot E(\omega) + \chi^{(3)} \cdot E(\omega) \cdot E(\omega) \cdot E(\omega) + \cdots \quad (5.1)$$

where $E(\omega)$ is the applied oscillating electric field, $P(\omega)$ is the induced polarization, and $\chi^{(n)}$ is the nth order susceptibility of the medium. The oscillating induced polarization is the source of the generated coherent light, which can appear at new frequencies because of the nonlinear terms in Eq. (5.1). The second-order term gives rise to the well-known processes of second harmonic generation, optical parametric oscillation, and sum and difference frequency mixing in crystals. All of these processes are very efficient, but to have a nonzero $\chi^{(2)}$, the nonlinear medium cannot have a center of symmetry, meaning that crystalline nonlinear materials must be used as nonlinear media for these processes. The birefringent crystals commonly available can only generate coherent light at wavelengths longer than about 1800 Å. It is unlikely that materials will be developed that will work at significantly shorter wavelengths, below about 1500 Å.

Generation of coherent light at shorter wavelengths requires a gaseous nonlinear medium, making the $\chi^{(3)}$ term the first nonzero nonlinear term. Since the third-order response is proportional to the product of three oscillating electric fields, at frequencies ω_1, ω_2, and ω_3, the light generated as a result of this term can be at any combination of the applied frequencies: $\omega_{new} = \omega_1 \pm \omega_2 \pm \omega_3$, with ω_{new} being the generated frequency. The new generated frequency means that four electric fields are oscillating in the nonlinear medium simultaneously, three fundamental and one generated, so this type of frequency mixing, mediated by the third-order nonlinearity, is called four-wave mixing (FWM). The simplest type of FWM uses only one fundamental frequency, and the generated wave is at three times the fundamental frequency: $\omega_{new} = 3\omega_1$. This process is called third harmonic generation, and it is commonly used because of its technical simplicity.

For this simplest case, the intensity of the generated third harmonic is described by

$$I_{3\omega} = N^2|\chi^{(3)}(3\omega)|^2 I_\omega^3 F(b\Delta k) \qquad (5.2)$$

where I_ω is the intensity of the fundamental, N is the number density of the nonlinear medium, and $\chi^{(3)}(3\omega)$ is the third-order susceptibility for third harmonic generation. The function $F(b\Delta k)$, where b is the path length and Δk the wavevector mismatch, depends on the macroscopic properties of the medium and the focusing of the fundamental, and represents the phase matching between the input light and the generated VUV. For third harmonic generation with focused lasers, $F(b\Delta k)$ is nonzero only if the refractive index at 3ω is less than the refractive index at ω. Since $F(b\Delta k)$ is pressure dependent because of the pressure dependency of Δk, one cannot simply increase N to improve $I_{3\omega}$, but the $N^2 F(b\Delta k)$ term as a whole must be optimized. Although Eq. (5.2) is valid only for third harmonic generation, similar equations apply to the other forms of four-wave mixing. The very strong intensity dependence indicated by the I_ω^3 term essentially restricts four-wave mixing to pulsed lasers, although it is possible to generate CW coherent VUV (14).

In this chapter, we shall be concerned exclusively with applications of VUV generated by three types of four-wave mixing (see Fig. 5.1): third harmonic generation (THG), two-photon resonant sum-frequency mixing (RSFM), and two-photon resonant difference-frequency mixing (RDFM). Since the simplest method of VUV generation is THG, it shall be described first. In principle, all that you need to do to generate third harmonic is to focus a pulsed laser into a low-pressure gas. For typical nanosecond laser pulses, focusing with a 15–30 cm focal length lens works well. In general, the more polarizable an atom is, the larger $|\chi^{(3)}|^2$ will be, so third harmonic generation should work best in metal vapors and rare gases such as Kr and Xe. The severe limitation on THG comes from the phase-matching term in Eq. (5.2), $F(b\Delta k)$, which forbids THG wherever the refractive index at 3ω is higher than the refractive index at ω. Since the refractive index normally increases with frequency, this behavior $[n(3\omega) < n(\omega)]$ is called negative dispersion. Regions of negative dispersion are found to the blue of resonance lines in gases, and this is where THG is most efficient. As an example, if we consider Kr gas as the nonlinear medium, the first resonance line is at 1235.85 Å, meaning

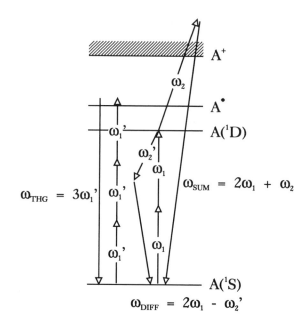

Figure 5.1. Schematic diagram of the three types of four-wave frequency mixing considered in this chapter. The nonlinear medium is represented by A, which has excited states and ionization limit as shown. From left to right, nonresonant third harmonic generation ($\omega_{output} = 3\,\omega_1'$), two-photon resonant difference frequency mixing ($\omega_{output} = 2\omega_2 - \omega_2'$), and two-photon resonant sum frequency mixing ($\omega_{output} = 2\omega_1 + \omega_2$) are illustrated.

that there is no THG possible in Kr for wavelengths longer than about 1230 Å. The region of negative dispersion extends to about 1205 Å, giving a tuning range for THG of 1205–1230 Å (15). The next resonance in Kr is at 1164.87 Å, and this gives rise to another range for THG from about 1160–1100 Å, and the next region is a narrow window just below the 1030.03 Å resonance. At shorter wavelengths the regions of negative dispersion are narrower and more numerous, but there is never an extended range of continuous tunability (16). This behavior is typical of all rare gases (12), which means that THG in rare gases does not produce broadly tunable coherent light. However, by choosing between Ar, Kr, or Xe, it is possible to generate third harmonic light over a broad range of wavelengths from 900 to 1470 Å with a few gaps. Mercury vapor is negatively dispersive, from 1310 to 1850 Å, so THG in Hg is possible over this range, but this requires dealing with the problems of using metal vapors for nonlinear media (see below). THG is widely used in spite of these limitations because it is technically nondemanding. All that is necessary is a gas cell or pulsed jet for the nonlinear medium and a tunable pulsed laser. Since only one VUV wavelength is generated by THG, there is no need for wavelength separation if the experiments you want to carry out are not affected by the fundamental radiation. For experiments in the VUV, a gas cell with a MgF_2 or LiF exit window can be used for the Kr or Xe, followed by a simple

MgF$_2$ or LiF lens for collimation of the VUV beam. For experiments in the XUV, where windows and lenses absorb, a pulsed jet can be used for the nonlinear medium (17), focusing the fundamental into the jet just after the nozzle, and a capillary can be used after the focus to collimate the XUV and reduce the gas load on the rest of the vacuum system (18). Using pure gas as the nonlinear medium, the density is restricted by the phase matching condition to around 100 torr, which results in a relatively low conversion efficiency, since $|\chi^{(3)}|^2$ is small for nonresonant THG. The expected conversion efficiency is about 1 part in 10^7, which still results in 10^9 photons per pulse for millijoule pulses of fundamental light, a sufficient flux for most experiments, if a broad tuning range is not required. THG can be optimized at a given wavelength by using mixtures of positively and negatively dispersive gases, such as a mixture of Kr and Ar for Lyman alpha (1216 Å) generation. In this way, the pressure of the negatively dispersive gas can be increased by using the positively dispersive gas to "counteract" the large negative Δk and preserve phase matching. In this way, one can increase the pressure of Kr from 50 torr, the phase matched pressure for pure Kr with a 20-cm focal length focusing lens, to about 500 torr, thus increasing the THG efficiency by about 100 times. The limitation of this method is that since the refractive index of Kr varies quickly with VUV wavelength, the phase matched tuning range in the VUV is very limited for gas mixtures compared with pure Kr.

Another possibility to increase efficiency and maintain the broad tunability of the coherent VUV is to increase the value of the $|\chi^{(3)}|^2$ term in Eq. (5.2). The equation for $\chi^{(3)}$ for third harmonic generation has the form:

$$\chi^{(3)} = \frac{e^4}{\hbar^3} \sum_{i,j,k} \frac{Z_{gi}Z_{ij}Z_{jk}Z_{kg}}{(\Omega_{ig} - \omega)(\Omega_{jg} - 2\omega)(\Omega_{kg} - 3\omega)} \qquad (5.3)$$

where Z_{xy} are dipole matrix elements, Ω_{xg} are the complex transition frequencies for the $x \leftarrow g$ transitions, and the summation is over all states. The resonant terms in Eq. (5.3) lead to a strong enhancement of $|\chi^{(3)}|^2$ when either ω, 2ω, or 3ω are resonant with a transition in the nonlinear medium. Since a resonance at ω or 3ω would lead to absorption of the fundamental or third harmonic light, a two-photon resonance is best for enhancing $|\chi^{(3)}|^2$. With only one input frequency, fixing ω such that 2ω corresponded to a resonance would not allow for tuning of the generated coherent light. To have both tunability and resonant enhancement, two fundamental frequencies must be used: ω_1 and ω_2. In this simplest (and most common) case of two fundamental frequencies, coherent light can be generated at either the sum or the difference frequency: $\omega_{VUV} = 2\omega_1 \pm \omega_2$. The rules for phase matching are the same for THG and RSFM (10), so this process will only work if the generated VUV is to the blue of a principal resonance in the nonlinear gas, or above the ionization limit. The phase matching conditions for RDFM are much less stringent (10), and both positive and negative Δk values can be tolerated, making it possible to do RDFM at essentially any wavelength. In practice, we have found RDFM to be generally more efficient than RSFM when the same two-photon resonance is used, and RSFM is more efficient than THG by a few orders of magnitude under similar conditions. This means one can expect to get 10^{11} photons per pulse from RSFM and 10^{11}–10^{12} photons per pulse with RDFM, when millijoule pulses of fundamental are used.

Initial work in RSFM was carried out in metal vapors (19–21), which are very efficient nonlinear media and useful over a broad range of VUV wavelengths. Two systems that have seen wide application in chemical physics are Mg vapor, which is useful in the 1400–1700 Å range (21), and Hg vapor, which is useful at wavelengths below 1300 Å (22,23), although the efficiency of Hg RSFM drops as the wavelength is reduced below 1100 Å. The main difficulty with using metal vapors, as opposed to noble gases, for nonlinear media is the need for a vapor over. Metal vapors do have significant advantages that may make the bother of oven construction worthwhile. One of the advantages is the potential of very high conversion efficiency, which has been demonstrated in Hg vapor by a number of groups, including a report of 5% conversion efficiency (24). The high conversion efficiency in Hg is found in the region of Rydberg resonances below the ionization limit (10.437 eV, 1188 Å) in the 1205–1290 Å range. By coincidence, one of these resonances is very close to 1251 Å, which can be produced by RSFM using a single dye laser at 6256 Å. The doubled output of this dye laser, at 3128 Å, is a two-photon resonant with the $7s \leftarrow 6s$ transition in Hg, meaning that by using 3128 Å for ω_1, and 6256 Å for ω_2, the sum frequency $(2\omega_1 + \omega_2)$ is generated at 1251 Å (9.91 eV). This method produces in excess of 10^{13} photons per pulse with Nd:YAG pumped dye lasers, more than enough for use as a photoionization detector of reaction products (25) or a photolysis light source for VUV photofragment spectroscopy (26,27). Another advantage of metal vapors is one of VUV line width, since in RSFM and RDFM, the VUV or XUV line width is given approximately by the number of visible dye laser photons used times the visible line width. Generating VUV for laser-induced fluorescence (LIF) detection of CO photofragments, which uses VUV around 1500 Å, can be done either by RSFM in Mg vapor, or RDFM in Kr or Xe. Although the Kr/Xe scheme is more efficient, it requires a total of five visible photons, compared with three visible photons for RSFM in Mg. This means the VUV line width will be approximately 60% larger, which is important if one wants to measure Doppler spectra of photofragments (28,29).

The most generally useful media for RSFM and RDFM are Kr and Xe, either in gas cells or pulsed jets (16,30,31). There are a wide variety of possible two-photon resonances possible, from 1927 to 2525 Å, and some of these are summarized in Table 5.1, along with data for Hg and Mg vapors. Our experience has been that in both Xe and Kr, the most efficient resonances correspond to $np[n/2, 0]$ levels, for both RSFM and RDFM (32). Although the tuning range for RSFM includes the region below the ionization limit, in practice this is a difficult region in which to work. In both Xe and Kr, the range of photon energies corresponding to the region between the two lowest ionization limits (Kr$^+$ or Xe$^+$: $^2P_{3/2}$ and $^2P_{1/2}$) corresponds to a series of autoionizing resonances, and the efficiency for generating XUV varies strongly with wavelength in this range, although XUV can be generated at all photon energies in this range (14.00–14.66 eV in Kr, 12.13–13.44 eV in Xe). Above the $^2P_{1/2}$ limit in both cases, the conversion efficiency for

Table 5.1. Examples of Four-Wave Frequency Mixing Techniques

Medium	Process	$2\omega_1$ Transition	ω_1 (cm^{-1}); λ_1 (Å)^a	Tuning Range (Å)	Intensity^b
Mg/He	$2\omega_1 + \omega_2$	$3s3d\ ^1D_2 \leftarrow 3s^2\ ^1S$	23,201.5; 4308.8	1430–1700	x
Hg/He	$2\omega_1 + \omega_2$	$6s6d\ ^1D_2 \leftarrow 6s^2\ ^1S$	35,666.6; 2802.8	1050–1200^c	
Hg/He	$2\omega_1 + \omega_2{}^d$	$6s7s\ ^1S_0 \leftarrow 6s^2\ ^1S$	31,964.1; 3127.7	1251.4	
Xe (jet)^e	$2\omega_1 + \omega_2$	$6p[0\frac{1}{2},0] \leftarrow 5p^6$	40,059.7; 2495.5	780–923^f	10
Xe (cell)^g	$2\omega_1 - \omega_2$	$6p[0\frac{1}{2},0] \leftarrow 5p^6$	40,059.7; 2495.5	1460–2000	
Xe (jet)	$2\omega_1 + \omega_2$	$6p'[0\frac{1}{2},0] \leftarrow 5p^6$	44,930.3; 2225.0	720–923^f	5
Xe (cell)	$2\omega_1 - \omega_2$	$6p'[0\frac{1}{2},0] \leftarrow 5p^6$	44,930.3; 2225.0	1175–2000	2x
Kr (jet)	$2\omega_1 + \omega_2$	$5p[0\frac{1}{2},0] \leftarrow 4p^6$	47,046.8; 2124.9	700–845^f	5
Kr (cell)	$2\omega_1 - \omega_2$	$5p[0\frac{1}{2},0] \leftarrow 4p^6$	47,046.8; 2124.9	1215–2000	20
Kr (jet)	$2\omega_1 + \omega_2$	$5p'[0\frac{1}{2},0] \leftarrow 4p^6$	49,427.9; 2022.5	1150–2000	5
Kr (cell)	$2\omega_1 - \omega_2$	$5p'[0\frac{1}{2},0] \leftarrow 4p^6$	49,427.9; 2022.5	1150–2000	15
Kr (jet)	$2\omega_1 + \omega_2$	$6p[0\frac{1}{2},0] \leftarrow 4p^6$	51,881.2; 1926.8	655–845^f	
Kr (cell)	$2\omega_1 - \omega_2$	$6p[0\frac{1}{2},0] \leftarrow 4p^6$	51,881.2; 1926.8	1090–2000	

^{a}Wavelength in air.
^{b}Relative intensity, see text for explanation.
^{c}Tuning range in a gas cell. Lower wavelength limit is 830 Å in windowless configuration.
^{d}Special case where $\omega_2 = \omega_1/2$, so only one tunable laser is needed.
^{e}The (jet) means the nonlinear medium is in a pulsed-free-jet.
^{f}Upper wavelength limit is Xe$^+$ or Kr$^+$ ($^2P_{1/2}$) ionization limit. Tuning to longer wavelengths is possible (see text).
^{g}The (cell) means that a gas cell is used to contain the nonlinear medium.

generating XUV does not vary strongly with photon energy. Below the lowest ionization limit, RSFM occurs most efficiently to the blue of resonance lines, no matter what the nonlinear medium is, although if the spectral density of lines is high enough, there will always be a nonzero conversion efficiency. The data given in Table 5.1 only represents a small fraction of possible $2\omega_1$ resonances, and does not include nonresonant third harmonic generation. The relative intensities are only intended as a rough guide, and represents what one can expect under standard operating conditions with a laser system comparable to the Nd:YAG pumped dye lasers used for the comparisons (Spectra-Physics GCR4 pumping two Lambda Physik 3002E). No attempt has been made to quantify XUV generation efficiency under equivalent input laser power conditions, rather, typical pulse energies available at different ω_1 were used. The comparison between RDFM in Xe and RSFM in Mg was made on a different apparatus, and the intensity from RDFM in Xe was slightly higher than RDFM in Kr, but no direct comparison has been made to date by us. The 1927 Å resonance has been described in the literature (33), but we have not yet tested it and quote no intensity.

In our experience, the most convenient methods for generating VUV and XUV are as follows. From 655 to about 1000 Å (12.4–18.9 eV), RSFM in pulsed jets of Kr or Xe works well, except for problems of variations in conversion efficiency from 923 to 1000 Å. For the range from 1090 to 1900 Å (6.5–11.4 eV), RDFM in a gas cell of Kr or Xe is the easiest and usually the most efficient method, although there can be some advantages to using RSFM in Hg or Mg vapor. The gap between the RSFM and RDFM tuning ranges, from 1000 to 1100 Å, presents some difficulties, although THG in pulsed jets or Kr or Ar can be used in this range, as can RSFM in pulsed jets of CO. Above 1900 Å, one can use birefringent crystals to generate coherent light. Thus, the experimenter has available efficient methods of generating coherent radiation continuously tunable over the entire 6.5–18.9 eV photon energy range, using standard pulsed laser systems to provide the input coherent light.

5.3. HIGH-RESOLUTION PHOTOIONIZATION SPECTROSCOPY

In the photoionization spectra of small molecules, sharp resonances due to auto-ionization are essentially universal in the photon energy range just above the lowest ionization threshold. To understand the spectroscopy and dynamics of the Rydberg states responsible for these resonances, it is vital to have a light source with sufficient resolving power to excite single quantum states of the autoionizing Rydberg series. This need for resolution can be easily seen by examining the photoionization efficiency (PIE) spectrum of almost any diatomic molecule. For example, the PIE spectrum of jet-cooled O_2 recorded with coherent XUV is shown in Fig. 5.2. Numerous Rydberg resonances can be seen in this spectrum, and as shown in the figure, several Rydberg series can be identified (34). The ionization thresholds for forming O_2^+ ions in various vibrational levels of the ground $X^2\Pi$ and excited $a^4\Pi$ states are shown. These thresholds will be discussed in the next section. In Fig. 5.3, a comparison is made between a section of the Fig. 5.2 spectrum and a photoionization mass spectrum with a resolution of 0.07 Å (10 cm^{-1}) recorded by Berkowitz (35). This comparison shows two advantages of using coherent XUV. The first, and most striking, is the factor of 10 difference in resolution. In addition to permitting better resolution of spectral features and expanded possibilities for determining lineshapes, this difference in resolution allows for more detailed dynamical studies of the autoionizing Rydberg states. Selective excitation of the very sharp resonances seen in Fig. 5.2 requires a narrow-band light source to avoid "dilution" by simultaneous excitation of the underlying continuum.

Another less obvious advantage is that the increased intensity of the coherent source allows for the measurement of excitation spectra of very "thin" targets, as is the case with collimated supersonic molecular beams. The ability to use supersonic beams means that it is possible to cool the target molecules through supersonic expansion, thus greatly simplifying the spectrum. An example of this simplification is shown in Fig. 5.4, where a small portion of the NO PIE spectrum is shown for two different supersonic beam conditions: pure NO, and 3% NO in Ne, both expanded from a pulsed supersonic source and collimated by a 0.5-mm skimmer. The rotational temperature drop from about 25 to about 3K results in a considerable simplification of the spectrum of the $7s\sigma(v = 2)$, $12s\sigma(v = 1)$, $11d\delta(v =$

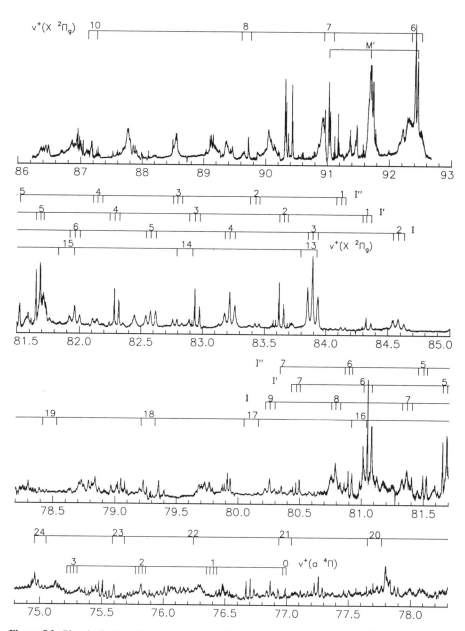

Figure 5.2. Photoionization efficiency spectrum of jet-cooled O_2 from 748 to 927 Å (13.4–16.6 eV). The ionization thresholds for forming O_2^+ X $^2\Pi_g$ ($v^+ = 6$–24) and O_2^+ a $^4\Pi_u$ ($v^+ = 0$–3) are marked on the spectrum, as are the Rydberg state assignments of Ref. 34.

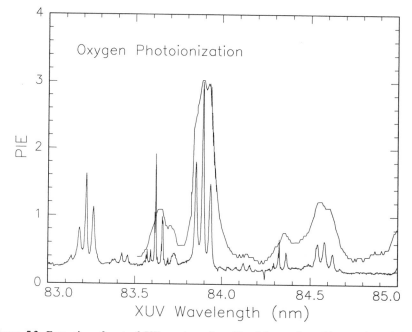

Figure 5.3. Expansion of part of PIE spectrum from Fig. 5.2, together with a portion of the PIE spectrum recorded by Berkowitz (35) (upper trace). The Berkowitz data is shifted upwards for clarity.

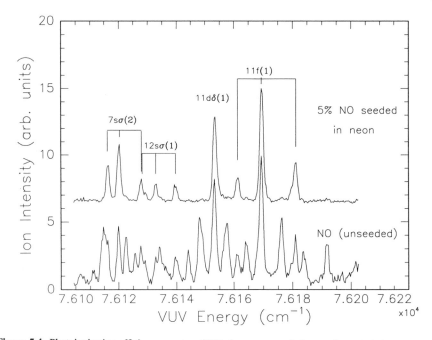

Figure 5.4. Photoionization efficiency spectra of NO, for a supersonic beam of pure NO (lower trace), and a beam formed by expanding 3% NO in Ne (upper trace).

1), and $11f(v = 1)$ Rydberg states. Even at the greatly reduced rotational temperature of 25 K, the spectral region shown in Fig. 5.4 is very congested. At 300 K, or even 78 K, assignment of the spectrum would be very difficult, given the number of overlapping resonances that would be present.

The PIE spectrum of supersonically cooled NO shown in Fig. 5.5 is a small portion of a spectrum recorded by us from the lowest ionization threshold to above the $v^+ = 2$ limit (36), corresponding to a photon energy range from 9.26 to 9.97 eV. The NO was in a collimated supersonic beam of 3% NO in Ne, and the coherent VUV was generated by FWDM in a gas cell of Kr. From the unambiguous assignment of the PIE spectrum, one can see that the most prominent features in the spectrum are due to the weakly predissociated nf states. There are, in fact, only very weak resonances due to the predissociating np states, which autoionize by "unfavorable" $\Delta v > -1$ transitions (37), in contrast with conclusions from previous, lower resolution studies (38). However, some of these states can still be seen in the spectrum, such as the $5p\sigma(v = 3)$ resonance at 75,880 cm^{-1}, allowing for a detailed comparison with multichannel quantum defect theory calculations on the interaction between autoionization and predissociation (39,40). Once a PIE spectrum, such as that of NO shown in Fig. 5.5, is recorded and detailed assignments are made, the intensity and resolving power of the coherent VUV makes it possible to study the autoionization dynamics for selected autoionizing Rydberg states. As an example, above the $v^+ = 2$ threshold, the $5f(v = 4)$, $8f(v = 3)$, $5p\pi(v = 5)$, and $9p\pi(v = 3)$ states can be selectively excited, and TOF photoelectron spectroscopy can be used to measure the NO$^+$ ($v^+ = 0$ to $v^+ = 2$) yields resulting from autoionization. Some of this work has been done in the past (41), and more experiments in this area are ongoing.

One example of the power of jet cooling and high resolution in photoionization measurements is provided by recent work on spin–orbit autoionization in HI. This work, done simultaneously by two different research groups (42,43), was the first rotationally resolved work on electronic autoionization, and was initiated because of detailed theoretical calculations on HI (44). In addition to providing an excellent test of ab initio theory, this work also demonstrates Hund's case (e) coupling (45), which had been described in a previous theoretical paper (46). This work on HI was followed in one of the groups by a series of elegant experiments, which measured asymmetry parameters and photoelectron spin polarization following rotationally resolved excitation of autoionizing levels in HI. These experiments are described briefly in section 5.5.

5.4. THRESHOLD PHOTOELECTRON SPECTROSCOPY

5.4.1. Introduction

One of the most exciting recent developments in laser spectroscopy has been the development of the technique of pulsed field ionization–zero kinetic energy (PFI–ZEKE) photoelectron spectroscopy. This technique, developed by Müller–Dethlefs et al. (47), allows one to record threshold photoelectron spectra at an

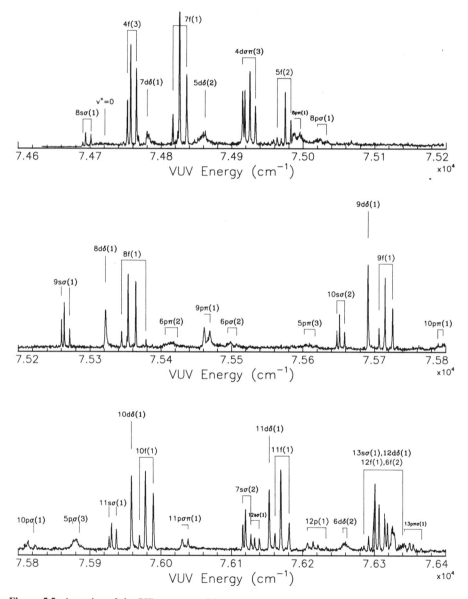

Figure 5.5. A portion of the PIE spectrum of jet-cooled NO, in the photon energy range just above the lowest ionization energy. The $8s\sigma(1)$ resonance below the $v^+ = 0$ limit results from Stark ionization. The states are labeled according to the Rydberg state, with the vibrational quantum number in brackets. Thus, $5f(2)$ is the $5f$ Rydberg state converging to the NO^+ ($v^+ = 2$) limit.

energy resolution comparable to laser bandwidths, about 1 cm^{-1} or better. The basis for this technique, which has been recently reviewed (48,49), is the excitation of very high principal quantum number Rydberg states that lie just below every ionization threshold, followed after a time delay by field ionization and detection of the field ionized electrons. This is illustrated schematically in Fig. 5.6, where excitation of Rydberg states converging to an ionization limit of the ion is shown. When the exciting laser is tuned to an energy just below threshold, where the density of Rydberg states is extremely high, a quasicontinuum of states within the laser bandwidth is excited. Simultaneously, degenerate ionization continua corresponding to lower levels of the ion are excited, resulting in direct photoionization and yielding what shall be referred to as prompt electrons. Under most circumstances, the high principal quantum number Rydberg states are metastable (50), and do not decay over periods as long as hundreds of microseconds (51). Thus, after a time delay of about 1 μs from excitation, one can expect all prompt electrons to have flown out of the ionization volume, following which one can detect the remaining high Rydberg states by field ionization.

It should be pointed out that although the simplest method for detecting this field ionization is electron detection, one can also detect the positive ions formed by field ionization. This variation on PFI–ZEKE, called MATI (for mass analyzed threshold ionization) by its inventor (52), is a very useful technique for cases where

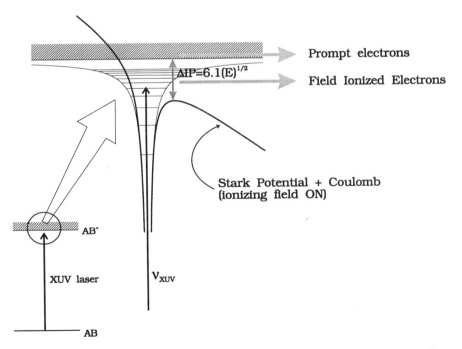

Figure 5.6 Schematic diagram of the PFI–ZEKE process. High principal quantum number Rydberg states are excited by the XUV laser, and after a time delay are field ionized by applying a Stark field. The adiabatic shift in the ionization energy is shown.

the carrier of the PFI–ZEKE spectrum might be ambiguous. This would be the case for van der Waals clusters, or free radicals formed in a photolysis source, where there may be several different species with similar ionization energies.

Field ionization of these high Rydberg states is caused by the Stark shift of the Coulomb potential, which results in a lowering of the ionization energy given by

$$\Delta \text{IE} = \kappa(E)^{1/2} \text{ cm}^{-1} \qquad (5.4)$$

where ΔIE is the lowering of the ionization energy from the field-free value, E is the applied electric field in volts per centimeter (V cm^{-1}) units, and κ is a constant between 4 and 6, depending on whether the field ionization is diabatic or adiabatic (53,54). Experimentally, the value of κ is normally about 4, meaning that a field of 1 V cm^{-1} will result in a 4 cm^{-1} lowering of the ionization energy. If a field of 1 V/cm^{-1} is applied after the delay period, Rydberg states between the ionization limit and 4 cm^{-1} below the ionization limit (between $n = \infty$ and $n \approx 170$) will be field ionized and detected. In practice, stray fields on the order of tens of millivolts per centimeter are usually present, which will result in field ionization immediately after excitation to about 0.5–1 cm^{-1} below the field-free ionization energy. Under these circumstances, the delayed 1 V cm^{-1} ionizing field will ionize Rydberg states between 4 and 1 cm^{-1} below the ionization limit. Combinations of fields can be used to isolate a narrow range of metastable Rydberg states and, hence, achieve better energy resolution. As a simple example, if a pulsed field of 0.5 V cm^{-1} is applied immediately after excitation, followed by a 0.6 V cm^{-1} field 1 μs later, detection of the electrons formed by the second field will detect Rydberg states between 3.1 and 2.8 cm^{-1} below threshold, resulting in an energy resolution of 0.3 cm^{-1}, which is 0.04 meV. This is more than 100 times better than the energy resolution of the best conventional photoelectron spectrometer, and is sufficient to resolve rotational structure in the photoelectron spectra of small molecules.

The technique of PFI–ZEKE spectroscopy was originally developed by multiply resonant multiphoton excitation to the thresholds, as illustrated in Fig. 5.7. While this approach is very powerful and has many advantages, it cannot be used universally. The first limitation is the requirement for a stable intermediate state whose spectroscopy is well understood. While this is not a problem in small molecules or well-characterized larger molecules, it can pose problems if one wishes to study unstable species whose spectroscopy is not well understood, or molecules with heavily predissociated excited states. Even in cases where there are suitable resonant states, there is still a limitation of the multiphoton PFI–ZEKE technique with respect to excited-state thresholds in the ion. This is shown in Fig. 5.7, using NO as the example. This molecule is an ideal candidate for multiphoton PFI–ZEKE since it has a wide range of possible intermediate neutral states and a low ionization threshold (9.26 eV). However, the first excited state of NO$^+$, the $a^3\Sigma^+$ state, has an IE of 15.67 eV, 6.41 eV above the NO$^+$ $X\,^1\Sigma^+$ limit. With conventional pulsed laser systems limited to photon energies below about 6.5 eV, it is impossible to reach this state by $1 + 1$ excitation, by far the best choice for PFI–ZEKE. Even the use of $2 + 1$ excitation will not work, since a 6.5-eV photon will not be

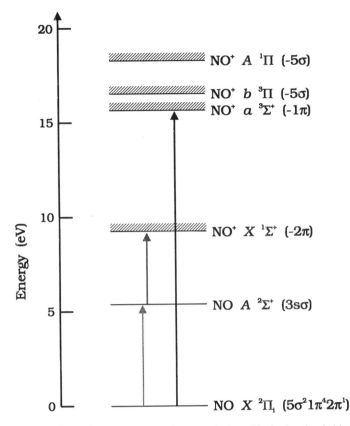

Figure 5.7. The use of one photon versus two photon excitation of ionization thresholds for NO. The NO$^+$ excited states shown cannot be reached by multiphoton excitation, but are all within range of the coherent XUV sources described in this chapter.

sufficient to reach 15.67 eV from the characterized states of NO, meaning that one has to somehow drive a two-photon excitation through the ionization continuum, a possible but unlikely process at moderate laser powers. A solution is to generate the necessary 15.67-eV coherent light by the frequency mixing techniques previously described, and use this for single-photon excitation of the excited-state threshold. With the tuning range of the laser systems described in Section 5.4.1, all of the NO$^+$ ionization thresholds shown in Fig. 5.7 can be accessed by a single coherent XUV photon. Single-photon excitation is also a completely general method of reaching any ionization threshold up to the energy limit of the coherent source, independent of the intermediate states that may or may not be available for multiresonant excitation.

The use of coherent XUV for PFI–ZEKE spectroscopy is now well established (55–58), and to date studies have been carried out on a wide range of small molecules and free radicals. This chapter will consider a few aspects of PFI–ZEKE spectroscopy, although none will be discussed in great detail. To begin, we shall look at the most obvious application of PFI–ZEKE: the determination of accurate ionization energies and ion spectroscopic constants. Following this, the use of coherent XUV in PFI–ZEKE spectroscopy of excited states of ions will be explored.

5.4.2. Determination of Ionization Energies; Ion Spectroscopy

As stated above, PFI allows for the accurate determination of ionization thresholds in a molecule, in principle with an accuracy at least as good as the energy resolution of the coherent XUV light source. The only other method for such accurate determination of ionization energies is extrapolation of Rydberg series to their $n = \infty$ limits, a very challenging spectroscopic task that has only been accomplished in a few molecules. The use of conventional photoelectron spectroscopy can only determine IE values to an absolute accuracy of several meV (several tens of cm^{-1}). Achievement of even this accuracy is not trivial, as accurate modeling of the rotational contour in the photoelectron bands is necessary to deconvolute the unresolved structure for an accurate assignment of the band origin. Consequently, for almost all molecules, including diatomics, the uncertainty on the ionization energy is several tens of wavenumbers or higher. The combined use of supersonic cooling and PFI allows for an improvement of one to two orders of magnitude in the precision of ionization energy determinations. Furthermore, by using coherent XUV for single-photon excitation, this method of IE determination is completely general.

The first step in the absolute determination of an IE is the calibration of the XUV photon energy, which is done by accurate calibration of the fundamental lasers used to generate the coherent XUV (ν_1 and ν_2). This should be done concurrently with the recording of the PFI–ZEKE spectrum. Using standard techniques, such as iodine LIF or optogalvanic spectroscopy, it is possible to produce an absolute XUV calibration accurate to ± 0.5 cm^{-1} or better. The principal uncertainty in determining an accurate ionization energy comes from the extrapolation to zero field of the PFI–ZEKE peak positions. This can be seen by considering the PFI–ZEKE spectrum at the Kr$^+$ ($^2P_{1/2}$) limit at various field strengths, as shown in Fig. 5.8. The general behavior of the PFI–ZEKE peak shifting to lower photon energies as the ionizing field is increased can be easily understood in terms of the Stark shift of the ionization threshold, which shifts the lower energy side of the PFI–ZEKE peaks according to Eq. (5.4). However, because the lower principal quantum number Rydberg states can be lost through a variety of predissociation and autoionization processes, determining the low-energy ''edge'' of the peak has some uncertainty associated with it. This can be seen by examining the peak shapes shown in Fig. 5.8, which do not have Gaussian low-energy edges. If only the initial part of the low energy rise is fitted to a Gaussian, using the known XUV lineshape, the Stark shifted ionization potential can be determined. If this is plotted as a function of $(E)^{1/2}$, the zero-field ionization energy can be extracted. For the Kr data given in Fig. 5.8, the resulting plot is shown in Fig. 5.9. The slope of the plot, -4.08 ± 0.12, is consistent with diabatic field ionization of high Rydberg states

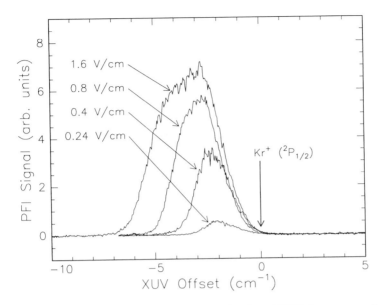

Figure 5.8. Stark ionized electron signal from Kr as a function of XUV photon energy for different values of the pulsed field. The XUV energy is measured with respect to the Kr^+ ($^2P_{1/2}$) limit.

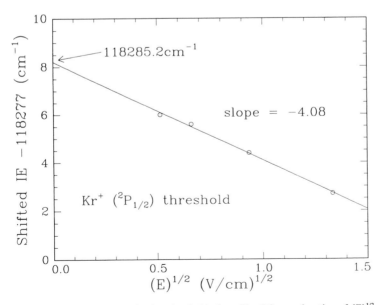

Figure 5.9. Plot of the ionization thresholds from Fig. 5.8, as a function of $(E)^{1/2}$.

(53,54), and the extrapolated ionization energy is in agreement with the known value within the accuracy of the PFI–ZEKE measurement. If one could be assured that all PFI were diabatic, the known Stark shift could be used to determine the zero-field ionization energies from a single PFI–ZEKE spectrum. However, in practice one finds varying degrees of adiabaticity in the field ionization, meaning a field shift between the limiting $-6.1\sqrt{E}$ and $-4\sqrt{E}$ for adiabatic and diabatic ionization, respectively. Since this can make a difference of 2 cm^{-1} for a 1-V cm^{-1} field, an accurate determination of the ionization energy requires an extrapolation to zero field.

Two somewhat surprising examples of molecules in which the IE values had not been accurately determined before PFI–ZEKE work are CO and HF. In the case of CO, conflicting assignments of the Rydberg series resulted in an uncertainty of 20 cm^{-1} in the IE (59), while for HF, the only accurate determination was based on a simulation of the photoelectron spectrum (60). In both cases, the determination of the adiabatic ionization energies by PFI–ZEKE was straightforward, using the calibrated coherent XUV source to measure the spectrum at various pulsed fields. The PFI–ZEKE spectrum of the $v^+ = 0$ state of HF$^+$ is shown in Fig. 5.10 (61). This spectrum could be assigned unambiguously, which means the HF$^+$ $X\,^2\Pi_{3/2}$ $(v^+ = 0, J^+ = \frac{3}{2})$ threshold can be assigned and measured with an absolute certainty of ± 1 cm^{-1}. Figure 5.10 also provides an indication of the resolving power of this

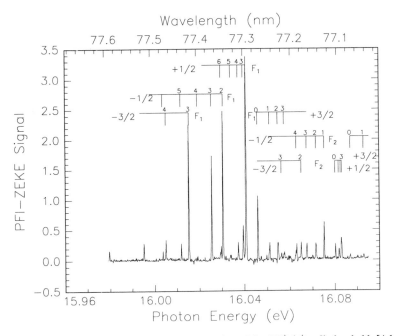

Figure 5.10. The PFI–ZEKE spectrum of HF in the region of the HF$^+$ ($v^+ = 0$) threshold. [Adapted from Ref. 61.]

type of threshold photoelectron spectroscopy, with the individual lines having a fwhm of 2 cm^{-1} (0.2 meV).

The PFI–ZEKE spectrum of CO (62), shown in Fig. 5.11, also shows clearly resolved rotational structure, allowing an accurate determination of the ionization energy of CO. From both spectra shown, one can make accurate measurements of the spectroscopy of the molecular ions. Although the spectroscopy of the HF$^+$ X $^2\Pi_i$ and CO$^+$$X$ $^2\Sigma^+$ states is well understood, this is not always true, as in the case of CH$_3^+$, which has also been investigated by PFI–ZEKE using coherent XUV (63). Another aspect of PFI–ZEKE spectroscopy, which has been discussed extensively in recent literature (56,64–67), is the line strengths for the rotational transitions observed. There have been several excellent articles published on this topic (68) and readers are referred to these for more detailed information. Of interest here are the models that can be used to describe the rotational line strengths in the PFI–ZEKE spectrum.

One model for rotational line strengths that is particularly useful for experi-

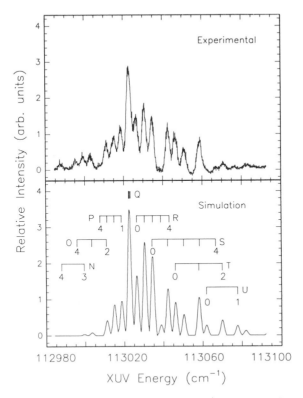

Figure 5.11. The PFI–ZEKE spectrum of jet-cooled CO at the $v^+ = 0$ threshold (top panel) along with the assignment and BOS simulation (bottom panel). The branch designation indicates the $N^+ \leftarrow J''$ change. [Adapted from Refs. 32 and 62.]

mentalists was developed initially by Buckingham et al. (69) and recently rederived in slightly different form by Xie and Zare (70). We shall refer to the Buckingham et al. formulation (the BOS model). The original BOS model was developed to describe rotational line strengths in photoelectron spectroscopy, where the molecule is excited directly into the ionization continuum. However, because of the continuity of oscillator strength (35) through the ionization threshold, the BOS model can also provide a description of the PFI–ZEKE process. The simplified formula for the partial ionization cross section is

$$\sigma_{N^{+}\leftarrow J''} \propto \sum_{\lambda} Q(\lambda;J'', N^{+})C_{\lambda} \qquad (5.5)$$

where $\sigma_{N^{+}\leftarrow J''}$ is the partial cross section for the ionization from rotational state J'' in the neutral to form N^{+} in the ion, $Q(\lambda; J'', N^{+})$ is an angular momentum coupling factor, and the C_{λ} are expansion coefficients, which can be treated as empirical fitting parameters. In the original formulation of the BOS model, the C_{λ} coefficients can be viewed as being related to the coefficients for the expansion of the one-electron molecular wave function in terms of hydrogenic atomic orbitals for the ionized electron. This simple picture ignores coupling between the various angular momentum continua caused by the nonspherical ion core. However, the BOS model, when used empirically, still provides a useful framework for interpreting observed rotational line strengths and allows for realistic spectral simulations. As an example, the simulated spectrum shown in Fig. 5.11 was calculated by assuming a CO rotational temperature of 10 K and then adjusting the C_{λ} coefficients for $\lambda = 0–4$ to fit the measured spectrum. By comparing the simulated and experimental spectra, one can immediately quantify the importance of "forced rotational autoionization (62)," which is responsible for the "extra" intensity observed in the P and O branches of the spectrum.

To avoid empiricism, ab initio methods can be used to calculate the rotational line strengths using ab initio initial wave functions and accurate quantum scattering treatment for the outgoing continuum waves in the presence of the nonspherical field of the resulting molecular ion (71,72). An example of this approach is given by recent work comparing the measured PFI–ZEKE spectra of CO and N_2 with theoretical calculations (62). This comparison for N_2 is shown in Fig. 5.12, where it can be seen that except for the "forced rotational autoionization," which is not included in the theoretical model, there is excellent agreement between theory and experiment. In the case of the isoelectronic CO molecule, the same theoretical treatment produces a result quite different from the experimental spectrum, as illustrated in Fig. 5.13. The ab initio calculation predicts a spectrum very similar to that observed for N_2, with a Q branch dominating, which is not at all what is observed. Interestingly, if the Q branch is ignored, there is good agreement between theory and experiment for the other branches, with the exception of the well-understood "forced autoionization." The reason for this discrepancy between ab initio theory and experiment in the case of the CO Q branch is not clear.

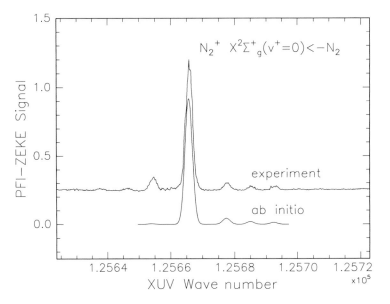

Figure 5.12. Experimental (upper trace) and ab initio calculated PFI–ZEKE spectrum of jet-cooled N_2. The experimental spectrum has been shifted up by 0.25 units for clarity. [Adapted from Ref. 62.]

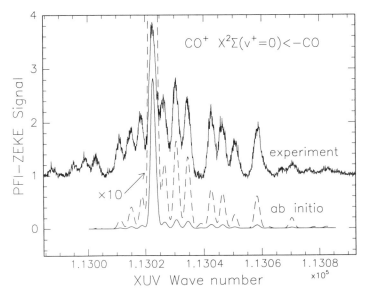

Figure 5.13. Experimental (upper trace) and ab initio PFI–ZEKE spectrum of jet-cooled CO. The experimental spectrum has been shifted up by 1 unit for clarity. The ab initio spectrum was calculated for a 10 K rotational temperature, and an ×10 expanded version of the theoretical spectrum is plotted as a dashed line. [Adapted from Ref. 62.]

5.4.3. Spectroscopy of Excited States of Ions

Although most work in PFI–ZEKE spectroscopy has investigated the lowest ionization thresholds in molecules, radicals, and clusters, the method is quite general and can be applied to the study of higher ionization thresholds. In our work in this area, states more than 6 eV above the lowest ionization threshold have been investigated successfully (73,74). This work shows that the PFI–ZEKE method is quite general, and that the long-lived Rydberg states necessary for PFI can exist even far above the onset of the ionization continua. While the spectroscopic information provided by these high-energy studies is just as reliable as work at the lowest thresholds, providing an accurate description of line intensities is more complicated due to the presence of other ionization and decay channels. Sometimes the deviation from standard theories of intensities in photoelectron spectra can lead to very useful results, as is the case in oxygen. Following the work done by White and co-workers on the lowest vibrational levels of O_2^+ (55,75) and inspired by results from synchrotron studies on the TPES of molecular oxygen (76–78), we investigated the PFI–ZEKE spectrum of oxygen in the energy range corresponding to the $v^+ = 6$ to $v^+ = 24$ thresholds of the O_2^+ X $^2\Pi_g$ state. What was being looked for was dramatic deviations from the band intensities predicted by Franck–Condon factors between the initial $v = 0$ state of O_2 and the final v^+ states of O_2^+.

In standard photoelectron spectroscopy, one expects the relative band intensities within a given electronic state of the ion to be well described by Franck–Condon factors (79), except in the cases when the ionizing wavelength coincides with an autoionizing resonance in the neutral molecule (80). Since PFI–ZEKE is a threshold process, relying on the existence of metastable Rydberg states, one should not necessarily expect to have the relative intensities of different vibrational bands well described by Franck–Condon factors. In fact, the situation in small molecules is often that the band intensities do not agree with Franck–Condon factors. In the case of oxygen, we have observed PFI–ZEKE spectra for the vibrational thresholds between $v^+ = 6$ and $v^+ = 24$ of the X $^2\Pi_g$ state of O_2^+, in spite of the Franck-Condon factors between these states and $v = 0$ of O_2 being negligible for all levels above $v^+ = 6$. An example of these spectra is given in Fig. 5.14, where the $v^+ = 6$, 15, and 21 bands are shown. The fact that the ionization thresholds are due to the X $^2\Pi_g$ ground electronic state of O_2^+ is proven by the simulated spectra also shown in Fig. 5.15. These spectra were calculated using the BOS model, with known (81) (for $v^+ = 6$) or extrapolated spectroscopic constants. A schematic plot of the complete PFI–ZEKE spectrum of oxygen over the 13.4–16.5 eV photon energy range is given in Fig. 5.15. As can be seen, the intensities of the Franck–Condon forbidden X state bands are comparable to the intensities of the allowed a $^4\Pi_u$ bands. In contrast with the earlier synchrotron TPES work, there were no "extra" peaks in the PFI–ZEKE spectrum coming from the strong autoionizing resonances in this energy range. Along with the very interesting ionization dynamics probed by these experiments, these data also are the first rotationally resolved spectra of X $^2\Pi_g$ vibrational levels higher than $v^+ = 11$.

Another example of spectroscopy of excited states of ions is found in the recent results on the HBr^+ A $^2\Sigma^+$ state (82). This excited state of HBr^+ is strongly pre-

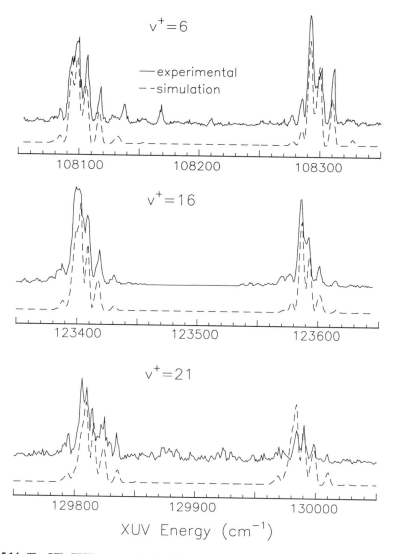

Figure 5.14. The PFI–ZEKE spectra for the v^+ = 6, 16, and 21 thresholds of the O_2^+ X $^2\Pi_g$ state. Experimental spectra (solid lines) are shifted up, and plotted along with the results of a simulation of the bands using the BOS model. [Adapted from Ref. 32.]

dissociated for all vibrational levels above v^+ = 1, and the photoelectron spectrum of HBr is one of the "textbook" examples of the effects of predissociation on line widths in photoelectron spectra (79). What is observed in the HBr photoelectron spectrum for the bands corresponding to the HBr$^+$ A $^2\Sigma^+$ state is a lifetime broadening for vibrational levels $v^+ \geq 3$, corresponding to predissociation lifetimes of 10^{-13} or less. The complete absence of fluorescence from the v^+ = 2 level shows

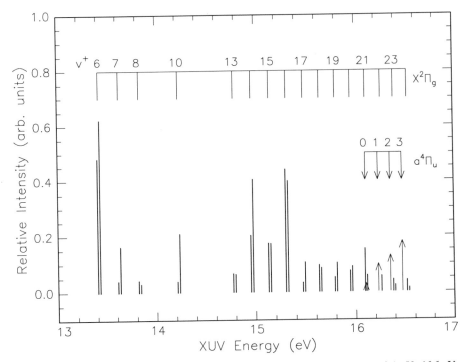

Figure 5.15. Schematic representation of the PFI–ZEKE spectrum of O_2 in the 13.4 eV–16.6-eV photon energy range. The relative intensities of the observed bands are indicated by the line heights, for the spin–orbit sublevels of the $X\ ^2\Pi_g$ state, and the overall bands of the $a\ ^4\Pi_u$ state. [Adapted from Ref. 32.]

that it is also completely predissociated, with an estimated lifetime of 10^{-10} s. In light of these facts, it is interesting to examine the PFI–ZEKE spectrum of HBr for ionization thresholds corresponding to predissociated states of HBr⁺.

The spectrum of one such level, the $v^+ = 2$ level, is shown in Fig. 5.16. This PFI–ZEKE spectrum shows no unusual features and can be simulated using the standard BOS model with spectroscopic constants that are reasonable extrapolations from the known $v^+ = 0$ and 1 constants. This simulation is also shown in Fig. 5.16. Interestingly, the parameters used for the BOS simulation were very similar to those used for the PFI–ZEKE spectrum of the undissociated $v^+ = 0$ level. The intensity of the $v^+ = 2$ band is approximately twice that observed for the $v^+ = 0$ band, in qualitative agreement with the Franck–Condon factors. Thus, the PFI–ZEKE spectrum shown in Fig. 5.16 is completely normal, in spite of the fact that the corresponding HBr⁺ ion dissociates in 10^{-10} s. What this means is that the ion core of the initially prepared Rydberg state has dissociated by the time the ionization pulse is applied, and the field ionization is Stark ionization of a high Rydberg state of Br. Although the lifetime broadening of the lines in Fig. 5.16 was too small to be observed at the 2 cm⁻¹ resolution, the PFI–ZEKE spectrum at the $v^+ = 3$ threshold

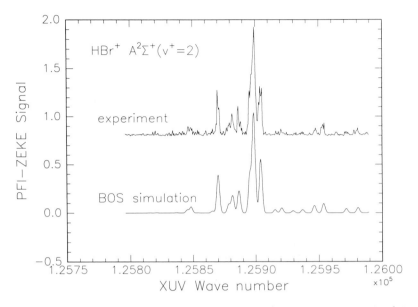

Figure 5.16. The PFI–ZEKE spectrum of the predissociating $v^+ = 2$ level of the HBr^+ A $^2\Sigma^+$ state, shown together with the spectral simulation using the BOS model. [Adapted from Ref. 82.]

does not show any rotational structure because of the very short lifetime ($\sim 10^{-13}$ s) of this state. What is observed is an unresolved band with a contour consistent with a 10^{-13} s lifetime.

5.5. SPIN POLARIZATION, ANGLE-RESOLVED PHOTOELECTRON SPECTROSCOPY

As previously discussed, the coherent VUV generated by frequency mixing is highly polarized, with a polarization determined by that of the input fundamental radiation. This feature has been exploited by the group of Heinzmann to study spin polarization of photoelectrons resulting from autoionization of hydrogen halides (83), using circularly polarized VUV and XUV generated from circularly polarized input radiation. The control over polarization is especially valuable in the case of circular polarized XUV, as it is difficult to generate circularly polarized XUV using incoherent sources. It is possible to collect circularly polarized synchrotron radiation from above or below the orbit plane of a storage ring, but this requires a specially adapted beam line (84), and one obtains the same photon flux and energy resolution described in the introduction for conventional bend magnet beam lines. To generate circularly polarized radiation from a synchrotron undulator, a helical undulator (85) or a crossed undulator (86) must be used. In the case of coherent VUV generated by frequency mixing, all that is needed is circularly polarized input beams. To change the sense of polarization from right to left handed, an electrooptic modulator can be used to reverse the sense of the input polarization. Although somewhat less efficient, linearly polarized ω_2 can be used with circularly polarized ω_1 to generate circularly polarized $2\omega_1 \pm \omega_2$, meaning only the ω_1 polarization needs to be controlled (87).

The power of circularly polarized coherent VUV has been demonstrated by Heinzmann and co-workers in a study of HI autoionization, where the resolution of the coherent source made possible excitation of selected rotational levels of autoionizing Rydberg states. By carrying out the excitation with circularly polarized VUV, they were able to measure the spin polarization of the resulting photoelectrons (83). This work showed that the spin polarization for the total photoelectrons varied strongly across Rydberg resonances and even changed sign within a given band when two adjacent rotational lines were compared. Since the rotational lines in HI Rydberg states are separated by about 12 cm^{-1}, these measurements would have been impossible without the coherent source. More recent work on HI has shown that it is possible to make angle-resolved spin polarization measurements using the coherent VUV source (87).

Linearly polarized XUV is required for angle-resolved photoelectron spectroscopy (ARPES). With conventional light sources, which can easily be linearly polarized, the photoelectron spectrometer is normally rotated with respect to the polarization vector of the XUV. However, coherent XUV can have its polarization trivially rotated simply by rotating the polarizations of the fundamental light. Since broadband UV half-wave rotators (Fresnel rhombs) are readily available, it is very simple to rotate the input polarization, thus altering the XUV polarization. A simple single-reflection polarization analyzer (88) can be employed after the ionization region to measure the orientation and degree of polarization of the coherent XUV in case there is a reduction in polarization from optical elements, such as the input window before the XUV generation region or the diffraction grating in the spectrometer after the XUV is generated.

For photoionization with polarized light, the photoelectron angular distribution can be expressed as (35):

$$\sigma(\theta) = \frac{\sigma}{4\pi} \left[1 + \frac{\beta}{4} (3\mathcal{P} \cos(2\theta) + 1) \right] \tag{5.6}$$

where σ is the total cross section, and β is the asymmetry parameter, which ranges from -1 to $+2$. The polarization of the XUV, $\mathcal{P}$, is defined in terms of a coordinate system with the XUV propagation direction defined as the z axis, the major polarization axis, the x axis, and the minor polarization axis, the y axis.

$$\mathcal{P} = \frac{I_x - I_y}{I_x + I_y} \tag{5.7}$$

where I_x is the intensity of the x-polarized component of the XUV, and I_y is the y-polarized component. The polarization varies between 0 and 1. Using the same coordinate system, θ is defined as the angle between the x axis and the photoelectron momentum, measured in the x–y plane. The value of β provides additional information about the photoionization dynamics (89), and can vary sharply as the photon energy is tuned across an autoionizing resonance.

This variation in β can be seen in Fig. 5.17, where the total photoionization cross section and β parameter for Kr are plotted as a function of wavelength in the region around the 8s and 6d autoionizing resonances in Kr (32). The experimental values of β vary from +0.4 to −0.6 across the narrow 8s resonance, a change of 3 meV in photon energy. The data shown in Fig. 5.17 could not be measured using current synchrotron sources, but are easily obtained with coherent XUV. Also plotted on the figure is the result of a recent ab initio calculation of

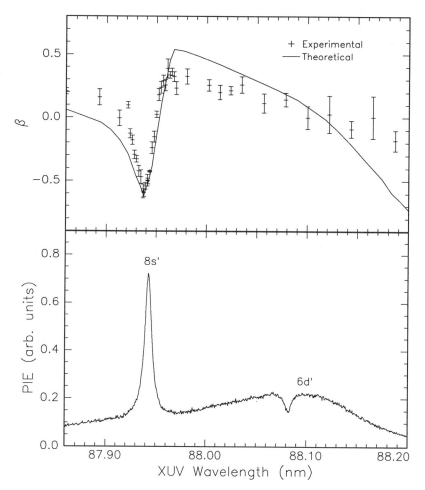

Figure 5.17. Photoionization efficiency (PIE) spectrum of Kr in the region of the 8s and 6d autoionizing resonances (bottom) and the measured photoelectron anisotropy parameters over the same wavelength range (top). The dip in the PIE spectrum at 880.0 Å is an artifact of the power normalization. Theoretical β values from Ref. 90.

the variation of β with wavelength (90), showing reasonable, but not perfect, agreement with the experimental data.

In molecules, the photoionization spectrum is much more complex, and generally there is a less dramatic change in the asymmetry as one tunes the photon energy across autoionizing resonances. Nonetheless, the β parameter is still a strong function of wavelength in the region of autoionizing resonances, which, as discussed in Section 5.3, can be exceedingly narrow in molecules. A striking example of this is given by recent work done by White and co-workers (91) on autoionizing states just above the lowest ionization threshold in oxygen. By using coherent XUV for excitation together with TOF photoelectron spectroscopy, they looked at the variation of the asymmetry parameter with wavelength over the photon energy range between the $v^+ = 1$ and $v^+ = 2$ limits in O_2, a range that includes a number of $H\ ^3\Pi_u$ autoionizing resonances. The results show highly structured, strong variations in the β parameter as the photon energy passes through an autoionizing resonance.

A further example of this variation in β parameter is provided by our results on nitrogen (32). Nitrogen has a complex autoionizing resonance close to the $v^+ = 1$ threshold (92), which results from the interaction between a low principal quantum number Rydberg state and high principal quantum number Rydberg states converging to the $v^+ = 1$ limit. The result is a very complex pattern of resonances in the PIE spectrum, as shown in Fig. 5.18. One can see in the inset that the resonance

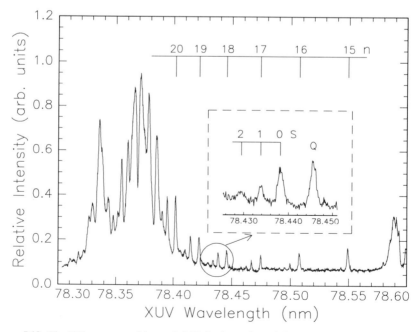

Figure 5.18. The PIE spectrum of jet cooled N_2 in the region of the complex resonance. Expanded version of one of the Rydberg resonances is shown as an inset, with the rotational assignment given.

structure is very sharp, and the rotational structure is resolved. Theoretical studies on complex resonances in CO have shown that the β parameter should show rapid oscillations in the region of complex resonances (93). Before the advent of narrow-band coherent XUV sources, observation of these oscillations in single-photon ionization was impossible. The data shown in Fig. 5.19, although preliminary, clearly shows the capability of making such measurements with a coherent XUV light source. The data in this figure represents the measured β parameter as a function of wavelength in the region of one of the Rydberg resonances shown in Fig. 5.18.

5.6. CONCLUSIONS

Hopefully, this brief glimpse at some of the work being done in Chemical Physics with coherent VUV and XUV light sources has provided some insight into the capabilities of these sources. This chapter is not meant as a review, but as examples

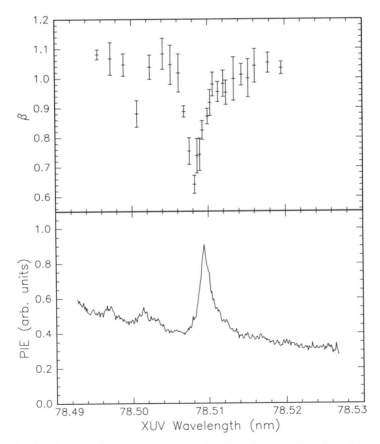

Figure 5.19. Photoelectron anisotropy parameter as a function of wavelength in the region of one of the autoionizing resonances in N_2 shown in Fig. 5.18.

of the current state-of-the-art in short wavelength laser applications drawn from work in our laboratory. At this point, I would like to gaze into the future and make some obvious predictions about coming developments in short wavelength coherent sources.

This chapter has been focused on nanosecond pulsed lasers, which currently represents the vast majority of work at short wavelengths. As stated earlier, femtosecond coherent sources with the same photon energy tuning range are possible in principle, although to the best of my knowledge have not yet been used. Picosecond work has been done, a nice example of which are experiments on time resolved semiconductor photoemission (94). Given the universal nature of photoionization, this would seem to be a natural area for development of detectors with femtosecond time resolution, which could be applied to a wide range of dynamics experiments.

As stated earlier, the tuning range of current four-wave mixing sources extends up to photon energies of 19 eV. Although this range can be extended through the use of higher order processes, such as six-wave mixing, our experience has been that the loss in photon flux is several orders of magnitude, compared with four-wave mixing. A more likely extension of the tuning range will come from developments in frequency mixing crystals, which promise to extend the tuning range of pulsed laser systems to 1500 Å. This extension of the tuning range would allow much higher values of ν_1 to be used in nonlinear media, such as Ar and Ne, and would result in a potential tuning range to higher than 24-eV photon energy with photon fluxes comparable to those used for the experiments discussed in this chapter. This seems a likely development within the next 5 years, and even better may be achieved if the crystal tuning range can be extended below 1500 Å.

The coherent sources discussed all produce intense pulses of XUV, at a low repetition rate. While this is good for the experiments covered, it makes coincidence experiments difficult, because of multiple ionization events occurring within a single light pulse. For coincidence experiments, a CW or pseudo-CW light source, such as a synchrotron, is much more useful. Ideally, one would like to preserve the photon flux (in photons per second), but greatly increase the repetition rate, meaning far fewer photons per pulse. For the experiments presented in this chapter, a repetition rate of tens of kilohertz or higher would allow for coincidence measurements following coherent XUV excitation, with 100 kHz to 1 MHz being ideal. High-power laser systems with these repetition rates are being developed, and this will provide for even more possibilities for coherent XUV experiments. An important area of application for high-repetition rate XUV laser systems will be in surface photoemission, where space charge effects are much worse than in the gas phase, but high average photon fluxes are still necessary.

Finally, developments in pulsed laser technology, including optical parametric oscillators, and convenient single mode lasers will have an important impact on this field, as tunable pulsed lasers will have improved performance, greater ease of operation, and potentially lower cost in the longer run. Given the demonstrated capabilities, and the possibilities for the technical improvements in the coming years, it seems certain that coherent sources will become the light source of choice in the photon energy range below 25 eV, and have the same impact on the VUV and XUV as visible and IR lasers have had at longer wavelengths.

REFERENCES

1. A good introduction to several aspects of applications of VUV to chemical physics is Ng, C. Y. (Ed.). *Vacuum Ultraviolet Photoionization and Photodissociation of Molecules and Clusters*, World Scientific, Singapore, 1991.

2. One of the classical texts describing experimental techniques in this spectral range is Samson, J.A.R. *Techniques of Vacuum Ultraviolet Spectroscopy*, Wiley, New York, 1967.

3. Radler, K. and Berkowitz, J. *J. Opt. Soc. Am.* **1978**, *68*, 1181. This comparison is not true for third generation synchrotron sources now becoming available.

4. These figures are based on our experience at the SA63 beamline at Super-ACO.

5. Data for ALS performance taken from *An ALS Handbook*, LBL publication PUB-643 (April, 1989), and: *ALS Beamlines for Independent Investigators*, LBL publication PUB-3104 (August, 1992).

6. McIlrath, T. J., in *Laser Techniques for Extreme Ultraviolet Spectroscopy*, McIlrath, T. J. and Freeman, R. R. (Eds.). AIP Conference Proceedings No. 90, American Institute of Physics, New York, 1982, pp. 9–18

7. Hepburn, J. W., in *Vacuum Ultraviolet Photoionization and Photodissociation of Molecules and Clusters*, Ng, C. Y. (Ed.). World Scientific, Singapore, 1991, pp. 435–486.

8. Kung, A. H. and Lee, Y. T., in *Vacuum Ultraviolet Photoionization and Photodissociation of Molecules and Clusters*, Ng, C. Y. (Ed.), World Scientific, Singapore, 1991, pp. 487–502.

9. Hanna, D. C., Yuratich, M. H. and Cotter, D. *Nonlinear Optics of Free Atoms and Molecules*, Springer-Verlag Berlin, 1979.

10. Vidal, C. R., in *Tunable Lasers*, L. F. Mollenauer and J. C. White (Eds.). *Topics in Applied Physics* series; Springer-Verlag Berlin, 1979.

11. Hepburn, J. W., *Isr. J. Chem.* **1984**, *24*, 273.

12. Hilbig, R., Hilber, G., Lago, A., Wolff, B., and Wallenstein, R. *Comm. At. Mol. Phys.* **1986**, *18*, 157.

13. At very high laser intensities, one can generate very high harmonics of the fundamental radiation. See Ivanov, M. Yu, and Corkum, P. B. *Phys Rev. A* **1993**, *58*, 580; Zuo, T., Chelkowski, S., and Bandrauk, A. *Phys. Rev. A* **1993**, *48*, 3837.

14. Timmermann, A. and Wallenstein, R., *Opt. Lett.* **1983**, *8*, 517.

15. Hilbig, R. and Wallenstein, R. *IEEE J. Quant. Electron.* **1981**, *QE-17*, 1566.

16. Lago, A. Ph.D. Thesis, Universität Bielefeld, Bielefeld, Germany, 1987.

17. Rettner, C. T., Marinero, E. E., Zare, R. N., and Kung, A. H. *J. Phys. Chem.* **1984**, *88*, 4459.

18. Tonkyn, R. G. and White, M. G. *Rev. Sci. Instrum.* **1989**, *60*, 1245.

19. Harris, S. E. and Bloom, D. M. *Appl. Phys. Lett.* **1974**, *24*, 229.

20. Hodgson, R. T., Sorokin, P. P., and Wynne, J. J. *Phys. Rev. Lett.* **1974**, *32*, 343.

21. Zdasiuk, G. and Wallace, S. C. *Appl. Phys. Lett.* **1976**, *28*, 449.

22. Hilbig, R. and Wallenstein, R. *IEEE J. Quant. Elect.* **1983**, *QE-19*, 1759.

23. Hermann, P. R. and Stoicheff, B. P. *Opt. Lett.* **1985**, *10*, 502.

24. Muller, C. H., Lowenthal, D. D., DeFaccio, M. A., and Smith, A.V. *Opt. Lett.* **1988**, *30*, 651.

25. Bartz, J. A., Barnhart, T. M., Galloway, D. B., Huey, L. G., Glenewinkel-Meyer, T., McMahon, R. J., and Crim, F. F. *J. Am. Chem. Soc.* **1993**, *115*, 8389.

26. Krautwald, H. J., Schnieder, L., Welge, K. H., and Ashfold, M. N. R. *Faraday Disc. Chem. Soc.* **1986**, *82*, 99.

27. Yen, M., Johnson, P. M., and White, M. G. *J. Chem. Phys.* **1993**, *99*, 126.

28. Sivakumar, N., Hall, G. E., Houston, P. L., Hepburn, J. W., and Burak, I. *J. Chem. Phys.* **1988**, *88*, 3692.

29. Burak, I., Hepburn, J. W., Sivakumar, N., Hall, G. E., Chawla, G., and Houston, P. L. *J. Chem. Phys.* **1987**, *86*, 1258.

30. Hilbig, R. and Wallenstein, R. *IEEE J. Quantum Elec.* **1983**, *QE-19*, 194.

31. Hilber, G., Lago, A., and Wallenstein, R. *J. Opt. Soc. Am. B* **1987**, *4*, 1753.

32. Kong, W. Ph.D. Thesis, University of Waterloo, 1993.

33. Srinivasan, T., Egger, H., Pummer, H., and Rhodes, C. H. *I.E.E.E. J. Quantum Electron.* **1983**, *QE-19*, 1270.

34. Dehmer, P. M. and Chupka, W. A. *J. Chem. Phys.* **1975**, *62*, 4525.

35. Berkowitz, J. *Photoabsorption, Photoionization, and Photoelectron Spectroscopy*, Academic, New York, 1979.

36. Guo, J. M.Sc. Thesis, University of Waterloo, 1994.

37. Berry, R. S. *Phys. Rev. A* **1970**, *1*, 383.

38. Ono, Y., Linn, S. H., Prest, H. F., Ng, C. Y., and Miescher, E. *J. Chem. Phys.* **1980**, *73*, 4855.

39. Giusti-Suzor, A. and Jungen, Ch. *J. Chem. Phys.* **1984**, *80*, 986.

40. Raoult, M. *J. Chem. Phys.* **1987**, *87*, 4736.

41. Milburn, D., Hart, D. J., and Hepburn, J. W., *Short Wavelength Coherent Radiation: Generation and Applications*, Vol. 2: OSA, Washington, 1988; pp. 384.

42. Hart, D. J. and Hepburn, J. W. *Chem. Phys.* **1989**, *129*, 51

43. Huth-Fehre, T., Mank, A., Drescher, M., Böwering, N., and Heinzmann, U. *Phys. Scr.*, **1990**, *41*, 454.

44. Lefebvre-Brion, H., Giusti-Suzor, A., and Raseev, G. *J. Chem. Phys.*, **1985**, *83*, 1557.

45. Mank, A., Drescher, M., Huth-Fehre, T., Böwering, N., Heinzmann, U., and Lefebvre-Brion, H. *J. Chem. Phys.* **1991**, *95*, 1676.

46. Lefebvre-Brion, H. *J. Chem. Phys.* **1990**, *93*, 5898.

47. Müller-Dethlefs, K., Sander, M., and Schlag, E. W. *Chem. Phys. Lett.* **1984**, *112*, 291.

48. Müller-Dethlefs, K. and Schlag, E. W. *Annu. Rev. Phys. Chem.* **1991**, *42*, 109.

49. Wang, K., Stephens, J. A., and McKoy, V. *J. Phys. Chem.* **1993**, *97*, 9874.

50. Chupka, W. A. *J. Chem. Phys.* **1993**, *99*, 5800.

51. Scherzer, W. G., Selzle, H. L., Schlag, E. W., and Levine, R. D. *Phys. Rev. Lett.* **1994**, *72*, 1435.

52. Zhu, L. and Johnson, P. M. *J. Chem. Phys.* **1991**, *94*, 5769.

53. Chupka, W. A. *J. Chem. Phys.* **1993**, *98*, 4520.

54. Pratt, S. T. *J. Chem. Phys.* **1993**, *98*, 9241.

55. Tonkyn, R. G., Winnizcek, J. W., and White, M. G., *Chem. Phys. Lett.* **1989**, *164*, 137.

56. Merkt, F. and Softley, T. P. *J. Chem. Phys.* **1991**, *96*, 4149.

57. Wiedmann, R. T., Grant, E. R., Tonkyn, R. G., and White, M. G. *J. Chem. Phys.* **1991**, *95*, 746.

58. Tonkyn, R. G., Wiedmann, R., Grant, E. R., and White, M. G. *J. Chem. Phys.* **1991**, *95*, 7033.

59. (a) Ogawa, M. and Ogawa, S. *J. Mol. Spectrosc.* **1972**, *41*, 393. (b) Huber, K. P. (private communication).

60. Wang, K., McKoy, V., Ruf, M. W., Yencha, A. J., and Hotop, H. *J. Electron. Spec. Relat. Phenom.* **1993**, *63*, 11.

61. Mank, A., Rodgers, D., and Hepburn, J. W. *Chem. Phys. Lett.* **1994**, *219*, 169.

62. Kong, W., Rodgers, D., Hepburn, J. W., Wang, K., and McKoy, V. *J. Chem. Phys.* **1993**, *99*, 3159.

63. Blush, J. A., Chen, P., Wiedmann, R. T., and White, M. G. *J. Chem. Phys.* **1993**, *798*, 3557.

64. Merkt, F. and Softley, T. P. *Phys. Rev. A.* **1992**, *46*, 302.

65. Tonkyn, R. G., Wiedmann, R. T., and White, M. G. *J. Chem. Phys.* **1992**, *96*, 3696.

66. Lee, M. T., Wang, K., McKoy, V., Tonkyn, R. G., Wiedmann, R. T., Grant, E. R., and White, M. G. *J. Chem. Phys.* **1992**, *96*, 7848.

67. Merkt, F., Fielding, H. H., and Softley, T. P. *Chem. Phys. Lett.* **1992**, *202*, 153.

68. A good review on this topic is given in Ref. 49.

69. Buckingham, A. D., Orr, B. J., and Sichel, J. M. *Philos. Trans. R. Soc. London Ser. A*, **1970**, *268*, 147.

70. Xie, J. and Zare, R. N. *J. Chem. Phys.* **1992**, *97*, 2891.

71. Dixit, S. N. and McKoy, V., *Chem. Phys. Lett.* **1986**, *128*, 49.

72. Wang, K. and McKoy, K. *J. Chem. Phys.*, **1991**, *95*, 4977.

73. Kong, W., Rodgers, D., and Hepburn, J. W. *J. Chem. Phys.* **1993**, *99*, 8571.

74. Kong, W., Rodgers, D., and Hepburn, J. W., in *Laser Techniques for State Selected and State-to-State Chemistry*, C. Y. Ng (Ed.). *SPIE Conference Proceedings No. 1858* (SPIE, 1993) pp. 207–216.

75. Braunstein, M., McKoy, V., Dixit, S. N., Tonkyn, R. G., and White, M. G. *J. Chem. Phys.* **1990**, *93*, 5345.

76. Ferreira, L. P., Ph.D. Thesis, Université Paris-Sud, 1984.

77. Merkt, F. Merkt, Guyon, P. M., and Hepburn, J. W. *Chem. Phys.* **1993**, *173*, 479.

78. Ellis, K., Ph.D. Thesis, University of Manchester, 1992.

79. Rabalais, J. W. *Principles of Ultraviolet Photoelectron Spectroscopy*, Wiley, New York, 1977.

80. Smith, A. L. *Philos. Trans. R. Soc. London*, **1970**, *A268*, 169.

81. Coxon, J. A. and Haley, M. P. *J. Mol. Spectrosc.* **1984**, *108*, 119.

82. Mank, A., Nguyen, T., Martin, J. D. D., and Hepburn, J. W. *Phys. Rev. A*, **1995**, *51*, R1.

83. Huth-Fehre, T., Mank, A., Drescher, M., Böwering, N., and Heinzmann, U. *Phys. Rev. Lett.* **1990**, *64*, 396.

84. Schäfers, F., Peatman, W., Eyers, A., Heckenkamp, Ch., Schönhense, G., and Heinzmann, U. *Rev. Sci. Instrum.* **1986**, *57*, 1032.

85. Halbach, K. *Nucl. Instrum. Methods*, **1981**, *187*, 109.

86. Kim, K. J. *Nucl. Instrum. Methods*, **1984**, *219*, 425.

87. Mank, A., Ph.D. Thesis, Fakultät für Physik, Universität Bielefeld, 1991.

88. Rabinovitch, K., Canfield, L. R., and Madden, R. P. *Appl. Opt.* **1965**, *4*, 1005.

89. Dill, D. *Phys. Rev. A*, **1973**, *7*, 1976.

90. Johnson, W. R., Chang, K. T., Huang, K. N., and Le Dourneuf, M. *Phys. Rev. A*, **1980**, *22*, 989.

91. Tonkyn, R. G., Winniczek, J. W., and White, M. G. *J. Chem. Phys.* **1989**, *91*, 6632.

92. Giusti-Suzor, A. and Lefebvre-Brion, H. *Phys. Rev. A*, **1984**, *30*, 3057.

93. Leyh, B. and Raseev, G. *J. Chem. Phys.* **1988**, *89*, 820.

94. Bucksbaum, P. H., Bokor, J., Haight, R., and Freeman, R. R., in *Optical Science and Engineering Series 7: Short Wavelength Coherent Radiation: Generation and Applications*, Attwood, D. T. and Bokor, J. (Eds.). AIP Conference Proceedings No. 147, AIP, New York, 1986, pp. 401–411.

Chapter **VI**

TIME-RESOLVED RESONANCE RAMAN IN THE VISIBLE AND ULTRAVIOLET: TECHNIQUES AND APPLICATIONS

Peggy A. Thompson and Richard A. Mathies
Department of Chemistry,
University of California at Berkeley, Berkeley, California

Laser Techniques In Chemistry, Edited by Anne B. Myers and Thomas R. Rizzo.
Techniques of Chemistry Series, Vol. XXIII.
ISBN 0-471-59769-4 © 1995 John Wiley & Sons, Inc.

6.1. INTRODUCTION

The most widely used experimental techniques for studying reaction dynamics in solution have been time-resolved absorption and emission spectroscopies. These techniques have provided a vast amount of kinetic information about electron-transfer, proton-transfer, and isomerization reactions by monitoring the excited-state lifetimes or rates of product formation. However, because the electronic spectra of most polyatomic molecules in solution are broad and structureless, it is difficult to obtain detailed structural information about the mechanistic intermediates. Also, the evolution of the electronic absorption spectrum does not easily distinguish between changes in the electronic structure and vibrational energy redistribution.

The high selectivity and structural specificity of time-resolved Raman spectroscopy makes it possible to obtain much more detailed information on reaction dynamics. In a time-resolved Raman experiment, a transient molecular species characterized by a unique set of vibrational modes is formed by a pump pulse and the time evolution of its vibrational spectrum is monitored. This technique provides a wealth of information about both the vibrational structure and dynamical properties of chemical and biological systems. For example, structural changes can readily be observed by monitoring the frequencies of vibrational modes. Kinetic information is obtained by resolving the intensity change of the vibrational bands as a function of time. In addition, microscopic molecular interactions can be examined by studying the dynamics of molecular vibrational modes under different environmental conditions. Comparison of time-resolved Stokes and anti-Stokes Raman scattering gives direct information about the population of excited vibrational states and provides a method for determining vibrational energy relaxation pathways. These factors have been the driving force behind the development of time-resolved Raman spectrometers.

Picosecond Raman spectroscopy was initially developed by Delhaye (1) in 1975, who used a single 25-ps laser pulse to generate Raman spectra of stable species. Resonance Raman spectra of short-lived transients were reported in 1980 where a single laser pulse was used for both photolyzing and probing the Raman scattering (2,3). Although a single-pulse experiment gives structural information about the species present during the laser pulse, kinetic information is only indirectly obtained. A two-pulse, pump–probe technique was developed by Gustafson et al. (4) in 1983 for exploring the dynamics of chemical reactions in solution. Since this time, several groups have designed Raman spectrometers with time resolutions ranging from a few picoseconds (5–13) to 500–800 fs (14,15).

There are several requirements that need to be met when designing a picosecond or subpicosecond Raman spectrometer. First, the laser must be able to produce a Raman spectrum with a high signal-to-noise (S/N) ratio. A high repetition rate helps to meet this requirement. Generally speaking, one desires a repetition rate that is as high as possible; however, the repetition rate must not be so high that the sample cannot be replaced (or relaxed) between pump–probe cycles. In addition, there must be enough energy in the pump pulses to initiate significant photochemistry. Care must be taken not to use pulses with too high peak powers since unwanted nonlinear processes can be initiated. An average probe beam energy of approximately 10–100 mW is typically necessary for nonresonance Raman scattering, and 1–10 mW is required when on-resonance with an electronic transition. Second, if resonance enhancement is taken advantage of, it is necessary to have a laser source that is tunable throughout a wide frequency range. For example, many organic molecules and aromatic amino acids have absorption transitions in the UV, whereas biological systems have chromophores with absorption transitions in the visible to near-infrared (IR). The fact that there are many molecular systems that can be studied in the UV puts an additional requirement on these laser systems to provide enough energy in the visible range to allow efficient frequency doubling and tripling. The last requirements are the spectral and temporal resolution of the laser. The bandwidth of the laser must not be broader than the bandwidth of the Raman bands. A gain in frequency resolution will result in loss of time resolution. For transform-limited, Gaussian pulses, the relationship between the spectral and temporal widths is given by

$$\Delta\omega * \Delta t \cong 0.44 \qquad (6.1)$$

where $\Delta\omega$ refers to the spectral width and Δt is the pulse width. Therefore, with 10-cm^{-1} resolution, this puts a limit of approximately 1.5 ps on the pulse width.

High time resolution is essential if the dynamics of ultrafast reactions in solution are to be examined. For instance, from time-resolved optical and vibrational spectroscopies it is known that the photoisomerization of *cis*- and *trans*-stilbenes in solution occurs within 150 fs and 80 ps, respectively; the initially prepared excited state in pericyclic reactions depopulates in tens of femtoseconds, vibrational energy relaxation can range from nanoseconds to subpicoseconds depending on the molecular system, the isomerization of retinal in rhodopsin is complete in 200 fs, and ligand photodissociation of heme proteins occurs in less than 50 fs. Many of the molecular systems just mentioned have recently been studied using time-resolved Raman spectroscopy. This chapter will survey the various techniques for performing time-resolved UV and visible resonance Raman spectroscopy and will present examples of the information that can be obtained from picosecond–subpicosecond Raman spectroscopy.

6.2. THEORY

6.2.1. Raman Scattering

Raman scattering occurs when some of the photon's energy is transferred to the molecule or vice versa. A frequency shift to lower energy is called a Stokes shift. Scattering from vibrationally excited molecules results in a frequency shift of the scattered photon to higher energy (an anti-Stokes shift). There are several good review articles that describe the theory of Raman scattering (16–20). The probability of Raman scattering from an arbitrary initial state I to a final state F corresponds to the molecule's Raman cross section $\sigma_{I \rightarrow F}$:

$$\sigma_{I \rightarrow F} = \frac{8\pi e^4}{9\hbar^4 c^4} E_s^3 E_L \left| \sum_{\rho\lambda} (\alpha_{\rho\lambda})_{I \rightarrow F} \right|^2 \quad (6.2)$$

where c is the velocity of light, E_s and E_L are the energies of the scattered and incident light, respectively, and $\alpha_{\rho\lambda}$ is the transition polarizability tensor with incident and scattered polarizations indicated by ρ and λ, respectively. Second-order perturbation theory gives the following expression for the polarizability:

$$(\alpha_{\rho\lambda})_{I \rightarrow F} = \sum_V \left(\frac{\langle F|m_\rho|V \rangle \langle V|m_\lambda|I \rangle}{E_V - E_I - E_L - i\Gamma} + \frac{\langle F|m_\lambda|V \rangle \langle V|m_\rho|I \rangle}{E_V - E_F + E_L - i\Gamma} \right) \quad (6.3)$$

where E_L is the incident photon energy, $|I\rangle$, $|V\rangle$, and $|F\rangle$ are the initial, intermediate, and final vibronic states, m_ρ and m_λ are the dipole moment operators, and Γ is the homogeneous line width of the electronic transition.

By invoking the Born–Oppenheimer and Condon approximations, the complete expression for the resonance Raman cross section is given by

$$\sigma_{i \rightarrow f} = 5.87 \times 10^{-19} M^4 E_s^3 E_L \left| \sum_V \frac{\langle f|v \rangle \langle v|i \rangle}{\varepsilon_v - \varepsilon_i + E_0 - E_L - i\Gamma} \right|^2 \quad (6.4)$$

where E_0 is the zero–zero energy separation between the lowest vibrational levels of the ground and excited states, and $|v\rangle$ and $|i\rangle$ are vibrational states with energies ε_v and ε_i (see Fig. 6.1). The energies E_L, E_s, E_0, ε_i, and ε_v are in reciprocal centimeters (cm^{-1}), M is in angstroms (Å), and σ is in angstroms squared per molecule (Å^2/molecule).

This is the Albrecht A-term expression, which is the dominant contribution to resonance enhanced scattering from allowed electronic transitions (21). Resonance enhancement occurs when the incident photon energy E_L approaches that of an allowed electronic transition ($E_0 + \varepsilon_v - \varepsilon_i$) and the first term in Eq. (6.3) becomes dominant. This effect is due to the coupling between the electronic and vibrational modes. The vibrational modes that get enhanced are localized on the group of atoms, that is, the chromophore, that gives rise to the electronic transition. Only totally symmetric modes can contribute to fundamental A-term scattering. The relative intensities depend on both the oscillator strength and the Franck–Condon overlaps.

Resonance enhancement can be very advantageous. First, Raman cross sections can be increased by orders of magnitude ($10^3 - 10^6$). This allows species to be detected with concentrations in the range of 10^{-6} M. Second, by choosing a wavelength that is on-resonance with an allowed electronic transition, vibrational modes that contribute to the absorption band are enhanced and undesirable scattering is rejected. Third, the resonance effect is very useful for investigating complex vi-

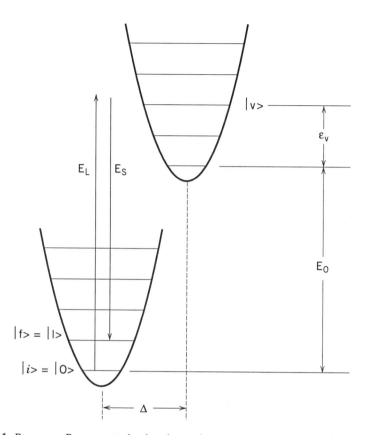

Figure 6.1. Resonance Raman scattering in a harmonic system. The initial system $|i\rangle$ is excited by incident light of energy E_L that is resonant with the set of virtual levels $|v\rangle$. Scattering of a photon of energy E_S raises the system from $|i\rangle$ to $|f\rangle$.

brational systems like proteins. Since there are $3N - 6$ normal modes of vibration, interpretation of protein spectra can be very difficult; however, resonance enhancement allows the vibrational modes due to selected chromophores to be enhanced and monitored.

6.2.2. Photoalteration Parameter

Picosecond and subpicosecond laser systems are being used more often for resonance Raman experiments because they provide the time resolution needed to monitor ultrafast photochemical reactions. In order to use high-energy laser systems as spectroscopic tools to study photochemically reactive systems, it is important to ensure that the molecule or state is not perturbed. This is especially important in resonance Raman spectroscopy since the probability that a molecule will absorb a photon can be 10^7 times greater than the probability of Raman scattering.

In the case of pulsed excitation the fraction of molecules photolyzed is given by the photoalteration parameter F, (22),

$$F = \frac{2303E\varepsilon\varphi}{\pi r^2 N_A} \tag{6.5}$$

where F is the fraction of molecules photolyzed by a single pulse with energy E (photons), ε is the molar extinction coefficient ($M^{-1}cm^{-1}$), φ is the photochemical quantum yield, and r is the radius of the focused laser beam (cm).

It is necessary to keep the probe pulse intensity low ($F < 0.2$) to ensure that no more than about 10% of the molecules in the illuminated volume absorb a photon, as given by Eq. (6.5). Also, for systems that do not rapidly relax back to their initial state, it is desirable to flow or rotate the sample so that it can be replaced between pump–probe cycles. For pump pulses, $F > 1$ is needed to initiate the photolysis. Better S/N can be achieved with a higher repetition rate; however, if the repetition rate is too high it becomes difficult to produce enough total energy in each pump pulse to sufficiently photolyze the sample.

6.3. INSTRUMENTAL

6.3.1. One-Pulse Raman Spectrometers

Picosecond time-resolved Raman experiments can be categorized as either single- or two-pulse experiments. In a single-pulse experiment, the same pulse photolyzes the sample and produces the Raman scattering. In this procedure, a Raman spectrum is first produced with a low power ($F < 0.2$) beam. Then a higher power ($F > 1$) beam is used to generate photoproduct. The "difference" spectra are generated by subtracting a low-power spectrum from a high-power spectrum. The single-pulse method has been used by van den Berg et al. (14,23) to determine the resonance Raman spectrum of early intermediates in the photocycle of bacteriorhodopsin and in the photodissociation of carbonmonoxy- and oxyhemoglobin. The pulse duration of their Raman spectrometer was 800–900 fs with a bandwidth ranging from 16–18 to 22–25 cm^{-1} for the two experiments, respectively.

There are several disadvantages with this technique. First, since the time delay between the "pump" and "probe" pulses cannot be adjusted, this method provides only Raman scattering from whatever is present during the pulse. Second, care must be taken not to generate artifacts when taking the high-power spectrum. This is due to the fact that unwanted interactions between molecules and the incident field can occur because of high peak powers from high F laser pulses. Possibilities that are likely to perturb the spectrum are thermal heating of the reactant, multiphoton absorption and ionization, and repumping the photoproduct. Third, a bandwidth of 22–25 cm^{-1} is approximately two times the transform-limited bandwidth of a 800–900-fs pulse. Therefore, even though this time scale comes close to monitoring the subsequent events after photoinitiation of the reaction, the broad bandwidth makes it difficult to resolve small frequency shifts of the Raman bands.

6.3.2. Two-Pulse Raman Spectrometers

The two-pulse configuration is the best method for obtaining the kinetic and structural detail needed for monitoring dynamical events in a photoinitiated reaction. In this case, the pump pulse photolyzes the sample initiating the reaction, and the probe pulse produces the Raman scattering of the photoproduct. The pump and probe pulses are temporally separate, allowing the time delay to be independently adjusted, as well as spectrally separate so that the wavelength can be changed to simultaneously optimize pump photolysis and resonance enhancement of the photoproduct.

An example of a two-pulse picosecond Raman spectrometer is shown in Fig. 6.2. (10). The design criteria were based on the desire to have about 1-ps pulses for probing ultrafast reaction dynamics and structural changes, and the ability to generate high intensity ultraviolet (UV) and visible tunable pulses. This system consisted of an extended cavity dye laser (Coherent 590) synchronously pumped

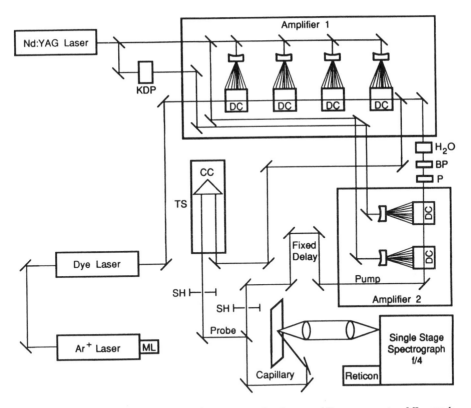

Figure 6.2. Schematic diagram of a two-pulse, pump–probe picosecond Raman apparatus. ML = mode locker, KDP = potassium dihydrogen phosphate doubling crystal, DC = prism dye cell, BP = band-pass filter, TS = translation stage, CC = corner cube, SH = shutter, P = polarizer. [Reprinted with permission from S. J. Doig, P. J. Reid, and R. A. Mathies, *J. Phys. Chem.* **1991**, *95*, 6372. Copyright © (1991) American Chemical Society.]

with 360–400 mW from a mode-locked argon ion laser (Spectra-Physics 2040E) operating at 80 MHz. The bandwidth and wavelength were controlled using a three-plate birefringent filter. The dye laser could be operated between 550 and 600 nm and generated 200 pJ per pulse (15–30 mW) with a pulse width of 2-3 ps (fwhm) and spectral width of 7 cm^{-1}.

The low-energy dye laser pulses were amplified in a four-stage dye amplifier pumped with a total energy of 170 mJ per pulse at 532 nm from a 50-Hz Q-switched Nd:YAG laser (Continuum YG581C-50). The gain cells were Bethune style cells constructed from 90° prisms of quartz or BK-7 glass (24). Amplified pulse energies of 1–2 mJ per pulse with a pulse amplitude jitter of less than 10% were obtained in the 565–600-nm range using different combinations of dyes. The output of the amplifier was split into a pump and probe beam with a 80:20 beam splitter and the probe beam was directed to an optical delay line.

This laser system can be operated in any of the following two-color pump–probe configurations: visible–visible, visible–UV, or UV–UV. By frequency doubling the dye amplifier output in a nonlinear crystal (KDP or β-BBO), UV pulses are generated for two-color visible-UV Raman spectroscopy. For two-color UV–UV experiments, the dye laser fundamental as well as visible pulses with a wavelength different from the fundamental are frequency doubled. Generation of visible pulses at different wavelengths is described below.

One method for creating visible pulses at wavelengths different from the dye laser fundamental is continuum generation. A variety of media have been used, (acetone, H_2O, D_2O, or $CHCl_3/CCl_4$ at a 6:5 v/v ratio), which generate emission from less than 400 nm to greater than 1 μm. Using 0.8-mJ incident pulses at 590 nm, the continuum intensity is peaked at the incident frequency and rapidly decays. We have measured 0.4 μJ per pulse per 10 nm at 490 nm and 3 μJ per pulse per 10 nm at 550 nm. The best spatial and focusing properties have been found using the $CHCl_3/CCl_4$ mixture. The wavelength of choice was selected with a 10-nm bandpass filter. These approximately 1-μJ pulses were subsequently amplified in a two-stage dye amplifier pumped with 50 mJ per pulse of residual 532 nm from the Q-switched Nd:YAG generating 0.6 mJ per pulse of amplified continuum. It should be noted that continuum generation results in spectrally broad pulses that cannot be used as probe pulses in a Raman experiment. However, these spectrally broadened pulses can be used as the pump source. Either the fundamental from the dye laser or its doubled harmonic can be used as probe pulses to maintain a spectral resolution of a few wavenumbers.

Raman shifting has also been used for generating two-color visible picosecond pulses. Stokes-shifted lines can be efficiently (~20% conversion) generated by Raman shifting in liquids. Anti-Stokes lines in liquids are generated with less conversion efficiency (2–5%) and with large divergence angles resulting in poor focusing qualities; therefore these pulses cannot be efficiently amplified. The divergence problem accompanying anti-Stokes generation can be overcome by Raman shifting in gases. Conversion efficiencies of approximately 1% in 1000 psi of H_2, 1% in 1000 psi of D_2, and 3% in 400 psi of CH_4 have been obtained. In addition to this, greater than 25% conversion efficiency was measured for the first

Stokes line in CH_4 gas. These efficiencies were obtained by focusing 1 mJ of 566 nm with a 0.5-m lens into a 0.5-m pipe. There was minimal spectral broadening in CH_4 gas for the first Stokes line (~5%), whereas for the first anti-Stokes line about 30% broadening of the bandwidth was observed.

In all experiments, the pump and probe beams were overlapped on the sample using a backscattering geometry. The samples were flowed using either quartz flow cells or a wire-guided jet to make a smooth film (25). Quartz flow cells were not used for UV probe experiments since there was substantial background scattering from the quartz. Raman scattering was collected with a 50-mm biconvex lens and imaged onto the entrance slit with a 200-mm plano convex lens f-matched to the $f/4$, 0.5-m single-stage spectrograph (Spex Model 500M, $f/4$). The spectrograph was equipped with a 1200 grooves per millimeter grating blazed at either 500 or 750 nm operating in second or third order, respectively. The dispersed light was detected with an intensified reticon (PAR Model 1421) cooled to $-35°C$. Calibrations were accurate to within 1 cm^{-1} and peak positions were accurate to ± 1 cm^{-1}.

This laser system has offered the flexibility of providing independently tunable UV–visible pump and probe pulses with high time and frequency resolution. This Raman spectrometer has been used to determine detailed kinetic and structural information on photochemical and photobiological systems as discussed below.

6.3.2A. Comparison of Two-Pulse Raman Spectrometers.

Table 6.1 summarizes the properties of the picosecond and subpicosecond two-pulse Raman spectrometers that have been developed thus far. The important characteristics of these laser systems are given along with the wavelength range and average energy that has been reported. A more detailed summary of these systems is given below along with a discussion of the advantages and disadvantages of each design.

The first report of a subpicosecond resonance Raman spectrometer was given by Petrich et al. (15) Pulses from a colliding pulse ring laser were amplified in a four-stage dye amplifier pumped by the second harmonic of a Nd:YAG, generating 1 mJ 100-fs pulses at 10 Hz. To obtain independently tunable pump–probe pulses, the technique of continuum amplification was used. This technique along with frequency doubling generates pump–probe pulses throughout the UV–visible region. Even though this Raman spectrometer provides probe pulses with 500-fs time resolution, which is desired for monitoring subpicosecond reaction dynamics, the high time resolution results in a bandwidth of 30 cm^{-1} and is much broader than the width of most Raman bands.

Recently, Iwata et al. (5) constructed a transform-limited picosecond time-resolved Raman spectrometer (see Fig. 6.3). The IR output of a mode-locked Nd:YAG was pulse compressed to about 5 ps, and frequency doubled for synchronously pumping a dye laser. The dye laser output was amplified using a CW Nd:YAG regenerative amplifier operating at 2 kHz as the pump source. Frequency doubling the amplified dye provided enough energy to be used as a UV pump source and the remaining visible fundamental could be used as the probe. This spectrometer provided transform-limited pulses with a 3-cm^{-1} bandwidth and a pulse duration of 3 ps. This system offers a tunable (UV–visible) laser source with

Table 6.1. Comparison of Two-Pulse Picosecond Raman Spectrometers

Laser System	Pulse Width	Spectral Width (cm^{-1})	Repetition Rate	Wavelength (nm)	Energy (mW)
Nd:YAG Amplified Dye Laser[a]	1.7–2.5 ps	10	50 Hz	680–726[b]	20–30
				566–600	50–100
				486–510	10–15
				283–300	5–10
				226–240[b]	1–2
Nd:YAG Amplified CPM[c]	≤500 fs	30	10 Hz	870	2
				570–620	10
				435	0.8
Nd:YAG Regen Amplified Dye Laser[d]	2–3 ps	3	2 kHz	588	20–30
				294	5–8
Nd:YLF Regen Amplified Dye Laser[e]	6 ps		500 Hz	590	10
				295	1.5–2
				278	0.4–0.6
Nd:YLF Amplified Dye Laser[f]	1.5–2 ps	15–16	1 kHz	585.5	4
				293	
Copper Vapor Amplified Cavity Dumped Dye Laser[g]	5–6 ps		4.46 kHz	550–610	4.5–54
				275–305	0.9–11
Copper Vapor Amplified Cavity Dumped Dye Laser[h]	5–6 ps		5 kHz	600	100–250
				300	10
Ar$^+$ Amplified Cavity Dumped Dye Laser[i]	3–8 ps		1 MHz	560–620	200–400
				577–650	30
				280–310	20–40
Cavity Dumped Dye Laser[j]	6–8 ps	~6	1–76 MHz	570–650	2–4
				560–570	10–15
				400–460	2–3
				287	2–3
Nd:YAG Regen Amplified Nd:YAG[k]	8 ps		1–2 kHz	532	300
				355	30

[a]Doig, S. J., Reid, P. J., and Mathies, R. A. *J. Phys. Chem.* **1991**, *95*, 6372
[b]Thompson, P. A., and Mathies, R. A. (unpublished data).
[c]Petrich, J. W., Martin, J. L., Houde, D., Poyart, C., and Orszag, A. *Biochemistry* **1987**, *26*, 7914
[d]Iwata, K., Yamaguchi, S., and Hamaguchi, H. *Rev. Sci. Instrum.* **1993**, *64*, 2140
[e]Phillips, D. L., Rodier, J. M., and Myers, A. B. *Chem. Phys.* **1993**, *175*, 1
[f]Qian, J., Schultz, S. L., Bradburn, G. R., and Jean, J. M. *J. Phys. Chem.* **1993**, *97*, 10638
[g]Matousek, P., Hester, R. E., Moore, J. N., and Parker, A. W. *Meas. Sci. Tech.* **1993**, *4*, 1090
[h]Hopkins, J. B. and Rentzepis, R. M. *Chem. Phys. Lett.* **1986**, *124*, 79
[i]Gustafson, T. L., Roberts, D. M., and Chernoff, D. A. *J. Chem. Phys.* **1983**, *79*, 1559
[j]Atkinson, G. H., Brack, T. L., Blanchard, D., and Rumbles, G. *Chem. Phys.* **1989**, *131*, 1
[k]Orman, L. K., Chang, Y. J., Anderson, D. R., Yabe, T., Xu, S., Yu, S. C., and Hopkins, J. B. *J. Chem. Phys.* **1989**, *90*, 1469

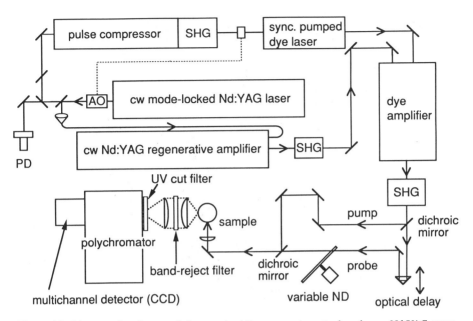

Figure 6.3. Diagram of a picosecond time-resolved Raman spectrometer based on a Nd:YAG regenerative amplifier. AO = Bragg cell, PD = photodiode, ND = neutral density filter. [Reproduced with permission from Ref. 5.]

enough energy to produce a Raman spectrum with high S/N and resolution to extract physically and chemically important information.

A complementary system was recently built by Phillips et al. (7), where a mode-locked Nd:YLF was used to synchronously pump a dye laser that was amplified by a CW Nd:YLF regenerative amplifier operating at 500 Hz. The dye laser can be operated with rhodamine dyes to generate a tunable visible source. The corresponding tunable UV pulses were produced by frequency doubling. Frequency mixing of the dye laser fundamental with the second harmonic of the Nd:YLF provided additional tunability in the UV region. The output of the mode-locked Nd:YLF was not pulse compressed resulting in a pulse duration of approximately 6 ps, slightly longer than what is needed for resolving ultrafast processes.

Three other picosecond Raman spectrometers have been reported based on a mode-locked Nd:YAG synchronously pumped dye laser. All systems can produce tunable visible and second harmonic generated UV pulses. The system designed by Qian et al. (8) amplified the dye laser output with a Q-switched Nd:YLF operating at 1 kHz as the pump laser. The pulse duration reported for this system was 1.5–2.0 ps; however, the spectral bandwidth was on the order of $15-16$ cm^{-1} indicating that the pulses were not transform limited. The two other laser systems were based on a similar design in which a train of pulses from a cavity-dumped dye laser were amplifed by a copper vapor laser-pumped dye amplifier operating at either 4.46 (6) or 5 kHz (13). Both of these systems reported pulse widths of

5–6 ps; again this does not offer the time resolution necessary for monitoring many ultrafast reaction dynamics.

A different approach for generating independently tunable pump and probe pulses has been to synchronously pump two dye lasers. This design for a picosecond Raman spectrometer has been reported by Gustafson et al. (4,9,26) and Atkinson et al. (12). In general, the laser systems consisted of a mode-locked Nd:YAG, which synchronously pumped two visible, independently tunable cavity-dumped dye lasers operating at 1 MHz. Gustafson et al. also amplified the output from one dye laser at 1 MHz with a cavity-dumped argon ion laser (9). These systems offer the advantage of having independently tunable pump–probe pulses both in the visible and UV regions. Pulse widths in the range of 3–8 ps have been reported for these systems. A disadvantage of the pump–probe pulses being derived from different dye lasers is that a timing jitter of greater than or equal to 5 ps was present between the two pulses, which ultimately limits the time resolution. Also, when operating at high repetition rates, care must be taken to make sure there is enough total energy to initiate the photolysis and that the photoproduct does not get repumped before being removed from the illuminated volume.

Another picosecond Raman spectrometer has been reported by Orman et al. (11) This system consisted of a mode-locked Nd:YAG laser that was amplified using a high repetition rate (2 kHz) Nd:YAG regenerative amplifier. Using an optical fiber and grating pair pulse compressor, pulses as short as 8 ps could be generated. Frequency doubling and tripling of the fundamental (1064 nm) generated pump or probe pulses at 532 and 355 nm. In addition to the relatively low time resolution, this experimental setup has the limitation of only providing pump–probe pulses at some harmonic of the fundamental.

Examples of how these laser systems have been applied to examine chemically and biologically important problems will be discussed in the following sections.

6.4. PHOTOCHEMICAL SYSTEMS

6.4.1. Ring-Opening Reactions

Electrocyclic ring-opening reactions are important for understanding polyene and vitamin D photochemistry (27–29). Recently, the excited-state dynamics and kinetics of the photochemical ring opening of 1,3-cyclohexadiene, 1,3,5-cyclooctatriene, and α-phellandrene have been studied (30–32). These reactions are characterized by a rapid (10–20 fs) (33,34) depopulation of the initially prepared excited state to a lower-lying optically dark excited state (35,36), followed by a 10-ps relaxation to the ground state. The interesting fact about these ring-opening reactions is that even though the stereochemistry changes depending on the number of double bonds in the ring, the product formation time is not significantly different.

The two-color UV Raman spectrometer that was used to obtain picosecond data on the photochemical ring opening of 1,3-cyclohexadiene (CHD) to *cis*-hexatriene (*cis*-HT)(30) has been presented in Fig. 6.2. A 2-nm bandpass filter centered at 550 nm was used to select the desired wavelength from a D_2O continuum. This was amplified in a two-stage dye amplifier and frequency doubled in a 1-mm piece of KDP generating 275-nm pulses that were used to initiate the photochemistry. The fundamental of the dye laser (568 nm) was frequency doubled in a 2-mm piece of KDP generating 284-nm light that was used to probe the photoproduct.

The Stokes Raman data from this experiment have been reproduced in Fig. 6.4. At time zero, ground-state depletion of CHD by the pump pulse was seen at 1578 and 1323 cm^{-1}. By 4 ps, scattering from the *cis*-HT photoproduct was observed at 390, 1236, and 1610 cm^{-1} corresponding to the skeletal deformations, C—C stretching, and C=C stretching frequencies, respectively. Between 8 and 22 ps, the ethylenic and single-bond stretching frequencies of the *cis*-HT photoproduct grew in and shifted to 1625 and 1249 cm^{-1}, respectively. The spectrum did not evolve further between 100 ps and 1 ns. A plot of the intensities of the photoproduct peaks as a function of time have shown that *cis*-HT was formed from CHD in 6 ± 1 ps.

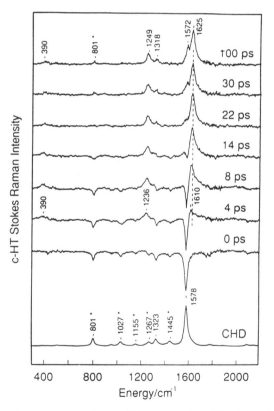

Figure 6.4. Two-color Stokes resonance Raman difference spectra of the 1,3-cyclohexadiene (CHD) to *cis*-hexatriene (*cis*-HT) photoconversion. Spectra were obtained with 625-μW irradiation in the pump (275 nm) and 180-μW irradiation in the probe (284 nm). The lines at 801, 1027, 1155, 1267, and 1445 cm^{-1} are due to the cyclohexane solvent. [Reproduced with permission from P. J. Reid, S. J. Doig, S. D. Wickham, and R. A. Mathies, *J. Am. Chem. Soc.* **1993**, *115*, 4754. Copyright © (1993) American Chemical Society.]

Shifts in the ethylenic and skeletal frequencies at later times suggested the formation of additional conformational intermediates.

Time-resolved resonance Raman anti-Stokes spectra were taken to provide insight into the role of molecular cooling and conformational relaxation in the initial photoproduct dynamics of CHD. Anti-Stokes difference spectra of the photoconversion of CHD to *cis*-HT are shown in Fig. 6.5. Two features close in frequency to that observed in the Stokes spectra were seen within 4 ps at 1614 and 1240 cm^{-1} indicating that the scattering was due to the ethylenic and single-bond stretching modes of *cis*-HT. Both the intensity and frequency of these two modes increased until 10 ps, representing production of the *cis*-HT photoproduct on the ground-state surface. Also at this time, intensity at 828 cm^{-1} was observed, which was a strong indication of the presence of the all-cis conformer. The ethylenic intensity decayed in about 9 ps, while the single-bond mode decayed in about 15

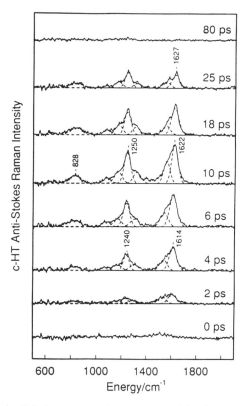

Figure 6.5. Time-resolved anti-Stokes resonance Raman spectra of the photoconversion of CHD to *cis*-HT. The spectra were obtained with 800-μW irradiation in the pump (284 nm) and 180-μW irradiation in the probe (284 nm). The least squares fit to the observed anti-Stokes intensity by a sum of Gaussian peaks is presented as the dashed lines. [Reproduced with permission from P. J. Reid, S. J. Doig, S. D. Wickham, and R. A. Mathies, *J. Am. Chem. Soc.* **1993**, *115*, 4754. Copyright © (1993) American Chemical Society.]

ps. The more rapid decay of the ethylenic stretching intensity was consistent with molecular cooling (30,37–39). The relationship between the ethylenic and single-bond anti-Stokes intensities gave an estimated initial molecular temperature of 1500 ± 500 K at 4 ps(30,37).

In summary, these data have shown that CHD, after photoexcitation, forms vibrationally hot all-*cis*-HT in the ground state within 6 ± 1 ps with a molecular temperature of 1500 ± 500 K, and cools within 9 ± 2 ps along with a 7 ± 1 ps conformational relaxation of the all-*cis*-HT to mono-*s-cis*-HT. Picosecond resonance Raman spectroscopy has provided the first determination of photoproduct formation and cooling of a pericyclic ring-opening reaction.

6.4.2. Vibrational Energy Relaxation

Vibrational energy relaxation processes are of great interest since they are important in reaction dynamics. Vibrational energy relaxation in polyatomic molecules in solution has mainly been studied by transient visible (40–44), and IR (45–50) absorption spectroscopies. Recently, several groups have shown that transient Raman spectroscopy can also provide valuable information for understanding vibrational energy relaxation in solution (5,7–9,30,51–56).

A major advantage of time-resolved resonance Raman spectroscopy is the ability to provide mode-specific information about the microscopic interactions between solute and solvent molecules. This has recently been demonstrated by Weaver et al. (9) where mode-specific changes in peak positions, bandwidths, and intensities of *trans*-stilbene were probed. *trans*-Stilbene was pumped to its first excited singlet state (S_1) with UV pulses in the 280–310-nm range generated by frequency doubling the amplified output of a cavity-dumped dye laser. The visible probe pulses were generated by a second independently tunable cavity-dumped dye laser operating at 577 nm.

Figure 6.6 presents transient Raman spectra of the first excited singlet state (S_1) of *trans*-stilbene in hexane. The primary observations were that the peak positions and bandwidths of the ethylenic stretching mode were sensitive to solvent polarity. The bandwidth decays exponentially with delay time and the peak positions shift to higher frequency after several picoseconds. These observations have been attributed, in part, to the fact that photoexcitation of *trans*-stilbene creates S_1 that was initially vibrationally excited. Very fast intramolecular vibrational energy relaxation occurs and generates vibrationally excited low-frequency modes that distort the molecular geometry. The intermolecular vibrational relaxation or "cooling" process involves energy transfer between the low-frequency modes of *trans*-stilbene and the reorienting solvent environment. This process results in an increasing stabilization of the excited S_1 state of *trans*-stilbene.

These mode-specific, solvent-dependent variations in the vibrational spectra of *trans*-stilbene have been observed in both protic and aprotic solvent environments (5,8,52–54), and in substituted *trans*-stilbenes (51). These transient Raman experiments have provided new information about the microscopic interactions between solute and solvent molecules.

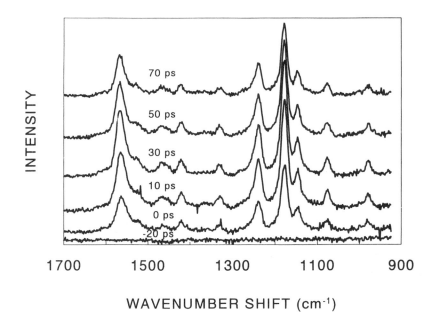

Figure 6.6. Transient Raman spectra of the S_1 excited state of *trans*-stilbene in *n*-hexane at delays from −20 to 70 ps after photoinitiation at 294.4 nm: probe = 577 nm, repetition rate = 1 MHz. Each delay represents the unsmoothed spectrum obtained by 20 minutes of total observation time. [Reproduced with permission from W. L. Weaver, L. A. Huston, K. Iwata, and T. L. Gustafson *J. Phys. Chem.* **1992,** *96,* 8956.

6.5. PHOTOBIOLOGICAL SYSTEMS

6.5.1. Bacteriorhodopsin

Bacteriorhodopsin (BR) is a retinal–protein complex found in the cell membrane of *Halobacterium halobium* (57). Since BR is a proton-translocating system, it can serve as a model for transmembrane ion pumps. The light sensitive part of BR is the all-*trans* retinal chromophore, which is bound to a lysine group in the interior of the protein via a protonated Schiff base. Upon absorption of a photon, the retinal chromophore undergoes a rapid all-trans to 13-cis isomerization about the C13=C14 double bond and about 30% (15 kcal/mol^{-1}) of the initial photon energy is stored (58,59). Bacteriorhodopsin then goes through a series of structurally different intermediates before returning to the initial state in approximately 5 ms. The photocycle of BR is shown in Fig. 6.7. Kinetic constants for the *L* through *O* intermediates have been obtained from absorption or Raman spectroscopic studies (58,60–62). Subpicosecond transient absorption spectroscopy has shown that the initial isomerization occurs within 500 fs producing the *J* intermediate, which converts to a second intermediate *K* within 3 ps (63–66). Even though these experiments provide valuable information about the photoproduct kinetics, they do not provide specific structural information.

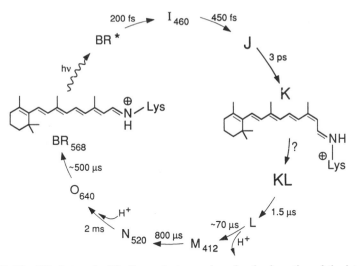

Figure 6.7. The BR photocycle. Kinetic constants are given for the formation of the intermediates during the photocycle.

To provide structural information on the mechanism of phototransformation, picosecond time-resolved resonance Raman spectra of the initial intermediates of BR have been taken using the laser system shown in Fig. 6.2 (10). The photocycle of BR was initiated with a pump pulse at 550 nm, generated by continuum amplification, and a probe pulse at 589 nm generated the Raman scattering from the photoproduct with a 2–3-ps pulse width. As determined by absorption spectroscopy, these two wavelengths are near the absorption maxima of ground-state BR (568 nm) and the K (590 nm) intermediate, respectively.

Stokes Raman spectra of the photoproduct were collected from 0 ps to 13 ns (Fig. 6.8) and used to determine the structure and kinetics of the $J \rightarrow K \rightarrow KL$ transitions (10). The ethylenic stretching frequency was at 1518 cm^{-1} for both the J and K intermediates indicating that they have a similar electronic structure. The Stokes Raman spectrum of the J intermediate at 0-ps delay exhibits strong hydrogen out-of-plane (HOOP) intensities at 1000 and 956 cm^{-1} suggesting that it contains a highly twisted 13-cis chromophore. Single-pulse subpicosecond (800–900 fs) (14) and single-pulse (67) or two-pulse (12) picosecond (7 ps) Raman experiments are consistent with this result. The HOOP frequencies lose intensity within 3 ps indicating that conformational relaxation occurs forming a planar K intermediate. At much longer times (~100 ps) the HOOP intensity slowly increased upon formation of the KL intermediate in which twists were reintroduced into the chromophore.

Anti-Stokes resonance Raman spectra of BR and its photoproduct are shown in Fig. 6.9 for time delays from 0 to 10 ps (10). Anti-Stokes data provide a direct measure of the rate of vibrational cooling and can provide important information on protein dynamics. At 0 ps, the ethylenic stretching frequency was very broad

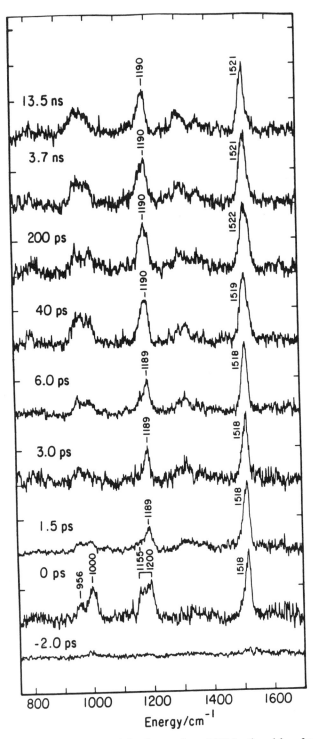

Figure 6.8. Resonance Raman spectra of the photoproduct of BR for time delays from −2 ps to 13.5 ns. Spectra were obtained by using probe and pump wavelengths of 589 and 550 nm, respectively. [Reproduced with permission from S. J. Doig, P. J. Reid, and R. A. Mathies. *J. Phys. Chem.* **1991,** *95,* 6372. Copyright © (1991) American Chemical Society.]

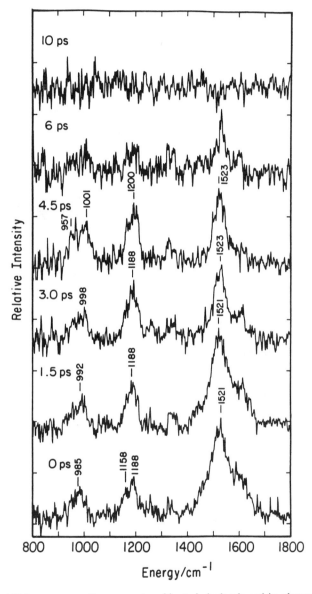

Figure 6.9. Anti-Stokes resonance Raman spectra of bacteriorhodopsin and its photoproduct for time delays from 0 to 10 ps (588 nm excitation). [Reproduced with permission from S. J. Doig, P. J. Reid, and R. A. Mathies *J. Phys. Chem.* **1991**, *95*, 6372. Copyright © (1991) American Chemical Society.]

with the frequency centered at 1521 cm^{-1}. The intensity of this band decreased and the bandwidth narrowed with a decay time of 2.5 ps. It was concluded that J was formed vibrationally hot and conformationally twisted and that the $J \rightarrow K$ transition was due to vibrational cooling and conformational relaxation of the chromophore. This rapid cooling rate was thought to be facilitated by the strong coupling between the chromophore and the protein bath allowing for more efficient energy transfer as compared to conventional solute–solvent interactions (68,69).

From these data, information about the configurations of the J, K, and KL intermediates were obtained and important information about the influence of protein dynamics on the chromophore reaction has been revealed. In summary, during the $J \rightarrow K$ transition, the highly twisted chromophore vibrationally cools and conformationally relaxes within 3 ps. On the 20–100-ps time scale, an isomerization-induced protein conformational change results in the formation of the KL intermediate. These experiments have shown the detailed information that can be obtained using picosecond time-resolved resonance Raman spectroscopy. This information has helped to elucidate both the chromophore and protein structural changes that occur during the complex BR photocycle.

6.5.2. Heme Proteins

Hemoglobin (Hb), a tetrameric protein, is in equilibrium between two different quaternary structures: a T (tense) structure that is stable in the absence of ligands, and an R (relaxed) structure that is stable when ligated. There is a factor of 60 difference in ligand-binding efficiency between the R and T structures (70). These two structures are the basis for the cooperative ligand binding that is communicated through structural changes in the heme pocket. The effect of a small ligand is to initiate extensive tertiary and quaternary structural changes. Several spectroscopic techniques have been used to study hemoglobin photochemistry and conformational changes on various time scales.

6.5.2A. Structural Dynamics.   Structural dynamics can be optically triggered by forcing a large number of ligands to photodissociate. The importance of studying structural dynamics on an ultrafast time scale was revealed by femtosecond transient absorption experiments, which have shown that a deoxy-like heme species was formed within 300–350 fs of photoexcitation (71–73). However, transient absorption experiments are not very sensitive to structural changes of the heme. In order to understand the transition between the R and T structures, heme–heme and heme–protein coupling have been studied using time-resolved resonance Raman spectroscopy, which provides a much more sensitive tool for observing structural changes.

The first subpicosecond resonance Raman spectra of heme proteins were reported by Petrich et al. (15). The femtosecond laser system generated 120-fs pump pulses at 575 nm that were used to photodissociate HbCO. Probe pulses (500 fs) at 435 nm were on resonance with the Soret band and corresponded to the absorption maximum of unligated hemoglobin. This spectrometer was designed to study the structural changes of the heme 0.2–95 ps after photodissociation of carbonmonoxide from hemoglobin. These experiments focused on monitoring the changes in the ν_4 vibrational mode because this mode is sensitive to both the ligation and oxidation state of the heme (74,75). Upon deligation, it shifts from 1370 to 1350 cm^{-1} (74,76–79). In the subpicosecond experiments, the ν_4 mode at 1373 cm^{-1} had completely disappeared by 0.2 ps (see Fig. 6.10) and had downshifted with a time constant of 30 ps (15). This downshift was thought to reflect the protein's response to doming of the heme. Raman experiments with 800–900-fs time resolution have reached the same conclusion (23).

6.5.2B. Vibrational Energy Relaxation.   In addition to providing a means to determine the early time structural dynamics, optical excitation produces large amounts of excess vibrational energy, which is deposited at the heme site. In order to interpret the dynamics of ultrafast laser experiments and compare these experiments to heme protein dynamics under biological conditions, it is necessary to know how the residual vibrational energy dissipates. Vibrational energy relaxation in heme proteins has been studied by a variety of spectroscopic techniques (80). These techniques have ranged from time-resolved IR spectroscopy (81–83), time-resolved thermal-phase grating spectroscopy (80,84–86), and time-resolved Raman

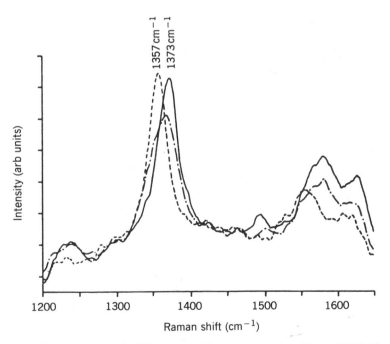

Figure 6.10. High-frequency portion of the resonance Raman spectra of equilibrium HbCO (–), equilibrium Hb (– –), and the HbCO photoproduct 0.2 ps subsequent to ligand dissociation (–·–) obtained with 435-nm, 500-fs, and 80-μJ probe pulses. [Reproduced with permission from J. W. Petrich, J. L. Martin, D. Houde, C. Poyart, and A. Orszag *Biochemistry* **1987**, *26*, 7914. Copyright © (1987) American Chemical Society.]

spectroscopy (15,23,37,87,88). Transient Stokes and anti-Stokes Raman spectroscopies provide the most direct method for monitoring vibrational energy relaxation.

The excited-state lifetime of the heme chromophore subsequent to photodissociation of HbCO is 300 fs (89). Therefore, a highly vibrationally excited ground electronic state is generated by rapid internal conversion with a large excess of energy localized on the heme porphyrin. Using a subpicosecond Raman spectrometer, Petrich et al. (15) monitored the asymmetric broadening and frequency Stokes shift of the ν_4 mode to investigate the vibrational energy relaxation of the heme following photodissociation of CO from hemoglobin. They estimated that by 0.9 ps the vibrational temperature of the heme had increased 218 K and that the heme had substantially cooled by 10 ps. It should be pointed out that interpretation of this experiment was complicated by the fact that a shift in the ν_4 mode has contributions from both structural changes and vibrational relaxation.

In order to separate structural dynamics from vibrational dynamics, Lingle et al. (37) studied deoxyhemoglobin using picosecond Stokes and anti-Stokes Raman spectroscopy. Vibrational energy relaxation was directly probed by comparing the Stokes and anti-Stokes data and was determined to occur with a time constant of 2–5 ps. This was consistent with the time constant of less than 10 ps proposed by Petrich et al. (15). These experiments estimated an internal temperature increase of about 36 K, consistent with vibrational energy relaxation on a picosecond time scale. Overall, these transient Raman experiments have provided a consistent picture for vibrational energy relaxation in heme proteins. The vibrational energy relaxes within a few picoseconds; during this time about 90% of the energy has dissipated into the protein.

6.6. PROSPECTS

In many of the photochemical or photobiological processes discussed here, it was seen that the use of time-resolved vibrational spectroscopy provided information about these systems that was difficult or impossible to obtain by absorption or emission spectroscopies. This success has prompted the designing of new or changing of existing Raman spectrometers to expand the accessible wavelength region such that new information about chemical and biochemical reaction dynamics can be obtained.

An area that is currently being pursued is generating tunable picosecond UV pulses in the 200–250-nm region. One method for generating these UV pulses is based on using the picosecond Raman spectrometer shown in Fig. 6.2. Eighty percent of the 1-mJ per pulse amplified dye laser output (568 nm) is used for Raman shifting in CH_4 gas. The first Stokes line generates 0.20–0.25 mJ per pulse at 680 nm and is amplified in a one-stage dye amplifier pumped with 50 mJ per pulse of 532 nm from a Q-switched Nd:YAG at 50 Hz. This results in 2-ps pulses with a spectral width of 10–11 cm^{-1} at 680 nm and an energy of 0.4–0.5-mJ per pulse. The amplified 680-nm pulses are then doubled and tripled using β-BBO crystals to generate 226-nm pulses. Tuning the dye laser from 568–590 nm can produce Stokes lines in the range of 680–712 nm, which upon doubling and tripling the amplified output can potentially generate UV pulses from 226–237 nm.

Generating picosecond UV pulses in the 220–250-nm region has also recently become feasible with the advancement of Ti:Sapphire laser technology. In this approach, a CW argon ion laser pumps a femtosecond Ti:Sapphire oscillator generating a wavelength range of 700–900 nm. Pulses at 800 nm (1 ps) will be amplified using a regenerative amplifier pumped by a Nd:YAG laser at 1 kHz and frequency doubled in a β-BBO doubling crystal generating 400-nm pulses. The 500–650-nm wavelength range can be accessed using an optical parametric gain (OPG), optical parametric amplifier (OPA) combination and should provide enough energy to be used as a visible pump source. Also, part of the 500–650-nm pulses can be mixed with the remaining 400-nm doubled fundamental in a β-BBO mixing crystal to produce 222–247 nm. A second option for this configuration is to operate with two OPG/OPA combinations. This would allow independent tuning of the pump and probe pulses. One issue that still needs to be resolved for this picosecond setup is the bandwidth, since there is some concern that the OPG/OPA might spectrally broaden the pulses. However, this design involves some of the latest developments in laser technology and offers an all solid state system that can potentially generate high-energy pulses throughout the UV–IR wavelength range.

One experimental area that this new technology will open up is picosecond resonance Raman spectroscopy in the UV. Probing in the 200–250-nm range will allow the excited-state structure and dynamics of many small molecules to be studied. Also, this wavelength range will allow the role of protein dynamics in controlling reaction rates to be determined and will provide unique information for characterizing chemical reactivity in proteins. In addition to the UV region, the recent advancement in Ti:Sapphire technology offers high-power visible and IR pulses as well. This opens up the possibility of studying an unlimited number of chemical and biophysical systems.

ACKNOWLEDGMENTS

This work was supported by grants from the NSF and NIH and by a NIH postdoctoral fellowship to PAT (GM15778-02). We also thank H. Hamaguchi, T. L. Gustafson, and J. L. Martin for permission to use previously published figures.

REFERENCES

1. Delhaye, M., in *Lasers in Physical Chemistry and Biophysics*, J. Joussot-Dubien (Ed.). Elsevier, New York, 1975, p. 213.

2. Terner, J., Spiro, T. G., Nagumo, M., Nicol, M. F., and El-Sayed, M. A. *J. Am. Chem. Soc.* **1980,** *102,* 3238.

3. Coopey, M., Valat, H. P., and Alpert, B. *Nature (London)* **1980,** *284,* 568.

4. Gustafson, T. L., Roberts, D. M., and Chernoff, D. A. *J. Chem. Phys.* **1983,** *79,* 1559.

5. Iwata, K., Yamaguchi, S., and Hamaguchi, H. *Rev. Sci. Instrum.* **1993,** *64,* 2140.

6. Matousek, P., Hester, R., Moore, J., and Parker, A. *Meas. Sci. Tech.* **1993,** *4,* 1090.

7. Phillips, D. L., Rodier, J.-M., and Myers, A. B. *Chem. Phys.* **1993**, *175*, 1.

8. Qian, J., Schultz, S. L., Bradburn, G. R., and Jean, J. M. *J. Phys. Chem.* **1993**, *97*, 10638.

9. Weaver, W. L., Huston, L. A., Iwata, K., and Gustafson, T. L. *J. Phys. Chem.* **1992**, *96*, 8956.

10. Doig, S. J., Reid, P. J., and Mathies, R. A. *J. Phys. Chem.* **1991**, *95*, 6372.

11. Orman, L. K., Chang, Y. J., Anderson, D. R., Yabe, T., Xu, X., Yu, S. C., and Hopkins, J. B. *J. Chem. Phys.* **1989**, *90*, 1469.

12. Atkinson, G. H., Brack, T. L., Blanchard, D., and Rumbles, G. *Chem. Phys.* **1989**, *131*, 1.

13. Hopkins, J. B., and Rentzepis, P. M. *Chem. Phys. Lett.* **1986**, *124*, 79.

14. van den Berg, R., Jang, D. J., Bitting, H. C., and El-Sayed, M. A. *Biophys. J.* **1990**, *58*, 135.

15. Petrich, J. W., Martin, J. L., Houde, D., Poyart, C., and Orszag, A. *Biochemistry* **1987**, *26*, 7914.

16. Myers, A. B., and Mathies, R. A., in *Biological Applications of Raman Spectroscopy: Vol. 2– Resonance Raman Spectra of Polyenes and Aromatics,* T. G. Spiro; (Ed.). Wiley, New York, 1987, p. 1.

17. Heller, E. J., Sundberg, R L., and Tannor, D. *J. Phys. Chem.* **1982**, *86*, 1822.

18. Champion, P. M., and Albrecht, A. C. *Annu. Rev. Phys. Chem.* **1982**, *33*, 353.

19. Spiro, T. G., and Stein, P. *Annu. Rev. Phys. Chem.* **1977**, *28*, 501.

20. Warshel, A. *Annu. Rev. Biophys. Bioeng.* **1977**, *6*, 273.

21. Albrecht, A. C. *J. Chem. Phys.* **1961**, *34*, 1476.

22. Mathies, R. A., in *Chemical and Biochemical Applications of Lasers,* Moore, C. B. (Ed.). Academic, New York, 1979, p. 55.

23. van den Berg, R. and El-Sayed, M. A. *Biophys. J.* **1990**, *58*, 931.

24. Bethune, D. S. *Appl. Optics* **1981**, *2011*, 1897.

25. Reider, G. A., Traar, K. P., and Schmidt, A. J. *Appl. Optics* **1984**, *23*, 2856.

26. Gustafson, T. L., Roberts, D. M., and Chernoff, D. A. *J. Chem. Phys.* **1984**, *81*, 3438.

27. Hudson, B. S., Kohler, B. E., and Schulten, K., in *Excited States*, Lim, E. C. (Ed.). Academic, New York, 1982, p. 1.

28. Havinga, E., de Kock, R. J., and Rappoldt, M. P. *Tetrahedron* **1960**, *11*, 276.

29. Rappoldt, M. P., and Havinga, E. *Recl. Trav. Chim. Pays-Bas* **1960**, *79*, 369.

30. Reid, P. J., Doig, S. J., Wickham, S. D., and Mathies, R. A. *J. Am. Chem. Soc.* **1993**, *115*, 4754.

31. Reid, P. J., Doig, S. J., and Mathies, R. A. *Chem. Phys. Lett.* **1989**, *156*, 163.

32. Reid, P. J., Doig, S. J., and Mathies, R. A. *J. Phys. Chem.* **1990**, *94*, 8396.

33. Trulson, M. O., Dollinger, G. D., and Mathies, R. A. *J. Am. Chem. Soc.* **1987**, *109*, 586.

34. Trulson, M. O., Dollinger, G. D., and Mathies, R. A. *J. Chem. Phys.* **1989**, *90*, 4274.

35. Share, P. E., Kompa, K. L., Peyerimhoff, S. D., and Van Hermet, M. C. *Chem. Phys.* **1988**, *120*, 411.

36. Buma, W. J., Kohler, B. E., and Song, K. *J. Chem. Phys.* **1990**, *92*, 4622.

37. Lingle, R., Xu, X., Zhu, H., Yu, S. C., and Hopkins, J. B. *J. Phys. Chem.* **1991**, *95*, 9320.

38. Schomacker, K. T., Bangcharoenpaurpong, O., and Champion, P. M. *J. Chem. Phys.* **1984**, *80*, 4701.

39. Schomacker, K. T., and Champion, P. M. *J. Chem. Phys.* **1989**, *90*, 5982.

40. Sension, R. J., Szarka, A. Z., and Hochstrasser, R. M. *J. Chem. Phys.* **1992**, *97*, 5239.

41. Sukowski, U., Seilmeier, A., and Elsaesser, T. *J. Chem. Phys.* **1990**, *93*, 4094.

42. Brito Cruz, C. H., Fork, R. L., Knox, W. H., and Shank, C. V. *Chem. Phys. Lett.* **1986**, *132*, 341.

43. Weiner, A. M. and Ippen, E. P. *Chem. Phys. Lett.* **1985**, *114*, 456.

44. Doany, F. E., Greene, B. I., and Hochstrasser, R. M. *Chem. Phys. Lett.* **1980**, *75*, 206.

45. Doorn, S. K., Stoutland, P. O., Dyer, R. B., and Woodruff, W. H. *J. Am. Chem. Soc.* **1992**, *114*, 3133.

46. Owrutsky, J. C., Kim, Y. R., Li, M., Sarisky, M. J., and Hochstrasser, R. M. *Chem. Phys. Lett.* **1991**, *184*, 368.

47. Hubner, H. J., Worner, M., and Kaiser, W. *Chem. Phys. Lett.* **1991**, *182*, 315.

48. Graener, H., Ye, T. Q., and Laubereau, A. *J. Chem. Phys.* **1989**, *90*, 3413.

49. Scherer, P. O. J., Seilmeier, A., and Kaiser, W. *J. Chem. Phys.* **1985**, *83*, 3948.

50. Seilmeier, A., Scherer, P. O. J., and Kaiser, W. *Chem. Phys. Lett.* **1984**, *105*, 140.

51. Butler, R. M., Lynn, M. A., and Gustafson, T. L. *J. Phys. Chem.* **1993**, *97*, 2609.

52. Hamaguchi, H. and Iwata, K. *Chem. Phys. Lett.* **1993**, *208*, 465.

53. Iwata, K., Toleutaev, B., and Hamaguchi, H. *Chem. Lett.* **1993**, *9*, 1603.

54. Hester, R. E., Matousek, P., Moore, J. N., Parker, A. W., Toner, W. T., and Towrie, M. *Chem. Phys. Lett.* **1993**, *208*, 471.

55. Iwata, K. and Hamaguchi, H. *Chem. Phys. Lett.* **1992**, *196*, 462.

56. Xu, X., Lingle, R., Yu, S. C., Chang, Y. J., and Hopkins, J. B. *J. Chem. Phys.* **1990**, *92*, 2106.

57. Oesterhelt, D. and Stoechenius, W. *Proc. Natl. Acad. Sci. USA* **1973**, *70*, 2853.

58. Mathies, R. A., Lin, S. W., Ames, J. B., and Pollard, W. T. *Annu. Rev. Biophys. Biophys. Chem.* **1991**, *20*, 491.

59. Birge, R. R. *Biochim. Biophys. Acta* **1990**, *1016*, 293.

60. Ames, J. B. and Mathies, R. A. *Biochemistry* **1990**, *29*, 7181.

61. Varo, G. and Lanyi, J. K. *Biochemistry* **1990**, *29*, 6858.

62. Milder, S. and Kliger, D. *Biophys. J.* **1988**, *53*, 465.

63. Mathies, R. A., Brito Cruz, C. H., Pollard, W. T., and Shank, C. V. *Science* **1988**, *240*, 777.

64. Dobler, J., Zinth, W., Kaiser, W., and Oesterhelt, D. *Chem. Phys. Lett.* **1988**, *144*, 215.

65. Sarkov, A. V., Pakulev, A. V., Chekalin, S. V. and Matveetz, Y. A. *Biochim. Biophys. Acta* **1985**, *808*, 94.

66. Polland, H.-J., Franz, M. A., Zinth, W., Kaiser, W., Koelling, E., and Oesterhelt, D. *Biophys. J.* **1986**, *49*, 651.

67. Stern, D. and Mathies, R. A., in *Time-resolved Vibrational Spectroscopy*, Stockburger, M. and Laubereau, A. (Eds.). Springer-Verlag, 1985, p. 250.

68. Zinth, W., Kolmeder, C., Benna, B., Irgens-Defregger, A., Fischer, S. F., and Kaiser, W. *J. Chem. Phys.* **1983**, *78*, 3916.

69. Wild, W., Seilmeier, A., Gottfried, N. H., and Kaiser, W. *Chem. Phys. Lett.* **1985**, *119*, 259.

70. Murray, L. P., Hofrichter, J., Henry, E. R., Ikeda-Saito, M., Kitagishi, K., Yonetani, T., and Eaton, W. A. *Proc. Natl. Acad. Sci. USA* **1988**, *85*, 2151.

71. Petrich, J. W., Houde, D., Poyart, C., Orszag, A., and Martin, J. L. *Photobiochem. Photobiophys. Suppl.* **1987**, 77.

72. Martin, J. L., Migus, A., Poyart, C., Lecarpentier, Y., Astier, R., and Antonetti, A. *Proc. Natl. Acad. Sci. USA* **1983**, *80*, 173.

73. Martin, J. L., Migus, A., Poyart, C., Lecarpentier, Y., and Antonetti, A., in *Ultrafast Phenomena*, Auston, P. H. and Eisenthal, K. B. (Eds.). Springer-Verlag, 1984, p. 447.

74. Rousseau, D. L., Tan, S. L., Ondrias, M. R., Ogawa, S., and Noble R. W. *Biochemistry* **1984**, *23*, 2857.

75. Choi, S., Spiro, T. G., Langry, K. C., Smith, K. M., Budd, D. L., and La Mar, G. N. *J. Am. Chem. Soc.* **1982**, *104*, 4345.

76. Yamamoto, T., Palmer, G., Gill, D., Salmeen, I. T., and Rimai, L. *J. Biol. Chem.* **1973**, *248*, 5211.

77. Spiro, T. G. and Burke, J. M. *J. Am. Chem. Soc.* **1976**, *98*, 5482.

78. Friedman, J. M., Rousseau, D. L., and Ondrias, M. R. *Annu. Rev. Phys. Chem.* **1982**, *33*, 471.

79. Asher, S. A. *Methods Enzymol.* **1981**, *76*, 371.

80. Miller, R. J. D. *Annu. Rev. Phys. Chem.* **1991,** *42,* 581.

81. Locke, B., Lian, T., and Hochstrasser, R. M. *Chem. Phys.* **1991,** *158,* 409.

82. Anfinrud, P. A., Han, C., and Hochstrasser, R. M. *Proc. Natl. Acad. Sci. USA* **1989,** *86,* 8387.

83. Anfinrud, P., Han, C., Hansen, P. A., Moore, J., and Hochstrasser, R. M., in *Ultrafast Phenomena VI,* Yajima, T., Yoshihara, K., Harris, C. B., and Shionoya, S. (Eds.). Springer-Verlag Springer Series in Chemical Physics, 1988, p. 442.

84. Genberg, L., Richard, L., McLendon, G., and Miller, R. J. D. *Science* 1991, *251,* 1051.

85. Genberg, L., Bao, Q., Gracewski, S., and Miller, R. J. D. *Chem. Phys.* **1989,** *131,* 81.

86. Genberg, L., Heisel, F., McLendon, G., and Miller, R. J. D. *J. Phys. Chem.* **1987,** *91,* 5521.

87. Alden, R. G., Schneebeck, M. C., Ondrias, M. R., Courtney, S. H., and Friedman, J. M. *J. Raman Spectrosc.* **1992,** *23,* 569.

88. Alden, R. G., Chavez, M. D., Ondrias, M. R., Courtney, S. H., and Friedman, J. M. *J. Am. Chem. Soc.* **1990,** *112,* 3241.

89. Petrich, J. W., Poyart, C., and Martin, J. L. *Biochemistry* **1988,** *27,* 4049.

Chapter **VII**

LOCAL ORDER AND ULTRAFAST DYNAMICS IN LIQUIDS: TRANSIENT GRATING OPTICAL KERR EFFECT EXPERIMENTS

Abhijit Sengupta and M. D. Fayer
Department of Chemistry,
Stanford University, Stanford, California

Laser Techniques In Chemistry, Edited by Anne B. Myers and Thomas R. Rizzo.
Techniques of Chemistry Series, Vol. XXIII.
ISBN 0-471-59769-4   © 1995 John Wiley & Sons, Inc.

7.8. Concluding Remarks
Acknowledgments
References

7.1. INTRODUCTION

Disordered condensed matter materials, such as glasses and liquids, are more common in nature than perfect crystals. Our understanding of disordered systems is poor compared to our knowledge of the properties of crystals. Recent advances in the theory of condensed matter and new spectroscopic techniques are providing the necessary tools for studying the dynamics and structure of disordered materials. In fact, the concepts of order and disorder are no longer absolute, but rather, depend on the distance and time scale associated with an observation made either in the time or frequency domains. In this chapter, we explore the extent of local order, the time scale of its persistence, and its influences on molecular dynamics in several systems in their isotropic liquid phase. Understanding molecular dynamics in liquids is important to the field of chemistry since the majority of chemical reactions occur in the liquids. Chemical reactions occur on the time scale on which molecular events, such as collisions, librations, or orientational relaxation, take place. The local environment has a strong influence on a chemical reaction. The evolution or fluctuation of a local environment is determined by the orientational dynamics of the molecules. In spite of the importance of liquids, a general description of liquid dynamics on all distance and time scales is lacking. The inhomogeneity in liquids on short time scales, combined with possible long-range orientational correlations, can give rise to important physical phenomena by changing the dynamic properties of a system. This chapter presents experimental studies of the orientational dynamics of molecules in complex liquids. The experiments are performed using the ultrafast transient grating optical Kerr effect (TG–OKE) technique, which is a powerful method for investigation of the relationship between dynamics and local structure. The TG–OKE method is described in Section 2.

This chapter examines the orientational dynamics in a variety of liquids. The first example discussed is biphenyl. This system is used to illustrate simple hydrodynamic behavior and to demonstrate the nature of the TG–OKE observables. Following ultrafast (fs) librational dynamics, the orientational relaxation of biphenyl is diffusive. The relaxation obeys the Debye–Stokes–Einstein (DSE) equation, that is, the orientation relaxation time (time for the decay of the orientational correlation function) is proportional to the temperature-dependent viscosity $\eta(T)$ and to the inverse of the temperature.

The main emphasis of this chapter is the illustration of nonhydrodynamic orientational relaxtion and its relationship to local structure in liquids. Five molecular liquid systems are described. These are 2-ethylnaphthalene (2EN), a common molecular liquid; the liquid crystals pentylcyanobiphenyl (5CB) and N-(methoxybenzylidene)butylaniline (MBBA) in their isotropic phases; and the polymers poly(methylphenylsiloxane) (PMPS) and poly(2-vinylnaphthalene) (2PVN). The remarkable feature of all of these systems is that they have a significant component of their orientational dynamics that is independent of temperature and viscosity. Although the temperature (T) and viscosity (η) of the liquids change substantially, the orientational relaxation dynamics are completely independent of T and η. These observations are related to the presence of local liquid structures that occur on short time and distance scales. The η/T independent dynamics are related to the existence of a local potential surface that gives rise to the local structure. On the time scale on which the local structures exist, the dynamics involve relaxation on the local potential surface rather than rotational diffusion.

Sometimes the complexity of the condensed phase cannot easily be extrapolated from the properties of its constituents without a knowledge of the nature of local order. Consider the formation of crystals of definite spatial symmetries by lowering the temperature of a homogeneous and isotropic liquid composed of spherical atoms. Atoms are spherical but the local packing can produce icosahedral clusters where orientational correlations are present. The first quasicrystal discovered in 1984 was a rapidly quenched alloy of 14% Mn atoms in Al (1). Sharp peaks in X-ray diffraction come from the long-range orientational order in the quasicrystal. The X-ray data suggest that before the system solidified it had short-range orientational order (i.e., icosahedral clusters) in the liquid. Local orientational structures have been observed in molecular liquids, such as CCl_4 (2), benzene (3–5), and naphthalene (3), using neutron scattering and Raman measurements. Section 7.7 presents the TG–OKE results for the molecular liquid 2-ethylnaphthalene, which show the influence of local structure on fast orientational dynamics.

Studies of phase transitions provide a perspective for understanding long-range order in terms of the correlations among spontaneous fluctuations. In many systems, critical phenomena are associated with a complex temperature-dependent long-range order. An example of long range ordering is the critical opalescence that occurs near the liquid–vapor critical point. Another example of thermodynamic criticality occurs near the isotropic-to-nematic phase transition in liquid crystals. The characteristic order parameter of this critical transition is a measure of orientational correlation. Therefore local order in the isotropic phase of liquid crystals is orientational in nature and dynamic orientational fluctuations show critical behavior. Many pretransitional phenomena found in the isotropic phase of liquid crystals have their origin in local orientational order. Section 7.5 discusses in detail the fast orientational dynamics of the liquid crystals 5CB and MBBA in their isotropic phases.

Local structure can also emerge near a point of dynamic criticality. At a dynamic critical transition, the heat capacity of the system does not diverge. In glass-forming liquids, a dynamic critical transition occurs at a temperature T_c ($T_c > T_g$, the glass transition temperature); decay of the density correlation function on a short-distance scale no longer decays to zero on a short time scale below T_c (6,7). The resulting local structure causes a bifurcation of the relaxation processes below T_c into a slow branch (α relaxation) and a fast branch (β relaxation). The side group dynamics in polymer glass-forming liquids contribute to fast temperature independent β relaxation. Section 6 discusses the results of TG–OKE investigations of the side group dynamics in P2VN and PMPS. Results for PMPS as a neat liquid (melt) and in dilute solution are presented. The side group dynamics of PMPS and P2VN in solution indicate that the short-range order, which exists on short time scales, is created by the restricted geometry of the backbones in polymer.

First, the experimental and theoretical aspects of the TG–OKE technique are described in Section 2. Section 3 introduces aspects of hydrodynamic and nonhydrodynamic orientational relaxation. Section 4 illustrates the typical features of TG–OKE data and the simple hydrodynamic model of orientational relaxation for liquid biphenyl. Section 4 also helps to establish a connection between the TG–OKE method and other methods used to study the dynamics of a simple hydrodynamic liquid. Sections 5–7 present detailed TG–OKE experiments for systems that display dramatically nonhydrodynamic behavior. The nonhydrodynamic orientational relaxation is a result of local order in the liquids, although the physical phenomena that give rise to the local order are distinct in the three types of systems, a molecular liquid, liquid crystals, and polymeric liquids, that are studied.

7.2. TRANSIENT GRATING OPTICAL KERR EFFECT MEASUREMENTS

The TG–OKE experiments are nonresonant four-wave mixing experiments (8–12). Two laser excitation pulses are crossed in space and time within the sample creating an optical interference pattern. If the pulse length is short compared to the characteristic response times of the sample, the optical interference pattern generates a spatially periodic modulation of refractive index, that is, a diffraction grating. The excitation pulses amplify a single Fourier component of the spontaneous fluctuations of the system. A probe beam with a variable time delay is Bragg diffracted from the induced grating. The diffracted signal is recorded versus the probe delay. This is shown schematically in Fig. 7.1(a).

Qualitatively, the TG–OKE experiment works in the following manner. The excitation electric field induces a dipole moment in the molecules. Because the molecules have anisotropic polarizabilities, the induced dipoles are not parallel to the field. The interaction of the field with the induced dipoles causes a torque to be exerted on the molecules. This interaction produces a slight alignment of the molecules along the field direction. Since the electric field is spatially periodic because of the interference of the two excitation pulses, the molecular alignment is also spatially periodic. This periodic alignment results in a periodic variation in the sample's index of refraction (susceptibility). The spatially periodic index acts as a volume holographic Bragg diffraction grating. The delayed probe pulse is brought in to meet the Bragg diffraction condition. The small component of the probe that is diffracted from the grating is the signal. As the molecular alignment decays by orientational relaxation, the grating decays, and the diffraction efficiency is reduced. Thus the time dependence of the diffracted signal is directly related to

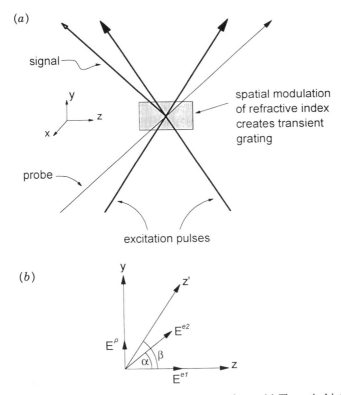

Figure 7.1. Schematic illustration of the transient grating experiment. (*a*) The optical interference pattern due to the crossed-excitation pulses produces a periodic spatial modulation of the refractive index of the sample and thus generates a transient diffraction grating. The probe pulse is Bragg diffracted off the grating producing a signal beam at the phase matched angle. The polarizations of all four beams are controlled in the experiment and are used to separate and identify contributions to the signal from various physical processes. (*b*) Illustration of the polarization configuration used in the experiments. The E fields of the two excitation beams and the probe beam are labeled *e1*, *e2*, and *p*, respectively. The parameter α is the angle between the polarizations of the excitation beams and β is the polarization of a polarizer in the signal beam. The angle between E^p and E^{e1} is 90°.

the orientational relaxation. This qualitative picture neglects the details of the quantum mechanical coupling of the radiation field to the material system. This is discussed in Section 7.2.1.

7.2.1. Experimental Technique

The transient grating experiments discussed in this chapter were performed using a subpicosecond dye laser system or a 100-ps mode-locked Q-switched Nd:YAG laser system, depending on the time scale of the dynamics studied. The subpicosecond system has been described in detail elsewhere (13). The 70-ps output pulses from a continuous wave (CW) mode-locked Nd:YAG laser are sent through an optical pulse compressor and frequency doubled. The resulting 2.5-ps pulses at 532 nm with an average power of 650–700 mW are used to synchronously pump a linear astigmatically compensated three mirror dye laser, which is tuned by a single-plate birefringent filter. With DCM, the dye laser delivers tunable 300-fs pulses with an average output of 70–220 mW over the wavelength range 605–705 nm. These dye pulses are amplified by pumping a three stage amplifier with frequency doubled pulses from a cavity-dumped, Q-switched and mode-locked Nd:YAG laser operating at a 1.75-kHz repetition rate. The cavity-dumped Nd:YAG laser produces 1-mJ IR pulses which, when frequency doubled, generate 700-μJ pulses at 532 nm to pump the three amplifier stages. The two Nd:YAG lasers are electronically synchronized by a common radio frequency (rf) master oscillator for both mode lockers, and the timing is set by a voltage controlled phase shifter for the rf. A saturable absorber jet between the second and third amplifier stage suppresses the unamplified dye pulses and shortens the width of the amplified pulse. The laser system is slightly different from the one described by Newell et al. (13). A 4X telescope has been added to the cavity to increase the spot size at the $LiNbO_3$ crystal reducing the rate of photorefractive damage. For the dye DCM, typical amplified pulse widths are 250 fs with an average pulse energy of 10–20 μJ. A laser wavelength of 665 nm was chosen to avoid two-photon absorption.

For the TG–OKE experiments, the amplified pulse is split into three pulses to yield the two excitation pulses and the probe pulse. The two excitation pulses, focused to 120-μm spot sizes, are crossed in the sample at an angle of 15° to produce an optical interference pattern (see Fig. 7.1). The optical interference pattern induces the OKE grating with spatial modulation of the refractive index that mimics the interference pattern. This grating is monitored by the 90-μm spot size probe pulse that is incident at the phase matching angle for Bragg diffraction (slightly noncollinear with the excitation beam to achieve spatial separation of the signal). The probe pulse is variably delayed by an optical delay line that is controlled by a 1-μm resolution stepper motor. This delay line is used for the short time scale measurements. The stepper motor delay line is mounted on a long optical rail and can be moved along the optical rail with a computer controlled direct current (dc) motor. This is used for the nanosecond time scale measurements (e.g., slowest dynamics in the isotropic phase of liquid crystals).

The intensity as well as the polarization of all three beams can be independently controlled by sets of half-wave plates and linear polarizers. A fourth linear polarizer is set in the signal path and permits any polarization of the diffracted beam to be monitored. A technique pioneered by Etchepare et al. (14) and later extended by Deeg and Fayer (12) is used to selectively probe the different dynamic processes contributing to the formation and relaxation of a phase grating. Following the notations of Deeg and Fayer, the configuration of the grating experiment is described as (0°/α/90°/β) where the four quantities within the parentheses stand for the angles of polarization of the two excitation beams, the probe beam and the detected signal beam, respectively. All the angles were in reference to the polarization of the excitation beam nearest to the probe beam. Near zero delay of the probe beam (t < 2 ps), the (0°/45°/90°/135°) configuration is used. It detects only nuclear $\chi_n^{(3)}$ processes by suppressing the overwhelmingly large electronic contribution. Before and after each of these nuclear scans, an electronic scan with the configuration (0°/45°/90°/56°) is taken. Since the electronic response is instantaneous, this configuration gives the instrument response. Any change in the pulse width during the nuclear OKE measurement can be detected. In addition, this instrument response is a three-pulse correlation, which provides a very accurate method for determining the pulse duration and shape (13). Since the instrument response is taken with all components of the experiment identical to the actual measurement (in contrast to using an autocorrelator) and it provides pulse shape information as well as information on the pulse duration, it makes possible very accurate deconvolutions to obtain quantitative information for times shorter than the pulse duration (15). At longer times ($t > 2$ ps), when the electronic OKE no longer contributes to the signal, the polarization grating configuration (0°/90°/90°/0°) is employed. The signal-to-noise (S/N) ratio is truly outstanding in this configuration as the polarizer in the signal beam eliminates all the (nondepolarized) scattered light from the nearby excitation beam. It also insures no formation of acoustic and thermal gratings that can contaminate the signal at longer times (11,12). A study of laser intensity dependence is important to assure that there are no artifacts. The signal varied linearly in the intensity of each beam with an overall I^3 dependence on the laser intensity. The individual beams were not depolarized passing through the sample.

One of the excitation beams or the probe beam is chopped and the signal is detected by a photomultiplier tube (PMT). On each laser shot, the output of the PMT is measured by a gated integrator. A computer controls the scanning of the delay line, performs analog-to-digital conversions of a gated integrator output, and averages the data over several scans. The intensity of each laser shot is also measured, and data from shots outside of a typical 10% window are rejected. The scattered light level is measured by blocking (chopping) a beam that does not contribute to the scattered light every other shot. The measured scattered light is normalized by the ratio of the laser intensity on the signal shot and the scattered light shot, and the normalized scattered light is subtracted from the amplitude measured on the signal shot.

In the isotropic phase of nematic liquid crystals orientational dynamics extends to many tens and hundreds of nanoseconds. A better S/N ratio is achieved in this regime by a two-color transient grating experiment with longer pulses. In the experiments performed by Stankus et al. (16), 100-ps pulses from a mode-locked Q-switched Nd:YAG laser were used. The sample is excited with 1.06-μm pulses and probed with a 532-nm pulse. The longer pulse creates a larger long time anisotropy, and the use of two colors makes it a truly zero background measurement since the PMT is not sensitive to the IR scattered light background from the excitation beams. A polarization grating was used to improve the S/N ratio and to eliminate acoustic wave generation (11,12). One of the IR excitation beams is chopped. The laser is run synchronously with this chopper. The output of the PMT is measured using a lock-in amplifier and digitized and stored on computer. The delay line is also computer controlled, which permits averaging of a number of delay line scans.

7.2.2. Theoretical Background of the TG–OKE Experiment

The TG–OKE signal comes at a phase matched angle; for the configurations used in the experiments (Figs. 7.1 and 7.2), the signal wavevector, $k_S = k_p + (k_{e1} - k_{e2})$, where p, e1, and e2 stand for the probe and the excitation beams, respectively. The third-order nonlinear polarizability at the phase matched angle is given by (17)

$$P_i(\omega) = \chi^{(3)}_{ijkl}(\omega,\omega_3,\omega_2,\omega_1)\, E_j(\omega_3)\, E_k(\omega_2)\, E_l^*(\omega_1) \qquad (7.1)$$

where P_i is the component of the induced polarization in the medium and the components of the electric fields are E_j, E_k, and E_l. This relationship is characterized by the properties of $\chi^{(3)}_{ijkl}$, the elements of the third-order nonlinear susceptibility tensor. For the evaluation of a time-resolved experiment, a description of the FWM process in the time domain

$$P_i(t) = \int_{-\infty}^{+\infty} dt_3 \int_{-\infty}^{+\infty} dt_2 \int_{-\infty}^{+\infty} dt_1\, G^{(3)}_{ijkl}(t - t_3, t - t_2, t - t_1)\, E_j(t_3)\, E_k(t_2)\, E_1(t_1) \quad (7.2)$$

is more convenient. Here the impulse response function $G^{(3)}_{ijkl}(t - t_3, t - t_2, t - t_1)$ is the Fourier transform of $\chi^{(3)}_{ijkl}(\omega)$ in Eq. (7.1). In general, one is interested in the

Polarization Selectivity in TGOKE Experiments

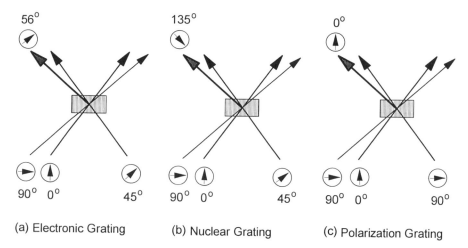

(a) Electronic Grating   (b) Nuclear Grating   (c) Polarization Grating

Figure 7.2. The three polarization configurations that are used in the transient grating optical Kerr effect experiments are illustrated. (a) Electronic grating: It provides the instrument response. (b) Nuclear grating: For $t < 5$ ps, it is useful for measurements of librational, interaction-induced, and orientational relaxations. (c) Polarization grating: Equivalent to the VH measurements of depolarized light scattering experiments. For $t > 2$ ps, the signal decay is due to orientational relaxations. No acoustic grating is formed at longer times.

time dependence of $G_{ijkl}^{(3)}$ $(t - t_3, t - t_2, t - t_1)$ because it can provide valuable information about the specific molecular dynamics of the sample under investigation. The symmetry properties of the $\chi^{(3)}$ tensor depend on the nature of the observed physical phenomena. Since different physical processes are characterized by different $\chi^{(3)}$ symmetries, polarization-dependent FWM experiments can be used to distinguish and separate time dependent signals arising from various physical processes.

There exist certain relationships among the elements of $\chi^{(3)}$ by virtue of the spatial symmetry class of the material under consideration. These relations have been tabulated by Butcher (18). For an isotropic sample, like a liquid or glass, there are 21 nonzero elements of $\chi^{(3)}$ of which only three are independent. These are $\chi_{1111}^{(3)}$, $\chi_{1122}^{(3)}$ and $\chi_{1212}^{(3)}$ with

$$\chi_{1221}^{(3)} = \chi_{1111}^{(3)} - \chi_{1122}^{(3)} - \chi_{1212}^{(3)} \tag{7.3}$$

Hellwarth (19) has shown that the Born–Oppenheimer approximation imposes additional restraints on the symmetry properties of $\chi^{(3)}$. For an isotropic sample the electronic $\chi^{(3)}(e)$ is a scalar and the relationship among the elements of $\chi^{(3)}(e)$ are given by

$$\chi_{1111}^{(3)}(e) = 3\chi_{1122}^{(3)}(e) = 3\chi_{1212}^{(3)}(e) = 3\chi_{1221}^{(3)}(e) \tag{7.4}$$

$\chi^{(3)}(n)$, which describes all processes associated with nuclear motions, contains only the two independent elements $\chi_{1111}^{(3)}(n)$ and $\chi_{1122}^{(3)}(n)$, and the relations among the elements are

$$\chi_{1221}^{(3)}(n) = \chi_{1212}^{(3)}(n) = [\chi_{1111}^{(3)}(n) - \chi_{1122}^{(3)}(n)]/2 \tag{7.5}$$

With these general relationships due to the space symmetry group and Born–Oppenheimer approximation, there are special symmetry properties of $\chi^{(3)}$, which arise from the symmetry of the underlying physical mechanism that generates $\chi^{(3)}$. The optical Kerr effect (OKE) has two contributions. The first is the electronic OKE, which is caused by the virtually instantaneous distortion of the molecular electron cloud by the applied optical field. The second is the nuclear OKE, which comes about by the slower reorientation of the nuclei in the optical field. The properties of $\chi^{(3)}(eOKE)$ for the electronic OKE are given by the relationships in Eq. (7.4). It can be shown for the nuclear OKE, that

$$\chi_{1111}^{(3)}(nOKE) = a + b \tag{7.6a}$$

$$\chi_{1122}^{(3)}(nOKE) = a \tag{7.6b}$$

and with Eq. (7.5)

$$\chi_{1212}^{(3)}(nOKE) = b/2 \tag{7.6c}$$

a and b are often referred to as the "isotropic" and "anisotropic" parts of the susceptibility (20) as they are the quantities that are responsible for polarized and depolarized light scattering, respectively. For liquids, theories that relate the third-order nonlinear susceptibility $\chi^{(3)}$ to the linear molecular polarizability α, have shown the equation

$$b \approx -3a \qquad (7.7)$$

should hold (19). In general, this has been confirmed by experiments. Inserting Eq. (7.7) into Eqs. (7.6) gives

$$\chi_{1111}^{(3)}(\text{nOKE}) = -2\chi_{1122}^{(3)}(\text{nOKE}) = \frac{4}{3}\chi_{1212}^{(3)}(\text{nOKE}) \qquad (7.8)$$

Using Eqs. 7.4 and 7.8 it is relatively straightforward to prove that the configuration (0°/45°/90°/135°) suppresses the electronic OKE, giving a signal that arises only from the nuclear OKE. The configuration (0°/45°/90°/56°) suppresses the nuclear OKE, giving a signal that is only from the electronic OKE. In the absence of absorption (including two photon absorption) by the sample, which is prevented by the choice of laser wavelength, these are the only gratings that will be formed at short time. At longer time, an acoustic grating may be formed. This is suppressed by the configuration (0°/90°/90°/0°) (12,21).

For the interpretation of the signals recorded in the TG experiments described here, it is necessary to consider the explicit time dependence. If we assume that the two excitation pulses are time coincident at the sample at $t' = t_1 = t_2$ and that we can separate excitation and probe processes, $\chi_{ijkl}^{(3)}(t - t_3) = \delta(t - t_3)$, Eq. (7.2) gives

$$P_i = E_j^p(t) \int_{-\infty}^{t'} dt' G_{ijkl}^{(3)}(t - t') E_k^{el}(t') E_l^{e2}(t') \qquad (7.9)$$

The Bragg scattered intensity $S(t)$ of a probe pulse delayed by time t with respect to the excitation pulses is given by

$$S(t) = \int_{-\infty}^{+\infty} dt'' I_p(t'' - t) \left[\int_{-\infty}^{t''} G^{\varepsilon\varepsilon}(t'' - t') I_e(t') dt' \right]^2 \qquad (7.10a)$$

where $G^{\varepsilon\varepsilon}(t)$ is the impulse response function (Green function) of the liquid's dielectric tensor for ultrashort optical pulse excitation, and $I_e(t)$, and $I_p(t)$ are the intensity profiles of the excitation and probe pulses, respectively. At longer time it is unnecessary to perform the convolutions, and

$$S(t) \propto [G^{\varepsilon\varepsilon}(t)]^2 \qquad (7.10b)$$

The impulse response function $G^{\varepsilon\varepsilon}(t)$ is related to $C^{\varepsilon\varepsilon}(t)$, the time correlation function of the collective polarizability of the liquid at a fixed $\mathbf{q}$ (grating wavevector) (22),

$$G^{\varepsilon\varepsilon}(t) = -\frac{\Theta(t)}{k_B T}\frac{\partial}{\partial t}\,C^{\varepsilon\varepsilon}(t) \tag{7.11}$$

$\Theta(t)$ is a unit step function, and the dielectric time correlation function is

$$C_{ijkl}^{\varepsilon\varepsilon}(t) = \langle\delta\varepsilon_{ij}(0)\delta\varepsilon_{kl}(t)\rangle$$

In Eq. (7.11), the $1/k_B T$ term normalizes the magnitude of the signal and does not influence the time dependence (23). Equation (7.11) can be applied to an arbitrary analytic function $G^{\varepsilon\varepsilon}(t)$ only if the complex function, $G^{\varepsilon\varepsilon}(\omega)$ [Fourier transform of $G^{\varepsilon\varepsilon}(t)$], is regular in the upper half-plane (23). In a situation where $C(t)$ is a sum of discrete exponentials:

$$C_{ijkl}^{\varepsilon\varepsilon}(t) \propto \sum_n A_n \exp\,(-f_n t) \tag{7.12a}$$

Eq. (7.11) gives

$$G_{ijkl}^{\varepsilon\varepsilon}(t) \propto \sum_n A_n f_n \exp\,(-f_n t) \tag{7.12b}$$

Equations (7.12a) and (b) are the observables for dynamic light scattering (DLS) and TG–OKE experiments. The factor f_n in the amplitude of the TG–OKE observable means the contribution of a relaxation is amplified by its rate. Thus, faster decays have inherently greater amplitude. This is one of the reasons that a TG–OKE experiment is better suited than a DLS experiment for observing fast dynamics.

7.2.3. Relationship to Other Experimental Techniques

Dynamics of a fluid system can be described in terms of the density correlation function and the current correlation function (24). Various types of intramolecular and intermolecular motions participate in the density fluctuation of a liquid. There are a variety of spectroscopic techniques that probe the scattering function or the density correlation function. Neutron scattering (NS) measures the Fourier transform of the density correlation function $\langle\rho_{-q}(0)\rho_q(t)\rangle$ in the frequency domain. The TG–OKE technique, DLS, and Raman spectroscopy examine the correlation of molecular polarizability associated with a dynamic mode or modes responsible for the density fluctuation. Low-frequency Raman spectroscopy probes single-molecule reorientation in liquids (25,26), while the DLS and the TG–OKE techniques can give additional information about collective and cooperative molecular reorientation. Within the framework of linear response theory (25), DLS and TG–OKE experiments contain the same information (22). In TG–OKE experiments, the signal comes from Bragg diffraction of the probe beam by a single Fourier component (amplified by the excitation pulses) of the spontaneous dielectric fluctuation of the medium. Decay of the signal corresponds to the quasielastic central peak in DLS and gives information about molecular reorientation. While in principle, DLS and the TG–OKE experiment provide the same information, in practice, the grating experiment has tremendous advantages for the measurement of fast and ultrafast phenomena. The TG–OKE measurements have been made over 6 decades of time and 12 decades of signal intensity (27). As with many time domain analogs of frequency domain experiments, there are large increases in the S/N ratio. In DLS it is difficult to resolve very fast components of multitime scale dynamics because the faster components can appear as an essentially flat baseline. The DLS measures the projections of the polarizability correlation function like $\langle \alpha_{ij}(0)\alpha_{ij}(t) \rangle$ ($i = j$ for VV configuration and $i \neq j$ for VH configuration). In TG–OKE experiments, polarization selectivity allows one to measure additional projections of the third-order nonlinear susceptibility $\chi^{(3)}$, such as χ_{1122} or χ_{1222}, which are proportional to $\langle \alpha_{11}(0)\alpha_{22}(t) \rangle$ or $\langle \alpha_{12}(0)\alpha_{22}(t) \rangle$, respectively (19,22).

7.3. ORIENTATIONAL RELAXATION

7.3.1. Hydrodynamic Model of Orientational Relaxation

A liquid is a dense medium where fluctuations occur spontaneously and continuously. Long-range modes of fluctuation with slow relaxation, that is, with characteristic decay time long compared to the molecular interaction time, are collective modes. These processes involve a large number of particles and their relaxation times are proportional to the square of their characteristic wavelength, which is large compared to the intermolecular distance (24). These slow modes are called hydrodynamic modes. In a "hydrodynamic" liquid, molecular orientation is coupled to the hydrodynamic modes. In the simple hydrodynamic model, a molecule is assumed to undergo orientational diffusion in a featureless continuous medium. Molecular motions in simple hydrodynamic liquids that occur on a picosecond or longer time scale can generally be described by the Debye-Stokes-Einstein (DSE) equation (28). The rotational diffusion time constant τ (decay time for the orientational correlation function) is given by

$$\tau = \frac{V\eta}{k_B T} \qquad (7.13)$$

where V is the volume of the rotating particle, η is the shear viscosity of the fluid, T is the temperature, and k_B is the Boltzmann constant. The application of the DSE equation to the experimental analysis of rotational diffusion of molecules in liquids has shown the necessity for the addition of a term, τ_0 (see below) (29).

This simple model for rotational diffusion is often accurate in the case of a dilute solution. One needs to consider the collective effects in high concentration solutions or neat liquids. The correlation function measured by the OKE and light scattering is a multiparticle correlation function. This is different from NMR and fluorescence experiments, for example, which measure single-particle correlation functions. Keyes and Kivelson (30) showed that although the single-particle and two-particle correlation functions may relax with different correlation times, the total correlation will relax with a single-correlation time. Specifically, if the single-particle correlation function relaxes as a single exponential with a time constant τ, then the total correlation function relaxes as a single exponential with a time constant τ_c, related by (30,31)

$$\tau_c = (g_2/j_2)\tau \qquad (7.14)$$

where g_2 is the static orientational correlation parameter, and j_2 is the dynamic orientational correlation parameter.

In general, it appears that the dynamic correlation parameter j_2 is approximately unity (28). The static correlation factor can be independently determined by integrated light scattering intensities (32). If the assumption that $j_2 \approx 1$ is made, then g_2 can also be determined by comparison of the results of an experiment that measures τ_c with another experiment that measures τ. This approach was pioneered by Alms et al. (33), using DLS and NMR. If one also makes the assumption that the hydrodynamic boundary conditions remain unchanged by dilution, then one can determine g_2 by comparison of τ_c for the neat liquid with τ_c for a dilute solution. This is the approach that was used by Greenfield et al. (34) for studying 2EN. Thus, the DSE equation in its final form is

$$\tau_c = g_2 \left(\frac{\lambda V \eta}{k_B T}\right) + \tau_0 \qquad (7.15)$$

where the Youngren and Acrivos frictional coefficient λ and the nonzero intercept τ_0 have been included (28). The correction parameter λ takes into account other factors that contribute to the rotational friction besides the viscosity. Sometimes λV is also referred to as V_{eff}, the effective volume.

7.3.2. Local Structure and Nonhydrodynamic Relaxation

Local orientational relaxation does not necessarily couple to the low-frequency hydrodynamic modes, and therefore does not always obey the DSE equation. This has been observed in several systems (27,34,35) described later in this chapter. The lack of η/T dependence can be attributed to the well-defined local structures that persist for a time much longer than the time scale of the observed relaxations. In a crystal, lattice vibrations, including those involving orientational motions, occur superimposed on a minimum energy background lattice that is fixed at all times. However, in liquids, the background structure itself is subject to change during the time a librational mode is excited due to the presence of other configurational states that are close in energy. (A libration is an orientational "vibrational mode" of the liquid. It has been shown that such modes of guest molecules in mixed-molecular crystals actually involve both orientational and translational components (36). This finding is certainly true in liquids as well. However, following common usage, we will refer to these modes as librations instead of the more general term pseudolocal modes.)

Recent theoretical calculations of the instantaneous normal modes of liquids (37) revive a concept first put forward by Maxwell (38), that is, for times shorter than a characteristic "Maxwell time" τ_m, a liquid behaves in many respects like a solid. Using a method analogous to the harmonic normal mode analysis of solid state physics, configuration averaged densities of states $\langle \rho(\omega) \rangle$ of liquids have been calculated (37). The calculated modes consist of both real and imaginary values of the frequency. The imaginary ω values imply that the structure is unstable on some time scale. The real ω values represent a spectrum of oscillatory modes that are analogous to the phonons of a fixed crystalline lattice. τ_m, the time scale on which the local structure is preserved, is determined by the density of unstable modes. When the density of unstable modes (imaginary ω values) becomes large at $|\omega| \leq |\omega_u|$, a significant structural rearrangement would occur at times $t \geq |\omega_u^{-1}| = \tau_m$ (37). For $t \geq \tau_m$, the local structure evolves, and the relaxations slower than τ_m should be sensitive to a change of temperature and viscosity.

Dynamics observed in TG–OKE experiments start with an anisotropy induced in the sample by the excitation beams. The coupling of the radiation field to the molecules involves stimulated Raman scattering exciting a subset of the real frequency spectrum of liquids, that is, librational modes (37,39–42) of the molecules. These essentially harmonic motions are modes of the potential surface associated with the fixed local structure of the liquid that exists on the ultrashort time scale of the excitation pulses. Because the subpicosecond pulses are short, they have a broad frequency bandwidth ($\sim$100 cm^{-1}). Thus, the pulses contain both frequencies necessary to excite the low-frequency (a few tens of cm^{-1}) librations by stimulated Raman scattering. Thus, in addition to the thermally populated librations, excitations are generated by stimulated Raman scattering. The orientation of the excited librations is determined by the polarization of the excitation radiation field. Therefore, the laser excited librations add to the isotropic distribution of thermally excited librations, making the ensemble of excited librations anisotropic. The initial ultrafast transient ($\sim$ 100 fs) seen in TG–OKE experiments consists of a rise in the signal as the overdamped librators begin to undergo angular displacement and the partial (or complete) decay of the signal as the librational motion dephases and damps. In a dense liquid, coherent motions of pairs, triplets, and so on, of molecules can also be driven to some extent and will lead to a very short-lived contribution to the signal through interaction induced polarizabilities (43,44).

In a crystal, stimulated Raman scattering excites optical phonons. Because of the well-defined lattice structure, damping of the optical phonons returns the molecules to their original positions, leaving no residual anisotropy. In a liquid, damping of the optically excited librations can result in orientational displacements from the initial isotropic configurations. This leaves a longer lived residual orientational anisotropy that will decay by some form of orientational relaxation. The orientational relaxation is nonhydrodynamic if the local liquid structure is preserved during the relaxation.

7.4.   A SIMPLE MOLECULAR LIQUID: BIPHENYL

In a simple liquid, such as biphenyl (45), fast librational motions and a bi-exponential decay of the residual anisotropy are observed. The librational dynamics are essentially temperature independent, but the long-lived anisotropy decays by orientational diffusion. Both components of the biexponential obey the DSE equation.

Figure 7.3 displays TG–OKE data taken on neat biphenyl liquid. Two experimental traces with fits are shown in Fig. 7.3(a). The curve labeled 1 in Fig 7.3(a) shows the electronic response from neat biphenyl. As discussed in Section 7.2.1, the intensity profile of the laser pulse $I(t)$ is obtained by observing the electronic OKE. $I(t)$ is found to be well described by a double sided exponential (13), as shown by the fit through the data. Once $I(t)$ is obtained, it is numerically convolved with an analytic response function, $G_n^{\varepsilon\varepsilon}(t)$, according to Eq. (7.10a) to fit the nuclear OKE data. The curve labeled 2 in Fig. 7.3(a) is the short time nuclear OKE data and the fit to the data including the convolution with the pulse shape determined by the fit to curve 1. At times greater than 2 ps, the data are taken using a polarization grating configuration, and the dynamics arise from orientational diffusion, as discussed below. For orientational relaxation having characteristic decay times much longer than the pulsewidth, that is, $t > 2$ ps, data are analyzed in a straightforward manner without having to numerically compute convolution integrals. In biphenyl, for $t > 2$ ps, the square root of the signal decay, $\sqrt{S(t)} \propto G^{\varepsilon\varepsilon}(t)$, is given by a biexponential function $A_1 \exp[-t/\tau_1] + A_2 \exp[-t/\tau_2]$. The dashed line through the data in Fig. 7.3(b), barely discernable from the data, is the square of this biexponential function. The general behavior described in connection with Fig. 7.3 is shared by all samples over the entire temperature range investigated, that is, the signal begins with a very fast (fs) librational component and then continues with a slower (picoseconds to nanoseconds) orientational decay of the residual anisotropy.

In a system in which the moment of inertia tensor and the molecular polarizability tensors coincide or almost coincide, which is the case for biphenyl, hydrodynamic theory predicts a biexponential decay of the orientational anisotropy (32). As can be seen from Fig. 7.3(b), in neat biphenyl, for $t > 2$ ps, the decay of $G^{\varepsilon\varepsilon}(t)$ is given by a biexponential characterized by two time constants τ_1 and τ_2. τ_1 characterizes the slower tumbling around the short axes, whereas τ_2 contains contributions from spinning and tumbling motion. The time constant for pure spinning motion is (45)

$$\tau_x = \frac{1}{6\Theta_\parallel} = \frac{2\tau_1\tau_2}{3\tau_1 - \tau_2} \tag{7.16}$$

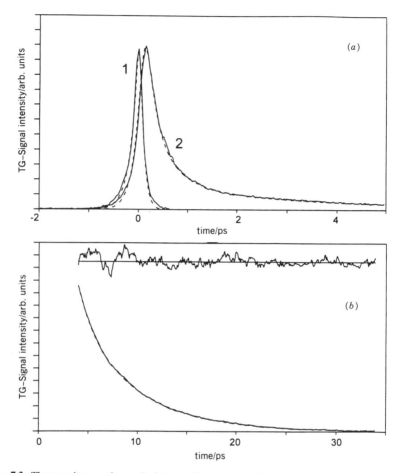

Figure 7.3. The transient grating optical Kerr effect signals $S(t)$ for a 0.77 M biphenyl/n-heptane solution at 23°C. (a) Ultrafast time scale behavior. Data labeled 1 and 2 are obtained using the electronic (0°/45°/90°/56°) and the nuclear (0°/45°/90°/135°) TG–OKE configuration, respectively. The dashed lines are the fits calculated by evaluating the convolution integrals of Eq. (7.10a). The response function from Eq. (7.18) is used to calculate the fit through the data set 2 (the nuclear Kerr response). Data set 1 (the electronic Kerr or instrument response) is fit with a $\delta(t)$ response function. (b) Slow time scale behavior of the signal, recorded using the polarization grating configuration (0°/90°/90°/0°). The dashed line is a fit with a biexponential [Eq. (7.10b)]. The trace at the top shows the residuals of the fit on a 10X magnified vertical scale. [Reprinted with permission from Ref 15.]

where $\tau_1 = 1/6\Theta_\perp$ and $\tau_2 = 1/(4\Theta_\| + 2\Theta_\perp)$. The parameters $\Theta_\|$ and $\Theta_\perp$ are the rotational diffusion constants for rotation around the long axis and two short axes of a biphenyl molecule, respectively. We plotted the values of τ_1 and τ_x versus η/T in Figs. 7.4 and 7.5, respectively. Data taken on biphenyl in n-heptane solutions are also displayed in the figures. In Figs. 7.4 and 7.5, τ_1 and τ_x show linear dependencies on η/T, as predicted by the DSE equation [Eq. (7.28)] (28). If there is

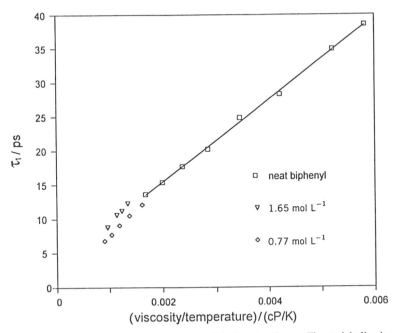

Figure 7.4. The η/T dependence of the slow exponential relaxation time τ_1. The straight line is a linear fit through the data for neat biphenyl only. Data taken in n-heptane solutions are also shown. [Reprinted with permission from Ref. 15.]

any orientational correlation between the biphenyl molecules, we would expect that this correlation, contained in the parameter $g_2 V_{\text{eff}}$, would change going from the neat liquid to the n-heptane solutions. Within experimental error, which is larger in the solutions than in the neat liquid, the slopes of the lines are unchanged. Thus $g_2 \approx 1$. The DSE equation is obeyed, and we consider neat biphenyl to exhibit similar hydrodynamic behavior. τ_x for biphenyl is smaller than the orientational relaxation times of benzene. Several possibilities were considered by Deeg et al. (45); they concluded that internal dihedral angle relaxation is the best description of the faster relaxation component.

Coherent librational motion and "incoherent" diffusive reorientational motion in a liquid are coupled. A correct response function has to describe the librational excitation and damping, the diffusive reorientational motion, and the coupling between the two kinds of motion. To take these into account, Deeg et al. (45) used a response function for a single libration/reorientation of the form

$$G^{\varepsilon\varepsilon}(t) = C_l[\exp(-t/\tau_b) - \exp(-t/\tau_a)] + C_r[1 - \exp(-t/\tau_b)]\exp(-t/\tau) \quad (7.17)$$

The C_l term describes the libration as an overdamped oscillator characterized by time constants τ_a and τ_b. τ_a characterizes the rise of the libration induced anisotropy, and τ_b characterizes the decay of the libration. The C_r term describes the diffusive

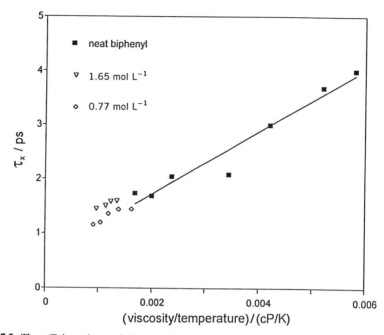

Figure 7.5. The η/T dependence of the relaxation time, τ_x as calculated using Eq. (7.16). The straight line is a linear fit through the neat biphenyl data only. Data taken in n-heptane solutions are also shown. [Reprinted with permission from Ref. 15.]

reorientation; the $[1 - \exp(-t/\tau_b)]$ factor assures the coupling between librational and reorientational relaxation and delays the onset of the diffusive reorientation by the librational damping of time. τ is the diffusive orientational relaxation time. The physical model embodied in Eq. (7.17) is as follows. The excitation pulses excite an anisotropic distribution of librations through stimulated Raman scattering. The excitation is impulsive, that is, the pulses are shorter than the period of the librational motion. Thus, at $t = \sim 0$ the librations have kinetic energy but have yet to displace from their initial positions. The signal depends on an orientational anisotropy in the sample. Therefore, the signal builds up as the librators displace. This is the rising edge of the signal that shows a significant shift from $t = 0$, as seen in Fig. 7.3(a) by comparing curves 1 and 2. The librations are overdamped. If the librators damped precisely to their initial configurations, then following damping, there would be no long-lived anisotropy in the sample and no long-lived signal. However, upon damping the molecules are left in somewhat different orientations than they had initially. This results in a residual anisotropy that, in a hydrodynamic system such as biphenyl, decays by rotational diffusion. Thus, the anisotropy that decays by diffusion is not present at $t = 0$, but rather builds in with the librational damping. Another factor that influences the signal on the time scale of the librational motion is inhomogeneous dephasing of the librators (46,47). Therefore the time dependence of the librational decay may not be due solely to damping. Nonetheless, the libration must thermalize; so the basic picture described above is essentially correct. Equation (7.17) provides a reasonable model for the short time dynamics that are important for understanding the longer lived orientational relaxation that follows the decay of the librational signal.

One immediate implication of the picture of molecular dynamics developed here is the fact that C_l and C_r in Eq. (7.17) are not independent parameters but are related. The ratio C_r/C_l depends on the relationship between the time scales for the librational damping and the solvent structure evolution. If we associate the time scale for the solvent structure randomization with the time constants of the orientational relaxation τ (this does not imply that they are identical), the C_r/C_l should be a function of τ_b/τ, the ratio of the librational damping time and the orientational relaxation time. If the librational damping does not depend strongly on temperature/viscosity, then we expect C_r/C_l to increase with increasing temperature/decreasing viscosity, that is, decreasing τ. Comparison of the intensity of the signal at its maximum (when it is dominated by the libration) with the intensity at $t = 3$ ps (when the diffusive reorientation is predominant), actually shows a trend confirming this idea.

Two librations, like the two diffusive reorientational components, corresponding to motions around the long axis and the short axes of the biphenyl molecule were considered for fitting the ultrafast time scale nuclear OKE data. The response function is

$$G^{\varepsilon\varepsilon}(t) = C_{l,1}[\exp(-t/\tau_{b,1})-\exp(-t/\tau_a)]+C_{r,1}[1 - \exp(-t/\tau_{b,1})]\exp(-t/\tau_1)$$

$$+C_{l,2}[\exp(-t/\tau_{b,2})-\exp(-t/\tau_a)]+C_{r,2}(1 - \exp(-t/\tau_{b,2})]\exp(-t/\tau_2) \quad (7.18)$$

The calculated curve (dashed line) through the data (curve 2) of Fig. 7.3(a) uses the parameters: $C_{l,1} = 90$, $C_{l,2} = 3.4$, $\tau_a = 40$ fs, $\tau_{b,1} = 50$ fs, $\tau_{b,2} = 480$ fs, $C_{r,1} = 1$, $C_{r,2} = 1.12$, $\tau_1 = 2.1$ ps, and $\tau_2 = 12.3$ ps. The agreement between the calculations using Eq. (7.18) and the data is excellent. Although there are a large number of parameters, many of them are determined independently, for example, the decay constants for the two components of the rotational diffusion are obtained from the fits to the data at long time and are independent of the model. The fact that the rotational diffusion decay of the residual anisotropy begins immediately following the decay of the librational signal and merges smoothly with it, as described by Eq. (7.18), lends further support to the basic physical model of the relationship between the librational damping and the development of the residual anisotropy that decays by orientational relaxation.

Another significant feature of the data presented in Fig. 7.3 and fit with Eq. (7.18) is its relationship to DLS experiments on the same system. Biphenyl in n-heptane was examined using DLS (48). In the DLS experiments, a single Lorentzian line was observed with a width that matches the slowest rotational diffusion decay constant. Where DLS observes a single decay component, the TG–OKE experiment reveals five components in the dynamics: the time constant for the initial displacement of the librators, the biexponential decay of the librational component, and the biexponential hydrodynamic decay of the residual anisotropy. In a DLS experiment, the faster components can be missed as part of a broad baseline.

7.5. THE ISOTROPIC PHASE OF LIQUID CRYSTALS

7.5.1. Criticality and Long-Range Order

The extent of order in a macroscopic system is often measured in terms of the thermodynamic function S, the entropy. The average fluctuation of the entropy in an equilibrium system is obtained by measuring the heat capacity, $C_p = \langle (\Delta S)^2 \rangle$ (49). Two phases can coexist under suitable temperature and pressure conditions. The transition that takes place through a coexistence curve is called first order; a latent heat L and a finite entropy change $\Delta S = L/T$ give a measure of the disorder (50). At a first-order transition, the first derivative of the Gibbs free energy G, with temperature is discontinuous, for example, the volume and the entropy (51). Higher order transitions do not display the sudden variations that are observed in the first-order transitions (51). The entropy changes steadily, as do the other state functions. However, the derivatives of these functions versus temperature, for example, the heat capacity display a divergence near the transition temperature, T_c, the critical temperature. Second-order transitions are those at which the second derivatives of G are discontinuous or infinite, while the first derivatives of G are continuous. Many important phenomena are observed around higher order phase transition temperatures. These are known in general as critical phenomena. A quantity called the order parameter, which is useful for the description of order–disorder transitions, can be identified for any critical transition. The order parameter is the fundamental macroscopic variable that makes it possible to formulate a unified description of the order of critical transitions in very different systems (52,53). Long-range correlations in the order parameter fluctuations are a general feature of critical behavior. Examples of such critical fluctuations occur in liquids above their critical points, ferromagnetic transitions, ferroelectric transitions, and Bose condensation in superfluids and superconductors. The order parameters in these examples are the density, the magnetization, the polarization, the wave function of ground level, and the wave function of Cooper pairs, respectively. The nematic–isotropic phase transition of liquid crystals is an important example in a chemical system in which criticality is superimposed on a weak first-order transition (54). It is a weak first-order transition because a small but finite entropy change ΔS is observed at the nematic–isotropic transition temperature T_{ni}, where a sharp transition in the system's symmetry occurs. The heat capacity shows divergence at T^* slightly below T_{ni}. This is a characteristic feature of second-order thermodynamic criticality. Many pretransitional phenomena are observed in the isotropic phase of nematic liquid crystals (54). They are explained as critical behavior, and orientational correlation is identified as the order parameter. The physical origin of critical anomalies is the increase in the correlation length of the order parameter on approaching the critical temperature from above. The increase in the range of correlated regions leads to a corresponding increase in the time scales for dynamic processes (critical slowing down) (55). The correlation extends over distances that are increasingly long and times that are much longer than those associated with the dynamics of microscopic elements of the system. Anisotropic intermolecular interactions and correlation of fluctuations cause cooperativity in the alignment of the permanent dipole moments of the liquid crystal molecules.

In this section, detailed TG–OKE experiments are presented that examine dynamics in the isotropic phase of liquid crystals in the pretransitional temperature regime. Two liquid crystal systems are studied in detail. These are pentylcyanobiphenyl (5CB) and N-(methoxybenzylidene)butylaniline (MBBA). Dynamics of the pseudonematic domains that exist in the isotropic phase are investigated. The temperature dependence of the orientational relaxation of the domains is followed over a greater temperature range than has been possible previously, and the temperature and correlation length at which the domain description ceases to be valid is found. The orientational dynamics of the molecules that make up the domains (intradomain dynamics) are investigated on fast time scales. It is found that the intradomain dynamics are independent of temperature and viscosity over wide ranges of η/T although over the same ranges, the orientational relaxation dynamics of the domains themselves, occurring on a much longer time scale, change dramatically.

7.5.2. Orientational Order in the Isotropic Phase of Liquid Crystals

Early work on the dynamics of the isotropic phase of nematic liquid crystals focused on the slow dynamics caused by long-range order parameter fluctuations. These slow dynamics are well described by Landau-de Gennes (LdG) theory and are the collective relaxation of the orientationally correlated regions often called pseudonematic domains (56–64). Flytzanis and Shen (65) pointed out that there should be additional orientational motions corresponding to the fast motions within the pseudonematic domains. Fast effects were shown to exist by both time (66) and frequency domain (67) experiments. These experiments had limited time resolution or temperature range, which did not permit quantitative analysis of the data. Quantitative analysis of these effects was made possible by using TG–OKE experiments, which increased the time resolution by four orders of magnitude over the bulk of previous experiments (57,59–62). These higher resolution TG–OKE experiments were performed over a wide temperature range on 5CB (68) and MBBA (27).

Landau–de Gennes theory applies to static parameters such as magnetic (56) and electric birefringence (59,69) and scattered light intensity (56,58) as well as dynamic effects seen in nuclear spin lattice relaxations (70,71), DLS (56,58), and the OKE (57,59–62) in the temperature region just above the nematic–isotropic phase transition. Deviations from LdG theory have been seen in TG–OKE experiments (16,68) and in magnetic and electric birefringence (72) at temperatures sufficiently above the nematic–isotropic transition.

Local orientational order in the isotropic phase of a nematic liquid crystal is characterized by a microscopic scalar order parameter, S,

$$S = \frac{1}{2} \langle 3\cos^2 \theta - 1 \rangle \qquad (7.19)$$

where θ is the angle between the axis of the rodlike molecule and a reference direction called the director (73,74). If the local order possesses some axial symmetry, one of the symmetry axes can be chosen as the director about which S has the highest value. For local nematic order, it is the local nematic axis. Sometimes it is not obvious whether the system has an axial symmetry; in that case the director is defined as the reference direction that maximizes S. In the isotropic phase of a nematic liquid crystal, a macroscopic symmetric traceless tensor order parameter of rank 2, $\underline{\underline{Q}}$ is used to describe the anisotropy of thermodynamic properties, for example, magnetic and dielectric anisotropy. The elements of $\underline{\underline{Q}}$ are determined by the values of the microscopic order parameter S.

Landau–de Gennes theory provides a good description of the phase transition in the pretransitional region. de Gennes extended the Landau theory of the second-order phase transitions to the weakly first-order nematic–isotropic phase transition (73,75). Landau–de Gennes theory as applied to liquid crystals defines the free energy as an expansion in powers of the order parameter. By neglecting the spatial variation of the director and the order parameter, the free energy is expanded as

$$F_n = F_i + AS^2 - BS^3 + CS^4 \tag{7.20}$$

where F_n is the free energy of the nematic phase, F_i is the free energy of the isotropic phase, A, B, C are constants, and S is the microscopic order parameter. Since the nematic–isotropic transition is a weak first-order transition, the magnitude of B is small. The constant A is inversely proportional to the birefringence when the sample is in a magnetic field, and goes to zero on approaching the transition temperature from above (54,73).

$$A = a(T - T^*)^\gamma \tag{7.21}$$

In this manner, the system produces large fluctuations in the order parameter near the temperature T^* with minimum expense to the free energy. This is the thermodynamic origin of pretransitional phenomena. The parameter T^* is the temperature where a second-order phase transition should occur, but it is actually slightly below the nematic–isotropic transition temperature T_{ni}. For systems where mean field theory is valid, $\gamma \approx 1$ (54).

The spatial variations of the local alignment contribute to the free energy through terms involving spatial derivatives of the order parameter. In the isotropic phase, these terms are important for fast time scale dynamics that occur on short distance scales. The total free energy in the isotropic phase can be expressed in terms of the fluctuations of the macroscopic order parameter $\underline{\underline{Q}}$:

$$F = F_0 + \sum_{\alpha\beta\gamma} A(Q_{\beta\gamma})^2 + L(\partial_\alpha Q_{\beta\gamma})^2 \tag{7.22}$$

where α, β, γ are indexes referring to the laboratory reference frame and $\partial_\alpha = \partial/\partial x_\alpha$. F_0 is the free energy at the minimum and L is the elastic constant in the isotropic phase; its magnitude is smaller than in the nematic phase and is only weakly dependent on temperature. The fluctuations can be expanded in terms of fluctuation modes.

$$Q(\mathbf{r},t) = \sum_{\mathbf{q}} Q(\mathbf{q},t)e^{i\mathbf{q}\cdot\mathbf{r}} \tag{7.23}$$

For simplicity the indexes are omitted. Thus, the free energy associated with a fluctuation of the order parameter at wavevector $\mathbf{q}$ is

$$f_q = (F - F_0)_q = AQ_q^2(t) + Lq^2Q_q^2(t) \tag{7.24}$$

Q_q can be expressed in terms of a microscopic order parameter S_q, which describes the molecular orientational order for various distance scales,

$$S_q = \langle S \rangle_q \tag{7.25}$$

where the averaging is done over a distance scale of q^{-1}. The relaxation of a fluctuation is determined by the slope of the free energy surface on the relevant distance scale. Then

$$\frac{\partial S_q}{\partial t} \propto \frac{\partial Q_q}{\partial t} = -\frac{1}{2\eta_q}\frac{\partial f_q}{\partial Q_q} \tag{7.26}$$

η_q is a viscosity coefficient on the corresponding length scale. For a parabolic surface, the time evolution is exponential with a decay constant Γ_q. From Eqs. (7.24) and (7.26).

$$\Gamma_q = \frac{1}{\eta_q}(A + Lq^2) \tag{7.27}$$

In the isotropic phase, just above the phase transition, there are regions of ori-entationally correlated molecules. In the LdG theory the correlations have an Ornstein–Zernike form (54):

$$\langle S(0)S(r) \rangle \approx \frac{1}{r}\exp\left(\frac{-r}{\xi}\right) \tag{7.28}$$

ξ is the correlation length, and r is the distance. ξ is related to the constants L and A, $\xi = (L/A)^{1/2}$. From Eq. (7.21), it is possible to see the manner in which the temperature dependence of the correlation length shows critical behavior (54)

$$\xi = \xi_0 \left(\frac{T^*}{T - T^*}\right)^{\gamma/2} \tag{7.29}$$

de Gennes identifies ξ_0 as a molecular dimension. Work by Courtens (76,77), Chu et al. (78,79), and Stinson and Litster (80) determined ξ_0 and found it to be in the range 5.5–7 Å. This is comparable to the cube root of the molecular volume (~8 Å) of a liquid crystal molecule, such as MBBA. For comparison the molecular length of MBBA is 18 Å.

The dispersion formula for Γ_q is the obtained by introducing the expression for ξ into Eq. (7.27).

$$\Gamma_q = \frac{L}{\eta_q}\,(q^2 + \xi^{-2}) \tag{7.30}$$

At short-distance scale where $q\xi \gg 1$, the relaxation depends on the fine structure of free energy surface and $\Gamma_q = Lq^2/\eta_q$. This is what we refer to as intradomain relaxation, the fast orientational dynamics. These dynamics have not been investigated previously. Results of the TG–OKE experiments presented below provide a very detailed picture of the intradomain dynamics. Since L is weakly dependent on temperature and the viscosity at high wavevector and high frequency is temperature independent (24,55), the relaxations in this limit are essentially temperature independent.

At $q\xi \ll 1$, orientational relaxation depends only on a critical length ξ, and the relaxation time is given by

$$\tau = \Gamma_q^{-1} \propto \frac{\eta(T)}{(T - T^*)^\gamma} \tag{7.31}$$

This is referred to as a modified DSE equation since it has a form analogous to Eq. (7.13) except that there is a critical temperature in the denominator. This slow relaxation can be visualized as orientational diffusive randomization of the pseudonematic domains. For liquid crystals, γ has been shown to be about 1, indicating the validity of the application of mean field theory (54). The most accurate measurement of γ is presented below.

At sufficiently high temperature, one would expect the domain size to become small enough that the isotropic phase of a liquid crystal would cease to exhibit behavior determined by the existence of pseudonematic domains. Early measurements on MBBA orientational relaxation were performed over a very limited temperature range that did not reach temperatures at which LdG theory no longer holds (61,81). As described in Section 7.5.3, recent measurements over a broad temperature range on MBBA (27) and on the liquid crystal 5CB (68) observed the deviations from the LdG theory well above T_{ni}.

7.5.3. Experimental Results

Figure 7.6 displays typical TG–OKE data taken on MBBA at 68.3°C. In panel (*a*), the electronic OKE and the nuclear OKE data are shown. All of the data sets begin with the very fast librational feature. The data from 1 to 4 ps are blown up

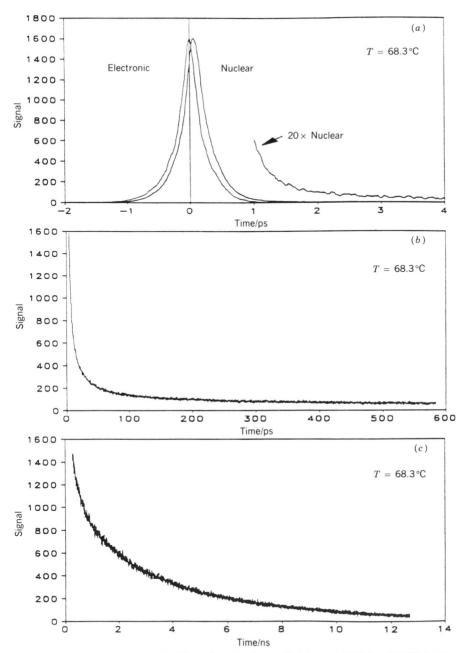

Figure 7.6. Transient grating optical Kerr effect data for the liquid crystal MBBA at 68.3°C. (*a*) Electronic and nuclear OKE. The electronic effect, centered at $t = 0$, gives the instrument response. The nuclear effect arises from the orientational dynamics of the MBBA. The short time behavior reflects librational dynamics. For $t > \sim 1$ ps the data are multiplied by 20. This is the beginning of the orientational relaxation. (*b*) The time scale is expanded. The steep part of the decay is the almost flat part (> 2 ps) in panel (*a*). (*c*) The time scale is expanded. The steep part of the decay is the almost flat part (> 30 ps) in panel (*b*). The data are nonexponential and span a very broad range of time scales and a very broad range of signal amplitude. [Reproduced with permission from J. J. Stankus, R. Torre, and M. D. Fayer, *J. Phys. Chem.* **1993**, *97*, 9478. Copyright © (1993) American Chemical Society.]

by a factor of 20. After 2 ps the decay is almost flat in this plot. In panel (b), the very fast decay is the continuation of the data that appear flat in panel (a). The data become almost flat after about 100 ps. In panel (c), the steep part of the data at short time is the almost flat data displayed in (b). Clearly these data are highly nonexponential. They span a very wide range of time [tens of femtoseconds (fs) to tens of nanoseconds (ns)] and a very broad dynamic range of signal intensity.

As seen from Fig. 7.6, the TG–OKE data of liquid crystals in their isotropic phase can be divided into three time ranges. At very short time (< 1 ps), the data reflect the librational dynamics discussed in Sections 7.3.2 and 7.4. The very short time behavior is fundamentally the same as discussed above and will not be considered here. The intermediate time scale data (~1 ps to ~ 1 ns) arise from the fast intradomain dynamics, and the long time scale data (>1 ns) arise from the slow randomization of the pseudonematic domains. The slow dynamics can be discussed in terms of the LdG theory.

7.5.3A. The Slow Dynamics.

The TG–OKE experiments were performed on both 5CB and MBBA over a broad range of temperatures. The long-time scale dynamics were fit to an exponential squared, since the observable is the square of the response function [see Eq. (7.10b)]. To analyze the faster dynamics (discussed below), the square root of the data was taken to eliminate cross terms in Eq. (7.10b). The slow LdG exponential decay is subtracted from the full data set, leaving the faster data.

The MBBA (T_{ni} = 46.9°C) slow orientational relaxation decay constant was measured as a function of temperature from 49.4–119.7°C. The 5CB (T_{ni} = 35.2°C) measurements were made between 36.4 and 120°C. The decay times for the pseudonematic domain randomization in MBBA and 5CB vary from 174 to 0.91 ns and 170 to 0.69 ns, respectively, over the temperature ranges. For MBBA, the decay times versus $\eta/(T - T^*)$ obtained at the lower temperatures and a fit to the LdG theory are shown in Fig. 7.7. The viscosity η of MBBA has been reported by Martinoty et al. (82). The temperature T^* was determined using a formula of de Gennes (10) followed by a slight variation in the fit and was found to be 319.6 K. Since at high enough temperature the data must deviate from the LdG theory, the LdG fit was performed over several low-temperature ranges to assure the theory is applicable. The LdG curve was fit for 49.4 → 60°C, 49.4 → 70.0°C, and 49.4 → 80.3°C. All three fits gave the same slope to within 1%, providing an accurate determination of the LdG curve.

All of the slow data (full temperature range) and the LdG curve obtained from Fig. 7.7 are plotted as rate versus temperature in Fig. 7.8. Similar data for 5CB are presented in Fig. 7.9. The fit to the LdG theory is good over a surprisingly wide temperature range with deviation beginning at about 90°C for MBBA and at about 70°C for 5CB. An accurate value of γ = 1.01 ± 0.01 was obtained. The deviations at higher temperatures cannot be accounted for by a change in γ. Previous light scattering and OKE experiments on MBBA covered a narrow temperature range of only a few degrees (61,81). These studies gave a value of $\gamma \simeq 1$. Because of the large temperature range covered here, this is the most accurate

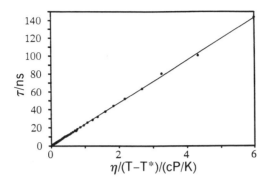

Figure 7.7. The critical behavior of the slow dynamics in MBBA. The slowest decay times versus the viscosity divided by the temperature minus the critical temperature T^*. The straight line is the fit to the LdG theory over the lower range of temperatures. [Reprinted with permission from J. J. Stankus, R. Torre, and M. D. Fayer, *J. Phys. Chem.* **1993**, *97*, 9478. Copyright © (1993) American Chemical Society.]

determination of the exponent γ, and it supports the applicability of mean field theory (54) over a broad range of temperatures.

A very important result of these measurements is that the deviation from LdG theory in both MBBA and 5CB occurs at the identical correlation length. Although the deviations begin at different temperatures in the two liquid crystals, in both the deviations begin when the correlation length is somewhat less than $3\xi_0$ (~$2.7\xi_0$). The best determination is that the onset of the deviation occurs at about $2.7\xi_0$ in both MBBA and 5CB. This fact may suggest a universal behavior.

When the correlation length in the isotropic phase falls below $3\xi_0$, the liquid crystals begin to make a transition from an ordered to a simple liquid. It is remarkable that the orientational dynamics follow the LdG theory to $\xi \approx 3\xi_0$. This

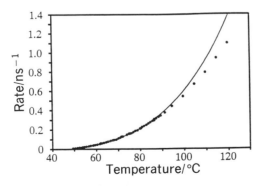

Figure 7.8. Rate versus temperature for the slowest decay in the isotropic phase of MBBA. The solid line is the LdG theory curve. Deviation from LdG theory begins at about 90°C, corresponding to a correlation length of about $3\xi_0$. [Reproduced with permission from J. J. Stankus, R. Torre, and M. D. Fayer, *J. Phys. Chem.*, **1993**, *97*, 9478. Copyright © (1993) American Chemical Society.]

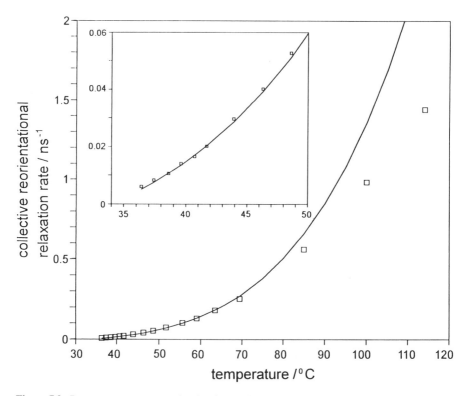

Figure 7.9. Rate versus temperature for the slowest decay in the isotropic phase of the liquid crystal 5CB. The solid line is the LdG theory fit calculated using $T^* = 34.0°C$ and an Arrhenius activation energy for the shear viscosity, $E_a = 34.3$ kJ mol^{-1}. Deviation from LdG theory occurs at about 70°C, corresponding to a correlation length of about $3\xi_0$, the same as observed for MBBA. [Reprinted with permission from Ref. 68.]

demonstrates the profound effect that local ordering in a liquid can have on liquid-state dynamics. The transition to a simple liquid will occur over a broad temperature range. If a sufficiently high temperature could be reached, the orientational relaxation should display the normal DSE behavior of a simple liquid. For temperatures at which $\xi < 3\xi_0$ but below the onset temperature for DSE behavior, MBBA and 5CB display dynamics that do not follow the LdG theory but are, nonetheless, strongly influenced by the details of the local liquid structure. In this range of temperatures, thermal fluctuations are not sufficient to totally overcome intermolecular interactions that affect orientational dynamics. The quantity $3\xi_0$ is the minimum correlation length for which the LdG theory applies. Therefore, it can be considered to be the minimum pseudonematic domain size. As mentioned above, ξ_0 for MBBA is between 5.5 and 7 Å (76–79). This value is approximately the length that gives rise to the (spherical) molecular volume. Within $3\xi_0$ there are about 30 molecules.

7.5.3B. The Fast Dynamics. Figure 7.10(*a*) shows a log–log plot of four short time scale data sets of MBBA taken at 52.6, 60.8, 68.5, and 78.2°C. Within experimental error, all of the data sets display identical highly nonexponential decays. The decays are viscosity/temperature (η/T) independent in contrast to the slow dynamics that change by a factor of about 170. Shear viscosity changes by a factor of 10 (82). The intradomain fast orientational relaxation is clearly not hydrodynamic.

Figure 7.10(*b*) compares the average of the lower temperature decays [all curves from Fig. 7.10(*a*)] with higher temperature decays. The faster dynamics become temperature dependent at about 90°C. This is the same temperature regime in which

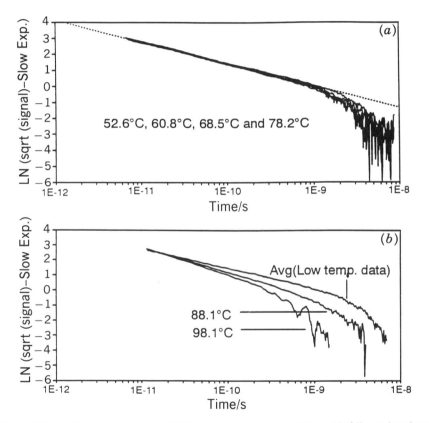

Figure 7.10. (*a*) Fast decay data sets of MBBA at four temperatures (52.6 → 78.2°C) are plotted versus time on a log–log plot. The decays are superimposable, showing that the fast MBBA dynamics are viscosity–temperature independent. The dotted line is a straight line through the data showing that the decays obey a power law ($t^{-\alpha}$) with $\alpha = 0.63 \pm 0.03$. (*b*) The average of the lower temperature data displayed in (*a*) and two data sets taken at higher temperatures. When the correlation length falls below $3\xi_0$, the LdG theory no longer describes the slow decay component, and the fast dynamics become temperature dependent. [Reproduced with permission from J. J. Stankus, R. Torre, and M. D. Fayer, *J. Phys. Chem.*, **1993**, *97*, 9478. Copyright © (1993) American Chemical Society.]

the slow dynamics of MBBA begin to deviate from LdG theory, that is, the temperature range in which the correlation length beocmes so small that pseudonematic domains no longer exist. The fast data for 5CB (1 ps to 1 ns) display the same behavior (68), that is, it is η/T independent from the phase transition temperature up to about 70°C. As long as the correlation length is sufficiently long for pseudonematic domains to exist, as demonstrated by the slow dynamics obeying the LdG theory, the fast intradomain dynamics are η/T independent. This is in spite of the fact that the relaxation times for the slow dynamics change by more than an order of magnitude over the temperature ranges of LdG theory applicability in both MBBA and 5CB. Thus, the fast, intradomain dynamics are not coupled to the hydrodynamic modes, and the results indicate that the dynamics are strongly influenced by the pseudonematic domain structure.

As seen in Fig. 7.10(a) for times shorter than 1 ns, the data fall on a straight line, corresponding to a power law decay.

$$G(t) = G_0 t^{-\alpha} \tag{7.32}$$

From the data, α is 0.63 ± 0.03. The 5CB data (Fig. 7.11) is also a power law with $\alpha = 0.63 \pm 0.02$. Thus, qualitatively and quantitatively, the orientational re-

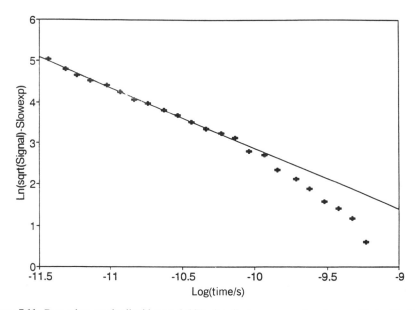

Figure 7.11. Data taken on the liquid crystal 5CB also display a power law decay with the identical exponent, $\alpha = 0.63 \pm 0.02$. The 5CB data are also viscosity/temperature independent until the correlation length falls below $3\xi_0$. The striking similarities in the temperature-dependent dynamics of MBBA and 5CB may indicate a universal behavior in the isotropic phase of liquid crystals. [Reproduced with permission from J. J. Stankus, R. Torre, and M. D. Fayer, *J. Phys. Chem.* **1993**, *97*, 9478. Copyright © (1993) American Chemical Society.]

laxation dynamics in the isotropic phases of both liquid crystals, MBBA and 5CB, are identical. The fast dynamics in both systems exhibit *power law behavior with the same exponent*, 0.63, and the slow dynamics, in both systems begin to deviate from LdG behavior at the same correlation length, about $3\xi_0$. When the correlation length becomes less than $3\xi_0$ the fast dynamics in both systems become temperature dependent. The remarkable similarities in the dynamical behavior of these systems may indicate the existence of a universal principle that governs the dynamics in the pretransitional isotropic phase of liquid crystals.

7.5.4. Models for the Temperature-Independent Fast Dynamics

As demonstrated above, MBBA and 5CB display temperature-independent dynamics over broad time and temperature ranges. In the liquid crystals, as long as the pseudonematic local structure lasts for a time longer than the intradomain relaxation, the dynamics are temperature independent. This demonstrates that local structures are responsible for the temperature-independent dynamics. The local structure is determined by the shape of the potential surface. The rate of relaxation is also determined by the shape of the potential surface. If the shape of the momentary, local potential surface is essentially independent of temperature, then the dynamics will also be independent of temperature. The dynamics involve relaxation on the potential surfaces defined by local structures rather than rotational diffusion, which displays a steep η/T dependence. On long time scales LdG-type orientational diffusion randomizes the local domain stuctures. On a time scale short compared to the domain randomizations, local structures are essentially fixed. As the temperature is increased, the time required for the domains to randomize becomes shorter. However, as long as this time is longer than the time scale for intradomain dynamics, the dynamics are η/T independent. (It is important to point out that the experiments might not detect a very weak dependence on the absolute temperature. For example, if there is a temperature dependence of $\sqrt{T}$, it is possible that it would not be observed since this would correspond to only a 4% change over the temperature range examined.)

The long-range correlation of initially excited librations is determined by the fringe spacing of the transient grating. Librational damping (dephasing) depends on the local molecular environment. Therefore, spatial variations in orientational correlation extend to high q (short distance scales) following librational damping (hundreds of femtaseconds). Thus the relaxation of the residual anisotropy produced by the excitation occurs over a broad range of q values that extend from the molecular length scale to the grating fringe spacing. Initial relaxation of residual anistropy is dominated by high q contributions (intradomain dynamics) so that $\Gamma_q = Lq^2/\eta_q$. The TG–OKE response function $G(t)$ is related to the time correlation of order parameter fluctuation as follows:

$$G(t) \propto \langle Q(\mathbf{r}, t_2)Q^*(\mathbf{r}, t_1) \rangle = \sum_q |a_q^2| \exp(-\Gamma_q t) \qquad (7.33)$$

where $t = |t_2 - t_1|$ and $|a_q^2|$ is the mean-square amplitude of the qth mode of the order parameter. The averaging is performed over all space. For $t < t_0$, the average depends on t_1. For $t > t_0$, the system is ergodic. Consider the slow dynamics first. For $t > t_0$, only one distance scale ξ becomes important as suggested by criticality, and therefore

$$|a_q^2| = |a_{q0}^2| = \text{constant}$$

where $|q_0| = (2\pi/\xi)$, the wavevector for the correlation length ξ. Therefore,

$$G(t) \propto \langle Q(\mathbf{r}, t)Q^*(\mathbf{r}, 0)\rangle \propto \exp[-\Gamma(T)t] \tag{7.34}$$

This is the hydrodynamic regime described by LdG theory, which gives $\Gamma(T)$. For $t < t_0$,

$$G(t) \propto \int_{q_c}^{q_m} q^2 dq \, |a_q^2| \exp(-\Gamma_q t) \tag{7.35}$$

where q_m is a molecular length scale and q_c corresponds to an average correlation length over the entire time scale of intradomain relaxation. From the equipartition theorem and the expression of f_q at high q [Eq (7.24)] one finds $|a_q^2| = (k_B T/Lq^2)$. In the high q limit $\Gamma_q = Lq^2/\eta_\infty$ [Eq. (7.30)], where η_∞ is a limiting high q and high-frequency viscosity, which is temperature independent (24). For an estimate we consider a simple limiting case where $q_c = 0$ and $q_m = \infty$. Then,

$$G(t) \propto \frac{k_B T}{L} \int_0^\infty dq \, \exp[-(L/\eta_\infty)q^2 t] \tag{7.36a}$$

$$G(t) \propto \frac{k_B T(\pi\eta_\infty)^{1/2}}{2L^{3/2}} t^{-1/2} \tag{7.36b}$$

This analysis suggests a power law with a temperature independent exponent of 0.5. The measured value of 0.63 may arise because of the critical correlation of fluctuations, which has been ignored in the above derivation.

The power law exponent in Eq. (7.36b) is obtained using a LdG continuum model for liquid crystals and is not far from the experimentally measured value. However, a power law can also be derived from molecular level models involving discrete processes. Some of the possible models are considered in the following. These models are quite general in the sense that they are applicable not only to liquid crystals but to other molecular systems as well. Also, they provide a general framework for understanding the molecular level origin of power laws in many disordered systems. A general form that yields a power law is

$$G(t) \propto \sum_i w_i(\Gamma_i)\exp(-\Gamma_i t) \qquad \text{(for discrete } \Gamma_i) \qquad (7.37a)$$

$$\text{or, } G(t) \propto \int d\Gamma_i w_i(\Gamma_i)\exp(-\Gamma_i t) \qquad \text{(for continuous distribution)} \qquad (7.37b)$$

Γ_i is the rate constant for the relaxation of microenvironments belonging to the ith subensemble of microenvironments inside a pseudonematic domain. All members of a subensemble have the same relaxation rate. The relaxation of the individual microenvironments is taken to be exponential, and $w_i(\Gamma_i)$ represent the weight of the ith component in the distribution of rates.

There are many types of physical mechanisms that can give rise to a power law. Very general, quantum mechanical descriptions have been given for the relaxation of a perturbation of a noncrystalline material that has some degree of local structure. See, for example, Dissado et al. (83) and references cited therein. Concepts presented in the quantum mechanical treatments apply here as well. Since the energies of the librational oscillators are small compared to kT, the dynamics will be discussed classically. Furthermore, the experiments only observe the anisotropy induced by the excitation fields. Therefore, the isotropic background of thermal excitations need not be considered. We will briefly discuss two well-known models that can give rise to power law decays. The models are a parallel process and a serial process (heirarchically constrained dynamics model). These models are considered because the physical nature of the pseudonematic domains suggests that they may be relevant.

7.5.4A. The Parallel Process Model.

The microscopic local structure is altered by the optical field (or by a thermal perturbation). The displacement occurs along a normal coordinate q_i associated with the one-dimensional potential surface for an instantaneous "normal mode" of the local structure. The relaxation dynamics are associated with the time evolution of the q_i and are determined by the slopes of the instantaneous normal modes' potential energy surfaces $V(q_i)$ values. For a single normal mode,

$$\frac{\partial q_i}{\partial t} = -\kappa \frac{\partial V(q_i)}{\partial q_i} \qquad (7.38)$$

κ is a proportionality constant. For a parabolic surface, the time evolution of q_i is exponential with a decay constant Γ_i. A distribution of curvatures of the parabolic surfaces for the normal modes in the various microenvironments gives rise to a distribution of Γ_i values. Then the observed decay of the TG−OKE signal will be a broad distribution of exponentials. The weight of a particular Γ_i occurring in the distribution is $w(\Gamma_i)$. Integrating over the distribution of decay constants, the observed power law decay will occur if

$$w(\Gamma_i) = \Gamma_i^{\alpha-1} \qquad (7.39)$$

that is,

$$G^{\varepsilon\varepsilon}(t) = \int_0^\infty (\Gamma_i^{\alpha-1}) e^{-\Gamma_i t} \, d\Gamma_i = \beta t^{-\alpha} \qquad t > 0 \qquad (7.40)$$

where β is a constant (84). For a finite range of Γ_i, $\Gamma_{min} < \Gamma_i < \Gamma_{max}$, the decay is a power law over a limited range of time, $\tau_{min} < t < \tau_{max}$. The crossover from a power law to a pure exponential occurs for $t > \tau_{max}$. To reproduce the observed data, the range of Γ values spans about 2.5 orders of magnitude. The experiment yields $\alpha = 0.63$. Therefore, the weighting function for the relaxation rates is $w(\Gamma_i) = \Gamma_i^{-0.37}$.

Distributions of rates of the form of Eq. (7.39) are seen in other condensed matter systems. For example, in low-temperature glasses, relaxation rates of glassy two level systems have been observed to have a distribution (85,86) $w(\Gamma) = \Gamma^{-1}$. Recent TG–OKE experiments on polysiloxane polymer melts also displayed a power law behavior (35b) (see Section 7.6.2). When the parallel process model is used to analyze the decays for this system, it gives $w(\Gamma) = \Gamma^{-0.18}$.

7.5.4B. The Serial Process Model.

In the serial process model, the dynamics of relaxation involve sequential release of constraints in a hierarchy of degrees of freedom (87,88). A condensed system can be nonergodic on the time scale of a fast measurement. This implies that simultaneous relaxation of all degrees of freedom does not happen, and the ensemble average of the relaxation of many microenvironments is not an accurate physical description. Fast dynamics within a smaller set of states bring the system into a configuration that permits somewhat slower dynamics within a larger set of configurations, which in turn release constraints that permit even slower dynamics within an even larger set of configurations. Here, the largest set of configurations corresponds to a correlation length of the microenviroment ξ, and the longest time scale is associated with an ergodic time τ_{max}.

Hierarchically constrained dynamics (87) have been proposed as a phenomenological model for non-Debye relaxations, especially in glassy systems. This model will give a stretched exponential decay or a power law decay depending on the distribution of states in the hierarchy of levels and the conditions for release of constraints to allow the subsequent set of motions (87,89). The constraints can dominate the relaxation dynamics over a wide time range. Palmer et al. (87) demonstrated a model of this type for a distinct series of levels of Ising spins. Here we will use the mathematical formulations of this model as a model for the response function for the relaxation dynamics of a perturbed microenvironment. For a series of n levels with N_n Ising spins or binary fluctuations in a given level, a constraint is imposed on the system such that a relaxation process in the succeeding level $(n + 1)$ may not occur unless μ_n elements have attained a particular configuration out of their 2^{μ_n} possible ones in the previous level (n). A number of molecules may generate a large number of orientationally disordered collective configurations. Configurations that are cooperatively interconvertible may constitute a level; and each single step of orientational distortion during interconversion may correspond to a binary fluctuation. (This should not be taken too literally since the process may involve more than a binary fluctuation). Assuming no intralevel correlation, the relationship between the average relaxation times of Ising spin flip in different levels is obtained.

$$\tau_{n+1} = 2^{\mu_n n}\tau_n \tag{7.41}$$

The microenvironment relaxation and the observable $G^{\varepsilon\varepsilon}(t)$ is determined by the constraint release dynamics, that is,

$$G^{\varepsilon\varepsilon}(t) = \sum_{n=0}^{\infty} w_n \left(\exp\left(\frac{-t}{\tau_n}\right)\right) \tag{7.42}$$

where $w_n = N_n/N$. w_n is the weight of the decay with decay constant τ_n in the distribution of decays. N_n is the number of states in the nth level; N is the total number of states.

A power law decay, as observed in the experiments, will only occur for certain conditions on μ_n and w_n. The weighting is geometric, with $N_{n+1} = N_n/\lambda$, where $\lambda > 1$. This weighting is allowed in that $N = \Sigma_{n=0}^{\infty}N_n$ is convergent. If the number of elements that must relax in each level is the same for all levels, $\mu_n = \mu_0$, $\tau_{\max}$ is not finite. There is a slowest decay in the isotropic phase of 5CB or MBBA. Therefore, we require $\tau_{\max}$ to be finite. Assuming an exponential distribution of the number of elements that must be relaxed to release constraints, $\mu_n = \mu_0 \exp(-\gamma n)$, gives a large but finite $\tau_{\max}$ for $\gamma \ll 1$. Using these conditions, $G^{\varepsilon\varepsilon}(t)$ is a power law for $t < \tau_{\max}$.

$$G^{\varepsilon\varepsilon}(t) = \beta t^{-(\ln\lambda)/\gamma} = \beta t^{-\alpha} \tag{7.43}$$

Experimentally, $(\ln\lambda)/\gamma = 0.63$ for 5CB and MBBA in their isotropic phase. $\gamma \ll 1$ and $\lambda \cong 1$. This implies that the number of states per level falls off very slowly with increasing n and therefore many levels participate in the process. The levels and the states are associated with the structure of the microenvironment. The optical (or thermal) perturbation modifies the local structure that extends over some correlation length, and the relaxation occurs through constraint release.

The above discussions show that either a parallel or a serial relaxation model can account for the observed power law decay. Based on these data there is no way to select between the two models or other possible processes that can give rise to power laws (83,90–92). However, in the constraint release picture, at least within its incarnation in terms of the Ising spin model, the observed value of $\alpha = 0.63$ implies that there are a great many levels with a large number of states per level. To have a very large number of levels each containing a large number of states suggests that the correlation length ξ is very large. The dynamics will be (essentially) temperature independent until the size of the pseudonematic domain is reduced to the maximum distance scale over which the constraints operate. Therefore, $\xi \cong 3\xi_0$ since this is the correlation length observed experimentally for both MBBA and 5CB at which the fast dynamics become temperature dependent. This corresponds to a length of approximately 20 Å. As mentioned above, this corresponds to about 30 molecules. While possible, it seems unlikely that there could be a sufficient number of states involving such a small number of molecules to yield the observed power law. In contrast, the parallel process model only requires that there is a broad inhomogeneous distribution of microenvironments. Local structure can create such inhomogeneity. Direct X-ray evidence for antiparallel local ordering in the isotropic phase of 5CB has been reported by Leadbetter et al. (93).

The results and discussion presented above established several important aspects of the influence of local order on the dynamics of liquid crystals in their isotropic phase. TG–OKE experiments demonstrate that the pseudonematic domains exist in the isotropic phase of liquid crystals far above the nematic–isotropic transition temperature. The shortest correlation length for the nematic order of $3\xi_0$ as observed in both 5CB and MBBA in their isotropic phases. The slow dynamics associated with the domain randomization show critical behavior and obey LdG theory in the temperature regime where the domain retains some nematic order. The intradomain dynamics, on time scales for which the nematic order persists, show a temperature-independent power law behavior. At higher temperautres, that is, $\xi < 3\xi_0$, the intradomain orientational dynamics become temperature dependent. Both MBBA and 5CB display power law intradomain dynamics with an exponent of 0.63 in the temperature range in which $\xi > 3\xi_0$.

7.6. POLYMER SIDE GROUP DYNAMICS IN LIQUIDS

In polymers, local order is created by the restricted geometry and connectivity of chemical bonds and local steric interactions. These restrictions create extended orientational correlations among side groups of the polymer over a longer distance scale than occurs because of close packing of monomers in a liquid. The side group dynamics in two very different polymers, poly(2-vinylnaphthalene) (P2VN) and poly(methylphenylsiloxane) (PMPS), are discussed in this section. Poly(2-vinyl-naphthalene) (T_g = 424 K) has a nonpolar backbone and is not very flexible, primarily due to steric interactions between the adjacent bulky naphthyl groups. On the other hand, PMPS (T_g = 228 K) is a very flexible polymer with a polar siloxane backbone. Local dynamics of a polymer in the amorphous glassy state, in a melt (neat liquid), or in solution depend on the extent of motional freedom that exists at the monomeric level. While local dynamics can result in fast orientational relaxation in polymeric systems, global relaxation, involving large scale backbone reorientation, occurs on much longer time scales (94). Unlike a small molecule isotropic liquid, in polymeric systems local restrictions of side groups give rise to cooperativity through the coupling of local and global modes (95–97). Although large scale motions are slow, observation of local dynamics, as demonstrated below, can require measurements on picosecond and femtosecond time scales. Like the dynamics of liquid crystals discussed in the previous section, polymer side group dynamics on short time scales are complex and are viscosity/temperature independent over a very broad range of η/T.

7.6.1. Poly(2-vinylnaphthalene)

This section focuses on the investigation of the femtosecond and picosecond orientational dynamics of the naphthyl side groups of the polymer P2VN both as an amorphous solid (glass) and in CCl_4 solution. The first direct ultrafast measurements of side group motion were performed on P2VN using the TG–OKE method (35a). The solid P2VN samples were made by either solvent casting or die pressing. The concentration of P2VN in CCl_4 solution was 1.5 M with respect to naphthyl groups at 25°C.

The TG–OKE data taken on P2VN glass (solid sample) and P2VN in CCl_4 solution are shown in Fig. 7.12(a) and (b) respectively. The signal in a TG–OKE experiment depends on the molecular polarizability anisotropy. For a polymer system, like P2VN, the polarizability anisotropy of the naphthyl side groups is very large compared to that of the polymer backbone. The backbone is a hydrocarbon chain. Long-chain hydrocarbon liquids yield measurable TG–OKE signals. However, the signals are extremely weak (98) compared to those obtained from P2VN. Thus, the signal is assumed to arise strictly from the aromatic side groups. This assumption will be confirmed in Section 7.6.2 on polysiloxanes. Therefore the TG–OKE experiments exclusively observe the orientational dynamics of the naphthyl side groups of P2VN.

Traces labeled 2 in Fig. 7.12(a) and (b) include fits to the short time scale nuclear OKE data. The response functions $G_n^{\varepsilon\varepsilon}(t)$ that give the best fit to the data are (15) for P2VN solid,

$$G_n^{\varepsilon\varepsilon}(t) = C_l[\exp(-t/\tau_b) - \exp(t/\tau_a)] \tag{7.44}$$

and for P2VN/CCl_4 solution,

$$G_n^{\varepsilon\varepsilon}(t) = C_l[\exp(-t/\tau_b) - \exp(-t/\tau_a)] + C_r[1 - \exp(-t/\tau_b)]\exp(-t/\tau_1) \tag{7.45}$$

The C_l term describes the ultrafast excitation and damping of the librations (15), as in Eq. (7.17) in Section 7.4. The parameters τ_a and τ_b are the rise and decay time constants, respectively. The C_r term describes the fast orientational relaxation. The prefactor $[1 - \exp(-t/\tau_b)]$ accounts for the delayed onset of the anisotropy that decays by orientational relaxation following the librational damping (15).

At longer times ($t > 1.5$ ps), the TG–OKE signal is proportional to the square of the sample response function [Eq. (7.10b)]. For P2VN/CCl_4 solution, we found

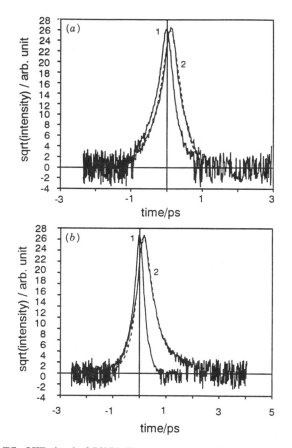

Figure 7.12. The TG–OKE signal of P2VN. Traces 1 and 2 are the square root of the signal. The signal was obtained using the $(0°/45°/90°/56°)$ and the nuclear $(0°/45°/90°/135°)$ TG–OKE configuration, respectively. Parts (a) and (b) show the ultrafast time scale behavior in P2VN glass and an 1.5 M (with respect to naphthyl groups) P2VN/CCl$_4$ solution, respectively, at 25°C. The dashed lines are the fits calculated by evaluating the convolution integral in Eq. (7.10a). The fit through trace 1 (the electronic Kerr or instrument response) was calculated with a $\delta(t)$ response function. The response function in Eq. (7.44) gives the fit through trace 2 in part (a) (the nuclear Kerr signal of P2VN glass), and the response function in Eq. (7.45) gives the fit through trace 2 in part (b) (the nuclear Kerr signal of the P2VN/CCl$_4$ solution). [Reproduced with permission from A. Sengupta, M. Terazima, and M. D. Fayer, *J. Phys. Chem.*, **1992**, *96*, 8619. Copyright © (1992) American Chemical Society.]

a sum of two exponentials to be an accurate expression for $G^{\varepsilon\varepsilon}(t)$

$$G^{\varepsilon\varepsilon}(t) = A_1 \exp(-t/\tau_1) + A_2 \exp(-t/\tau_2) \qquad (7.46)$$

τ_1 in Eq. (7.46) is the same as in Eq. 7.45. Figure 7.13 shows that the square of Eq. (7.46) fits the data extremely well.

Using Eqs. (7.10), (7.44–7.46), and fits to the data such as those shown in Figs.

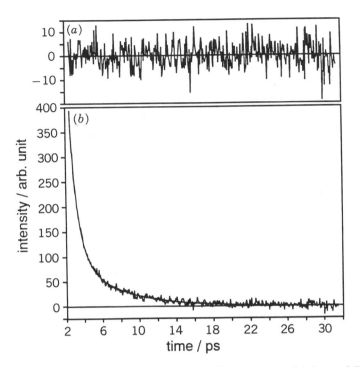

Figure 7.13. The slow time scale behavior of an 1.5 M (with respect to naphthyl groups) P2VN/CCl₄ solution: The TG–OKE signal at $t > 2$ ps was obtained using a polarization grating configuration (0°/90°/90°/0°). The solid line through the data is a fit using a bioexponential decay response function [Eq. (7.46)]. The residuals of the fit are shown on the top. [Reproduced with permission from A. Sengupta, M. Terazima, and M. D. Fayer, *J. Phys. Chem.*, **1992**, *96*, 8619. Copyright © (1992) American Chemical Society.]

7.12 and 7.13, the time constants associated with the dynamics of the solid and solution systems were accurately determined. The P2VN die pressed pellets and the solvent cast P2VN films showed no difference in their nuclear response. The librational damping constant $\tau_b = 140$ fs. The librational damping of the P2VN/CCl₄ solution is very similar to the solids with $\tau_b = 190$ fs. The estimated error in the τ values is ± 20 fs. However, because of the duration of the pulses used here, only an estimate of $\tau_a = 20$ fs can be made. In stark contrast to P2VN solid, P2VN in solution shows a residual anistropy following the ultrafast librational decay. As shown in Fig. 7.13, the residual anisotropy decays as a biexponential. The two time constants are significantly different so it is not difficult to obtain their values unambiguously. The values obtained for these slower decay constants are $\tau_1 = 0.95 \pm 0.1$ ps and $\tau_2 = 9.0 \pm 0.5$ ps.

There was one other consideration in the analysis of the ultrafast dynamics in the P2VN/CCl₄ solution. Neat CCl₄ gives a relatively weak but nonnegligible signal ($\sim \frac{1}{10}$ of the signal from P2VN/CCl₄ solution) due to interaction induced processes

(99–102). Therefore, the nuclear OKE data of pure CCl_4 were subtracted from the P2VN data before analysis. However, because the decay of the neat CCl_4 signal is extremely fast, it does not influence the τ_1 and τ_2 values, and at most introduces a small error in the value of τ_b for the solution.

Figure 7.14 shows that the two orientational relaxation components of the dynamics of the naphthyl groups of P2VN in CCl_4 are independent of temperature over the range studied, -3 to $+55°C$. The bulk shear viscosity of P2VN/CCl_4 solution changes by a factor of 20 and the viscosity of pure CCl_4 changes by a factor of 2.5 over this range of temperature. The orientational dynamics are not affected by the change in viscosity. Thus, the observed orientational relaxation is η/T independent and does not arise from rotational diffusion, that is, it is nonhy-

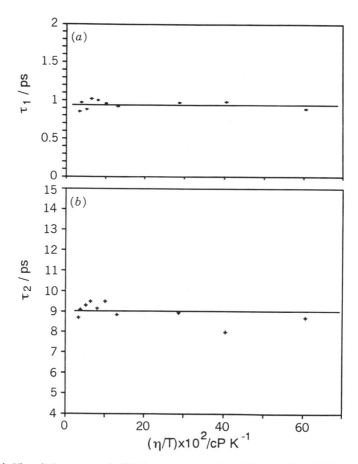

Figure 7.14. Viscosity/temperature (η/T) independent orientational relaxation in 1.5 M (with respect to naphthyl groups) P2VN/CCl_4 solution from -3 to $+55°C$. (*a*) Fast component: τ_1 does not show any η/T dependence. (*b*) Slow component: τ_2 does not show any η/T dependence and η/T changes by a factor of 20. [Reproduced with permission from A. Sengupta, M. Terazima, and M. D. Fayer, *J. Phys. Chem.*, **1992**, *96*, 8619. Copyright © (1992) American Chemical Society.]

drodynamic. The side group orientational relaxation is decoupled from the hydro-dynamic modes.

To make sure that the lack of temperature and viscosity dependence is not due to some anomalous behavior of either the naphthyl moiety or the CCl_4 solvent, the orientational relaxation of 2-ethylnaphthalene (2EN) in CCl_4 was studied. Following the ultrafast librational dynamics, a $1M$ solution of 2EN in CCl_4 also displays pure biexponential dynamics. Figure 7.15 shows the behavior of both the fast and the slow orientational relaxation components of the biexponential decay in 2EN

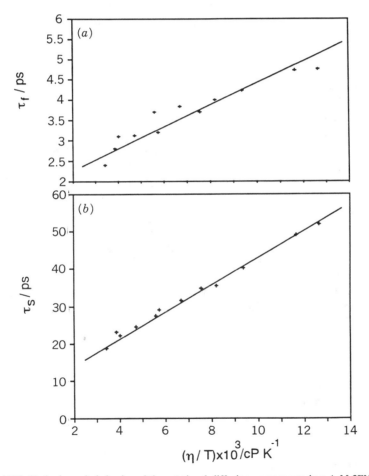

Figure 7.15. Hydrodynamic behavior of the rotational diffusion components in a 1 M 2EN/CCl_4 solution. (a) Fast component: τ_1 from -30 to $+ 25°C$. The straight line fit through the τ_f versus η/T demonstrates the hydrodynamic behavior. (b) Slow component: τ_s from $- 30$ to $+ 25°C$. A straight line also fits the τ_s versus η/T data and establishes hydrodynamic behavior. [Reproduced with permission from A. Sengupta, M. Terazima, and M. D. Fayer, *J. Phys. Chem.*, **1992**, *96*, 8619. Copyright (1992) American Chemical Society.]

solution. Both decays change considerably as η/T of the solution is varied. For $1M$ solution they display changes of greater than a factor of 2 when (η/T) changes a factor of 3. The behavior of 2EN shown in Fig. 7.15 is hydrodynamic in that the DSE equation is obeyed. In contrast, the data in Fig. 7.14 display no change although η/T of P2VN/CCl$_4$ solution varies by a factor of 20.

Both in the glassy solid and in the CCl$_4$ solution, P2VN displays ultrafast librational decays, which are 140 and 190 fs, respectively. The small difference in the decay time reflects a small change in the librational damping in the two environments. More important than the small difference in the decay times is the fact that solid P2VN displays a single decay while P2VN/CCl$_4$ solution shows two additional long-lived decays of 0.95 and 9 ps. These decays in solution are associated with the loss of residual anisotropy that remained after the librational decay. The amorphous glass of P2VN behaves much like a crystal. Following the librational damping, the naphthyl side groups return to their original configurations leaving no residual anisotropy and no signal.

Considering the structural constraints on the packing of a polymer glass, one might expect that the local potential surface that determines the orientational configurations would involve multiple nearly isoenergetic minima [see Fig. 7.16(a)]. In such a situation, after the ultrafast decay of libration is over, the naphthyl groups could return to any of a variety of local minima. This would result in a residual anisotropy. Slower orientational decays would be observed in the data in a manner analogous to the situation observed in the P2VN/CCl$_4$ solution. The ultrafast single-exponential decay seen in P2VN glass demonstrates that a single well constitutes the local potential energy surface with relatively high barriers separating the minimum from other orientational local minima. The picture in Fig. 7.16(a) does not apply to the solid; the packing around the naphthyl side groups permits the naphthyl groups to have essentially a single orientation, with even very small angular variations forbidden. Only librational (vibrational) motions are possible.

The P2VN/CCl$_4$ solution displays a residual anisotropy following the damping of the anisotropically distributed librations generated by the excitation field. The addition of solvent has loosened the packing to the point where damping of the librations can leave the naphthyl groups with a distribution of orientations. This is the same as the behavior of small molecules in dilute solution and of many neat liquids. However, in such systems the residual anisotropy decays because of orientational diffusion. They obey the DSE equation. An example of this is the behavior of 2EN, as shown in Fig. 7.15 or biphenyl discussed in Section 7.4. Even orientational relaxation of internal degrees of freedom of moderate size molecules can be diffusive, for example, the phenyl rings in biphenyl (15) or in *trans*-stilbene (103). In contrast to small molecules in solution, the naphthyl side groups do not obey the DSE equation. Figure 7.14 shows that over a very substantial range of η/T, the relaxation of the residual anisotropy is independent of η/T.

To gain insight into the relaxation of the naphthyl side groups of P2VN in CCl$_4$ solution, it is useful to examine the relative geometries and the nonbonded interactions of the naphthyl groups. The naphthyl group is rather bulky, which results in significant van der Waals interactions with the neighboring side groups as well

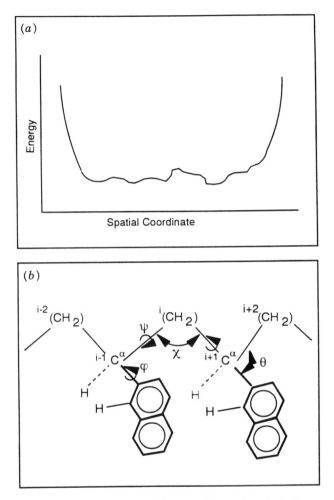

Figure 7.16. (*a*) Qualitative illustration of an arbitrary section of a hypothetical local potential energy surface for the orientational fluctuation of a naphthyl group in P2VN glass. (*b*) Schematic diagram of the isotactic portion of an atactic P2VN chain. It shows the different degrees of freedom that might be important for orientational relaxation of the residual anisotropy left after librational decay in the P2VN/CCl₄ solution. [Reproduced with permission from A. Sengupta, M. Terazima, and M. D. Fayer. *J. Phys. Chem.*, **1992**, *96*, 8619. Copyright © (1992) American Chemical Society.]

as with the methine-*H* and the methylene-*H*. For a given configuration, the steric energy of the naphthyl group arises mainly from the difference between the non-bonded interaction energy and the bending energy associated with the bond angles. The details of these interactions depend on the tacticity of the chain. Figure 7.16(*b*) shows the isotactic portion of an atactic P2VN molecule. The C^α–H bond and the C^α–C^2(naphthyl) bond define a plane. In the *o* conformation, the naphthyl plane is perpendicular to the H–C^α–C^2 plane, and in the π conformation the naphthyl plane is parallel to the $H-C^\alpha-C^2$ plane. De Schryver et al. (104) pointed out that the results of excimer formation experiments on P2VN solutions suggest the presence of oo and $\pi\pi$ conformations for adjacent naphthyl groups when the naphthyl groups are in their ground state. Molecular force field calculations (105) for P2VN show that $oo\pi$ conformations for isotactic and $\pi\pi o$ conformations for heterotactic sequences are the most stable. Frank et al. (106) concluded that the preferred backbone conformations for isotactic chains are trans-gauche or gauche-trans, and for syndiotactic chains it is trans–trans. Thus the preferred conformations do not have trans naphthyl groups. The isotactic portion of the P2VN chain may also form a 3_1 helix with the naphthyl groups stacked parallel to each other (107).

This information indicates that in all the preferred conformations, independent of the tacticity (which is unspecified for the sample studied), significant rotation of the naphthyl group about the $C^\alpha-C^2$ (naphthyl) bond will be highly hindered because of steric interactions. The data on the solid sample demonstrates that these steric interactions define a single minimum on the potential energy surface for the naphthyl group orientation. The librational states are determined by this surface. Excitation and damping of the libration returns the system to the initial minimum.

In the solution, there is more freedom for small changes in the backbone conformation. The librational motion is coupled through steric interactions to the backbone conformation. The experimental results suggest that librational excitation results in small conformational changes in the local geometry. Once the perturbation of the backbone caused by the excited libration is gone, the deformed local structure relaxes back to its initial geometry. The deformed local structure is responsible for the residual anistropy observed on the picosecond time scale. Relaxation back to the initial structure causes the observed decays.

Following the librational damping, P2VN in solution displays two distinct picosecond time scale orientational decays: 0.95 and 9 ps. While it is not possible to determine the nature of the motions involved from these measurements, the following is a possible scenario. The experimental results do prove that in solution there is relatively more freedom than in the solid for orientational change. The calculations of Seki et al. (107) show that the minimum energy of the trans–trans state for the meso configuration appeared over a range of θ (114°–120°). Due to connectivity of the $-CH_2-CH_2-$naphthyl groups in P2VN, the libration could cause elastic deformation of the angles θ and χ. After the libration damps, the χ, φ, and θ angles could be left with altered values, no longer corresponding to the minimum energy configuration. Relaxation of the angles θ and χ does not involve the solvent structure significantly and, therefore, could be temperature independent. Because of the inertial nature of this relaxation, it may be associated with the 0.95-ps decay. Following the relaxation of the angles θ and χ, the angle φ could relax by interrupted rotation of the naphthyl group about the $C^\alpha-C^2$ (naphthyl) bond. The distance between two naphthyl rings in meso dyad for $\psi_i\ \psi_{i+1} = 0°,0°$ is 2.60–2.77 Å (107). Due to the proximal alignment of the naphthyl groups, the intrapolymer interactions probably contribute more significantly to the local potential energy than interactions with the solvent, leading to temperature independent relaxations on a potential energy surface that is fixed on the time scale of the observed dynamics.

7.6.1A. Comparison to Other Methods. TG–OKE experiments have aspects that are fundamentally different from other methods that examine orientational dynamics of polymers or small molecules. To describe these aspects, it is necessary to briefly discuss the femtosecond TG–OKE experiment as it contrasts to slower OKE experiments and other techniques.

The transition energies for the librational modes will be in the range of a few tens of cm^{-1}. Since the thermal energy kT is large compared to the mode energy, the librational modes are constantly being thermally excited and are constantly relaxing. In a femtosecond TG–OKE experiment, the radiation field couples to the orientational degrees of freedom of a molecule via stimulated rotational (librational) Raman scattering. Thus, the orientationally anisotropic distribution of libration is created purely by quantum mechanical excitation.

Now consider the situation for the standard Kerr effect in which a static electric field is applied. Molecules in solution are constantly undergoing rotational diffusion, that is, an orientational random walk. With a static electric field, which couples to the molecules through either their permanent dipole moment or an induced dipole moment if there is no permanent moment, the molecules execute a biased random walk. Orientational steps in all directions are not equally probable, but rather have a greater probability in directions that align the molecular dipoles with the field. The extent of the alignment will depend on the strength of the field–dipole interaction relative to kT and the duration of the application of the electric field. When the field is turned off, the induced orientational anisotropy will relax at the rate of rotational diffusion. The OKE with long laser pulses (10 ns) is very similar to this. The coupling of the optical field to the molecule is through an induced dipole moment. Since the transform limited bandwidth of the optical field is much smaller than the librational frequencies, stimulated Raman scattering does not occur. Instead, as with the case of a static field, the molecules execute a biased walk, and an anisotropy is generated in the sample. When the optical field is removed, the anisotropy will relax by rotational diffusion.

In all three types of Kerr effect experiments (fs OKE, long-pulse OKE, or static field KE), an orientational anisotropy is induced in the sample. However, only the femtosecond OKE experiment actually excites librations. In translational diffusion, collisions excite a ballistic translation that is rapidly damped. This ballistic translation is the fundamental step in translational diffusion. Librational excitation and damping is the fundamental step in the orientational random walk that leads to rotational diffusion in liquids. In slow OKE experiments and in KE experiments, thermal excitation of librations, and therefore rotational diffusion, is an inherent part of generating the orientational anisotropy. In femtosecond OKE experiments, the pulses are sufficiently short to excite librations, and the generation of orientational anisotropy is not directly dependent on rotational diffusion.

Experiments such as time resolved fluorescence anisotropy (or equivalent pump–probe and excited-state transient grating experiments) or magnetic resonance line shape experiments do not induce an actual mechanical orientational anisotropy in a sample. Fluorescence depolarization measurements tag an orientationally anisotropic distribution of molecules by electronically exciting them and then observe the time-dependent anisotropy in the distribution of tagged molecules as they relax by rotational diffusion. Magnetic resonance experiments examine a line shape (T_2) or population relaxation (T_1). The T_1 and T_2 experiments are indirectly related to rotational diffusion through a model that relates orientational motions to spin dynamics. Like slow OKE and KE experiments, fluorescence depolarization and magnetic resonance experiments do not excite librations.

Orientational relaxation measurements on polymers involve a vast range of time scales from the ultrafast librational motions of side groups to very slow motions involving large scale backbone reorientation. What will be observed depends on the time scale and the nature of the measurement. An OKE experiment using long optical pulses, or a KE experiment in which the static field is applied for a relatively long time, will result in anisotropy developing due to the slow polymer degrees of freedom undergoing a biased random walk. If the application of the field lasts for a time scale that is on the same order as the time scale for diffusion of the slow degrees of freedom, substantial anisotropy will be generated. Observations following the application of the field (optical or static) will reveal the diffusive relaxation of the slow degrees of freedom. In contrast, a femtosecond OKE experiment will excite librations. The librations damp, leaving a residual anisotropy as observed in the polymer in solution experiments discussed above. However, on the ultrafast time scale, there is essentially no diffusive motion of the slow degrees of freedom and, therefore, there is essentially no buildup of anisotropy in the slow degrees of freedom.

From the preceding discussion, it is clear why the experiments presented here directly observe ultrafast dynamics in polymers while previous experiments have observed much slower time scale motions although some magnetic resonance experiments have indicated the existence of fast dynamics. Fluorescence depolarization experiments observed nanosecond time scale decays in polystyrene (108,109) and PMMA (110) labeled with anthracene in the middle of the chain. The decays showed solvent viscosity and temperature dependence. Excited-state transient grating experiments using 35-ps pulses on polyisoprene (111,112), similarly labeled with anthracene, detected backbone reorientation occurring from several hundreds of picoseconds to a few nanoseconds. Temperature and solvent viscosity dependencies of an average nanosecond correlation time were studied in these experiments.

Nuclear magnetic resonance (NMR) studies of side group motions were performed on polystyrene (113–115), PMMA, (116), poly(p-fluorostyrene) (117), and other polymers (118). Correlation times for internal motions of side groups are obtained using theoretical models (116,119–121) that relate T_1 and T_2 measurements to orientational relaxation. In some instances the interpretations of the experiments change dramatically with the model that is applied (114,122). There are reports of very fast dynamics. Polyisoprene experiments were interpreted as yielding a 100–300-fs orientational relaxation of the methyl group, and the results were found to be independent of temperature (123). This could be a methyl libration. In contrast, rotational correlation times of methyl and phenyl groups in polycarbonates are reported to have the same value, 10 ps, and a large activation energy (124).

Very slow relaxations, slower than 10^{-7} s, have been observed in dielectric relaxation and transient birefringence measurements. These involve large displacements of backbone segments comprised of many monomeric units. In a transient birefringence measurement, Pecora and co-workers (125) used electric field pulses with durations ranging from microseconds to milliseconds. They observed an enhancement of the amplitude of slower modes with longer pulses. This is consistent with the above discussion stating that the time scale of the application of the electric field must be on the order of the time scale of the diffusional relaxation of a degree of freedom for significant anisotropy to develop in that degree of freedom.

7.6.2. Poly(methylphenylsiloxane)

This section describes experiments examining polymer side group dynamics in polymer melts of poly(methylphenylsiloxane) (PMPS) and in PMPS in dilute solution (35b). The PMPS melt is a glass-forming liquid. The PMPS samples had a $M_w = 2600$ and a room temperature viscosity of 500 cP. The results provide insight into the role of side group dynamics in the dynamical criticality of a polymer glass-forming liquid (6,7). The TG–OKE experiments were conducted over a broad temperature range, 25–143°C. Over this broad temperature range, the ratio of the viscosity to the temperature η/T changes by a factor of 40. However, the orientational dynamics of the phenyl side groups are independent of temperature throughout the entire temperature range. Experimental results on poly(dimethylsiloxane) (PDMS) are also reported. Comparison of the PMPS and PDMS data demonstrate that the observed temperature-independent orientational dynamics of PMPS arise from motions of the phenyl side groups.

In liquids, local structural relaxation occurs over a broad range of time scales. It is well established that the liquid-to-glass transition is dynamic in origin (6,7,126). With respect to the time window of most spectroscopic techniques, structural disorder in a glass-forming liquid freezes in at the glass transition temperature (7). Especially in a fragile glass-forming liquid (127), a bimodal distribution in the relaxation spectrum exists over a wide temperature range above T_g (6,7,126). On approaching T_g from higher temperature, the relaxation process bifurcates into slow and fast relaxation branches below a temperature T_c. This is referred to as a dynamic critical transition. The criticality shows up in dynamical behavior only; the heat capacity C_p does not exhibit divergence at T_c. The low-frequency (slow) structural relaxation is known as α relaxation (7). The time scale for α relaxation diverges, becoming infinitely slow near T_g. This divergence is responsible for freezing the spatial inhomogeneity in a glass. Besides the quasiharmonic phonon-like modes, there exists a high-frequency relaxation branch known as β relaxation (7). The α relaxation is slow, highly temperature-dependent and **q**-dependent (wavevector dependent) structural relaxation (6,126), while β relaxation is fast, only weakly temperature-dependent, and **q**-independent local fluctuation (7). These features of β relaxation are in accord with the mode coupling theories and are experimentally observed in some fast time scale measurements, for example, neutron scattering and depolarized light scattering experiments (126–129). At low frequency, for example, dielectric relaxation measurements, the β process also appears as a broad distribution of relaxation rates. The low-frequency β process depends on temperature. However, the temperature dependence is very mild relative to that of the α relaxation observed in the same experiment (130).

In a polymer glass, structural disorder originates from innumerable combinations of backbone conformations and side group orientations. Due to connectivity of the bonds, restricted bond angles, and steric interactions of the side groups, backbone orientational diffusion is orders of magnitude slower than pure side group motions. Backbone motions are responsible for α relaxation in polymer melts (131,132). They are very sensitive to temperature change and often yield dynamical observables that display stretched exponential time dependencies (131,132). Some experiments (127) suggest that localized motions of polymer side groups are responsible for high-frequency β relaxation and act as precursors to α relaxation.

Figure 7.17 compares the TG–OKE signal from a PMPS melt (*a*) and a PDMS melt (*b*) at short times. The polarizations used give only nuclear OKE data; the electronic contribution to the OKE is suppressed (12). The PMPS signal is 400 times stronger than the PDMS signal near $t = 0$. While PMPS (*a*) displays a substantial long-lived signal running out to many tens of picoseconds, PDMS (*b*) displays no detectable signal after about 1 ps. The PDMS exhibits zero residual anisotropy following the librational damping. In contrast, the residual anisotropy is very large after the librational damping in PMPS, and its decay is highly nonexponential. The phenyl side groups in PMPS are very polarizable and have a large anisotropy in their polarizability. The observations that the PMPS signal is 400 times stronger than the PDMS signal and PMPS displays residual anisotropy for times longer than 1 ps demonstrate that the PMPS signal comes predominately from the phenyl side groups at short times and exclusively from the phenyl side groups at longer times. When the PDMS librations damp, the system is left essentially in its original configuration. However, following the damping of the PMPS librations, the phenyl side groups have not returned to their original orientations. The phenyl side group residual orientational anisotropy decays on the time scale greater than 1 ps. (These observations support the assignment of the signal in P2VN to naphthyl side group dynamics, as discussed in the previous section.)

To improve the S/N ratio, at $t > 3$ ps, the polarization grating configuration is used (12,34,35). The polarization grating configuration is equivalent to the frequency domain VH measurements in depolarized light scattering experiments. For small molecules in dilute solutions or in low viscosity neat liquids, VH measurements of DLS give rotational diffusion times that obey the DSE equation (28,32). The DSE equation is obeyed in the viscosity range where the shear relaxation is much faster than the orientational relaxation. In dense fluids, the VH spectra become complicated due to coupling of reorientation, local translation, and shear relaxation (24,32,133–135). Figure 7.18 displays the PMPS long-lived anisotropy decay at two temperatures, 25 and 90°C. The data are shown from 3 to 300 ps. In the inset, the data are displayed from 20 to 300 ps, and the data have been amplified. The data sets taken at all temperatures are absolutely superimposable in spite of the fact that the measured zero-frequency shear viscosity changes by a factor of greater than 20 over this range of temperatures, and η/T changes by more than a factor of 40.

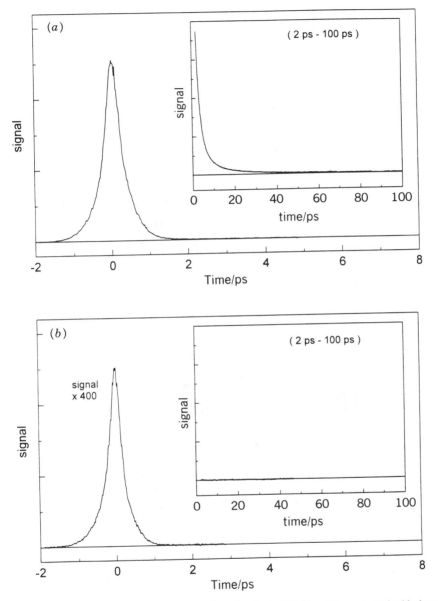

Figure 7.17. Transient grating optical Kerr effect signal of a PMPS melt is contrasted with that of a PDMS melt. The data reflect molecular orientational dynamics only; the electronic contribution to the OKE is suppressed by polarization selectivity. (a) Typical TG–OKE signal of PMPS melt. The dynamics at short time (< 1 ps) arise from librational excitation and damping (dephasing). The inset (the data are multiplied by 1000 relative to the main figure) shows the decay of a large residual anisotropy left after librational damping. (b) The TG–OKE signal of PDMS melt is 400 times weaker than that of PMPS melt. [The data are multiplied by 400 relative to the data in part (a).] The data in the inset are multiplied by 1000 relative to the main part of Fig. 7.18(b). There is no detectable signal, that is, no residual anisotropy, after 1 ps. [Reprinted with permission from Ref. 35b.]

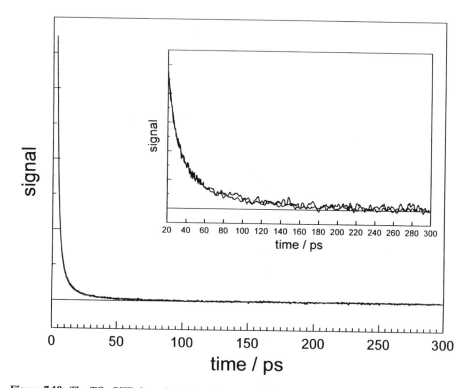

Figure 7.18. The TG–OKE data of PMPS melt at two different temperatures, 25 and 90°C, are perfectly superimposable. This shows that the orientational relaxation of the phenyl side groups is viscosity/temperature independent in spite of the large change in viscosity over this temperature range. [Reprinted with permission from Ref. 35b.]

To obtain a more quantitative description of the influence of temperature and viscosity on the phenyl group orientational dynamics, the data at a variety of temperatures were decomposed into a sum of exponentials. The data for $t > 3$ ps can be fit with the response function $G^{\varepsilon\varepsilon}(t)$, expressed as a sum of three exponentials,

$$G^{\varepsilon\varepsilon}(t) = A_f \exp(-t/\tau_f) + A_i \exp(-t/\tau_i) + A_s \exp(-t/\tau_s) \qquad (7.47)$$

Equation (7.47) is a fitting function used to permit a detailed comparison among data sets taken at different temperatures. A discussion of the actual functional form of the data is given below. The data sets were fit to the square of Eq. (7.47) [see Eq. (7.10b)]. Also, exponential decomposition of the data sets was employed. The fast and intermediate decay components can be followed for 9 and 6 factors of e of signal decay. The slowest component does not have as large a dynamic range as the intermediate and fast components and was determined most accurately by the full tri-exponential fit to the data. The decay times obtained from the tri-exponential fits are (f ≡ fast; i ≡ intermediate; s ≡ slow) $\tau_f = 2.3 \pm 0.2$ ps, $\tau_i =$

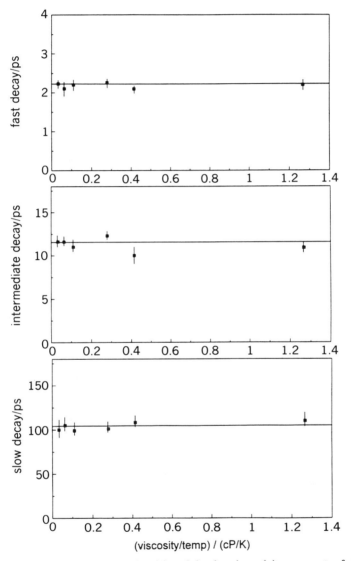

Figure 7.19. The fast (τ_f), the intermediate (τ_i), and the slow decay (τ_s) components of phenyl group orientational relaxation in PMPS melts as a function of viscosity/temperature (η/T). The results clearly demonstrate that the dynamics are completely independent of η/T. The DSE equation for hydrodynamic rotation diffusion predicts a change of a factor of about 40 over the range of η/T displayed. [Reprinted with permission from Ref. 35b.]

11.5 $\pm$ 1 ps, and τ_s = 105 $\pm$ 10 ps. Error bars were determined by the spread of fits over many separate data sets.

In Fig. 7.19, the fast, intermediate, and slow decay components of the tri-exponential fit are plotted versus η/T. This figure clearly shows the total lack of dependence of the dynamics on η/T on all the observed time scales. The viscosity data obey a Vogel–Fulcher–Tammann (VFT) equation, $\eta = \eta_0 \exp[DT_0/(T - T_0)]$ (7,136). The parameter T_0 is $(T_g - 20°)$, typical of many glass-forming liquids. For glass-forming liquids, the dynamical phase transition, that is, the bifurcation of relaxation times, occurs below T_c, which is well above T_g. Measurement of T_c is difficult. For many small molecule organic supercooled liquids, it is found that $T_c/T_g \approx 1.2 - 1.3$ and $\eta > 10^2$P in the bifurcation regime below T_c (137). The temperature range of the PMPS experiments is $1.3T_g - 1.82\ T_g$ and the viscosity regime is 10 P > η > 10^{-1}P. Since PMPS is a polymer, it is possible that T_c is much higher than for small molecule liquids. The observed VFT temperature dependence of the viscosity is a characteristic feature of small molecule liquids in the regime where the relaxation processes are bifurcated into α and β relaxations (137).

The lack of temperature dependence of the observed TG–OKE decays suggests that these are related to the β process. Over the temperature range studied, PMPS is viscoelastic (138,139), and the width of the Brillouin peak (the rate of acoustic damping) varies substantially with temperature (138). The fact that the orientational dynamics of the phenyl side groups are independent of temperature and viscosity over such a broad range of η/T (a factor of 40) demonstrates that the dynamics are decoupled from the low-frequency hydrodynamic modes. The lack of viscosity and temperature dependence of the PMPS phenyl side group dynamics in the pure polymer melt is analogous to the η/T independent orientational dynamics of the naphthyl side groups of P2VN discussed above.

The tri-exponential function was used as a fitting function and does not correspond to the true functional form of the TG–OKE decay. Figure 7.20 shows a log–log plot of the data from 3 to 300 ps. A straight line (dashed line) is drawn through a portion of the data. The data appear linear on the log–log plot from about 10 to 300 ps, spanning two decades of signal decay. A straight line on a log–log plot indicates that the functional form of the data is a power law, $t^{-\alpha}$. A faster component is observed for times less than 10 ps. The data at all temperatures examined in these experiments have the identical shape within a very small experimental error (see Figs. 7.18 and 7.19). In Fig. 7.21, the same data are plotted with a fit (dashed curve) to the function

$$S(t) \propto [G^{\varepsilon\varepsilon}(t)]^2 = [Ae^{-t/\tau} + Bt^{-\alpha}]^2 \qquad (7.48)$$

that is, the sum of an exponential and a power law. As seen from the figure and the residuals (labeled power law), the fit over the entire time range is essentially perfect. In the fit, τ = 1.95 ps and α = 0.82. At the long time limit of the data, there is no tailing off. This implies that the power law behavior extends to times

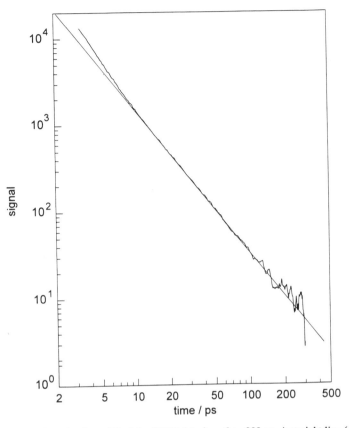

Figure 7.20. A log–log plot (base 10) of the PMPS data from 3 to 300 ps. A straight line (dotted line) is drawn through a portion of the data. The data appear linear on the log–log plot from about 10 to 300 ps, indicating that the functional form of the data is a power law, $t^{-\alpha}$. A faster component is observed for times less than 10 ps. [Reprinted with permission from Ref. 35b.]

longer than 300 ps. While the tri-exponential fit to the data is very useful for displaying the lack of η/T dependence of the fast, intermediate, and slow portions of the data, it does not yield nearly as good a fit as Eq. (7.48). The residuals of the tri-exponential fit are shown at the top of Fig. 7.21. It is clear that the tri-exponential does not reproduce the functional form of the data as well as Eq. (7.48). Only the high quality of the TG–OKE data makes the distinction possible.

Since Si–O bonds are polar, the backbone reorientation in PMPS is dielectrically active. The dipole components add up in the direction perpendicular to the backbone (140). Therefore, dielectric measurements will not be able to observe pure side group reorientation; there will be contributions from backbone reorientation. The polarization grating TG–OKE signal depends on the anisotropy of the polarizability. The data (Fig. 7.17) demonstrate that the response to the optical excitation of phenyl groups is much greater than that of the Si—O—Si— backbone or the CH_3

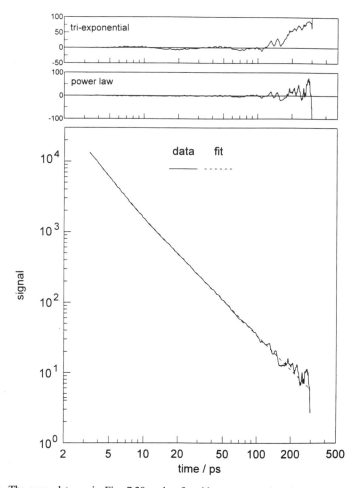

Figure 7.21. The same data as in Fig. 7.20 and a fit with a response function of the form of an exponential plus a power law [Eq. (7.48)]. The high quality of the data permits a very accurate fit as indicated by the residuals shown at the top (labeled power law). The exponential decay constant, $\tau = 1.95$ ps, and the power law exponent, $\alpha = 0.82$. The residuals are calculated as percent deviation of the data from the fit, $100 \times (\text{data} - \text{fit})/\text{data}$. The residuals of the fit to the same data with the tri-exponential response function [Eq. 7.47] (labeled tri-exponential) are also shown for comparison. The tri-exponential does not reproduce the functional form of the data. [Reprinted with permission from Ref. 35b.]

groups. This makes the study of pure side group orientational relaxation in PMPS melts possible. Because of the ultrashort duration of the excitation pulses, negligible anisotropy is induced in the slow dynamic modes involving backbone reorientation (see discussion in Section 7.6.1A). Therefore, the residual anisotropy left after the librational relaxation is due to orientational anisotropy of the phenyl groups. The TG–OKE measurements were performed at low q ($q \sim 2.5 \times 10^{-4}$ Å^{-1}). Like DLS experiments, these TG–OKE measurements may include contributions from collective reorientation of the phenyl groups (24,32). For small molecule liquids, experiments on dilute solutions yield the single molecule reorientation dynamics of the solute molecules. For a polymer in a random dilute solution, side group orientational relaxation will not be influenced by intermolecular interactions with side groups situated on different chains. Collective reorientation of nearby side groups may still contribute to the TG–OKE signal in dillute solution.

To determine if the observed phenyl side group dynamics in the melts involve interchain interactions, TG–OKE experiments were performed on a dilute solution of PMPS in CCl_4 at 25°C. Following the librational decay, the orientational dynamics are very similar to the melt, but not identical. Figure 7.22 shows a log–log plot of the dilute solution data with a straight line (dashed line) drawn through it. As in the melt, the decay is a power law over a significant portion of the decay. However, there are differences between the data in the melt and the data in solution. The exponent α in the power law is 0.87 in solution, while it is 0.82 in the melt. These are very close, but given the high quality of the data, they are not identical. A more obvious feature of the solution data is the tailing off at long time. By 100 ps, the data from PMPS in solution are clearly decaying faster than the power law and have the appearance of going over to a slowest exponential component at long time. This is in contrast to the melt data. Finally, the data in solution do not display the same fast exponential component at very short time. In the melt, there is a fast exponential decay ($\tau = 1.95$ ps) that does not merge into the power law until about 10 ps. In the solution data, there is also a fast component ($\tau = \sim 0.7$ ps) that merges into the power law by about 5 ps (141). While the differences between the melt data and the solution data are not dramatic, they indicate that interchain interactions play a role in the dynamics of the phenyl side groups in the melt. The extension of the power law to longer times in the melt demonstrates that the continuous range of time scales of dynamic processes extends to longer times. This is probably due to a more complex set of local structures in the melt involving the interaction of the phenyl side groups on one chain with the side groups and backbone of other chains. The differences in the nature of local structures in the melt versus the solution can also be responsible for the observed difference in the α values.

Through stimulated Raman scattering, the excitation pulses excite a subset of the real frequency spectrum of the PMPS liquid. The potential surface for a particular oscillation of a phenyl side group or groups is determined by local backbone structures and side group steric interactions. Only on a much longer time scale, that is, the time scale for backbone structure to evolve, will the local structure change significantly. Because of the large number of backbone conformations and chain–chain interactions, there will be a wide variety of phenyl group microenvironments. These will have different potential surfaces, and each microenvironment may have a variety of instantaneous normal modes existing on the short time scale of the experiment.

The data are analyzed in terms of two models described earlier in Section 7.5.4. To reproduce the observed data with a parallel process model, the range of Γ's spans about 2.5 orders of magnitude. The experiment yields $\alpha = 0.82$. Therefore, the

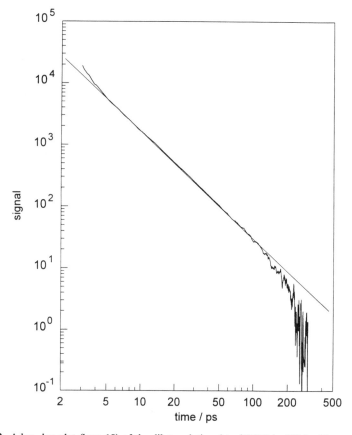

Figure 7.22. A log–log plot (base 10) of the dilute solution data (PMPS in CCl$_4$) with a straight line (dotted line) drawn through it. As in the melt, the data are a power law over a significant portion of the decay. There are differences between the data in the melt and the data in solution, which indicate that interchain interactions play a role in the phenyl side group dynamics in the melt. [Reprinted with permission from Ref. 35b.]

weighting function for the relaxation rates is $w(\Gamma_i) = \Gamma_i^{-0.18}$. Such distributions of rates are seen in other condensed matter systems (Section 7.5.3B, Figs 7.10 and 7.11).

Alternatively, the ultrafast side gorup dynamics of PMPS can be analyzed as a serial process (see Section 7.5.4B). In PMPS, a number of phenyl groups may generate a large number of orientationally disordered collective configurations. Configurations that are cooperatively interconvertible may constitute a level, and each single step of orientational distortion during interconversion may correspond to a binary fluctuation. (This should not be taken too literally since the process may involve more than a binary fluctuation.) When a serial process model is applied, experimentally, $(\ln \lambda)/\gamma = 0.82$ for the PMPS melt. Since $\gamma \ll 1$, $\lambda \cong 1$.

This result implies that many levels participate in the process, that is, a number of phenyls, on and off chain, interact over some correlation length to form a micro-environment. The optical (or thermal) perturbation modifies the local structure and the relaxation occurs through constraint release.

As in the description of the power law dynamics observed in the liquid crystal systems (Section 7.5.C), either a parallel or a serial relaxation model can account for the observed PMPS power law decay. Based on these data there is no way to select between the two models or other possible processes that can give rise to power laws (83,90–92). However, in the constraint release picture, the observed value of $\alpha = 0.82$ implies that there are a great many levels with a large number of states per level. Since the backbone structure is fixed on the time scale of the experiments, this suggests that the levels and states involve configurations of the side groups. To have a very large number of levels each containing a large number of states indicates that the correlation length ξ is very large, and therefore fast (100 ps) correlated dynamics are occurring over long distances. In contrast, the parallel process model only requires that there is a broad inhomogeneous distribution of microenvironments. Since the backbone is very flexible because of the large ionic character of the Si—O bonds, it is reasonable to assume that the phenyl side groups will exist in many structural configurations either in the melt or in solution. The experiments on P2VN in CCl_4 solution (see Section 7.6.1) over a wide η/T range displayed no η/T dependence, just like the PMPS. However, the P2VN data, following the ultrafast librational dynamics, are a true biexponential decay. The decay constants are 0.95 and 9.0 ps. Both components have large amplitudes. The P2VN is a carbon backbone polymer that does not have the structural flexibility of PMPS. The bonds are strictly covalent and the geometry is tetrahedral. The biexponential observed in P2VN was described as relaxation on the local potential surface determined by the local backbone geometry and naphthyl side group steric interactions. If this is correct, then it suggests that the power law observed in PMPS arises from parallel dynamics but that PMPS has a very broad distribution of local structures, in contrast to P2VN.

In the PMPS melts, following the ultrafast librational dynamics, an exponential decay ($\tau = 1.95$ ps) can be noticed at short time that merges (~ 10 ps) into a power law decay. Equation (7.48) gives a remarkably good fit to the data as shown in Fig. 7.21. Samios and Dorfmuller (26) performed room temperature Raman line shape measurements of the reorientation of the C_6 axis of the phenyl group in PMPS ($M_w = 2600$, viscosity 500 cP). They fit the observed line shape to a single Lorentzian and obtain a relaxation time of 2.8 ps. While this is similar to the exponential decay component of the TG–OKE data, the difference seems to be too great to assume that the two correspond to the same motion. The Raman lineshape provides the decay of the single-particle orientational correlation function. The TG–OKE experiments on PMPS are sensitive to single and collective phenyl orientational relaxations involving any motion that affects the polarizability anisotropy.

Dielectric relaxation spectroscopy and quasielastic light scattering studies of PMPS (131,142) observed a temperature-dependent collective relaxation (α relaxation), which can be fit to a stretched exponential form ($\tau_{kww} \sim 10^{-9}$ s). Since dielectrically active relaxation in PMPS originates primarily from the dipole moments of Si—O bonds, α relaxation is believed to involve large-scale backbone reorientation. Neutron scattering (NS) studies from polybutadiene (128), di-2-ethylhexylphthalate (DOP) (129), PMPS (127), and low-frequency (2–100 cm^{-1}) Raman spectroscopy of GeSBr$_2$ (126) observed a high-frequency q-independent, temperature-independent β process. In most cases, the accuracy of the measurements of β relaxation is limited by the time window of the spectroscopic techniques. In any specific case, elucidation of the β relaxation in terms of molecular level dynamics (distance scale of a few Å) is very difficult. Neutron scattering measurements on PMPS (127) indicate coupling of α relaxation to phenyl reorientation above 331 K. Neutron scattering is unable to observe pure side group motion isolated from the contribution of backbone motions occurring in the same wavevector and time windows. At lower temperatures, the NS experiments (127) observed a q-independent and temperature-independent relaxation (β process). The data are fit to a single Lorentzian and yield an exponential correlation decay time of 30 ± 3 ps. The TG−OKE results for the phenyl side group dynamics are a power law from 10 to beyond 300 ps. This is the system's impulse response function, which is related to the decay of the correlation function. Clearly, there are dynamics on the 30-ps time scale, but the TG−OKE experiments expose more complex dynamics in this time window.

If the observed η/T independent dynamics are identified with the β relaxations of glass-forming liquids and are ascribed to relaxations on local potential surfaces fixed on the time scale of the measurements, then what is the connection to the slow α relaxations? We suggest the following picture that is based on concepts found in the results of simulations of glass-forming liquids and in mode coupling theory. Mode coupling theory and molecular dynamics methods indicate that β relaxation involves only a small number of atoms (molecules) (143). The description given above involves the motions of single and small numbers of phenyl side groups relaxing on a local potential surface. It is fluctuations of the local environment about the minimum energy structure that leaves the system displaced from the minimum following the librational dampling. A classical harmonic oscillator, overdamped in a viscous medium, if displaced from the origin will relax back to the origin as a biexponential (144). The two decay rates are related to the oscillator frequency and the viscosity. The decay of the residual phenyl group anisotropy is the equivalent of the slow component of the classical system with the local molecular environment replacing the viscosity. The environment of a set of phenyl groups involved in a particular motion is also composed of molecular librators, that is, other phenyls, methyls, and components of the backbone. In most cases, a displaced oscillator will relax, reforming the initial local structure. Occasionally, however, the simultaneous motions of several adjacent oscillators will block the return path and shift the local structure. This is like the sudden jumps that are seen in molecular dynamics simulations (143) and represent a single step in the evolution of the structure. The collection of these steps, which will have a broad distribution of step rates, is the α relaxation.

As the temperature is reduced, the β relaxation does not change its time dependence since it is relaxation on local potential surfaces that are fixed on the β relaxation time scale. However, the probability that a step in the α relaxation will occur is greatly reduced since a single step requires the simultaneous excitation of a number of oscillators with the proper correlation of motions to couple to a mode responsible for the structure change. At low enough temperature, the probability of a significant number of such steps occurring on any reasonable experimental time scale becomes vanishingly small, and α relaxation ceases to occur. The temperature at which this occurs is to some extent dependent on the time scale of the experimental observable. At sufficiently high temperature, the time scale for the loss of the local structure becomes comparable to the time for relaxation on the potential surface (27,34,68); all dynamics are coupled. Even the fast relaxations becomes temperature dependent, and β relaxation ceases to exist.

Unlike many simple liquids, the PMPS phenyl group orientational relaxation does not follow the DSE equation and the dynamics are decoupled from the low-frequency hydrodynamic modes. On a time scale longer than that examined in these experiments, PMPS displays temperature dependent dynamics, for example, the temperature dependence of Brillouin line width and the observed Vogel−Fulcher−Tammann temperature dependence of the viscosity.

The temperature-independent nature of the dynamics and the local character of the dynamics are distinctive features of β relaxation in glass-forming liquids (7,126). In scattering experiments on glass-forming liquids near the glass transition temperature, a Rayleigh peak is observed, a Boson peak is observed, and a broad spectral feature between the peaks is seen (42,126). A narrow Rayleigh peak arises from the slow orientational motions of the liquid; the Boson peak comes from low-frequency phonon modes, which include librations (42,126,145,146). The spectrum of the broad feature is temperature independent, and is ascribed to β relaxation (126,128). In transient grating experiments, the equivalent of the Rayleigh peak is slow temperature-dependent decays as observed in biphenyl and other simple liquids. The equivalent of the Boson peak is the ultrafast (<1 ps) librational dynamics. Because of the large dynamic range and broad time scale examined in the TG−OKE experiments, the ultrafast librational dynamics do not interfere with the observation of the fast time scale dynamics. It is proposed that the temperature-independent orientational dynamics of PMPS observed with the TG−OKE experiments result from relaxation on local potential surfaces that are fixed on the time scale of the dynamics under observation. The observed dynamics contribute to the broad temperature independent β relaxation feature seen in various scattering experiments.

7.7.   A COMPLEX MOLECULAR LIQUID: 2-ETHYLNAPHTHALENE

In liquids, the probability of finding another molecule at a distance R from a reference molecule is expressed in terms of the radial distribution function or pair correlation function $g(R)$. Structure exists even in a hard sphere simulated liquid (147). There are maxima and minima in $g(R)$. In real molecular liquids, the maxima are found close to the molecular positions determined by X-ray or neutron diffraction from the corresponding crystal lattice. Some of the order present in the crystal is preserved in the liquid on a small distance scale. The structure becomes progressively less correlated with increasing distance from a reference molecule. The function $g(R)$ depends on the potential energy $U(R)$ associated with intermolecular interactions. For small nonspherical molecules that have anisotropic polarizabilities, the van der Waals interaction potential $U(\mathbf{R})$ depends on the angle. Molecules with permanent dipole moments have additional, strong angle-dependent intermolecular interactions. The angular dependence of intermolecular interactions produces local orientational structure. Triplet $g^{(3)}$ and higher order correlation functions contain informations about orientational structure, but experimental measurements of $g^{(3)}$ are very difficult (148).

To illustrate the influence of local liquid structure on molecular orientational dynamics in liquids that do not have the clear cut structural aspects of liquid crystals or polymers, TG–OKE experiments performed on 2-ethylnaphthalene (2EN) are presented (34). The slowest dynamics are composed of a biexponential decay that obeys the DSE equation (hydrodynamic). However, there is an additional fast relaxation process that does not display any temperature dependence over a significant range of temperatures. This behavior is described in terms of relaxation of local structures rather than orientational diffusion. Dilution of 2EN in CCl$_4$ eliminates the local structures in the liquid, and a hydrodynamic biexponential decay is observed.

Figure 7.23 shows TG–OKE data for neat 2EN out to 200 ps at 25°C. The data are displayed following the decay of the ultrafast librational dynamics. A fit to the data that is virtually indistinguishable from the data is also shown. The data are well fit by a tri-exponential model, as seen by the residuals shown in the inset. This decay, as with all of the postlibrational data, is scanned out until the signal reaches the baseline using a polarization grating. Because the data in these scans are not analyzed until several picoseconds after $t = 0$ (many times the optical pulse length), the electronic OKE response does not affect the data, and the use of a polarization grating is justified. Similarly, the finite instrument response does not affect the data at these times, so convolutions are unnecessary.

Any attempt to fit the data with less than three exponentials was unsuccessful, producing poor residuals. Since any decaying function can be fit to the sum of enough exponentials, it is essential to establish that the data is truly the sum of three distinct exponentials. Figure 7.24 shows semi-log plots (natural logarithm) of the polarization grating data at 25°C. The log of the slowest component is shown in Fig. 7.24(a). The data are clearly single exponential for seven factors of e, with a time constant τ_s (s $\equiv$ slow). As the measured signal is proportional to the square of the sample response function, the square root of the data is taken to eliminate cross-terms before the slow decay is subtracted off. A semi-log plot showing the square root of the data with the slow component properly subtracted off is shown in Fig. 7.24(b). It is single exponential for seven factors of e of the signal (3.5 factors of e of the square root of the signal), with a time constant τ_i (i $\equiv$ intermediate). Figure 7.24(c) shows the log of the square root of the data with both the

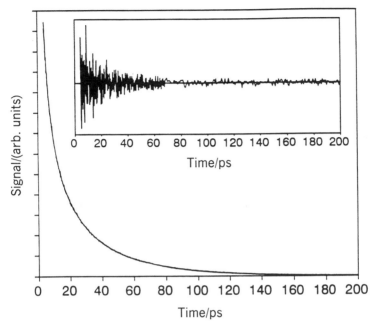

Figure 7.23. The TG–OKE signal for neat 2EN following the ultrafast librational dynamics at 25°C. The fit to the triple exponential model is also shown. The greatly magnified residuals of the fit are shown in the inset. [Reproduced with permission from S. R. Greenfield, A. Sengupta, J. J. Stankus, M. Terazima, and M. D. Fayer, *J. Phys. Chem.*, **1993**, *98*, 313. Copyright © (1993) American Chemical Society.]

slow and intermediate components properly subtracted off. It is single exponential for eight factors of *e* of the signal (four factors of *e* of the square root of the signal), with a time constant τ_f (f ≡ fast). At 25°C, the decay constants are τ_s = 68.8 ps, τ_i = 13.9 ps, and τ_f = 3.5 ps. The substantial amplitudes and clear separation of time scales of the three decays provide a high level of confidence in the use of the triple exponential function to describe the postlibrational data in neat 2EN.

It is worth pointing out that the data analysis is made possible by the outstanding S/N ratio that is obtained from TG–OKE experiments. From the peak of the librational signal, the decay was followed for 12 decades of signal intensity. This corresponds to a decay of about 28 factors of *e*. It is this incredible dynamic range of the measurements that makes it possible to observe the full extent of the dynamics in the 2EN system and the other systems discussed in the previous sections.

The temperature dependence of the slow decay of neat 2EN from 2 to 100°C is shown in Fig. 7.25(*a*). The dynamics are clearly coupled to the hydrodynamic modes, following a DSE equation of the form $\tau = C\eta/T + \tau_0$. A linear least-squares fit to the data gives τ_s = (8090 ± 80) (ps K/cP) × (η/T) + (1.9 ± 1.9) ps. The temperature dependence of the intermediate decay from 2 to 80°C is shown in Fig. 7.25(*b*). The rotational diffusion time constant again obeys the DSE equation. The

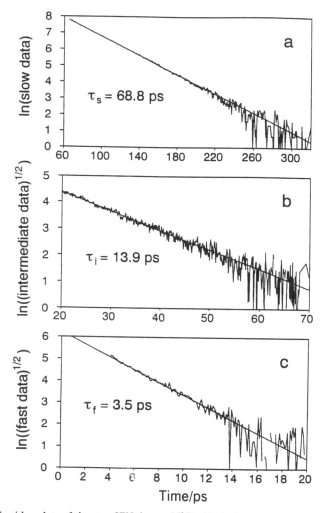

Figure 7.24. Semi-log plots of the neat 2EN data at 25°C with single exponential fits to each of the three components. The y-axis scale shows factors of e. (a) Slow component: natural log of the raw data showing the single exponential tail of the decay over 7 factors of e. (b) Intermediate component: natural log of the square root of the data minus the fit to the slow component. This component spans 7 factors of e of the signal. (c) Fast component: natural log of the square root of the data minus the fit to the slow and intermediate components. This component spans 8 factors of e of the signal. [Reproduced with permission from S. R. Greenfield, A. Sengupta, J. J. Stankus, M. Terazima, and M. D. Fayer, *J. Phys. Chem.*, **1993**, *98*, 313. Copyright © (1993) American Chemical Society.]

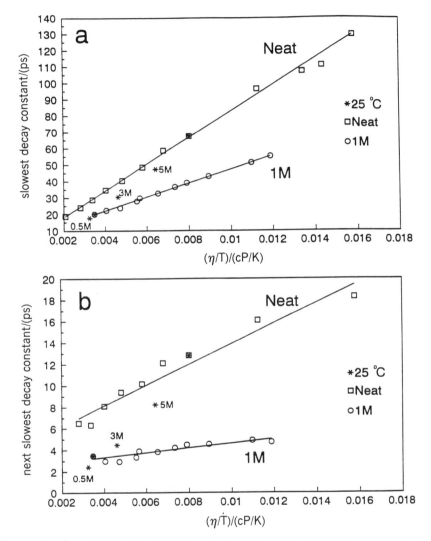

Figure 7.25. Viscosity/temperature (η/T) dependencies of the rotational diffusion components of the neat 2EN and various 2EN/CCl$_4$ solutions. (*a*) The slow component of neat 2EN (τ_s) and its solutions (τ_{ss}). (*b*) The intermediate component of neat 2EN (τ_i) and the next slowest component of the solutions (τ_{sf}). The parameter τ_s of neat 2EN is shown from 2 to 100°C and τ_i is shown from 2 to 80°C. The 1M 2EN/CCl$_4$ temperature range was -30 to 25°C for both (*a*) and (*b*). The data indicated by * show the concentration dependence of the relaxation components at 25°C. [Reproduced with permission from S. R. Greenfield, A. Sengupta, J. J. Stankus, M. Terazima, and M. D. Fayer, *J. Phys. Chem.*, **1993**, *98*, 313. Copyright © (1993) American Chemical Society.]

fit for the intermediate data gives τ_i = (960 ± 47) (ps K/cP) × (η/T) + (4.3 ± 1.6) ps.

Figure 7.25 also presents the results of a dilution study. Four different concentrations of 2EN in CCl_4 were used in the TG–OKE dilution experiments: 0.5, 1, 3, and 5 M (neat 2EN is 6.35 M). The mole fraction of 2EN in these solutions is 0.050, 0.10, 0.36, and 0.70, respectively. Polarization grating experiments were performed on these solutions at 25°C. Following the ultrafast librational dynamics, the data at the lowest three concentrations are cleanly fit with a biexponential model function. Only the 5 M sample, which is close to the pure liquid in concentration, displays a triexponential decay. The 1 M solution was examined at 11 temperature points between −30 and + 25°C. The temperature dependence of the solution data is shown in Fig. 7.25 along with the neat data for τ_s and τ_i. Two observations about the 1 M data are immediately obvious. The first is that both components are clearly hydrodynamic, being well fit by straight lines. To avoid confusion with the neat data components, the solution data rotational diffusion time constants are referred to as τ_{ss} (ss ≡ solution slow) and τ_{sf} (sf ≡ solution fast). A linear least-squares fit to the slower decay for the 1 M 2EN data in Fig. 7.25(a) gives τ_{ss} = (4220 ± 120) (ps K/cP) × (η/T) + (5.2 ± 1.5) ps. The fit to the faster decay shown in Fig. 7.25(b) gives τ_{sf} = (218 ± 40) (ps K/cP) × (η/T) + (2.4 ± 0.5) ps. It should be noted that the hydrodynamic volume for the slowest component in the solution is approximately one-half of the volume in the neat liquid. The second observation from Fig. 7.25 is that the neat data and the solution data do not have the same slope. The most likely explanation for the differences in the slopes (effective hydrodynamic volumes) is that there are multiparticle contributions to the neat 2EN TG–OKE observable [$g_2 \neq 1$ in Eq. (7.15)]. On the time scale of the data in Fig. 7.25, the solvent does not produce a signal in the 1 M solution. The concentration is low enough that the signal arises from essentially isolated 2EN molecules undergoing hydrodynamic rotational diffusion. In the neat liquids, the slow and intermediate decays also indicate hydrodynamic rotational diffusion. However, the molecules are not isolated. Using the 1 M data, the g_2 values can be obtained for the neat liquid [Eq. (7.15)]. For the slow decay (tumbling), $g_2 \geq 2$, and for the faster decay (a combination of spinning and tumbling) $g_2 \geq 4$. With $g_2 \approx 1$ for the 1 M data, then the single-particle "effective" hydrodynamic volume (λV) of the tumbling rotation can be calculated with Eq. (7.15) to be 58 ± 2 Å^3.

The ability of a biexponential to fit the dilute 2EN solutions is very significant. In general, the orientational correlation function decay by rotational diffusion of an anisotropic rigid rotor is given by the sum of five exponentials (32). If the principal axis systems for the moment of inertia tensor and the polarizability tensor coincide or almost coincide, the number of exponentials is reduced to two (32). The two principal axis systems almost coincide in 2EN. Therefore, it is expected and observed that the TG–OKE data should be reproduced by a biexponential model (following the librational dynamics). The existence of the third exponential component in neat 2EN is the key point.

Figure 7.26 shows the temperature dependence of the fast component of neat 2EN from 2 to 80°C. The behavior of this component of the orientational anisotropy relaxation is nonhydrodynamic. The value for τ_f is temperature independent from the lowest temperature at which we made measurements (2°C) to 40°C. Above 40°C, τ_f becomes temperature dependent. In the temperature-independent regime, the viscosity of the liquid changes by a factor of 2.1, and τ_s and τ_i change by factors of approximately equal to 3 and approximately equal to 2, respectively. Thus τ_f is associated with orientational relaxation that is decoupled from the hydrodynamic modes.

Because the 1 M 2EN data are well described by a biexponential decay, the additional component, τ_f, seen in the neat data must be associated with a dynamic phenomena that is not anisotropic rigid rotor rotational diffusion. Greenfield et al. (34) proposed that τ_f is the time for local structure relaxation. In liquid naphthalene (as in benzene), pairs of molecules are known to form a crooked ''T'' shaped

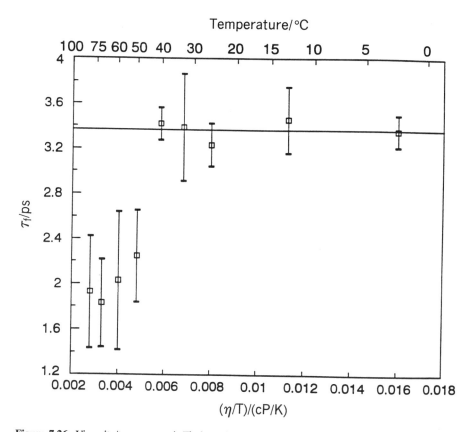

Figure 7.26. Viscosity/temperature (η/T) dependence of the fast component (τ_f) of neat 2EN from 2 to 80°C. The straight line shows the average of the 2–40°C data ($\tau_f = 3.36$ ps). The data are viscosity/temperature independent from 2 to 40°C, showing the nonhydrodynamic behavior of the orientational dynamics. [Reproduced with permission from S. R. Greenfield, A. Sengupta, J. J., Stankus, M. Terazima, and M. D. Fayer, *J. Phys. Chem.*, **1993**, *98*, 313. Copyright © (1993) American Chemical Society.]

structure (3). It is reasonable to assume that 2EN forms similar structures. Dilute 2EN in CCl_4 will not have these structures, and τ_f is not observed in the dilution experiments until the concentration is almost that of the pure liquid.

On a time scale fast compared to the rotational diffusion times (time scale for randomization of orientations), local structures in the neat liquid are associated with local, momentary minima on a free energy surface. The optical field perturbs the local structure. With the electric field on, the potential surface is changed, and local structure is no longer at the potential minimum. The local structure evolves toward the new minimum. After the optical pulse has passed through the sample, the electric field is gone, and the local structure has evolved away from the zero-field potential minimum. The local structure will now relax toward the (zero-field) potential minimum. It is this relaxation time that we associate with τ_f. Evidence for similar low-frequency nondispersive collective modes in simple liquids has recently been observed in benzene and pyridine (149). The polarizability of 2EN is greatest along the length of the molecule (150). Thus, the result of the optical pulse is to slightly align the molecules, primarily along their long axis. Such alignment involves a disturbance of the crooked "T" shape of neighboring 2EN molecules. After the relaxation back to the "crooked T's," one is left with a collection of partially aligned crooked T's. This leaves a residual anisotropy. It is the randomization of this residual anisotropy that is responsible for the anisotropic rigid rotor rotational diffusion components (τ_i and τ_s) of the decay.

The experiments on 2EN demonstrate the influence of local structure on orientational dynamics in a conventional molecular liquid. The dynamics, like those observed in the liquid crystals and the polymers, are η/T independent on the short time scale on which the local structure exists.

7.8. CONCLUDING REMARKS

In this chapter, we have examined the nature of orientational relaxation in complex molecular liquids. The fundamental concept that has been illustrated by the experimental examples is the relationship between local liquid structure and orientational dynamics. In a simple liquid, such as biphenyl, orientational relaxation occurs by rotational diffusion. This is characterized by a strong dependence of τ, the decay time of the orientational correlation function, on the temperature and the viscosity. Rotational diffusion in simple liquids is described by the Debye–Stokes–Einstein (DSE) equation, which states that τ is proportional to viscosity/temperature (η/T). In the transient grating optical kerr effect (TG–OKE) experiments, the signal gives τ for simple liquids following the decay of the librational dynamics that occur on an ultrafast time scale. In examples such as biphenyl or 2-ethylnaphthalene in dilute solution, it was shown that the dynamics do obey the DSE equation.

In complex liquids, the DSE equation is not obeyed. In fact, TG–OKE measurements on a variety of systems showed that the orientational relaxation on short time scales is independent of temperature and viscosity. A number of examples were discussed in detail. These are liquid crystals in their isotropic phase, polymer side groups for pure polymer liquids and polymers in solution, and a conventional molecular liquid. In each of these systems, the short time scale dynamics is controlled by the existence of local structures in the liquids. On the time scale on which the local structure exists, dynamics involve relaxation on the potential surface that gives rise to the local structure. A thermal or optical perturbation moves the system away from the potential minimum. The system then relaxes back to the minimum, reforming the local structure, rather than undergoing rotational diffusion that randomizes the local structure. Rotational diffusion will occur on some longer time scale. For example this was observed in the liquid crystals. The temperature and viscosity determine how long the local structures exist. As long as the time scale for structural randomization is long compared to the time scale for relaxation on the local potential surface, the dynamics are not influenced by viscosity and temperature. Once the temperature becomes high enough so that the structures are destroyed on the same time scale as the relaxation on the potential surface, even the short time dynamics become viscosity and temperature dependent. This was observed in the liquid crystals and in liquid 2-ethylnaphthalene (2EN). This same concept was used to explain the difference between fast β relaxation and slow α relaxation in glass-forming liquids such as poly(methylphenylsiloxane) (discussed above).

In addition to the general features of the relationship between local structure and dynamics illustrated by the experiments, the specific details of the orientational relaxation of the liquid crystal systems are remarkable. In the two liquid crystal systems studied, both ceased to obey the Landau–de Gennes (LdG) theory describing the slow time scale dynamics when the correlation length of the pseudonematic domains fell below $3\xi_0$, where ξ_0 is a molecular length scale. In addition, in both systems the fast, intradomain dynamics decayed as power laws with identical exponents of 0.63. We are currently investigating a third liquid crystal. If the same behavior is observed, it will imply that there are fundamental physical principals that control the dynamics in the isotropic phase of liquid crystals. By applying the methods of LdG theory to the fast, intradomain dynamics, we were able to derive a power law with an exponent 0.5 in the limit of an infinite correlation length. We are extending the theory to include the concepts of critical systems. It is possible that this will lead to the experimentally observed results.

The experiments on the polymer systems provided the first measurements of the fast orientational dynamics of side groups. The difference in the observed dynamics of the carbon based polymer and the polysiloxane is significant. While the naphthyl side groups of poly(2-vinylnaphthalene) (2PVN) exhibit exponential dynamics, the phenyl side groups of poly(methylphenylsiloxane) (PMPS) displays power law decays. This difference may arise from the difference in the flexibility of the polymer backbones. However, it will be necessary to examine more systems to see if the nature of the dynamics is controlled by the chemical composition of the polymer backbone.

The influence of local structure on dynamics can be of fundamental importance, not only in understanding the dynamics of liquids, but also in understanding other phenomena, like reaction chemistry. In many chemical reactions in solution, the reacting species must obtain the appropriate relative geometry before the reaction can proceed. Once the reactants have come into close proximity, the orientational dynamics may be controlled by the details of anisotropic intermolecular interactions and local structures, rather than by rotational diffusion.

Our understanding of dynamics in liquids is accelerating rapidly because of advances in both theoretical and experimental methods. The femtosecond TG–OKE method and other ultrafast nonlinear techniques are providing the necessary observables for examining liquids on the time scales of key dynamical phenomena. Liquid dynamics at the molecular level is emerging from the realm of speculation into the light of understanding.

ACKNOWLEDGMENTS

Over the last few years, several co-workers have directly contributed to the TG–OKE experiments described in this chapter. We would like to thank Dr. V. J. Newell, Dr. F. W. Deeg, Dr. S. R. Greenfield, Dr. J. J. Stankus, Dr. M. Terazima, and Dr. R. Torre for their important contributions to both the experimental and theoretical components of the material presented here. This work was supported by the National Science Foundation, Division of Materials Research (Grant No. DMR90-22675).

REFERENCES

1. (a) Shechtman, D., Blech, I., Gratias, D., and Cahn, J. W. *Phys. Rev. Lett.* **1984**, *53*, 1951. (b) Levine, D. and Steinhardt, P. J. *Phys. Rev. Lett.* **1984**, *53*, 2477.

2. Misawa, M. *J. Chem. Phys.* **1989**, *91*, 5648.

3. Misawa, M. and Fukunaga, T. *J. Chem. Phys.* **1990**, *93*, 3495.

4. Pinan-Lucarre, J. P., Loisel, J., Berreby, L., Dayan, E., and Dervil, E. *J. Raman Spectrosc.* **1992**, *23*, 67.

5. Arunan, E. and Gutowsky, H. S. *J. Chem. Phys.* **1993**, *98*, 4294.

6. (a) Leutheusser, E. *Phys. Rev. A* **1984**, *29*, 2765. (b) Bengtzelius, D. *Phys. Rev. A* **1986**, *34*, 5059. (c) Gotze, W. and Sjogren, L. *J. Phys. Condens. Matter* **1989**, *1*, 4183; 4203.

7. (a) Sjolander, A., In *Static and Dynamic Properties of Liquids*; Springer Proceedings in Physics *40*, Davidovic, M. and Soper, A. K. (Eds.). Springer-Verlag: Berlin, 1989; p. 90. (b) Elmroth, M., Börjesson, L., and Torell, L. M. Springer Proceedings in Physics **1981**, *40*, 118.

8. Eichler, H. J., Guenter, P., and Pohl, D. W. *Laser-Induced Dynamic Gratings*, Springer, Berlin, 1986.

9. Fayer, M. D. *Annu. Rev. Phys. Chem.* **1982**, *33*, 63.

10. Eichler, H. J. (Ed.). *Special issue on Dynamic Gratings and Four-Wave Mixing, IEEE J. Quantum Electron.*, 1986.

11. Eyring, G. and Fayer, M. D. *J. Chem. Phys.* **1984**, *81*, 4314.

12. Deeg, F. W. and Fayer, M. D. *J. Chem. Phys.* **1989**, *91*, 2269.

13. Newell, V. J., Deeg, F. W., Greenfield, S. R., and Fayer, M. D. *J. Opt. Soc. Am. B* **1989**, *6*, 257.

14. Etchepare, J., Grillon, G., Chambaret, J. P., Hamoniaux, G., and Orszag, A. *Optics Comm.* **1987**, *63*, 329.

15. Deeg, F. W., Stankus, J. J., Greenfield, S. R., Newell, V. J., and Fayer, M. D. *J. Chem. Phys.* **1989**, *90*, 6893.

16. Stankus, J. J., Torre, R., Marshall, C. D., Greenfield, S. R., Sengupta, A., Tokmakoff, A., and Fayer, M. D. *Chem. Phys. Lett.* **1992**, *194*, 213.

17. Shen, Y. R. *The Principle of Nonlinear Optics,* Wiley, New York, 1984.

18. Butcher, P. N. *Nonlinear Optical Phenomena Bulletin 200*, Engineering Experiment Station, Ohio State University, Columbus, Ohio, 1965.

19. Hellwarth, R. W. *Prog. Quant. Electr.* **1977**, *5*, 1.

20. Hellwarth, R. W., Owyoung, A., and George, N. *Phys. Rev. A* **1971**, *4*, 2342.

21. Fayer, M. D. *IEEE J. Quant. Electr.* **1986**, *QE-22*, 1437.

22. (a) Yan, Y. X. and Nelson, K. A. *J. Chem. Phys.* **1987**, *87*, 6240.; 6257. (b) Yan, Y. X., Cheng, L. F., and Nelson, K. A. *Adv. Infrared Raman Spectrosc.* **1987**, *16*, 299.

23. (a) Landau, L. D. and Lifshitz, E. M. *Statistical Physics*; Pergamon Press, Oxford, 1980; 3rd ed., Part 1, pp. 377–389. (b) Reichl, L. E. *A Modern Course in Statistical Physics*. University of Texas Press, Austin, TX, 1984, pp. 545–556.

24. Boon, J. P. and Yip, S. *Molecular Hydrodynamics*. Dover, New York, 1991.

25. Dattagupta, S. *Relaxation Phenomena in Condensed Matter Physics*; Academic, London, 1987.

26. Samios, D. and Dorfmuller, T. *Chem. Phys. Lett.* **1985**, *117*, 165.

27. Stankus, J. J., Torre, R., and Fayer, M. D. *J. Phys. Chem.* **1993**, *97*, 9478.

28. Kivelson, D. In *Rotational Dynamics of small and macromolecules,* Dorfmuller, T. and Pecora, R., (Ed.). Springer, Berlin, 1987, p. 1.

29. Alms, G. R., Bauer, D. R., Brauman, J. I., and Pecora, R. *J. Chem. Phys.* **1973**, *58*, 5570.

30. Keyes, T. and Kivelson, D. *J. Chem. Phys.* **1972**, *56*, 1057.

31. Kivelson, D. and Madden, P. A. *Annu. Rev. Phys. Chem.* **1980**, *31*, 523.

32. Berne, B. J. and Pecora, R. *Dynamic Light Scattering*, Wiley, New York, 1976.

33. Alms, G. R., Bauer, D. R., Brauman, J. I., and Pecora, R. *J. Chem. Phys.* **1973**, *59*, 5310.

34. Greenfield, S. R., Sengupta, A., Stankus, J. J., Terazima, M., and Fayer, M. D. *J. Phys. Chem.* **1993**, *98*, 313.

35. (a) Sengupta, A., Terazima, M., and Fayer, M. D. *J. Phys. Chem.* **1992**, *96*, 8619. (b) Sengupta, A. and Fayer, M. D. *J. Chem. Phys.* **1993**, *100*, 1673.

36. Olson, R. W., Meth, J. S., Marshall, C. D., Newell, V. J., and Fayer, M. D. *J. Chem. Phys.* **1990**, *92*, 3323.

37. (a) Xu, B. C. and Stratt, R. M. *J. Chem. Phys.* **1990**, *92*, 1923. (b) Seeley, G. and Keyes, T. *J. Chem. Phys.* **1989**, *91*, 5581. (c) Wu, T. M. and Loring R. J. *J. Chem. Phys.* **1992**, *97*, 8568.

38. Frenkel, J. *Kinetic Theory of Liquids*. Dover, New York, 1955.

39. Ruhman, S., Williams, L. R., Joly, A. G., Kohler, B., and Nelson, K. A. *J. Phys. Chem.* **1987**, *91*, 2237.

40. Kohler, B. and Nelson K. A. *J. Phys. Condens. Matter* **1990**, *2*, SA109.

41. Ito, M. and Shigeoka, T. *Spectrochimica Acta* **1966**, *22*, 1029.

42. Gochiyaev, V. Z., Malinovsky, V. K., Novikov, V. N. and Sokolov, A. P. *Philos. Mag. B* **1991**, *63*, 777.

43. Madden, P. A. In *Molecular Liquids-Dynamics and Interactions*, Barnes, A. J., Orville-Thomas, W. J., and Yarwood, J. (Eds.). Reidel, Dordrecht, 1984, p. 431.

44. Geiger, L. C. and Landanyi, B. M. *Chem. Phys. Lett.* **1989**, *159*, 413.

45. Deeg, F. W., Stankus, J. J., Greenfield, S. R., Newell, V. J., and Fayer, M. D. *J. Chem. Phys.* **1989**, *90*, 6893.

46. Ruhman, S., Kohler, B., Joly, A. G., and Nelson, K. A. *IEEE J. Quant. Electron.* **1988**, *QE 24*, 470.

47. Kalpouzos, C., McMorrow, D., Lotshaw, W. T., and Kenney-Wallace, G. A. *Chem. Phys. Lett.* **1988**, *150*, 138.

48. Dorfmuller, Th. In *Rotational Dynamics of Small and Macromolecules*, Dorfmuller, Th. and Pecora, R. (Eds.). Springer, Berlin, 1987.

49. Landau, L. D. and Lifshitz, E. M. *Statistical Physics.* Pergamon, Oxford, 1980; 3rd ed. Part 1, p. 341.

50. Denbigh, K. *The Principles of Chemical equilibrium,* Cambridge University Press, Cambridge, 1983; 4th ed., p. 201.

51. Lerner, R. G. and Trigg, G. L. (Eds.). *Encyclopedia of Physics,* VCH Publishers, Inc. New York, 1991, 2nd ed., p. 879.

52. Careri, G. *Order and Disorder in Matter;* Benjamin / Cummings, Menlo Park, CA, 1984.

53. Stanley, H. E. *Introduction to Phase Transitions and Critical Phenomena.* Oxford, New York, 1971.

54. (a) De Gennes, P. G. *The Physics of Liquid Crystals.* Oxford, Clarendon, 1974. (b) Chandrasekhar, S. *Liquid Crystals.* Cambridge, New York, 1992, 2nd ed.

55. Brenig, W. *Statistical Theory of Heat,* Springer Verlag, New York, 1989.

56. Stinson III, T. W. and Litster, J. D. *Phys. Rev. Lett.* **1970**, *25*, 503.

57. Coles, H. *J. Mol. Cryst. Liq. Cryst. (Lett.)* **1978**, *49*, 67.

58. Litster, J. D. and Stinson III, T. W. *J. Appl. Phys.* **1970**, *41*, 996.

59. Wong, G. K. L. and Shen, Y. R. *Phys. Rev. Lett.* **1973**, *30*, 895.

60. Wong, G. K. L. and Shen, Y. R. *Phys. Rev. Lett.* **1974**, *32*, 527.

61. Wong, G. K. L. and Shen, Y. R. *Phys. Rev. A* **1974**, *10*, 1277.

62. Hanson, E. G., Shen, Y. R., and Wong, G. K. L. *Phys. Rev. A* **1976**, *14*, 1281.

63. Gierke, T. D. and Flygare, W. H. *J. Chem. Phys.* **1974**, *61*, 2231.

64. Alms, G. R., Gierke, T. D., and Flygare, W. H. *J. Chem. Phys.* **1974**, *61*, 4083.

65. Flytzanis, C. and Shen, Y. R. *Phys. Rev. Lett.* **1974**, *33*, 14.

66. Lalanne, J. R., Martin, B., Pouligny, B., and Kielich, S. *Optics Commun.* **1976**, *19*, 440.

67. Amer, N. M., Lin, Y. S., and Shen, Y. R. *Solid State Commun.* **1975, 16**, 1157.

68. Deeg, F. W., Greenfield, S. R., Stankus, J. J., Newell, V. J., and Fayer, M. D. *J. Chem. Phys.* **1990**, *93*, 3503.

69. Filippini, J. C. and Poggi, Y. *J. Phys. Lett. (Paris)* **1974**, *35*, L-99.

70. Cabane, B. and Clarke, G. *Phys. Rev. Lett.* **1970**, *25*, 91.

71. Gosh, S., Tettamanti, E., and Indovina, E. *Phys. Rev. Lett.* **1973**, *29*, 638.

72. Filippini, J. C. and Poggi, Y. *J. Phys. Lett. (Paris)* **1976**, *37*, L-17.

73. de Gennes, P. G. *Mol. Cryst. Liq. Cryst.* **1971**, *12*, 193.

74. Frank, F. C. In *Liquid Crystals*, Chandrasekhar, S. (Ed.). Heyden, London, 1980.

75. Vertogen, G. and de Jeu, W. H. *Thermotropic Liquid Crystals, Fundamentals,* Springer-Verlag: Berlin, 1988.

76. Courtens, E. and Koren, G. *Phys. Rev. Lett.* **1975**, *35*, 1711.

77. Courtens, E. *J. Chem. Phys.* **1977**, *66*, 3995.

78. Chu, B., Bak, C. S., and Lin, F. L. *Phys. Rev. Lett.* **1972**, *28*, 1111.

79. Gulari, E. and Chu, B. *J. Chem. Phys.* **1975**, *62*, 798.

80. Stinson, T. W. and Litster, J. D. *Phys. Rev. Lett.* **1973**, *30,* 688.

81. Prost, J. and Lalanne, J. R. *Phys. Rev. A* **1973**, *8*, 2090.

82. Martinoty, P., Candau, S., and Debeauvais, F. *Phys. Rev. Lett.* **1971**, *27*, 1123.

83. Dissado, L. A., Nigamatullin, R. R., and Hill, R. M. *Adv. Chem. Phys.* **1985**, *63*, 253.

84. Weast, R. C. *Handbook of Chemistry and Physics;* CRC Press, Florida 1986–1987; 67th ed., p. A-61.

85. Littau, K. A., Dugan, M. A., Chen, S., and Fayer, M. D. *J. Chem. Phys.* **1992**, *96*, 3484.

86. Littau, K. A. and Fayer, M. D. *Chem. Phys. Lett.* **1991**, *176*, 551.

87. Palmer, R. G., Stein, D. L., Abrahams, E., and Anderson, P. W. *Phys. Rev. Lett.* **1984**, *53*, 958.

88. Ajay, Palmer, R. G. *J. Phys. A.: Math. Gen.* **1990**, *23*, 2139.

89. Klafter, J. and Shlesinger, M. F. *Proc. Natl. Acad. Sci.* **1986**, *83*, 848.

90. Huberman, B. A. and Kerszberg, M. *J. Phys. A Gen. Phys.* **1985**, *18*, L331.

91. Kumar, D. and Shenoy, S. R. *Sol. St. Commun.* **1986**, *57*, 927.

92. Teitel, S. and Domany, E. *Phys. Rev. Lett.* **1985**, *55*, 2176.

93. Leadbetter, A. J., Richardson, R. M., and Colling, C. N. *J. Phys.* (Paris) **1975**, *36*, C1-37.

94. Bauer, D. R., Brauman, J. I., and Pecora, R. *Macromolecules* **1975**, *8*, 443.

95. Brueggeman, B. G., Minnick, M. G., and Schrag, J. L. *Macromolecules* **1978**, *11*, 119.

96. Fixman, M. and Evans, G. T. *J. Chem. Phys.* **1978**, *68*, 195.

97. Nomura, H. and Miyahara, Y. *Polym. J.* **1976**, *8*, 30.

98. Sengupta, A. and Greenfield, S. R. (unpublished results).

99. Frenkel, D. and Mc Tague, J. P. *J. Chem. Phys.* **1980**, *72*, 2801.

100. Madden, P. A. In *Molecular Liquids—Dynamics and Interactions*. Barnes, A. J., Orville-Thomas, W. J., Yarwood, J. (Eds.). Reidel, Dordrecht, 1984, p. 431.

101. Geiger, L. C. and Landanyi, B. M. *J. Chem. Phys.* **1987**, *87*, 191.

102. Deb, S. K. *Chem. Phys.* **1988**, *120*, 225.

103. Doany, F. E., Greene, B. I., and Hochstrasser, R. M. *Chem. Phys. Lett.* **1980**, *75*, 206.

104. De Schryver, F. C., Demeyer, K., van der Auweratr, M., and Quanten, E. *Ann. N.Y. Acad. Sci.* **1981**, *366*, 93.

105. Seki, K., Ichimura, Y., and Imamura, Y. *Macromolecules* **1981**, *14*, 1831.

106. Frank, C. W. and Harrah, L. A. *J. Chem. Phys.* **1974**, *61*, 1526.

107. Seki, K. and Imamura, Y. *Bull. Chem. Soc. Jpn.* **1982**, *55*, 3711.

108. Valeur, B. and Monnerie, L. *J. Polym. Sci., Polym. Phys. Ed.* **1976**, *14*, 11.

109. Viovy, J. L. and Monnerie, L. *Macromolecules* **1983**, *16*, 1845.

110. Sasaki, T., Yamamoto, M., and Nishijima, Y. *Makromol. Chem. Rapid Commun.* **1986**, *7*, 345.

111. Hyde, P. D., Waldow, D. A., Ediger, M. D., Kitano, T., and Ito, K. *Macromolecules* **1986**, *19*, 2533.

112. Waldow, D. A., Johnson, B. S., Hyde, P. D., Ediger, M. D., Kitano, T., and Ito, K. *Macromolecules* **1989**, *22*, 1345.

113. Heatley, F. and Wood, B. *Polymer* **1978**, *19*, 1405.

114. Allerhand, A. and Hailstone, R. K. *J. Chem. Phys.* **1972**, *56*, 3718.

115. Laupretre, F., Noel, C., and Monnerie, L. *J. Polym. Sci., Polym. Phys. Ed.* **1977**, *15*, 2127.

116. Heatley, F. and Begum, A. *Polymer* **1976**, *17*, 399.

117. Matsuo, K., Kuhlmann, K. F., Yang, H. W. H, Geny, F., Stockmayer, H., and Jones, A. A. *J. Polym. Sci., Polym. Phys. Ed.* **1977**, *15*, 1347.

118. Glowinkowski, S., Gisser, D. J., and Ediger, M. D. *Macromolecules* **1990**, *23*, 3520.

119. Woessner, D. E. *J. Chem. Phys.* **1962**, *36*, 1.

120. Valeur, B., Jarry, J. P., Geny, F., and Monnerie, L. *J. Polym. Sci., Polym. Phys. Ed.* **1975**, *13*, 667.

121. Schaefer, J. *Macromolecules* **1973**, *6*, 882.

122. Gronski, W. *Makromol. Chem.* **1979**, *180*, 1119.

123. Denault, J. and Prud'homme, J. *Macromolecules* **1989**, *22*, 1307.

124. Tekely, P. *Macromolecules* **1986**, *19*, 2544.

125. Lewis, R. J., Pecora, R., and Eden, D. *Macromolecules* **1986**, *19*, 134.

126. Kruger, M., Soltwisch, M., Petscherizin, I., and Quitmann, D. *J. Chem. Phys.* **1992**, *96*, 7352.

127. Meier, G., Fujara, F., and Petry, W. *Macromolecules* **1989**, *22*, 4421.

128. Richter, D., Frick, B., and Farago, B. *Phys. Rev. Lett.* **1988**, *61*, 2465.

129. Floudas, G., Higgins, J. S., and Fytas, G. *J. Chem. Phys.* **1992**, *96*, 7672.

130. Angell, C. A. In *Relaxations in Complex Systems*. Ngai, K. and Wright, G. B. (Eds.). National Technical Information Service, U.S. Department of Commerce, Springfield, VA, Vol. 1, 1985; Angell, C. A. *J. Non-Cryst. Solids* **1985**, *73*, 1.

131. Boese, D., Momper, B., Meier, G., Kremer, F., Hagenah, J. U., and Fischer, E. W. *Macromolecules* **1989**, *22*, 4416.

132. Frick, B. *Progr. Colloid Polymer Sci.* **1989**, *80*, 164–171.

133. Anderson, H. C. and Pecora, R. *J. Chem. Phys.* **1971**, *54*, 2584; *55*, 1496.

134. Wang, C. H., Ma, R. J., Fytas, G., and Dorfmuller, T. *J. Chem. Phys.* **1983**, *78*, 5863.

135. Wang, C. H. and Zhang, J. *J. Chem. Phys.* **1986**, *85*, 794.

136. (a) Vogel, H. *Phys. Z* **1921**, *22*, 645. (b) Fulcher, G. S. *J. Am. Ceram. Soc.* **1925**, *8*, 339.

137. Rossler, E. *Phys. Rev. Lett.* **1990**, *65*, 1595.

138. Wang, C. H., Fytas, G., and Zhang, J. *J. Chem. Phys.* **1985**, *82*, 3405.

139. Wang, C. H., Ma, R. J., and Liu, Q. L. *J. Chem. Phys.* **1984**, *80*, 617.

140. Kremer, F., Boese, D., Meier, G., and Fischer, E. W. *Prog. Colloid Polymer Sci.* **1989**, *80*, 129–139.

141. The 0.7-ps decay is fast enough to have a contribution from interaction induced polarizability. The slowest component of the librational decay is about 70 fs. Therefore the 0.7-ps decay is distinct from the librational decay. The 1.95-ps melt decay, which runs out to 10 ps, is probably too slow to arise from interaction induced effects.

142. Fytas, G., Lin, Y. H., and Chu, B. *J. Chem. Phys.* **1981**, *74*, 3131.

143. Barrat, J. L. and Klein, M. L. *Annu. Rev. Phys. Chem.* **1991**, *42*, 23.

144. Landau, L. D. and Lifshitz, E. M. *Mechanics*, 3rd ed., Pergamon Press, Oxford, 1988, p. 76.

145. Shuker, R. and Gammon, R. W. *Phys. Rev. Lett.* **1970**, *25*, 222.

146. Martin, A. J. and Brenig, W. *Phys. Status Solidi B* **1974**, *64*, 163.

147. Hansen, J. P. and Mc Donald, I. R. *Theory of Simple Liquids*, 2nd ed., Academic, New York, 1991.

148. March, N. H. and Tosi, M. P. *Atomic Dynamics in Liquids*. Dover, New York, 1991.

149. McMorrow, D. and Lotshaw, W. T. *Chem. Phys. Lett.* **1993**, *201*, 369.

150. Ruessink, B. H. and MacLean, C. *Mol. Phys.* **1987**, *60*, 1059.

Chapter **VIII**

ELECTRIC FIELD EFFECTS IN MOLECULAR SYSTEMS STUDIED VIA PERSISTENT HOLE BURNING

Bryan E. Kohler, Roman I. Personov[‡] and Jörg C. Woehl
Department of Chemistry, University of California, Riverside, California

8.1. **Introduction**
8.2. **Principles**
    8.2.1.   Inhomogeneous Broadening and Hole Burning
    8.2.2.   Hole Profile and Electric Field Effect
    8.2.3.   Fixed Frequency Experiments
    8.2.4.   Experimental Considerations
8.3. **Results**
    8.3.1.   Introduction
    8.3.2.   Molecular Orientation: Experiments on Langmuir–Blodgett Films
    8.3.3.   Molecular Structure: Experiments on Metalloporphyrins
    8.3.4.   Single Molecules: Stark Experiments
    8.3.5.   Internal Field: Experiments on Octatetraene
8.4. **Concluding Remarks**

8.1. INTRODUCTION

Most of our understanding of the electronic structure of organic solid solutions is derived from optical spectroscopy. It is now widely appreciated that the amount of unambiguous information that can be extracted from a given optical spectrum

---

[‡]1994 Regents Professor. Institute of Spectroscopy of the Academy of Sciences of Russia, Troitzk, 142092 Moscow, Russia

Laser Techniques In Chemistry, Edited by Anne B. Myers and Thomas R. Rizzo.
Techniques of Chemistry Series, Vol. XXIII.
ISBN 0-471-59769-4 © 1995 John Wiley & Sons, Inc.

may be substantially increased by increasing the homogeneity of the ensemble under investigation. This may be done by one or another of the methods of high-resolution site selective (more strictly speaking, energy selective) spectroscopy. The first experiments on fluorescence line narrowing (1–3) and persistent spectral hole burning (4,5) showed that these methods could increase spectral resolution in organic molecular systems by a factor of 10^3 (up to 10^5 in subsequent experiments). Since these early investigations these techniques have been broadly applied to measure vibronically resolved spectra of many different compounds; to investigate relaxation processes (in particular, optical dephasing and spectral diffusion) and energy transfer in crystalline and amorphous organic solids; to characterize the photophysics of biological systems; to improve sensitivity and selectivity in the spectrochemical analysis of complex organic products; to optical data storage and holography, and so on. Many different applications of fluorescence line narrowing and hole burning have been discussed in review articles (6–13).

This chapter focuses on the use of energy selective spectroscopy to investigate the influence of an external field on complex molecules in solids. Initial work on electric field or Stark (14–18) and magnetic field or Zeeman (19,20) effects was followed by the publication of a number of papers in this area, some of which are summarized in reviews (21,22). However, during the last few years novel and important results have been obtained and new methods for interpreting the effects of external fields have appeared. This chapter concentrates on these new developments, especially hole-burning investigations of the Stark effect. Special attention is given to the Stark effect in the case of centrosymmetric molecules and to the problem of the internal field in organic molecular systems.

Before going to the latest results in this field it is useful to briefly review some general principles of hole-burning spectroscopy and the special advantages of this technique for studying the effects of external fields.

8.2. PRINCIPLES

8.2.1. Inhomogeneous Broadening and Hole Burning

Two conditions are necessary for hole-burning spectroscopy: (1) the presence of inhomogeneous spectral broadening and (2) some process that can selectively change the spectroscopic properties of the initially excited molecules. When both these conditions are fulfilled, as they are in a great number of cases, irradiation with monochromatic light will selectively reduce absorption at the irradiation wavelength producing a persistent hole in the inhomogeneous absorption profile.

As is now well known, the band profiles seen in the low-temperature spectra of many organic molecules in solution are dominated by inhomogeneous broadening to such an extent that (especially in glasses and polymers) much of the fine structure is concealed. In these cases any broad vibronic band is the superposition of many narrow zero-phonon lines (accompanied by relatively broad phonon wings) whose frequencies are distributed because of differences in local conditions. In the case of molecules in amorphous organic matrices the bandwidth due to inhomogeneous broadening is usually $\Delta \nu \approx 10^2 - 10^3$ cm^{-1}. In the special case of a mixed crystal (including some organic molecules in *n*-alkane Shpol'skii matrices) inhomogeneous broadening is less so that at liquid helium temperatures $\Delta\nu \approx 2-10$ cm^{-1}, but even in this case the bandwidth is mainly determined by inhomogeneous broadening. Spectral hole burning is a technique that makes the narrow features that are hidden in the inhomogeneously broadened band visible.

Hole-burning mechanisms can be very different. It is conventional to divide hole-burning mechanisms into two types (see Ref. 9 and references cited therein) called photochemical and photophysical.

Photochemical hole burning generally refers to those cases where the excited molecule is transformed into a new species with a different chemical identity, often with a shift in excitation energy that is larger than the inhomogeneous bandwidth. Characteristic photochemical processes include proton transfer (as in free-base porphyrins and some dyes in biological macromolecules); electron transfer; isomerization; and dissociation. Some of these processes can be reversible.

Photophysical hole burning generally refers to those cases where the excited molecule retains its chemical identity but photoexcitation leads to changes in the local environment that shift the excitation energy. Usually this shift is smaller than the width of the inhomogeneous band. Photophysical hole burning is characteristic of low-temperature solutions of photostable molecules. It is nonspecific and can be observed in many polar, glassy, and polymer matrices. The description of photophysical hole burning in glasses in terms of the coupling of the chromophore molecule to so-called two-level systems (TLS) is now widespread (7,9,13).

Phototransformations induced by narrow band laser irradiation in different materials are schematically diagrammed in Fig. 8.1. The spectroscopic consequences of these phototransformations, photochemical and photophysical hole burning, are shown in Fig. 8.2. Of course, the simple idea that the signature of photochemical hole burning is a spectral shift that is larger than the inhomogeneous broadening has exceptions. For example, in the case of symmetrical free-base porphyrins the hole-burning process is photochemical, namely, the phototautomerization of the two central protons (23,24). Since the chemical change in this case is a subtle one (it is equivalent to a 90° rotation of the molecule in the matrix), the spectral shift is within the inhomogeneous band (25). On the other hand, there are cases where hole-burning processes have been assumed to be photophysical but show peculiarities that come from photochemistry. Examples of photochemical hole-burning processes are shown in Fig. 8.3.

One important process that leads to persistent spectral holes is sequential two-photon photochemistry. In this case absorption of the first photon produces an excited state that can absorb a second photon. When the metastable state produced by the absorption of the first photon has low probability for decaying to photoproduct and the photochemistry follows mainly from the absorption of the second photon, a type of hole burning called "photon-gated hole burning" results. An example is (c) of Fig. 8.3. In this case the efficiency of hole burning is enhanced by irradiation in the region of the $T_n \leftarrow T_1$ absorption of the anthracene–tetracene dimer (27). The irreversible photodecomposition of dimethyl-*s*-tetrazine (29) is a second example of photon gated hole burning. There are many others (see, e.g., Refs. 30–32 and references cited therein).

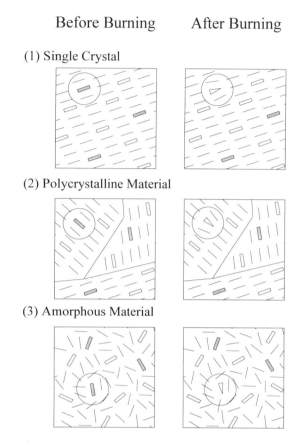

Figure 8.1. Scheme of phototransformations in different materials. Dopant molecules are indicated by rectangles that are filled in if the dopant molecule absorbs at the irradiation frequency. Dopant molecules that originally absorbed at the irradiation wavelength but after excitation now absorb at a different frequency are denoted by triangles. (1) In single crystals the impurity molecules absorb at different frequencies due to small differences in local environment. A fraction of the excited molecules (circles) are phototransformed into products with a different structure and/or local environment. Each of these "burned" molecules makes the same contribution to the hole area. (2) In the case of polycrystalline material that consists of randomly oriented, microscopic single crystals the fraction of burned molecules and the contribution of one burned molecule to the hole area depends on the orientation of the micro-crystal. (3) In amorphous materials the dopant molecules are randomly oriented. This case is similar to case (2) but the inhomogeneous broadening is much larger due to larger differences in local environments.

8.2.2. Hole Profile and Electric Field Effect

Before going to the experimental data on the Stark effect we will discuss the behavior expected for a persistent hole when an external field is applied.

For low-laser intensity and short burning time for the case where the inhomogeneous distribution is so broad as to be effectively constant, the hole profile is

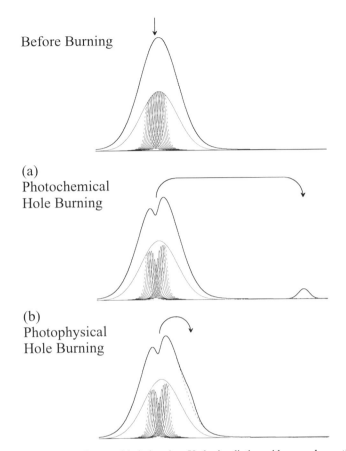

Figure 8.2. Different types of spectral hole burning. Under irradiation with monochromatic light, photobleaching of the sample takes place, that is, excitation transforms some molecules to photoproducts that absorb at a different frequency. This leads to a depletion of the number of molecules that absorb at the irradiation frequency (arrow) and a hole appears in the spectrum. Depending on the shift of the photoproduct relative to the burning frequency one can distinguish two different types of hole burning: photochemical (*a*) and photophysical (*b*).

described by the expression

$$-\Delta D(\nu) = K \iint (\mathbf{m} \cdot \mathbf{n_p})^2 (\mathbf{m} \cdot \mathbf{n_b})^2 \epsilon(\nu - \nu_0)\epsilon(\nu_b - \nu_0) \, d\Omega \, d\nu_0 \qquad (8.1)$$

Here $\Delta D(\nu)$ is the change in the optical density at frequency ν in the region of the hole, ν_b is the burning frequency (assumed to have a delta function distribution), $\mathbf{m}$, $\mathbf{n_b}$, and $\mathbf{n_p}$ are unit vectors in the directions of the transition dipole moment of the molecule and the polarization of the burning and probing light, respectively. The parameter $\epsilon(\nu - \nu_0)$ is the homogeneous optical absorption band contour (proportional to the probability that a molecule with transition frequency ν_0 absorbs at ν), which we expect to be the same for all the dopant molecules. Integration is

Unimolecular Photoreactions

(a)

(b)

Bimolecular Photoreactions

(c)

(d)

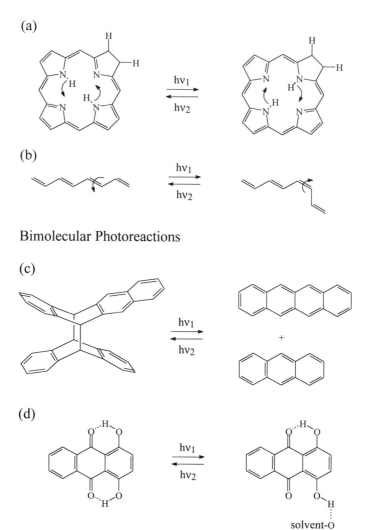

Figure 8.3. Examples of photoprocesses that can lead to spectral holes. (*a*) Phototautomerization of chlorin (23,24); (*b*) Photoinduced cis,trans-isomerization of trans,trans-1,3,5,7-octatetraene (26); (*c*) Photodissociation of the anthracene–tetracene dimer (27). (*d*) Hydrogen-bond photorearrangement in quinizarin (28). Processes (*a*) and (*b*) are unimolecular while processes (*c*) and (*d*) can be classified as bimolecular (guest + guest and guest + host, respectively). As the arrows indicate, these reactions can be reversed when light of different frequency is used.

over the transition frequency ν_0 and, in the case of randomly oriented molecules, the set of Eulerian angles Ω that define molecular orientation in the laboratory coordinate frame. The constant K depends on, among other things, the absorption cross section, the quantum yield for phototransformation, and various factors relating to experimental sensitivity.

At low temperatures the homogeneous optical absorption band $\epsilon(\nu - \nu_0)$ consists of a narrow zero-phonon line (ZPL) and broad phonon wing (PW) on the high-energy side of the zero-phonon line. Photochemical or photophysical removal of molecules via zero-phonon line absorption generates a hole profile similar to the homogeneous band profile: that is, a narrow resonant hole with a broad phonon wing extending to higher energy. The contribution to the hole profile from molecules burnt via phonon wing absorption is broad lying mainly to the long-wave side of the narrow hole. The most informative electric field effect experiments are those for which the deepest part of the hole profile is the narrow feature generated by zero-phonon line absorption. Therefore, in what follows we will assume that $\epsilon(\nu - \nu_0)$ is a narrow zero-phonon line. From Eq. (8.1) it follows that if $\epsilon(\nu - \nu_0)$ is Lorentzian with width Γ_{ZPL}, the hole width Γ_h is related to it by $\Gamma_h = 2\Gamma_{ZPL}$.

An electric field applied to an ideal single site single-crystal sample will produce identical shifts for the zero-phonon lines of all dopant molecules and, in consequence, the photochemical hole will shift without broadening. This is not the case for the randomly oriented ensembles found in polycrystalline samples or solutions in glasses and polymers. In these orientationally disordered samples shifts will be different for the various molecules, which means that the hole profile will broaden as the field strength is increased. (In the case of polymer and glassy hosts the electrostatic environment also varies. Determining the relative contributions of orientational and local field disorder to external electric field induced hole broadening is an area of current research.)

Classically, the energy of the ith electronic state of a molecule in an electric field E to second order in the field is written

$$W_i = W_{0i} - \mathbf{E} \cdot \boldsymbol{\mu}_i - \tfrac{1}{2} \mathbf{E} \cdot \underline{\underline{\alpha}}_i \cdot \mathbf{E} \qquad (8.2)$$

where $\boldsymbol{\mu}_i$ is the permanent dipole moment vector and α_i is the polarizability tensor. For a noncentrosymmetric site $\mathbf{E}$ is the vector sum of the field at the molecule created by the externally applied field (laboratory field times the appropriate local field factor) plus an internal field. In the presence of the field the frequency for a transition between the ith and jth states is

$$\nu = (W_j - W_i)/h = \nu_0 - 1/h(\mathbf{E} \cdot \Delta \boldsymbol{\mu} + \tfrac{1}{2} \mathbf{E} \cdot \Delta \underline{\underline{\alpha}} \cdot \mathbf{E}) \qquad (8.3)$$

where ν_0 is the unperturbed transition frequency, $\Delta \boldsymbol{\mu} = \boldsymbol{\mu}_j - \boldsymbol{\mu}_i$ and $\Delta \alpha = \alpha_j - \alpha_i$. If the principal axes of the polarizability tensor are the same in the ith and jth states, then with the appropriate choice of axes $\Delta \alpha$ will be diagonal. In this case six parameters (three components of $\Delta \boldsymbol{\mu}$ and the three diagonal elements of $\Delta \underline{\underline{\alpha}}$)

suffice to completely specify the shift of transition frequency in an arbitrary electric field $\mathbf{E}$. The frequency shift is given by

$$\Delta \nu = -1/h(\mathbf{E} \cdot \Delta \boldsymbol{\mu} + \tfrac{1}{2}\mathbf{E} \cdot \Delta \underline{\underline{\alpha}} \cdot \mathbf{E}) \tag{8.4}$$

where the first term corresponds to the linear effect and the second term to the quadratic effect.

If the internal field is either known or can be neglected, then the effect of an external field on the hole profile can be calculated using Eq. (8.1) and (8.4). The results depend on the angle between $\Delta \boldsymbol{\mu}$ and the transition dipole, on the orientation of the axes of the tensor $\Delta \underline{\alpha}$, on the polarization of the burning and reading light, and on the geometry of the experiment. This kind of theoretical analysis of the hole profile as a function of applied electric field has been performed for the most important cases in a number of papers (see, e.g., Ref. 15,21,22,33–36). For an isotropic system it has been shown that in the case of the linear Stark effect the hole profile broadens symmetrically, but the hole shifts and becomes asymmetric in the case of quadratic effects.

Alternatively, the shift in transition frequency may be formulated in terms of a two-level model (37). If $\mathcal{H}_0$ is the unperturbed Hamiltonian matrix with diagonal elements $h\nu_{01}$ and $h\nu_{02}$ ($\nu_{01} < \nu_{02}$) and $\boldsymbol{\mu}$ is the matrix of the dipole moment operator, then diagonalizing

$$\mathcal{H} = \mathcal{H}_0 - \mathbf{E} \cdot \boldsymbol{\mu} \tag{8.5}$$

and dividing the difference of the eigenvalues by h leads to the energies expressed as frequencies

$$\nu_1 = \{(\nu_{01} + \nu_{02}) - \mathbf{E} \cdot (\boldsymbol{\mu}_1 + \boldsymbol{\mu}_2)/h - [(\nu_0 - \mathbf{E} \cdot \Delta\boldsymbol{\mu}/h)^2 + (2\mathbf{E} \cdot \boldsymbol{\mu}_{12}/h)^2)]^{1/2}\}/2 \tag{8.6}$$

$$\nu_2 = \{(\nu_{01} + \boldsymbol{\mu}_{02}) - \mathbf{E} \cdot (\boldsymbol{\mu}_1 + \boldsymbol{\mu}_2)/h + [(\nu_0 - \mathbf{E} \cdot \Delta\boldsymbol{\mu}/h)^2 + (2\mathbf{E} \cdot \boldsymbol{\mu}_{12}/h)^2]^{1/2}\}/2 \tag{8.7}$$

where $\Delta \boldsymbol{\mu} = \boldsymbol{\mu}_2 - \boldsymbol{\mu}_1$ is again the difference between the dipole moments of the states labeled 1 and 2, $\nu_0 = \nu_{02} - \nu_{01}$ is the zero-field transition frequency and $\boldsymbol{\mu}_{12}$ is the transition dipole. If instead of exactly diagonalizing Eq. (8.5) the energies are evaluated by Rayleigh–Schrödinger perturbation theory carried to second order, the new energies expressed as frequencies are

$$\nu_1 = \nu_{01} - \mathbf{E} \cdot \boldsymbol{\mu}_1/h - (\mathbf{E} \cdot \boldsymbol{\mu}_{12}/h)^2/\nu_0 \tag{8.8}$$

$$\nu_2 = \nu_{02} - \mathbf{E} \cdot \boldsymbol{\mu}_2/h + (\mathbf{E} \cdot \boldsymbol{\mu}_{12}/h)^2/\nu_0 \tag{8.9}$$

where $\nu_0 = \nu_{02} - \nu_{01}$. The shift calculated from Eqs. (8.8) and (8.9) is exactly analogous to the classical expression in Eq. (8.4) with $\Delta\alpha_{xx} = -4(\boldsymbol{\mu}_{12x})^2/(h\nu_0)$, and so on. As is the case for the classical formulation there are six parameters (three components of $\Delta \boldsymbol{\mu}$ and three components of $\boldsymbol{\mu}_{12}$) that completely specify the shift of transition frequency in an arbitrary external field. A three level quantum mechanical model can describe the most general case.

Representative hole profiles calculated with the two-level quantum mechanical model for randomly oriented centrosymmetric molecules are shown in Fig. 8.4 and 8.5. The profiles for the case where there is a large internal field (upper panel of Fig. 8.4 and both panels of Fig. 8.5) are equivalent to profiles calculated classically for molecules that have permanent dipole moments (15,21,22,33). The Stark effect is linear and the hole profile is symmetric. The profile for the case where there is

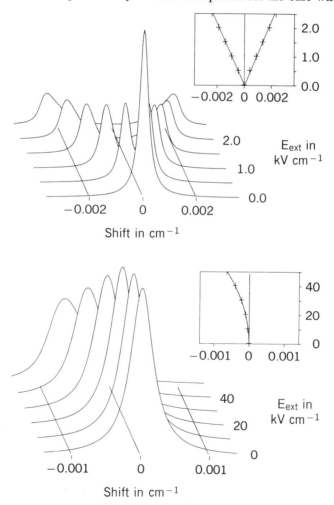

Figure 8.4. Linear and quadratic effects of an electric field on a hole profile. These profiles were calculated from the two-level quantum mechanical model using parameters appropriate for octatetraene (for the upper panel $\Delta\mu = 0$, $\nu_0 = 3500$ cm^{-1}, $\mu_{12} = 2$D, E_{int} parallel μ_{12} and $|E_{int}| = 1.6 \times 10^6$ V cm^{-1}: for the lower panel the same values except E_{int} was set to zero). These calculations are for the laser beam polarized parallel to the applied field assuming that all possible rotations from the molecular to the laboratory frame have equal weight. Further details are given in Section 8.3.5.

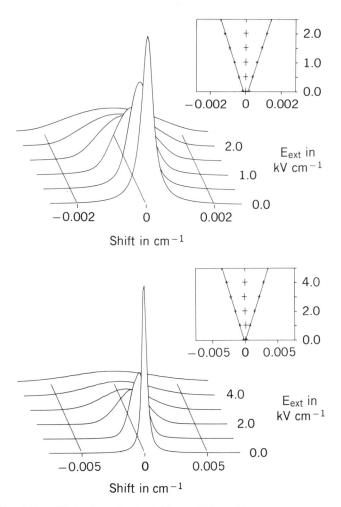

Figure 8.5. Broadening effects of an electric field on a hole profile calculated from the two-level quantum mechanical model using parameters appropriate for octatetraene. The parameters for the upper panel are the same as for the upper panel of Fig. 8.4 except that the laser beam is polarized perpendicular to the applied field. The parameters for the lower panel are the same as for the upper panel of Fig. 8.4 except that E_{int} was distributed randomly ($< | E_{int} | > = 1.6 \times 10^{6}$ V cm^{-1}). Further details are given in Section 8.3.5.

no internal field (lower panel of Fig. 8.4) is likewise equivalent to profiles calcu-lated from the classical expression for molecules that have no permanent dipole moments (15,21,22,33). The Stark effect is quadratic and the hole profile is asymmetric.

While the classical expressions are capable of concisely summarizing the results of electric field experiments, the quantum mechanical formulation has advantages for determining the local electric field. Perhaps the most important advantage is that the effect of an internal field can be evaluated without explicit knowledge of the polarizability. This will be discussed further below.

8.2.3. Fixed Frequency Experiments

So far our discussion of the effects of external electric fields on persistent spectral holes has focused on the frequency domain, that is, we have considered absorption intensity versus frequency at fixed field strength. It is also possible to do these experiments in the field domain, that is, to look at absorption intensity versus field at fixed laser frequency. This possibility was first discussed in 1982 (15) and has been used in several subsequent studies (17,21,22). This technique takes advantage of the broader availability of stabilized fixed frequency lasers and the lower demands on the accuracy with which the electric field is known (while the laser frequency must be stable to parts in 10^8, field stabilities of 1 part in 10^3 are more than adequate). In the field scan method the optical density as a function of the external field $\Delta D(E)$ at fixed frequency $\nu = \nu_{\text{burn}}$ is recorded. In the case of a linear Stark effect the resulting "field curve" is symmetrical with a full-width at half maximum (fwhm) Γ_E that is linearly related to the homogeneous line width Γ (38). Specifically,

$$\Gamma_E = \frac{hc}{f_e \Delta \mu} B \Gamma \qquad (8.10)$$

where $\Delta \mu$ is the magnitude of the difference between ground- and excited-state dipole moments, f_e is the local field factor, and B is a coefficient that depends on the experimental geometry and the angle γ between the dipole moment difference $\Delta \mu$ and the transition dipole. With Eq. (8.1) and (8.4) B is readily calculated for different cases. For example, if the laser polarization is the same for burning and readout and the light propagates along the applied electric field then $B = 8$ for $\gamma = 0$ and $B = 3.8$ for $\gamma = \pi/2$ (38).

Since the width of the field curve is proportional to Γ, this technique may be used to determine the dependence of the width of the zero-phonon line Γ on temperature even if the values of the Stark parameters $\Delta \mu$ and γ are not known. Furthermore, it is possible to determine the ratios of the homogeneous line widths for the same molecule in different matrices by measuring the widths of the field curves in these systems. For example, Fig. 8.6 shows three field curves obtained for chlorin in three different polymers. The ratios of the widths of the three field curves were found to correspond closely to ratios of the homogeneous line widths determined independently by hole burning (16).

8.2.4. Experimental Considerations

In this section, we briefly consider some of the experimental constraints and special considerations that apply to hole-burning spectroscopy. Specifically, we discuss the need for low temperature and extremely monochromatic light for burning and measuring the persistent holes, the magnitude of the external field that will lead to an observable shift, and techniques for sensitive detection.

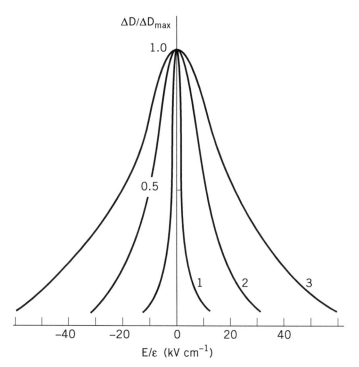

Figure 8.6. Field curves for chlorin in different polymer matrices (1 = polyvinylbutyral; 2 = polyme-thylmethacrylate; 3 = polystyrene) measured at the He–Ne laser frequency (λ = 6328 Å) at 4.2 K. The ratios of the measured value for Γ_E in these three polymers are approximately 1:4:8 (38). Measurements of the homogeneous linewidths by hole-burning spectroscopy gave values of Γ = 0.03, 0.15, and 0.25 cm^{-1}, respectively (16). The ratios of these line widths are the same as the Γ_E ratios to within experimental error. [Adapted from Ref. 38].

Temperature is a very important factor. As temperature increases the intensity of the zero-phonon line decreases with the consequence that at higher temperatures the hole will appear as a broad feature whose width is related to the width of the phonon wings rather than as a narrow feature whose width is related to the width of the zero-phonon line. In most cases in organic solids the zero-phonon line practically disappears at temperatures of 40–60 K [see, e.g., reviews (6–13)]. In some special cases, it has been possible to observe what is called high-temperature photochemical hole burning (39,40), where approximately 10 cm^{-1} wide holes could be recorded at liquid nitrogen temperature. Such broad holes are not useful for the kind of high-resolution spectroscopy that is needed for high information content electric field studies. In order to clearly see the effects of electric fields on persistent holes it is necessary to work at liquid helium temperatures (1.4–4.2 K), where the hole widths for organic systems are typically 10^{-1}–10^{-2} cm^{-1} [in the case of octatetraene in n-hexane at 1.4 K holes as narrow as 0.0002 cm^{-1} are observed (26)].

Because the hole width cannot be narrower than the frequency width of the light used for burning and observation, it is necessary to use single mode lasers. Even though the demands on bandwidth and frequency stability are quite impressive (0.0001 cm^{-1} at optical frequencies corresponds to < 1 part in 10^8), actively stabilized tunable laser systems whose line widths are approximately 10^{-4} cm^{-1} are commercially available.

The order of magnitude of the external field that has to be applied for Stark experiments may be estimated using Eq. (8.4). To be significant, the electric field induced shift of the zero-phonon line should be comparable to the width of the persistent hole, typically 0.1 cm^{-1}. In the case of a linear Stark effect, where $\Delta\mu$, the magnitude of the difference between the ground- and excited-state dipole moments, is 1 D, a shift of 0.1 cm^{-1} requires an external field of about 6 kV cm^{-1}. To observe a significant quadratic effect, much stronger electric fields are needed. With a typical value for $\Delta\alpha$, the difference between ground- and excited-state polarizability, of 20 Å^3, the same shift of 0.1 cm^{-1} requires a 300 kV cm^{-1} electric field. Of course, what is seen as the shift or broadening of a persistent hole for an orientationally disordered sample differs from the maximum possible shift of the zero-phonon line, but is comparable. For those systems where persistent holes can be narrower, smaller fields are needed to obtain significant effects. For example, as seen in Fig. 8.4, to obtain a shift of the maximum of 0.001 cm^{-1} in the case of a linear effect requires an electric field of 1 kV cm^{-1}: the same shift in the case of a purely quadratic effect needs an applied field of more than 50 kV cm^{-1}. At the needed field strengths the demands on accuracy and stability are quite modest: 1 part in 1000 is more than adequate in most cases.

A persistent hole results when the number of absorbers with a narrow range of zero-phonon line frequencies is reduced. Among the methods for detecting this selective loss of absorptivity, the two that are most direct and common are absorption spectroscopy and fluorescence excitation spectroscopy. Absorption spectroscopy requires samples that have good optical quality and whose product of thickness times concentration gives a significant diminution in the intensity of the probe beam. In the case of fluorescence excitation spectroscopy absorption is detected by the fluorescence that it generates. Since the emitted photons can be collected over a wide range of frequency and angle and can be directly counted, fluorescence excitation spectroscopy enjoys some important advantages over straight absorption. It is intrinsically more sensitive and can be used on very dilute and strongly scattering samples. The high sensitivity permits experiments on samples whose optical density is vanishingly small (e.g., the dye doped Langmuir–Blodgett films discussed in Section 8.3.2) and strongly scattering samples (e.g., the dye doped n-alkane polycrystals, discussed in Section 8.3.5).

A block diagram of a typical setup (which we are using in our laboratory) for hole-burning experiments is shown in Fig. 8.7.

8.3. RESULTS

8.3.1. Introduction

Persistent spectral holes can be three to five orders of magnitude narrower than inhomogeneously broadened absorption bands. This translates into enormous gains

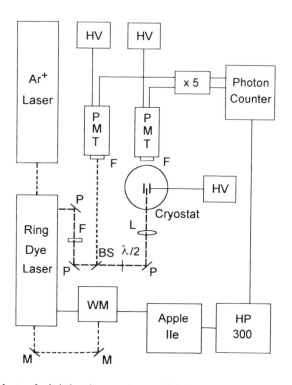

Figure 8.7. Typical setup for hole-burning experiments. Light from a tunable single frequency dye laser that is actively stabilized to obtain a bandwidth of less than 1 MHz irradiates a sample in a helium cryostat (P = prism, L = lens, M = mirror, BS = beam splitter). The sample is between electrodes connected to a high-voltage HV power supply. Polarization is determined by a half-wave plate (λ/2). For hole burning the laser is kept at a single frequency and the beam is unattenuated. For readout by fluorescence excitation the laser is scanned at lower intensity (typically reduced by a factor of 100). The fluorescence is detected with a photomultiplier tube through a filter chosen to transmit as much of the fluorescence spectrum as possible while blocking scattered excitation light. A second photomultiplier monitors the intensity of the exciting light. For samples whose fluorescence is weak, normalization is critically important. In our setup the photon counts from both photomultipliers are collected and processed by a computer (HP 300), which is also interfaced to the laser control system (wavemeter WM plus Apple IIe).

in sensitivity, which can be increased even more through holographic detection of the hole (41) or by using modulation techniques (42). It is not an exaggeration to say that during the last decade persistent hole-burning experiments have produced a renaissance in Stark effect spectroscopy making possible a number of high information content experiments on a variety of complex organic systems.

As discussed in Section 8.2.2 the shift of excitation energy (marked by the persistent spectral hole) is expected to be linear for noncentrosymmetric molecules since they will have nonzero ground- and excited-state dipole moments. This expectation has been borne out in a number of experiments on dilute dopants in organic matrices (most often polymers). In particular, electric field effect measurements have been reported for the polar molecules resorufin (14,43), chlorin

(15,16,18,34,38), octaethylchlorin (42), isobacteriochlorin (44,45), oxazine (36), and cresylviolet (36). These measurements yielded estimates of $|\Delta\mu|$, the magnitude of the difference between ground- and excited-state dipole moments and γ, the angle between $\Delta\mu$ and the transition dipole. The measured $|\Delta\mu|$ values range from 0.2 D for resorufin to 2 D for cresylviolet.

If a molecule is centrosymmetric in both ground and excited states, neither state can have a nonzero dipole moment and $\Delta\mu$ vanishes. In this case the shift of excitation energy with electric field is expected to be quadratic, that is, to be proportional to the square of the field strength. The effect of electric field on persistent hole profile has been determined for a number of centrosymmetric molecules including perylene (17,38,46,47), octaethylprophyrin, tetrazaporphin, Zn-tetrabenzoporphin, and phthalocyanine (33,34,48–51). In all of these cases, to field strengths of up to 5×10^4 V cm^{-1} there was no detectable shift in the center frequencies of the holes and, to within experimental error, the hole profile remained symmetrical. Our interpretation of this is that an internal electric field has broken centrosymmetry to such an extent as to make the quadratic Stark effect negligible. That is, because of the presence of a large internal field, the quadratic Stark effect is seen in the high-field limit where it appears linear over the field ranges explored. This point is discussed further in Section 8.3.5.

From Stark experiments on centrosymmetrical molecules in solid matrices one can straightforwardly determine the average value of the induced dipole moment difference $\Delta\mu$ (found to range from 0.05 to 0.5 D) and from $\Delta\mu$ and an estimate of the dopant molecule polarizability derive a value for the internal matrix field. These early experiments together with later ones on the Stark effect in media with different dielectric constants (51,52) and on the dependence of $\Delta\mu_{ind}$ on burning frequency in the inhomogeneously broadened band (52,53) suggest that persistent hole-burning studies of electric field effects can lead to an exquisitely detailed characterization of the electrostatic properties of molecular sites in organic solids. The experiments on octatetraene in n-alkanes that are described in Section 8.3.5 show that this is indeed the case.

In the rest of this chapter we focus on studies of the effects of external electric fields on persistent holes in the cases of dye molecules incorporated into ultrathin Langmuir–Blodgett films, metalloporphyrins in polymer matrixes, octatetraene in n-alkane polycrystals, and will also examine some recent results on the effect of an external electric field on single molecules in organic solids (not, strictly speaking, hole burning, but closely related to the picture of the local electric field that is derived from the hole-burning experiments). In discussing these data we will concentrate on those aspects of the Stark effect in condensed phase molecular systems that we feel have the most leverage on the questions of microscopic structure including the properties of the internal electric field.

8.3.2. Molecular Orientation: Experiments on Langmuir–Blodgett Films

Most of the work on hole-burning spectroscopy has been on different crystalline and amorphous solids. Recently, this method has also been used to spectroscopically characterize ultrathin molecular systems in the form of Langmuir–Blodgett films (54–58) [see also the recent review (12)]. Special attributes of Langmuir–Blodgett films, such as their extreme thinness, well-defined molecular structure, and controllable molecular architecture (which makes possible the creation of superlattices), make them very interesting systems for Stark spectroscopy and for future applications in molecular electronics (59–62). In principle, Stark measurements can give information about the magnitude and direction of the dipole moment difference between excited and ground states of the probe molecule, the internal electric field, and the orientation and distribution of the dopant molecules in the Langmuir–Blodgett film. We focus on these points in our brief review of experiments where hole-burning spectroscopy has been used to determine the orientation of dye molecules incorporated into Langmuir–Blodgett films (63,64).

The films used in these experiments were prepared by doping the cadmium salt of arachidic acid (CdA) with small amounts of the ionic cyanine dye N,N'-dioctadecylselenacarbocyanine (S20, Fig. 8.8). Here we consider only a few aspects of these experiments: Details about sample preparation, experimental setup, and background information on the solution spectroscopy of S20 are given in Ref. 64.

In this case the hole-burning experiments used more or less standard techniques. The excitation source was a single-mode ring dye laser (operating with rhodamine 6G and pumped by an Ar^+ laser). The laser line width was about 1 MHz, the intensity during hole burning was approximately 10 μW cm^{-2}, and the burning time ranged from 1 to 10 min. The hole was read by fluorescence excitation with the laser intensity decreased by a factor of 10–100 [this method is very sensitive and can be used even in the case of monomolecular layers (55–57)]. The laser beam was polarized perpendicular to the applied electric field for both burning and scanning.

The Stark parameters $\Delta\mu$ and γ for S20 in bulk CdA solution that were needed for the analysis of the Langmuir–Blodgett film data were determined from hole-burning studies. The fluorescence excitation and fluorescence spectra of S20 in CdA at room temperature and at liquid helium temperature are broad and structure-

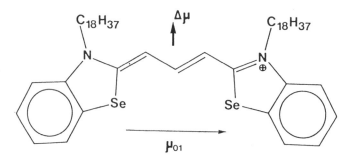

Figure 8.8. Structure of the ionic dye S20. Also shown are the transition dipole moment, μ_{01}, pointing along the main chromophore backbone, and the dipole moment difference between excited and ground states $\Delta\mu$, which coincides with the C_2 symmetry axis of the molecule.

less (Fig. 8.9), but holes can be burned in the region where the absorption and fluorescence bands overlap (583–598 nm). Electric field experiments were performed at $T = 2$ K and $\lambda \approx 593$ nm. Figure 8.10 shows that in an applied field the hole broadens symmetrically as expected for a linear Stark effect. Analysis showed that the hole profile is very nearly Lorentzian, even under a 15–20 kV cm^{-1} external field (64). (The hole broadens without the splitting that would be expected for $\gamma = \pi/2$ for the laser polarized perpendicular to the external field. This is, in part, because in bulk CdA there is strong depolarization due to scattering and, in part, due to the fact that there is a distribution of the induced dipole moment that adds to the static dipole moment. For details, see Ref. 64.) While in principle values for both of the Stark parameters $\Delta\mu$ and γ can be obtained from measurements on randomly oriented chromophores in bulk solution, in this special case, because of the symmetry of S20, $\gamma = \pi/2$. From a theoretical analysis and fitting of the hole profiles $|\Delta\mu|$ was determined to be 0.35 D.

Stark experiments were performed on Langmuir–Blodgett films doped with S20 with two different architectures, symmetric and alternating (Fig. 8.11). In the case of the symmetric sample (dye chains pointing both towards and away from the indium tin oxide electrode) the hole broadens without shifting when the field is applied (63). This is not the case for the alternating film (64), where all of the dye chains point towards the indium tin oxide electrode. Because of the anisotropy of

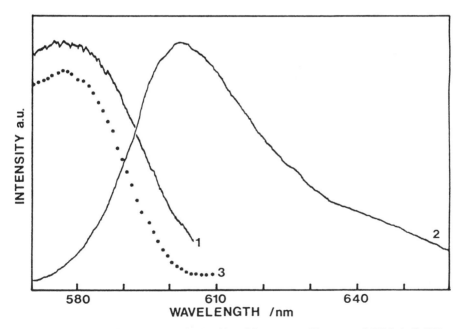

Figure 8.9. Broadband fluorescence excitation (1) and fluorescence (2) spectra of S20 in bulk CdA at 4.2 K. For comparison, a section of the excitation spectrum of S20 in a Langmuir–Blodgett film multilayer (3) is also shown. [Adapted from Ref. 64].

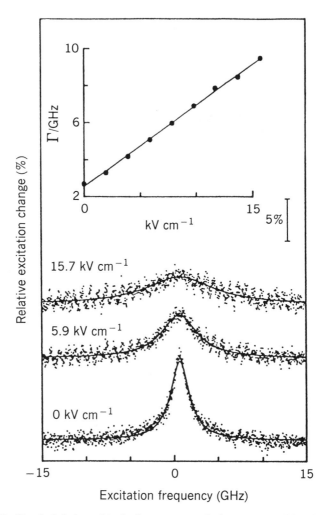

Figure 8.10. Profile of a hole burned in the fluorescence excitation spectrum of S20 in bulk CdA at 2 K as a function of the applied field (lower half). Hole broadening versus applied field is plotted in the upper half. [Adapted from Ref. 64].

the alternating architecture the hole is expected to both broaden and shift with increasing external field. The shift comes from the nonrandom average orientation of the chromophore with respect to the external field and the broadening reflects the distribution of this orientation. The experimental results for the alternating film are shown in Fig. 8.12. A comparison of these spectra to those measured for the bulk sample (Fig. 8.10) shows that the broadening in the case of the Langmuir–Blodgett film is much smaller (compare the spectrum at 20 kV cm^{-1} in Fig. 8.12 to the equally broadened spectrum at 5.0 kV cm^{-1} in Fig. 8.10). This confirms that the short axis of the chromophore is not randomly oriented in the Langmuir–

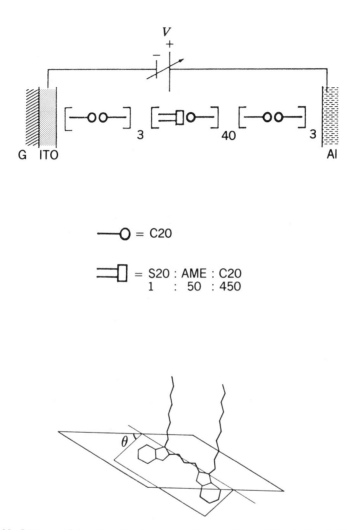

Figure 8.11. Structure of the alternating Langmuir–Blodgett film and the Stark cell. The voltage is applied between the transparent indium–tin oxide (ITO) electrode on a glass (G) surface, and the thick aluminum (Al) electrode. C20 indicates cadmium arachidate, S20 is the ionic dye, and AME (arachidic methyl ester) was added to improve the film packing. Also shown is the angle θ between the chromophore's short axis and the layer. [Adapted from Ref. 64].

Blodgett film. As seen in Fig. 8.12, the shift is rather small compared to the broadening. This means that the average angle of the chromophore's short axis with the layer plane (angle θ in Fig. 8.11) is much smaller than the width of its distribution. Quantitative analysis of the data (64) shows that the average angle $\langle\theta\rangle$ is about $+3°$ and that the average of the absolute value of θ, $\langle|\theta|\rangle$, is about $16°$. Thus, the chromophore orientation is primarily fixed by strong forces in the hydrophilic region of a bilayer unit and is not influenced by the almost perpendicular orientation

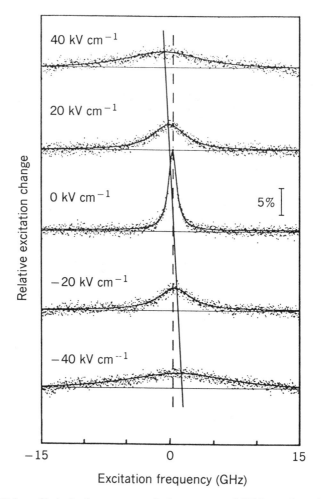

Figure 8.12. Hole profile in the fluorescence excitation spectrum of S20 in an alternating Langmuir–Blodgett film of CdA at 2 K as a function of external field. The broadening of the hole is much bigger than the shift. [Adapted from Ref. 64].

of the flexible chains. The conclusion that the rigid plane of the chromophore is roughly parallel to the layer plane is quite different from the picture determined by measuring the electrochromism of the inhomogeneously broadened spectra of a similar system (65), where it was concluded that the short chromophore axis was perpendicular to the layer plane. Since the electrochromism measurements were made on much more concentrated samples, this difference could be due to the formation of aggregates in which the short molecular axis was vertical. A simple explanation for the rather large distribution of the angle between the short chromophore axis and the layer plane is the existence of cadmium ions which, in some cases, might be located between the chromophore plane and the layer surface, thereby tilting the chromophore plane by $10-20°$. This picture would also account for the much smaller distribution of the long chromophore axis angle [as found by polarization experiments (66)] since the tilt effect along the longer axis of the chromophore plane due to cadmium ions would be much smaller for geometrical reasons. These first results show that the Stark effect on the hole is a very sensitive method to estimate the orientation of the chromophore even in the case where the dipole moment difference is almost perpendicular to the applied electric field. It will be very interesting to perform similar experiments on different systems. In the case of centrosymmetric dye molecules this method sems to be especially promising since it opens the possibility of obtaining information about the internal electric fields in Langmuir–Blodgett films in experiments analogous to those described in Section 8.3.5.

8.3.3. Molecular Structure: Experiments on Metalloporphyrins

Here we discuss another example of the application of the Stark effect technique in combination with persistent spectral hole burning to study aspects of molecular structure. Specifically, we examine some recent results relating to the excitation induced displacement of the metal ion in metalloporphyrin molecules (67). Metalloporphyrins (68) derived from the tetrapyrollic pigment porphin are an important class of biologically active chromophores that includes the light receptor in the photosynthetic reaction center (chlorophyll, a magnesium complex of porphin) and the functional part of hemoproteins (heme, a porphyrin iron complex).

Since porphyrin rings are planar due to delocalization of the π-electrons one might expect that in most cases metalloporphyrins would have D_{4h} symmetry with the metal ion in the molecular plane (69,70). This is not the case. X-ray crystallography shows that the metal ion is actually located slightly out of the plane of the porphyrin ring ($0.2-1$ Å) for different metalloporphyrins with and without additional ligands (68,71–74). Molecular orbital calculations also indicate that the out-of-plane configuration of the metal ion is more stable than the in-plane configuration (75,76).

While X-ray crystallography is satisfyingly direct, the determination of subtle structural details of metalloporphyrins by this technique can only be done for ground states of those systems where high quality crystals can be prepared. It has been suggested that Stark effect experiments on spectral holes in metalloporphyrins are sensitive enough to obtain some information about excitation induced geometry changes without the necessity of preparing single crystals (67). In this section we review how comparing the response of a persistent spectral hole to an external electric field in the case of a free-base porphyrin to that of the corresponding zinc complex can lead to conclusions about metalloporphyrin geometry changes following excitation.

For the experiments described in Ref. 67, samples of octaethyl-porphin (H_2–OEP) and its zinc complex (Zn–OEP) at concentrations of 10^{-3} mol L^{-1} in $100-200$ µm thick polyvinylbutyral (PVB) polymer films were prepared. The films were placed between two transparent electrodes (glass plates coated with SnO_2). Typical absorption spectra of these samples are shown in Fig. 8.13. Persistent spectral holes

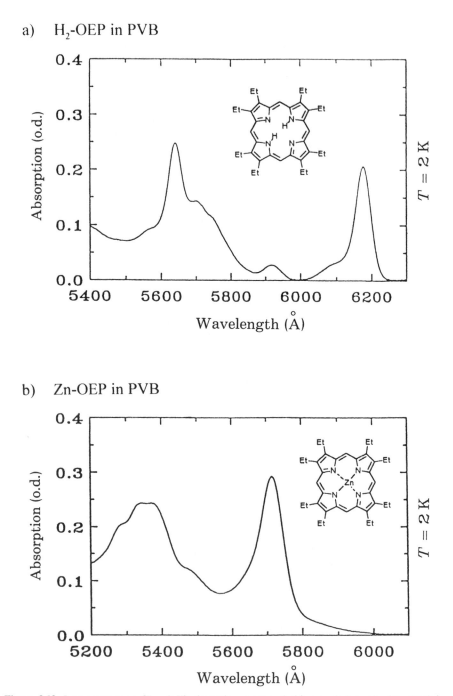

Figure 8.13. Low-temperature ($T = 2$ K) absorption spectra of: (a) octaethylporphin (H_2–OEP) in polyvinylbutyral ($d = 144$ μm, $c = 10^{-3}$ mol L^{-1}); (b) zinc–octaethylporphin (Zn–OEP) in polyvinylbutyral ($d = 161$ μm, $c = 1.2 \times 10^{-3}$ mol L^{-1}). [Adapted from Ref. 67].

were burned near the intensity maxima of the long-wavelength absorption bands using a single mode tunable dye laser with an average burning intensity of about 4 μW cm^{-2} in the case of the free-base porphyrin (photochemical hole burning due to tautomerization of the inner protons) and about 80 μW cm^{-2} in the case of the zinc complex (photophysical hole burning that requires higher burning intensities). The burning times ranged from a few dozens of seconds for free-base porphyrin versus a few dozen minutes for the zinc complex. The hole profiles were determined by absorption with the laser beam propagating parallel to the applied field. Upon application of an external electric field the holes broadened symmetrically in both cases as for a linear Stark effect. This behavior, similar to that observed for many other centrosymmetric molecules in polymer matrices, reflects the fact that internal fields in the polymer matrix induce significant dipole moments (which are absent in the free molecules).

In order to compare the Stark effect for H$_2$−OEP to that for Zn−OEP the additional field induced broadening $\Delta\Gamma = \Gamma_e - \Gamma_0$ (Γ_e = hole fwhm with field, Γ_0 = hole fwhm without field) was plotted against the applied electric field (Fig. 8.14). This additional broadening is directly related to the value of the dipole moment difference $|\Delta\mu|$. From Fig. 8.14 it is apparent that the field induced broadening of the hole in the case of Zn−OEP is approximately two times larger than that for the free-base porphyrin (similar results have been obtained for the free-base and zinc complex of tetrabenzoporphin in polyvinylbutyral and for H$_2$−OEP and Zn−OEP in polystyrene). In Ref. 67 it was suggested that it should be reasonable to assume that the matrix induced dipole moments in H$_2$−OEP and Zn−OEP are

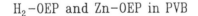

H$_2$−OEP and Zn−OEP in PVB

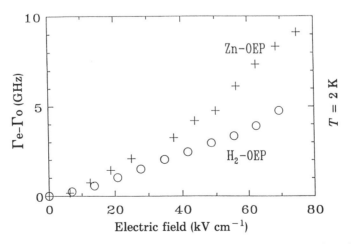

Figure 8.14. Broadening versus electric field strength for persistent holes burned into absorption spectra of octaethylporphin (H$_2$−OEP) in polyvinylbutyral (circles) and zinc−octaethylporphin (Zn−OEP) in polyvinylbutyral (crosses). [Adapted from Ref. 67].

similar since both systems contain the same polarizable π-electron system. In this case the larger Stark broadening of the zinc complex would then be connected to the additional dipole moment (with different values for the ground and excited states) that is associated with the out-of-plane position of the metal ion in at least one of the two states.

To analyze these data more quantitatively, the following model was applied. First, it was assumed that the internal electric field induces the same dipole moment in both Zn–OEP and H_2–OEP. The difference of this dipole moment between excited and ground states is

$$\Delta \mu_{ind} = \Delta \underline{\underline{\alpha}} \cdot E_{int} \tag{8.11}$$

where $\Delta \alpha$ is the polarizability tensor for the excited state minus that for the ground state. This follows from the idea that since Zn–OEP and H_2–OEP both have similar π-electron systems the polarizability difference will be the same for both molecules. Because of symmetry $\Delta \alpha_{zz} = 0$ so that the induced dipole moments lie in the molecular planes. For Zn–OEP symmetry also requires that $\Delta \alpha_{xx} = \Delta \alpha_{yy}$: This was assumed to be true for H_2–OEP as well. Second, since the statistical distribution of the magnitude of the internal field $|E_{int}|$ can be assumed to be the same for Zn–OEP and H_2–OEP in the same polymer matrix (17), the induced dipole moment will follow the same distribution. (In the calculation in Ref. 67 a two-dimensional random orientation of the induced dipole moment vectors in the molecular plane was assumed.) Finally, it was assumed that the only difference between H_2–OEP and Zn–OEP is an additional dipole moment difference perpendicular to the molecular plane $\Delta \mu_0$, which is due to the out-of-plane position of the metal. This additional dipole moment difference was assumed to be the same for all Zn–OEP molecules. (Of course, a more comprehensive analysis should include some distribution of this dipole moment difference as well, but for calculational simplicity it was reasonable to neglect this effect.) With these assumptions and Eqs. (8.1) and (8.4), hole profiles in external fields have been calculated and compared to the experimental results (67). For free-base octaethylporphin, the root-mean-square value of the induced dipole moment difference was determined to be $|\Delta \mu_{ind}| = 0.09$ D. To analyze the data of Zn–OEP, the parameter

$$p = \frac{|\Delta \mu_0|}{|\Delta \mu_{ind}|} \tag{8.12}$$

was introduced. The best agreement between the calculations and experimental data have been obtained for $p = 1.6$. From this a value for the dipole moment difference due the out-of-plane position of Zn in Zn–OEP in at least one of the two states $|\Delta \mu_0| = 0.15$ D was deduced.

These data demonstrate (even with the simple model that was described) the new and interesting possibility of using Stark effects on spectral holes to investigate subtle structural changes in organic molecules under electronic excitation. Porphyrin molecules are especially interesting in this respect since in important cases their photoexcitation is directly connected to their biochemical functions.

8.3.4. Stark Experiments on Single Molecules

Some very recent papers report measurements of the effect of an external electric field on the spectral lines of single dopant molecules in a solid matrix. While these experiments do not use the hole-burning technique, they are a logical extension that significantly adds to our current understanding of external field effects in molecular systems.

It is obvious that hole-burning techniques cannot remove all kinds of inhomogeneity from spectroscopic measurements. This follows from the fact that molecules that are selected due to their transition energy may differ from each other in many other ways: they may have different electron–phonon couplings, different line widths, different lifetimes, and so on. Because molecules are selected only on the basis of their excitation energies, high-resolution hole-burning spectroscopy only provides information about the average behavior of the ensemble of selected molecules. In the case of a single molecule, there is no ensemble to average over. The pioneering papers in single-molecule spectroscopy demonstrate that spectral lines of single molecules may be measured in absorption (77) and fluorescence excitation (78). Thus it is possible to obtain information about the behavior of single molecules without ensemble averaging.

To selectively irradiate only a single molecule, the number of molecules in the irradiated volume must be small enough that the lines of the individual molecules do not overlap. To meet this requirement both the irradiated volume and the concentration must be small. An estimate for the upper limit for the number of molecules that can be in the irradiated volume is the inhomogeneous width of the absorption band divided by the homogeneous line width of a single-dopant molecule. For a typical case ($\Gamma_h = 0.001$ cm^{-1}, $\Gamma_i = 100$ cm^{-1}) this corresponds to 10^5 molecules. If the irradiated volume is a cylinder 5 μm in diameter by 10 μm in thickness this is a concentration of 0.8×10^{-6} mol L^{-1}. With a typical extinction coefficient $\epsilon = 50{,}000$ L mol^{-1} cm^{-1} the optical density would be only 4×10^{-6}: Because of the need for high sensitivity, the fluorescence excitation technique (78) enjoys a number of advantages over absorption detection for single-molecule spectroscopy.

Single-molecule spectroscopy is a new and rapidly developing branch of solid state spectroscopy. Of the many fascinating results that have been obtained in this field (reviewed in Refs. 12 and 79) we only consider the Stark effect measurements. At this point, Stark effect measurements on single molecules have been reported for two systems: terrylene in polyethylene (Tr/PE) (80), a dye-doped amorphous polymer, and pentacene in p-terphenyl (Pc/pTP) (81), a mixed crystal. We first discuss the case of terrylene in polyethylene.

In contrast to the situation for hole-burning experiments, the application of an external electric field to a single molecule should shift the spectral line without broadening. For an ensemble of randomly oriented molecules the shift will be different for molecules that have different orientations with respect to the applied field. These expectations are borne out by experiment. As seen in Fig. 8.15, which summarizes the electric field induced shifts of five lines for Tr/PE, the sign and the magnitude of the shift are different for different lines. In Fig. 8.16, which shows the shift of single-molecule lines versus applied electric field for six different molecules, it is obvious that in all these cases the Stark effect is strictly linear to within the experimental errors, even though terrylene is a centrosymmetric molecule. This is expected to be the case. Since the electric field an angstrom away from a C–H bond is about 10^6 V cm^{-1}, in any system where the dopant molecule environment is not strictly centrosymmetric, the local field will induce significant dipole moment into the ground and excited states. Similar linear behavior is seen for all of the 60 single molecules investigated. The slope of a plot of shift versus field strength gives the magnitude of the projection of the dipole moment difference $\Delta\mu$ onto the applied electric field $\Delta\mu_E$. Analysis of these data (80) shows that the $\Delta\mu_E$ values are appreciable: The average of the absolute value of $\Delta\mu_E$, $< \mid \Delta\mu_E \mid >$, was found to be 0.5 D, while the largest observed projection of the dipole moment difference onto the electric field was $\mid \Delta\mu_E \mid = 2.5$ D. This is a very striking result: The average magnitude of the effective dipole moment difference for the similar system of perylene in polyethylene was found from hole-burning experiments to be $\mid \Delta\mu^* \mid = 0.06$ D (46), at least one order of magnitude smaller than in the case of Tr/PE. The reason for such a large discrepancy, which indicates that two of the internal fields must be quite different in the two cases, is not yet completely understood. Experiments on octatetraene in n-alkane matrixes, discussed further in Section 8.3.5, have shown that internal fields at a molecular site depend much more sensitively on local geometry than on the bulk properties of the host material.

In contrast to Pr/PE, the system of pentacene in p-terphenyl crystals exhibits a qualitatively different Stark behavior. It is interesting to note that this system was

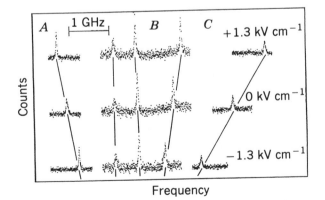

Figure 8.15. The shift of fluorescence excitation line with applied electric field of five different terrylene molecules in polyethylene at 1.8 K (*A* and *B*, four molecules at 575 nm; *C*, one molecule at 571 nm). Plus and minus 1.3 kV cm^{-1} top and bottom, respectively, zero field in the middle. [Adapted from Ref. 80].

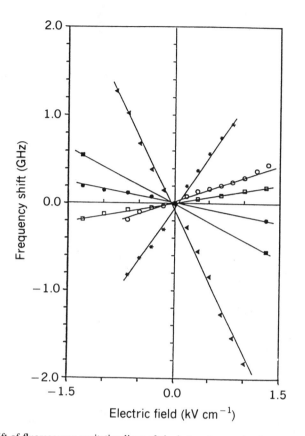

Figure 8.16. Shift of fluorescence excitation lines of single terrylene molecules versus applied electric field. The lines shift without broadening and the Stark effect is linear. The slope for each molecule is different because of differences in the induced dipole moments and/or differences in the orientation with respect to the external field. [Adapted from Ref. 80].

investigated in the earliest papers on Stark effect in doped solids (82,83) where, without energy selective spectroscopic techniques, it was shown that the shift in transition energy with applied electric field was purely quadratic. In the Stark experiments on single molecules in the same system, Pc/pTP, a second-order effect was also obtained (81). In this case, however, due to the extremely high resolution of this technique, it was possible to get information concerning a smaller linear contribution as well. Figure 8.17 shows the results of Stark experiments on four single-molecule lines. The shifts for molecules A, B, and D are mainly quadratic effects: The shift has the same value and direction when the electric field direction is reversed (compare this behavior with the behavior of the single-molecule lines in Fig. 8.15 when the electric field direction is changed). The shift of molecule C changes when the field is reversed demonstrating the superposition of a linear and quadratic effect. The measured Stark shifts $\Delta\nu$ were fit by a second-order poly-

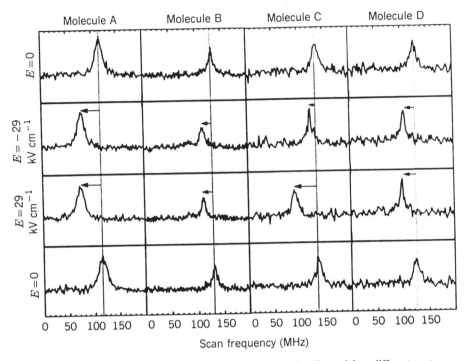

Figure 8.17. Electric field induced shifts of fluorescence excitation lines of four different pentacene molecules in p-terphenyl at 1.8 K in the 592.3 nm region. The shift for the lines of each of the molecules, A, B, and D is the same for positive and negative electric fields indicating mainly quadratic behavior. In contrast, the line for molecule C shifts differently when the field is reversed indicating a significant linear contribution. [Adapted from Ref. 81].

nomial, $\Delta\nu = \beta E + \gamma E^2$ [in principle, a simplified version of Eq. (8.4)], and the constants β and γ were obtained for each single molecule. The results of these fits to data for four single molecules are shown in Fig. 8.18. The dominance of the quadratic effect is clearly seen (compare this to the behavior of the terrylene molecules in the amorphous polyethylene matrix, Fig. 8.16), but the small linear contribution, represented by the constant β, is well defined. According to the analysis (81) the linear contribution is very different for different molecules indicating differences in the local environments. For example, for molecule C, the coefficient β is an order of magnitude larger than for molecule A. Since pentacene is a centrosymmetric molecule, the linear shift coefficient β is solely connected with the internal electric field induced dipole moment. From the experimental data the authors estimated a value for the internal field in p-terphenyl crystals of an order of magnitude of 10 k cm^{-1} which, in their case, was less than the applied electric field. This explains the observability of the second-order Stark effect. This internal field value is two orders of magnitude less than what has been seen so far in other amorphous and crystalline solids (see Section 8.3.5).

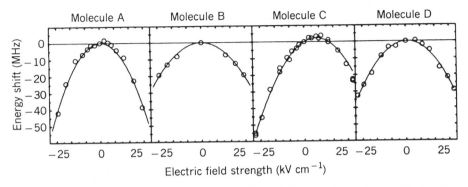

Figure 8.18. Shifts of lines for the four molecules of Fig. 8.17 versus the applied electric field. The smooth curves show the best-fit second-order polynomial. The quadratic effect dominates in all four cases: Only molecule C shows a significant linear contribution. [Adapted from Ref. 81].

8.3.5. Internal Field: Experiments on Octatetraene

Experiments on the linear polyene octatetraene have greatly clarified our understanding of local electric fields. Since the advances that have come from these experiments are due to a number of special properties of octatetraene's electronic structure we begin with a brief review of the key elements of linear polyene singlet states. [Additional details can be found in papers that review linear polyene electronic structure from theoretical (84,85) and experimental (85,86) perspectives.] Then the recent experiments will be discussed emphasizing the detailed characterization of the internal field that they provide. This leads naturally to the conclusion of this chapter in which we indicate what we think are the most interesting areas for future research.

Linear polyenes are conjugated chains with the general formula $H—(CH=CH)_N—H$. There are significant barriers to rotation about the double bonds so that for a given chain length there are a number of isomers that can be isolated and studied as distinct chemical compounds. The most stable of these is the conformation in which all double bonds are trans, in which case the molecule is a planar zigzag chain with C_{2h} symmetry (Fig. 8.19).

Linear polyene absorption spectra are dominated by a strongly allowed transition that gives the longer chains their characteristic colors (e.g., Fig. 8.20). It is now well established that in all polyene hydrocarbons the $1\ ^1B_u$ symmetry excited state that is responsible for this strong absorption is S_2, the second excited singlet state, and that S_1, the lowest energy excited singlet state, is the $2\ ^1A_g$ state to which electric dipole transition from the $1\ ^1A_g$ ground state is symmetry forbidden. In the shorter polyenes the energy difference between the S_1 and S_2 is small compared to the excitation energy of either state: The situation for octatetraene is summarized in Fig. 8.20.

There are a number of reasons why experiments on the conjugated linear polyene octatetraene in n-alkane crystals are especially informative with respect to re-

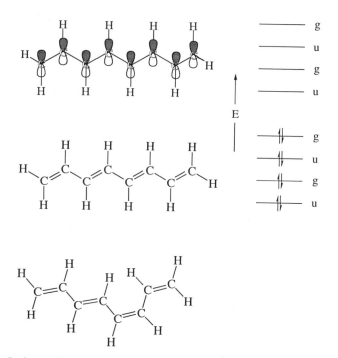

Figure 8.19. Conjugated linear polyenes. On the left side the *sp²* hybridized chain with the 2*p* π atomic orbitals on which the delocalized π-molecular orbitals are based is shown above two isomers of octatetraene, all-trans-octatetraene (middle) and cis-octatetraene (bottom). The description of the ground state in terms of the π-electron molecular orbitals (labeled by their inversion symmetry) is schematically indicated at the right.

fining our picture of the local electric field in an organic solid. The most important of these include:

- Octatetraene has well-characterized photochemical activity under conditions that are suitable for burning narrow holes (87). A number of experiments have established that the cis and trans isomers of octatetraene may be photochemically interconverted even when the molecules are guests in an n-alkane crystal cooled to liquid helium temperature. Furthermore, the $2\ {}^1A_g$ origin excitation energies for the cis and trans isomers differ by many times their inhomogeneous bandwidths. This means it is possible to burn persistent holes.

- There is significant intensity in the zero-phonon component of the $1\ {}^1A_g \rightarrow 2\ {}^1A_g$ 0–0 band. Furthermore, even though the approximately 4 cm^{-1} inhomogeneous width is nearly 10^5 wider than the lifetime limited homogeneous width, it is so much narrower than the envelope of the phonon side band that excitation can be completely restricted to the zero-phonon component. This insures that the widths of the persistent holes that are burned into the spectrum

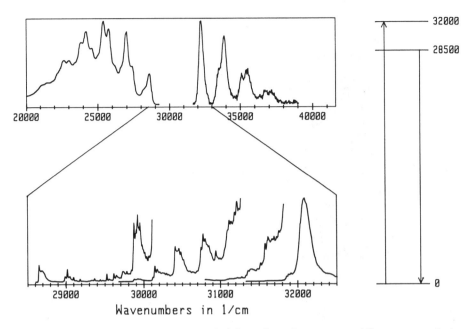

Wavenumbers in 1/cm

Figure 8.20. Octatetraene singlet states. At the left are shown fluorescence and fluorescence excitation spectra for n-octane solutions at 77 K (top), and an expanded view of the fluorescence excitation spectrum at 4.2 K (bottom). The energy level scheme is shown at the right. The $1\ ^1B_u$ state responsible for the strong absorption (0–0 at 32,000 cm^{-1}) is well described by a single-molecular orbital configuration that differs from the ground-state configuration by the promotion of one electron from the highest energy occupied molecular orbital (HOMO) to the lowest energy unoccupied molecular orbital (LUMO). Absorption to the $2\ ^1A_g$ state responsible for the fluorescence, is only seen under high-resolution conditions.

of this system will be determined by the zero-phonon line and, thus, will be narrow.

- In the case of the centrosymmetric all-trans isomer the $S_0 \rightarrow S_1$ transition is symmetry forbidden. Even when isomerization or local perturbations break down the strict symmetry forbiddeness, the transition remains very weak. The S_1 lifetimes for octatetraene in n-alkane matrixes at liquid helium temperatures range from 100 to over 250 ns (88), which means that the lifetime limited homogeneous widths are 1.6 MHz and narrower. For octatetraene in n-hexane at 1.4 K (the temperature used in the electric field experiments) hole widths less than 6 MHz may be directly measured (89,90). Extrapolation of a plot of hole width versus hole area to zero area gives a Lorentzian hole width of 4 MHz for the persistent hole (89) which, considering that the lifetime limited hole width of 3.2 MHz should be convolved with a laser bandwidth of order 1 MHz, means that, to within experimental error, the limiting hole width is set by the lifetime of the excited state. The ability to work with very narrow holes means that the effects of external electric fields can be determined with great precision.

- Octatetraene has a well-understood electronic structure that can be successfully described with relatively simple models (91). This greatly facilitates the translation of spectroscopic measurements into a detailed picture of local structure.

- Octatetraene is, at least to a first approximation, a one-dimensional molecule. For a one-dimensional molecule the transition dipole moments that determine the response to external electric field are all parallel to the chain. This can greatly simplify the analysis of observed shifts.

- Finally, the relative closeness of the $2\,^1A_g$ and $1\,^1B_u$ states in octatetraene (in n-hexane the 0–0 excitation energies differ by 3500 cm^{-1}) means that, at least as a conceptual guide, the shift in $2\,^1A_g$ excitation energy with electric field can be treated with the very simple and informative two-level quantum mechanical model described in Section 8.2.2.

If inversion is a valid symmetry operation for both ground and excited states, neither state can have a nonzero dipole moment and any electric field effect must be second order in the field strength. Since high-resolution spectroscopic experiments on all-trans-octatetraene seeded into supersonic rare gas expansions or substituted in crystalline n-octane have established that in their relaxed equilibrium geometries the octatetraene $1\,^1A_g$ and $2\,^1A_g$ states are both centrosymmetric (92,93), one expects a quadratic Stark effect similar to that shown in Fig. 8.4. Even if steric effects induce a noncentrosymmetric geometry, there still should be no significant permanent ground- and excited-state dipole moments because of the expected charge neutrality of the carbon atoms in a linear polyene (94). Thus, the observation that a photochemical hole burned into the zero-phonon component of the $1\,^1A_g \rightarrow 2\,^1A_g$ 0–0 band of all-trans-octatetraene in n-hexane splits symmetrically when an electric field is applied and that this splitting is linear in field strength was at first very surprising. However, these observations were completely reconciled with the current understanding of linear polyene electronic structure once it was postulated that the octatetraenes occupied sites in the n-hexane crystals where there was a well-defined local electric field (95).

In the initial analysis the effect of electric field on the octatetraene molecule was treated in the point dipole approximation. That is, it was assumed that the electric field perturbation could be written as $\mathcal{H}_1 = -\,\mathbf{E}\cdot\boldsymbol{\mu}$, where $\mathbf{E}$ is the vector sum of the internal and external fields and $\boldsymbol{\mu}$ is the electric dipole operator. Hole profiles for a given laboratory field were calculated by solving the 3×3 eigenvalue problem

$$\mathcal{H} = \begin{pmatrix} \overset{1\,^1A_g}{0} & \overset{2\,^1A_g}{0} & \overset{1\,^1B_u}{-\mathbf{E}\cdot\boldsymbol{\mu}_{13}} \\ 0 & 28500\ \text{cm}^{-1} & -\mathbf{E}\cdot\boldsymbol{\mu}_{23} \\ -\mathbf{E}\cdot\boldsymbol{\mu}_{13} & -\mathbf{E}\cdot\boldsymbol{\mu}_{23} & 32{,}000\ \text{cm}^{-1} \end{pmatrix} \begin{matrix} 1\,^1A_g \\ 2\,^1A_g \\ 1\,^1B_u \end{matrix} \qquad (8.13)$$

for all possible vector sums of the internal and external fields, weighting each excitation energy by the square of the transition dipoles for burning and readout, and convolving the resulting distribution of excitation energies with the Lorentzian that best fit the zero-field hole profile. In the orientation averaging for the poly-crystalline samples it was assumed that the internal field was the same for all octatetraene molecules in the ensemble and that two transition dipoles $\mathbf{\mu}_{13}$ and $\mathbf{\mu}_{23}$ were parallel. If the transition dipole magnitudes are taken from theory (96), the only adjustable parameter is the projection of the internal field on the transition dipoles. When this is set to 1.6×10^6 V cm^{-1} in the three level analysis, the agreement between the calculated and measured profiles is within experimental error (Figs. 8.21 and 8.22). Results for octatetraene in n-heptane are qualitatively similar but the splitting that is seen when the laser light is polarized parallel to the applied field is approximately one-fourth as large as in n-hexane. In the point dipole model this implies that the internal field is four times larger in n-hexane than in n-heptane. This is expected on the basis of the microscopic modeling discussed

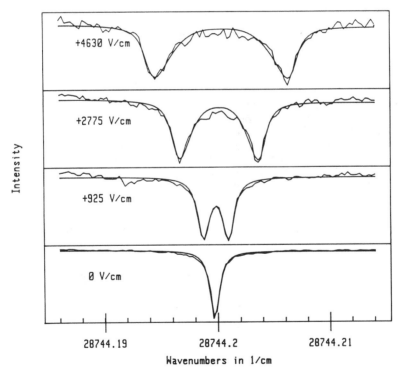

Figure 8.21. Field dependence of the photochemical holes burned into the zero-phonon component of the 0–0 band of the $1\,^1A_g\text{–}2\,^1A_g$ transition of octatetraene in n-hexane. The noisy lines are the fluorescence excitation scans of the hole at 1.4 K, the smooth curves are the profiles calculated from Eq. (8.13) assuming an internal field of 1.6×10^6 V cm^{-1}. Because of the large local field the splitting is very nearly linear with applied field over the field range investigated.

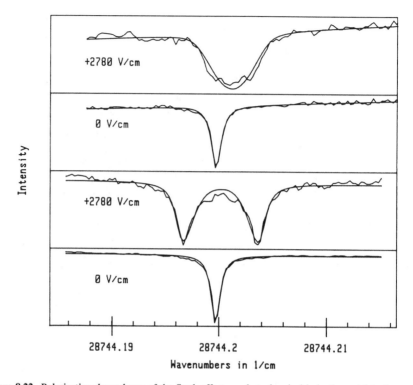

Figure 8.22. Polarization dependence of the Stark effect on photochemical holes burned into the zero-phonon component of the 0–0 band of the 1 ^1A_g–2 ^1A_g transition of octatetraene in n-hexane. The lower pair of curves are for the burning and scanning laser beams polarized parallel to the applied field, the upper pairs of curves are for the burning and scanning laser beams polarized perpendicular to the applied field. The noisy lines are the fluorescence excitation scans of the hole at 1.4 K, the smooth curves are the profiles calculated from Eq. (8.13) assuming an internal field of 1.6 × 10⁶ V cm⁻¹. The apparent shift in the upper curve is due to laser frequency drift.

below (the larger internal field in n-hexane just reflects the greater asymmetry of the dopant site where two host molecules must be removed to accommodate the guest). That internal fields should be so different in hosts that have virtually identical bulk properties shows that bulk parameters are not directly relevant to determining the internal field. How the different splittings seen in the n-hexane and n-heptane hosts are related to the strength of the internal field is shown in Fig. 8.23.

As seen in Fig. 8.24, the experimental results obtained for octatetraene in n-hexane and n-heptane are qualitatively the same as the results obtained for perylene in n-heptane (97). Because of this qualitative similarity it is clear that, although a different and somewhat more complicated higher order model has been invoked to explain the perylene results (98), the observed profiles can be completely accounted for by the internal field model that we have described (Fig. 8.23).

That there should be local fields in the 10⁶ V cm⁻¹ range in a crystal made up of symmetric electrically neutral hydrocarbon molecules may also seem surprising,

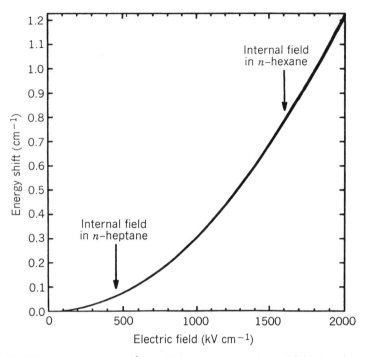

Figure 8.23. Shift of the octatetraene $2\,^1A_g$ excitation energy versus internal field strength as calculated by Eq. (8.13). The slope of this curve at a given internal field strength gives the shift per unit field strength that would be seen in a Stark experiment. The internal fields that give the slopes experimentally observed for octatetraene in n-hexane and octatetraene in n-heptane are indicated by labeled arrows.

but such large fields are to be expected whenever a molecular site does not have strict centrosymmetry. Chemical bonds are, in general, polar: This includes the C—H bonds in an n-alkane molecule. Although there is controversy as to its exact magnitude (98), the generally accepted value is 0.4 D with hydrogen positive (100). Since the field 1 Å from a 1 D dipole moment is 3×10^8 V cm^{-1}, this means that even though an n-hexane molecule is completely neutral at distances that are large with respect to its internal dimensions, at distances comparable to the size of the molecule electric fields are very strong. The consequences of this for charge separation and transport in molecular systems should not be underestimated: In any situation where the local environment is anisotropic for any reason, it is not realistic to expect to understand photoinduced charge separation solely in terms of chromophore electronic structure.

    In Ref. 95 the local fields for two dopant sites were calculated by summing the contributions from the C—H bond dipoles over the lattice. While this calculation showed that large local fields were reasonable, it also showed that these fields were far from constant, thus raising the problem of modeling the response of a molecule to an electric field that varies significantly over its length. In the case of the linear polyenes where excitation energies may be accurately reproduced by a simple quan-

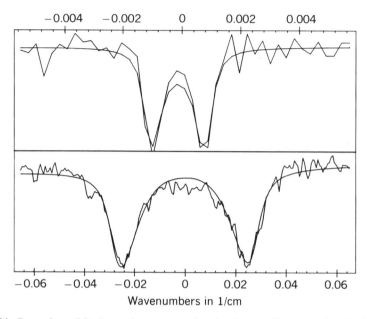

Figure 8.24. Comparison of Stark experiments on perylene in n-heptane (lower panel) to Stark experiments on octatetraene in n-heptane (upper panel, burning and reading light polarized parallel to the applied field). The experimental curve in the lower panel was digitized from Fig. 8.1 of Ref. 97. The noisy lines are the experimental data, the smooth curves were calculated for burning and reading light polarized parallel to the applied field using Eq. (8.13) in the case of octatetraene ($|\, \mathbf{E}_{int}\, | \sim 0.4$ MV cm^{-1}) and using the two-level analog of Eq. (8.13) (energy gap 22,500 cm^{-1}, $\mu_{12} = 2.46$ D, $|\, \mathbf{E}_{int}\, | \sim$ 2 MV cm^{-1}) in the case of perylene.

tum mechanical model (91), we found that this may be done in a conceptually simple way with surprising accuracy. The ability to realistically account for the effect of an electric field that varies over the length of the linear polyene opens the door to characterizing internal fields in molecular systems with unprecedented detail and precision. It is not an exaggeration to say that these experiments establish methods for using octatetraene as an eight element array detector for internal electric fields.

We now turn to the question of modeling electric field effects for linear polyenes. It is already established that a simple model based on Hückel theory (the tight-binding Hamiltonian for linear chains) accurately reproduces all of the linear polyene $2\,^1A_g$ and $1\,^1B_u$ 0–0 excitation energies that have been measured in high-resolution experiments on Shpol'skii systems (91). All that is done to generalize this model to describe linear polyenes in an arbitrarily complicated static electric field is to replace α_i, the Hückel coulomb integral (site energy) for the ith carbon, by $\alpha_i - e\mathbf{E}_i \cdot \mathbf{r}_i$, where $\mathbf{E}_i$ is the total electric field at $\mathbf{r}_i$, the location of the ith carbon atom and e is the electronic charge. The calculational protocol is then extremely simple.

For a given orientation of laboratory axes relative to molecular axes, $\mathbf{E}_i$ is calculated as the vector sum of internal and external fields. Then the $2 \, ^1A_g$ excitation energy for this orientation is calculated from the simple model and weighted by the squares of the amplitudes for burning and readout. For this weighting the $1 \, ^1A_g - 2 \, ^1A_g$ transition dipole is assumed to be parallel to the $1 \, ^1A_g - 1 \, ^1B_u$ transition dipole. (In the simple model the dipoles for all the allowed transitions are parallel, so any transition dipole by mixing A_g and B_u levels must have the same direction.) These operations are then repeated for all possible orientations of the laboratory and molecular frames.

The internal field at a given polyene carbon atom is calculated by superposing the fields at that point from all of the C—H bond dipoles in the lattice. The local structure around the linear polyene guest is calculated with a standard molecular dynamics program (101). A paper giving a detailed description of this model is in preparation: For now we only want to call attention to the facts that; first, the external electric field induced shifts calculated with this model are extremely sensitive to local structure in a way that allows one to choose between possibilities that are nearly isoenergetic in the molecular dynamics calculations and, second, that for one of the possible site identified by the molecular dynamics calculation this model gives calculated profiles that are reasonably close to experiment. This may be seen in Fig. 8.25. Two points deserve special emphasis. First, there are no adjustable parameters in this calculation. Second, the agreement between calculated and observed profiles is well within the uncertainty in the value of the C—H bond dipole. (It is reasonable to expect that these experiments will lead to a refinement of this important quantity.)

8.4. CONCLUDING REMARKS

This chapter has dealt with the basic principles and new results on persistent hole-burning studies of the effects of external electric fields on electronic transition energies in condensed phase molecular systems. We are well aware of the fact that this brief review does not discuss all of the new data in this field. Omissions include any discussion of practical applications of the electric field effect for optical data storage (102,103), for electrooptical modulation of laser beams (104), and, we are sure, other things. Our selection of topics to discuss was based on our desire to concentrate on phenomena that have special relevance for understanding basic microscopic structure, especially the issue of the internal field at a molecular site.

The most important thing in photoinduced charge transfer in molecular systems may well be the microscopic electric field at the photoactive molecule. Up to this point these fields have been neglected, in part because of lack of knowledge of how large they could be and in part because there were no methods for measuring them. Photochemical hole burning provides the sensitivity necessary to spectroscopically observe the consequences of these local fields and the simulation techniques described in Section 8.3.5 open the way to converting these spectroscopic observations into a detailed and accurate mapping of the electrostatic environment around a dopant molecule. We feel that there is no doubt that the application of these methods is going to undergo enormous expansion and that the experimental

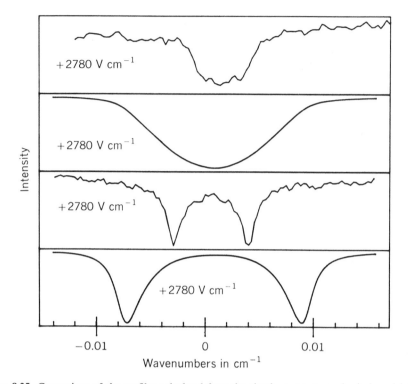

Figure 8.25. Comparison of the profiles calculated from the simple quantum mechanical model for linear polyene electronic structure using the internal field components at the octatetraene carbons calculated by superposing the fields from all the n-hexane C—H dipoles in the lattice to experimental measurements. The calculation is for one of the four site configurations identified in a molecular dynamics calculation. The bottom and second from the top panels show the profiles calculated for burning and probing light polarized parallel and perpendicular to the applied field, respectively. The second from the bottom and top panels show the corresponding profiles measured for octatetraene in n-hexane at 1.4 K. Further details will be given in a paper now in preparation.

results are going to have a major impact on our understanding of intermolecular interaction. It's an exciting time to be doing high-resolution condensed phase spectroscopy!

REFERENCES

1. Personov, R. I., Al'shits, E. I., and Bykovskaya, L. A. *Opt. Commun.* **1972**, *6*, 169; *JETP Lett.* **1972**, *15*, 431.

2. Personov, R. I., Al'shits, E. I., Bykovskaya, L. A., and Kharlamov, B. M. *JETP* **1974**, *38*, 912.

3. Abram, I., Auerbach, R. A., Birge, R. R., Kohler, B. E., and Stevenson, J. M. *Chem. Phys.* **1974**, *61*, 3857.

4. Kharlamov, B. M., Personov, R. I., and Bykovskaya, L. A. *Opt. Commun.* **1974**, *12*, 191.

5. Gorokhovskii, A. A., Kaarly, R. K., and Rebane, L. A. *JETP Lett.* **1974**, *20*, 216.

6. Personov, R. I. Site Selection Spectroscopy of Complex Molecules in Solutions and its Applications. In *Spectroscopy and Excitation Dynamics of Condensed Molecular Systems*; Agranovich, V. M. and Hochstrasser, R. M. (Eds.). North-Holland: Amsterdam, 1983, Chapter 10.

7. Small, G. J. Persistent Nonphotochemical Hole Burning and the Dephasing in Impurity Electronic Transitions in Organic Glasses. In *Spectroscopy and Excitation Dynamics of Condensed Molecular Systems*; Agranovich, V. M. and Hochstrasser, R. M. (Eds.). North-Holland, Amsterdam, 1983, Chapter 9.

8. Friedrich, J. and Haarer, D. *Angew. Chem., Int. Ed. Engl.* **1984**, *23*, 113.

9. Moerner, W. E. (Ed.). Persistent Spectral Hole Burning: Science and Applications. Springer, Berlin, 1988.

10. Völker, S. *Annu. Rev. Phys. Chem.* **1989**, *40*, 499.

11. Personov, R. I. *J. Photochem. Photobiol. A: Chem.* **1992**, *62*, 321.

12. Orrit, M., Bernard, J., and Personov, R. I. *J. Phys. Chem.* **1993**, *97*, 10256.

13. Jankowiak, R., Hayes, J. M., and Small, G. J. *Chem. Rev.* **1993**, *93*, 1471.

14. Marchetti, A. P., Scozzafava, M., and Young, R. H. *Chem. Phys. Lett.* **1977**, *51*, 424.

15. Samoilenko, V. D., Rasumova, N. V., and Personov, R. I. *Opt. Spectrosc. (USSR)* **1982**, *52*, 346.

16. Burkhalter, F. A., Suter, G. W., Wild, U. P., Samoilenko, V. D., Rasumova, N. V., and Personov, R. I. *Chem. Phys. Lett.* **1983**, *94*, 483.

17. Bogner, U., Schätz, P., Seel, R., and Maier, M. *Chem. Phys. Lett.* **1983**, *102*, 267.

18. Dicker, A. I. M., Johnson, L. W., Noort, M., and van der Waals, J. H. *Chem. Phys. Lett.* **1983**, *94*, 14.

19. Kharlamov, B. M., Al'shits, E. I., Personov, R. I., Nizhankovsky, N. I., and Nazin, V. G. *Opt. Commun.* **1978**, *24*, 199.

20. Dicker, A. I. M., Noort, M., Völker, S., and van der Waals, J. H. *Chem. Phys. Lett.* **1980**, *73*, 1.

21. Maier, M. *Appl. Phys. B.* **1986**, *41*, 73.

22. Personov, R. I. *Bull. Acad. Sci. USSR, Phys. Ser.* **1988**, *52*, 1.

23. Solov'ev, K. N., Salesskii, N. E., Kotlo, V. N., and Shkirman, S. F. *JETP Lett.* **1973**, *17*, 332.

24. Völker, S. and van der Waals, J. H. *Mol. Phys.* **1976**, *32*, 1703.

25. (a) Kharlamov, B. M., Al'shits, E. I., and Personov, R. I. *Bull. Acad. Sci. USSR, Phys. Ser.* **1984**, *48*, 65; (b) Kharlamov, B. M., Al'shits, E. I., and Personov, R. I. *J. Appl. Spectrosc.* **1986**, *45*, 1024.

26. Adamson, G., Gradl, G., and Kohler, B. E. *J. Chem. Phys.* **1989**, *90*, 3038.

27. Iannone, M., Scott, G. W., Brinza, D., and Coutler, D. R. *J. Chem. Phys.* **1986**, *85*, 4863.

28. Graf, F., Hong, H. K., Nassal, A., and Haarer, D. *Chem. Phys. Lett.* **1978**, *59*, 217.

29. de Vries, H. and Wiersma, D. A. *Phys. Rev. Lett.* **1976**, *36*, 91.

30. Macfarlane, R. M. *J. Luminescence* **1987**, *38*, 20.

31. Moerner, W. E. *Jpn. J. Appl. Phys.-S.* **1989**, *28*, 221.

32. Ambrose, W. P. and Moerner, W. E. *Chem. Phys.* **1990**, *144*, 71.

33. Meixner, A. J., Renn, A., Bucher, S. E., and Wild, U. P. *J. Phys. Chem.* **1986**, *90*, 6777.

34. Kador, L., Haarer, D., and Personov, R. I. *J. Chem. Phys.* **1987**, *86*, 5300.

35. Schätz, P. and Maier, M. *J. Chem. Phys.* **1987**, *87*, 809.

36. Renn, A., Bucher, S. E., Meixner, A. J., Meister, E. C., and Wild, U. P. *J. Luminescence* **1988**, *39*, 181.

37. Gradl, G., Kohler, B. E., and Westerfield, C. *J. Chem. Phys.* **1992**, *97*, 6064.

38. Ivanov, V. K., Personov, R. I., and Rusumova, N. V. *Optics Spectroscopy* **1985**, *58*, 2.

39. Furasawa, A. and Horie, K. *J. Chem. Phys.* **1991**, *94*, 80.

40. Bashé, Th. and Bräuchle, C. *J. Phys. Chem.* **1991**, *95*, 7130.

41. Renn, A., Meixner, A. J., Wild, U. P., and Burkhalter, F. A. *Chem. Phys.* **1985**, *93*, 157.

42. Korotaev, O. N., Surin, N. N., Yurchenko, A. I., Glyadkovsky, V. I., and Donskoi, E. I. *Chem. Phys. Lett.* **1984**, *110*, 533.

43. Renn, A., Bucher, S. E., Meixner, A. J., Meister, E. C., and Wild, U. P. *J. Luminescence* **1988**, *39*, 181.

44. Johnson, L. W., Murphy, M. D., Pope, C., and Foresti, M. *J. Chem. Phys.* **1987**, *86*, 4335.

45. Caro, C. D., Renn, A., Wild, U. P., and Johnson, L. W. *J. Luminescence* **1991**, *50*, 309.

46. Gerbinder, J., Bogner, U., and Maier, M. *Chem. Phys. Lett.* **1987**, *141*, 31.

47. Kanaan, Y., Attenberger, T., Bogner, U., and Maier, M. *Appl. Phys. B* **1990**, *51*, 336.

48. Sesselmann, Th., Kador, L., Richter, W., and Haarer, D. *Europhys. Lett.* **1988**, *5*, 361.

49. Kador, L., Jahn, S., Haarer, D., and Silbey, R. *Phys. Rev. B* **1990**, *41*, 12215.

50. Agranovich, V. M., Ivanov, V. K., Personov, R. I., and Rasumova, N. V. *Phys. Lett. A* **1986**, *118*, 239.

51. Meixner, A. J., Renn, A., and Wild, U. P. *Chem. Phys. Lett.* **1992**, *190*, 75.

52. Altmann, R. B., Renge, I., Kador, L., and Haarer, D. *J. Chem. Phys.* **1992**, *97*, 5316.

53. Hartmannsgruber, N. and Maier, M. *J. Chem. Phys.* **1992**, *96*, 7279.

54. Bogner, U., Röska, G., and Graf, F. *Thin Solid Films* **1983**, *99*, 257.

55. Orrit, M., Bernard, J., and Möbius, D. *Chem. Phys. Letts.* **1989**, *156*, 233.

56. Bernard, J., Orrit, M., Personov, R. I., and Samoilenko, A. D. *Chem. Phys. Letts.* **1989**, *164*, 377.

57. Bernard, J. and Orrit, M. *J. Luminescence* **1990**, *45*, 70.

58. Romanovskii, Yu. V., Personov, R. I., Samoilenko, A. D., Holliday, K., and Wild, U. P. *Chem. Phys. Lett.* **1992**, *197*, 373.

59. Kuhn, H., In *Proceedings of International Symposium on Future Electronic Devices*, Tokyo, 1985.

60. Möbius, D. (Ed.). *Langmuir–Blodgett Films*, Thin Solid Films **1988**, *159;* **1988**, *160.*

61. Blinov, L. M. *Sov. Phys.—Usp.* **1988**, *31*, 623.

62. Ulman, A. In *Introduction to Ultrathin Organic Films;* Academic P: New York, 1991.

63. Orrit, M., Bernard, J., Mouhsen, A., Talon, H., Möbius, D., and Personov, R. I. *Chem. Phys. Lett.* **1991**, *179*, 232.

64. Bernard J., Talon, H., Orrit, M., Möbius, D., and Personov, R. I. *Thin Solid Films* **1992**, *217*, 178.

65. Grewer, G. and Lösche, M. *Makromol. Chem. Macromol. Symp.* **1991**, *46*, 79.

66. Orrit, M., Möbius, D., Lehmann, U., and Meyer, H. *J. Chem. Phys.* **1986**, *85*, 4966.

67. Altmann, R. B., Haarer, D., Ulitsky, N. I., and Personov, R. I. *J. Luminescence* **1993**, *56*, 135.

68. Smith, K. M. (Ed.). *Porphyrins and Metalloporphyrins;* Elsevier: Amsterdam, 1975.

69. Schaffer, A. M., Gouterman, M., and Davidson, E. R. *Theoret. Chim. Acta* **1973**, *30*, 9.

70. Gouterman, M. In *The Porphyrins*, Vol. 3. Dolphin, D. (Ed.). Academic: New York, 1971, p. 1.

71. Timkovich, R. and Tulinsky, A. *J. Am. Chem. Soc.* **1969**, *91*, 4430.

72. Cullen, D. L. and Meyer, E. F. Jr. *Acta Crystallogr.* **1976**, *B32*, 2259.

73. Scheidt, W. R. and Dow, W. *J. Am. Chem. Soc.* **1977**, *99*, 1101.

74. Ball, R. G., Lee, K. M., Marschall, A. G., and Trotter, J. *Inorg. Chem.* **1980**, *19*, 1463.

75. Maggiora, G. M. *J. Am. Chem. Soc.* **1973**, *95*, 6555.

76. Bersuker, I. B. and Stavrov, S. S. *Chem. Phys.* **1981**, *54*, 331.

77. Moerner, W. E., Kador, L. *Phys. Rev. Lett.* **1989**, *62*, 2535.

78. Orrit, M. and Bernard, J. *Phys. Rev. Lett.* **1990**, *65*, 2716.

79. Moerner, W. E. and Basché, Th. *Angew. Chem. Int. Ed. Engl.* **1993**, *32*, 457.

80. Orrit, M., Bernard, J., Zumbusche, A., Personov, R. I. *Chem. Phys. Lett.* **1992**, *196*, 595; *199*, 408.

81. Wild, U. P., Güttler, F., Pirrota, M., and Renn, A. *Chem. Phys. Lett.* **1992,** *193,* 451.

82. Meyling, J. H. and Wiersma, D. A. *Chem. Phys. Lett.* **1973,** *20,* 383.

83. Meyling, J. H., Hesselink, W. H., and Wiersma, D. A. *Chem. Phys.* **1976,** *17,* 353.

84. Baeriswyl, D., Campbell, D. K., and Mazumdar, S. In *Conjugated Conducting Polymers*, Kiess, H. (Ed.). Springer-Verlag: Berlin, 1992.

85. Hudson, B. S., Kohler, B. E., and Schulten, K. *Excited States* **1982,** *6,* 1.

86. Kohler, B. E. In *Conjugated Polymers: The Novel Science and Technology of Highly Conducting and Nonlinear Optically Active Materials*; Brédas, J. L. and Silbey, R. (Eds.). Kluwer: Dordrecht, The Netherlands, 1991, p. 405.

87. Kohler, B. E. *Chem. Rev.* **1993,** *93,* 41.

88. (a) Ackerman, J. R., Huppert, D., Kohler, B. E., and Rentzepis, P. M. *J. Chem. Phys.* **1982,** *77,* 3967; (b) Kohler, B. E. and Snow, J. B. *J. Chem. Phys.* **1983,** *79,* 2134.

89. Adamson, G., Gradl, G., and Kohler, B. E. *J. Chem. Phys.* **1989,** *90,* 3038.

90. Gradl, G., Kohler, B. E., and Westerfield, C. *J. Luminescence* **1990,** *45,* 83.

91. Kohler, B. E. *J. Chem. Phys.* **1990,** *93,* 5838.

92. Granville, M. F., Holtom, G. R., and Kohler, B. E. *J. Chem. Phys.* **1980,** *72,* 4671.

93. Petek, H., Bell, A. J., Choi, Y. S., Yoshihara, K., Tounge, B. A., and Christensen, R. L. *J. Chem. Phys.* **1993,** *98,* 3777.

94. See, for example, Salem, L. In *The Molecular Orbital Theory of Conjugated Systems* Benjamin Press: Reading, MA, 1972.

95. Gradl, G., Kohler, B. E., and Westerfield, C. *J. Chem. Phys.* **1992,** *97,* 6064.

96. Soos, Z. G. and Ramasesha, S. *J. Chem. Phys.* **1989,** *90,* 1067.

97. Attenberger, T., Bogner, U., and Maier, M. *Chem. Phys. Lett.* **1991,** *180,* 207.

98. Gerblinger, J., Bogner, U. and Maier, M. *Chem. Phys. Lett.* **1987,** *141,* 31.

99. Reed, A. E. and Weinhold, F. *J. Chem. Phys.* **1986,** *84,* 2428.

100. For example, see Smyth, C. P. In *Dielectric Behavior and Structure*. McGraw-Hill, New York, 1955, pp. 243–244.

101. *Discover*, version 2.9.0/3.1.0, San Diego: Biosym Technologies, 1993.

102. Wild, U. P., Bucher, S. E., and Burkhalter, F. A. *Appl. Opt.* **1985,** *24,* 1526.

103. Bogner, U., Beck, K., and Maier, M. *Appl. Phys. Lett.* **1985,** *46,* 534.

104. Schätz, P., Bogner, U., and Maier, M. *Appl. Phys. Lett.* **1986,** *49,* 1132.

Chapter **IX**

EXCITED ELECTRONIC STATE PROPERTIES FROM GROUND-STATE RESONANCE RAMAN INTENSITIES

Anne B. Myers
Department of Chemistry,
University of Rochester, Rochester, New York

Laser Techniques In Chemistry, Edited by Anne B. Myers and Thomas R. Rizzo.
Techniques of Chemistry Series, Vol. XXIII.
ISBN 0-471-59769-4   © 1995 John Wiley & Sons, Inc.

9.1.   INTRODUCTION

Resonance Raman is gaining increasingly wide use as a spectroscopic probe for the structure and dynamics of both ground and excited electronic states of molecules in both vapor and solution phases. Raman spectroscopy in its nonresonant form is simply a type of vibrational spectroscopy similar to infrared (IR) absorption but with different, often complementary, selection rules. Quasimonochromatic light incident on a sample at frequency ω_L is inelastically scattered into other frequencies ω_S and the energy difference between the photons at ω_L and ω_S goes to excite ($\omega_S < \omega_L$, Stokes scattering) or deexcite ($\omega_S > \omega_L$, anti-Stokes scattering) a molecular vibration. A plot of scattered intensity versus the Raman shift, $\omega_L - \omega_S$, gives the vibrational spectrum, which is usually dominated by Stokes fundamentals with intensities determined by the derivatives of the molecular polarizability along each vibrational normal coordinate.

Raman scattering becomes "preresonant" or "resonant" when the laser frequency approaches or coincides with an electronic transition frequency of the material (Fig. 9.1). On or near resonance, the Raman scattered intensity is greatly increased, and the intensities become determined principally by the properties of the resonant electronic state. Resonance Raman intensities contain essentially the same information about the excited-state potential energy surface as do vibrationally resolved absorption or fluorescence spectra, but allow this information to be extracted even when the direct electronic spectra are completely vibrationally unresolved (1). Furthermore, since the shape of the excited-state potential surface dictates the initial dynamics of nuclear motion on that surface, there is an immediate connection between resonance Raman intensities (as well as ordinary electronic absorption spectra) and excited-state dynamics. This connection can be emphasized by Fourier transforming the sum-over-states expression for the resonance Raman amplitude into the time domain and interpreting the resulting equations as involving time-dependent overlap integrals between the final state in the Raman process and a moving wavepacket that represents the initial vibrational wave function propagated on the excited-state potential surface (1–3). This interpretation, originally put forth by Lee and Heller in 1979 (2), has spurred many subsequent studies in which resonance Raman intensities have been analyzed, either qualitatively or quantitatively, to probe the dynamics of fast photochemical reactions of a variety of small and large molecules in both vapor and condensed phases.

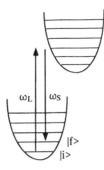

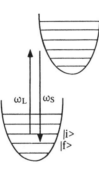

Figure 9.1. Resonance Raman scattering, Stokes (top) and anti-Stokes (bottom).

The resonance Raman intensities of different transitions, and their excitation profiles, carry vibrationally mode-specific information about the excited-state potential energy surface in the Franck–Condon region (and, in favorable cases, somewhat outside the Franck–Condon region) even if the direct electronic absorption spectra are completely diffuse due to fast photochemistry in the resonant state or severe solvent-induced electronic spectral broadening. In particular, the effect of the solvent environment on the potential surfaces may be examined. The absolute scattering intensities depend on the rate of dephasing of the electronic transition due to both population decay and solvent-induced pure dephasing, while the total emission intensity (Raman plus fluorescence) depends on the population lifetime alone. Thus, in an oversimplified sense, the ratio of Raman-like to fluorescence-like emission gives the ratio of the total electronic dephasing time to the lifetime, while knowledge of the total dephasing time allows the solvent-induced absorption spectral breadth to be separated into its "homogeneous" and "inhomogeneous" components. This statement is oversimplified because the classification of broadening mechanisms into homogeneous and inhomogeneous depends on the time scale on which the observation is made, and at least in liquids there are unlikely to be any truly static contributions to the breadth; rather, the solvent-induced breadth is better treated as a quantity having some characteristic time scale that may range from essentially instantaneous (the "fast modulation" or "motional narrowing" limit) to the nearly static "inhomogeneous" limit. The excitation frequency dependence of the Raman and fluorescence yields can be used to deduce this time scale. Finally, Raman spectra obtained on resonance with electronic states that are dissociative or have very large geometry changes can provide vibrational frequencies and line widths for highly excited vibrations of the electronic ground state, allowing the ground-state potential surfaces, the nature of highly excited vibrations, and vibrational dephasing rates at high vibrational energies to be explored in both isolated and solvated molecules.

The experimental and analytical methods discussed in this chapter should not be confused with those of true time-resolved resonance Raman scattering, discussed in Chapter VI in this volume. In that technique, an actinic pump pulse is used to initiate a photochemical or photophysical process followed by a probe pulse to obtain the resonance enhanced spectrum of a transient intermediate. In this manner the conformational dynamics and vibrational energy relaxation of excited electronic states and/or ground-state photoproducts can be examined on time scales from hundreds of femtoseconds on up. However, the time resolution that pump–probe spontaneous Raman can achieve is determined by the acceptable spectral bandwidth of the probing pulse, which limits the spectral resolution of the Raman spectrum it excites. For most applications the maximum useful bandwidth is probably in the range from 15 to 30 cm^{-1}, corresponding to a transform-limited pulse duration of approximately 1.0–0.5 ps (for Gaussian or sech^2 pulses).

Although both experimental and theoretical aspects of resonance Raman spectroscopy have been reviewed numerous times (4–10) most of the reviews that focus on extraction of excited-state information from intensities are by now somewhat outdated. The goal of this chapter is to summarize the current (early 1994) state of the art in experimental methods for obtaining resonance Raman intensities and excitation profiles as well as theoretical and computational methods for interpreting them. Selected applications of these methods to a variety of molecular systems, including both small and large molecules in both vapor and condensed phases, are cited, with an emphasis on work performed since the mid-1980s. Finally, prospects for further experimental and theoretical developments and future applications are discussed briefly.

9.2. EXPERIMENTAL METHODS

9.2.1. Light Sources

Nonresonant or "normal" Raman spectroscopy requires a narrow-bandwidth light source of relatively high average power, since normal Raman scattering is very weak. Furthermore, in order to obtain good spectral resolution the scattered light must be imaged through a relatively narrow spectrograph entrance slit, requiring that the incident light be focused to a small spot size at the sample. The most commonly used light sources for normal Raman are continuous wave (CW) blue-green argon ion lasers, although there has been considerable recent interest in the use of far-red or near IR excitation from krypton ion, titanium–sapphire, or Nd:YAG lasers, often combined with Fourier transform detection, to circumvent problems with interfering fluorescence (see Section 9.2.9) (11–13). For resonance Raman spectroscopy the requirements are somewhat different. Most obviously, the excitation frequency must be resonant with the molecular electronic transition of interest, which often lies in the blue or ultraviolet (UV) region. Because of the resonance enhancement of the scattering, high average powers are no longer as necessary, and generally cannot be used even if available because the sample will undergo multiphoton absorption, photochemistry, local heating, or other undesirable processes.

The intensity of spontaneous Raman scattering is linear with the incident laser intensity, whereas some of the competing processes have a higher than linear dependence. Additionally, problems with heating and photochemistry can often be mitigated by physically moving the sample (see Section 9.2.4) so that the same part of the sample is not continuously illuminated. Therefore the ideal light source for almost any ground-state resonance Raman experiment would be a CW one. For experiments requiring visible to near-IR excitation, ion, dye, and/or titanium–sapphire lasers fill the bill nicely. When UV excitation is required the situation becomes more difficult. Ion lasers have only a few near-UV lines, and extracavity frequency doubling of visible ion or dye lasers into the UV provides average powers that are too low (microwatts) to be generally useful (14). Accordingly, Raman excitation in the UV is most commonly obtained from pulsed excimer or Nd:YAG lasers and/or the dye lasers they pump, together with the nonlinear optical techniques of frequency doubling (15,16), mixing (17,18) and stimulated Raman shifting. The last is a particularly useful technique for shifting an incident pulsed laser output to shorter wavelengths; a Q-switched Nd:YAG-pumped dye laser combined with a Raman shifter allows simple, reliable, and reasonably inexpensive generation of continuously tunable radiation from the near-IR to somewhat below 200 nm (19–23). However, the low pulse repetition rate of such systems (typically 20–50 Hz) requires that the average power be kept very low in order to avoid saturation or nonlinear effects within a single pulse (24). There are several recent promising developments in UV Raman sources. Intracavity frequency doubled Ar^+ and Kr^+ ion lasers, which provide tens to hundreds of milliwatts of CW power at a number of discrete UV frequencies, are now being offered commercially. Alternatively, mode-locked and cavity-dumped picosecond lasers operating at high repetition rates (typically 1 MHz or so) have sufficiently high peak powers to allow reasonably efficient frequency doubling, tripling, and/or mixing, but low enough pulse energies that much higher average powers can be tolerated (25,26).

9.2.2. Sample Handling

Resonance Raman spectroscopy can be carried out on virtually any state of matter (gases, liquids, solids, supercritical fluids, surfaces and interfaces, etc.) over an enormous range of temperatures and pressures. While many of the considera-

tions involved in handling samples for resonance Raman are sample-specific and/
or are similar to those encountered in other optical spectroscopies, specific aspects
of the resonance Raman process do impose a few special constraints.

Normal nonresonant Raman spectroscopy on gases and liquids is usually carried
out in a 90° geometry [Fig. 9.2(a)], that is, the Raman emission is viewed propagating at right angles to the path of the laser through the sample. This configuration
can also be used on resonance, particularly for gases and for some solutions with
visible excitation. However, its utility for solutions is limited by the frequently high
optical densities of the samples used in resonance Raman, causing the incident
laser light to be absorbed within a very short path length and the Raman emission
trying to escape the sample to be reabsorbed severely. High optical densities are
often required in order to make the solute's Raman signal competitive with that of
the solvent, even when only the solute's scattering is resonance enhanced. This
problem is minimized in solvents that are themselves weak Raman scatterers, water
being a prime example; it becomes far more serious with UV excitation, where
Raman from all solvents is much stronger. Thus backscattering [Fig. 9.2(b)] is
often the optimal geometry for resonance Raman in solutions, as it is for any Raman
experiment in solids or other materials of poor optical quality. The backscattering
geometry also minimizes the distortion of the relative intensities of different Raman
lines due to reabsorption, as discussed in Section 9.2.6.

Unfortunately, the backscattering geometry also tends to enhance unwanted stray
laser light and Raman from any windows of the sample cell. Stray laser light can
be minimized by choosing the illumination and collection geometry such that the
direct reflection of the laser from the front of the sample misses the collecting
optics; however, in samples that scatter diffusely it may not be possible to avoid
collecting a large amount of laser light, and then the rest of the detection system
must have very good stray light rejection capabilities, as discussed further in Section 9.2.3. Interference from cell windows becomes particularly serious with UV
excitation due to the preresonance Raman enhancement of all common optical
materials. While in principle the window contribution can be subtracted from the
rest of the spectrum, this compromises the signal-to-noise (S/N) ratio and is feasible

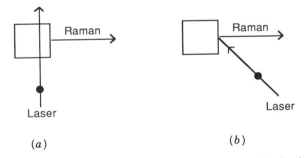

Figure 9.2. (a) The 90° scattering geometry. (b) A backscattering geometry. The dots indicate the usual
configuration with the laser polarization perpendicular to the plane of the paper.

only if the contribution is small. Since the Raman from the windows and from the solution come from physically different portions of the sample, it is often possible to greatly reduce the window contribution by carefully imaging only the light from the desired portion of the sample into the spectrograph. Alternatively, one can eliminate the sample cell altogether, and a variety of open-jet and guided-flow systems for achieving this with liquid samples have been described (17,20,27).

Normal Raman spectroscopy is generally performed on stationary samples. In resonance Raman it is usually necessary to move liquid or solid samples to avoid undesirable photochemistry or local heating at the laser powers needed to gather good Raman data. Liquid samples are usually flowed or spun past the excitation beam. Solid samples may be spun or translated. Moving a sample within a cryostat at low temperature presents special challenges.

9.2.3. Detection

Raman scattering, even on resonance, is a weak effect; the ratio of the Raman scattered power (integrated over all frequencies, directions, and polarizations) to the incident laser power rarely exceeds 10^{-6}. Detection of the signal therefore requires efficient collection of the emitted light, separation into its frequency components with very good rejection of the much more intense light at the laser frequency, and detection of the Raman photons with a high quantum efficiency and good linearity.

Raman scattering is emitted into all spatial directions, although usually not isotropically, and therefore an efficient collection system must gather light over a large solid angle (small f/number for collection). It should then focus the Raman light onto the entrance slit of the monochromator or spectrograph with a range of input angles that matches the f/number of the spectrograph, typically $f/4$ to $f/7$. This necessarily magnifies the image reaching the spectrograph, and the magnification of the optical system combined with the spectrograph slit width (which is dictated by the desired resolution of the Raman spectrum) determines the optimal size of the focused laser spot at the sample, although at a given laser power the focused spot often cannot be made this small due to photoalteration considerations (see Section 9.2.4). The collection optics may be either transmissive (glass or fused silica condenser lens) or reflective (usually an aluminum coated ellipsoidal, spherical, or parabolic mirror). Lenses tend to be less expensive and more resistant to damage, and generate less severe spatial aberrations for off-axis rays, but even "achromats" have some chromatic aberration (dependence of focal length on wavelength). This causes different scattered frequencies to be collected with different efficiencies, which can severely distort, in a way that is not readily reproducible, the apparent relative intensities of lines widely separated in frequency. Therefore reflective optics, which have a small and readily calibrated wavelength dependence of their reflectivity, are preferred when accurate intensity information is required, particularly in the UV where chromatic aberrations of lenses are most severe (21).

Dispersion of the emitted light into its frequency components is typically achieved by using a grating monochromator or spectrograph. Traditional Raman systems formerly consisted of scanning double monochromators with photomultiplier detectors. While such systems provide high spectral resolution and very good stray light rejection, they are not very efficient due to the reflection losses at a minimum of six mirrors and two diffraction gratings, and the need to scan one emitted wavelength interval at a time through the monochromator exit slit to a single detector. Most modern Raman systems employ a spectrograph to disperse a large portion of the spectrum at once onto a multichannel detector. For highly scattering samples such as solids, or when Raman lines at very low frequencies (close to the laser line) are of interest, a triple spectrograph is generally used, providing excellent stray light rejection at the expense of poor light throughput. A single spectrograph can provide very good throughput and often has adequate stray light rejection for samples of reasonably good optical quality if care is taken to avoid imaging the scattered laser light into the spectrograph. A variety of both spectral (solution filters, long-pass color filters, atomic and molecular vapor absorption filters, and interference filters) (12,13,28) and spatial filtering approaches have also been described to improve the stray light rejection of single spectrographs without the cost and severely reduced throughput of a multiple spectrograph. Spectral filters can be very efficient at blocking the laser frequency, but often also attenuate the low-frequency part of the Raman spectrum, and are generally tunable either not at all or over a very limited range of laser wavelengths.

Traditional single-channel detectors, usually cooled photomultiplier tubes with photon counting electronics, are still quite useful with CW light sources, particularly when very high spectral resolution is needed and/or only a small region of the Raman spectrum is of interest. However, multichannel detectors have become fairly standard in modern Raman instrumentation because of the great reduction in data accumulation times (or, equivalently, the ability to obtain spectra with much lower incident laser powers or from much weaker Raman scatterers) afforded by detection in many frequency channels simultaneously. The common multichannel detectors in current use are intensified diode arrays (IDAs) (29) and charge-coupled devices (CCDs) (13,30–32).

An IDA consists of a photocathode similar to that on a photomultiplier, which converts incident photons to photoelectrons, followed by a microchannel plate intensifier that multiplies the photoelectrons, a phosphor that converts the electrons back to photons, and a linear array of photodiodes that convert the photons to stored charge. The array usually consists of either 512 or 1024 pixels, each 25 μ wide. Charge can be integrated on the detector for up to several minutes before being read out, minimizing noise in the digitizing electronics. Thermoelectric cooling of the array renders the dark noise of the diode array essentially negligible compared with that from the photocathode. When used with low repetition rate pulsed lasers the intensifier can be gated with a gate width of tens to hundreds of nanoseconds, allowing a large reduction in the photocathode dark noise as well as rejection of some forms of stray light. The IDAs provide performance similar to that of a linear array of photomultipliers: comparable quantum efficiencies, response throughout the visible and UV (and vacuum-UV, with appropriate window materials), reasonably good linearity and dynamic range, gatability, ease of damage by excessive light levels (accidentally setting the spectrograph such that the laser line falls on the detector can destroy the intensifier), and significant dark noise that can be greatly reduced by gating, or, if only the UV region is of interest, choosing a nearly solar blind photocathode. Spectral resolution is also fairly poor with IDAs due to the approximately three channels of crosstalk within the intensifier.

The more recently introduced CCDs have several advantages, and a few disadvantages, compared with IDAs. They are two-dimensional detectors useful in various imaging applications, and even if only one-dimensional spectral information is needed, the ability to observe the distribution of light in the direction perpendicular to the wavelength axis (i.e., along the height of the spectrograph slit) and to select the channels that will be digitized is often helpful in excluding spurious light without losing real signal. The number of pixels in each dimension varies from a few hundred to over 1000, with pixel sizes of 20–30 μ. If unintensified and cooled to near liquid nitrogen temperature, CCDs have virtually no true dark noise, although they also cannot be gated; an intensifier can be added if gating is required, but then the dark noise and spectral response become comparable to that of an IDA. Unintensified CCDs have very high peak quantum efficiencies, from 40% to as high as 75%, and are far more responsive in the red- and near-IR than IDAs or photomultipliers, making them the clear detectors of choice in this region. Ultraviolet response is generally mediocre and achieved by coating with a UV scintillator, although this is changing with improved technologies for fabricating thinned, backilluminated chips. Unintensified CCDs also provide very good resolution (roughly one pixel), and are quite resistant to damage by high light levels. The main disadvantage is the sensitivity of such devices to cosmic rays, which impinge on the chip in random positions to generate very large peaks in a single or a few contiguous pixels. The rate varies widely with geographical location and possibly other factors, but typically averages several per minute. While these features can be removed in software, they destroy any real spectral information present in those pixels, and greatly limit the length of time during which signal can be accumulated on the otherwise very low-noise detector.

The development of new types, and improvement of existing types, of solid-state multichannel detectors for spectroscopic applications continues to move forward quite rapidly. Particular emphasis is now being placed on the near-IR region of the spectrum (11,33).

9.2.4. Photoalteration

In resonance Raman spectroscopy the incident light is necessarily absorbed by the sample, and since resonance Raman quantum yields are low, typically 10^6 or more molecules will be photoexcited for each one that scatters a Raman photon. The generation of excited states can have several undesirable effects on an experiment whose goal is to measure the resonance Raman spectrum of the ground state. If the excited state itself is long lived, or if it forms a photoproduct, triplet, or other long-lived state, the concentration of ground-state molecules will be reduced from its nominal value. This ground-state depletion or saturation is a problem if one wishes to measure absolute intensities. If the excited state or photoproduct has any resonance enhancement at the incident laser frequency it will also contribute to the resonance Raman spectrum, causing interferences even if one wants only the spectrum and not absolute intensities. Finally, even if photoexcitation is followed by very fast ground-state recovery with negligible photochemistry or long-lived transient formation, the resulting local heating can cause the temperature to differ from the nominal sample temperature. Here we use the term "photoalteration" to refer to any light-induced change in the sample, whether temporary or permanent.

Photoalteration can be mitigated by limiting the incident laser power and moving the sample such that the majority of molecules in the illuminated volume at any given time are in the ground state. In order to quantify how photoalteration depends on the experimental parameters, we distinguish between CW excitation, in which the sample is flowed, spun, or otherwise moved through a continuous excitation beam, and pulsed excitation, in which the sample is assumed to be effectively stationary for the duration of a single laser pulse, but the illuminated volume is completely replaced between pulses. With CW excitation, the fraction F of molecules that undergo photoconversion during a single pass through the beam is given by

$$F_{CW} = 2303 \; P \; \varepsilon \; \phi/[N_A \; \pi^{1/2} \; r \; v] \tag{9.1}$$

for $F \ll 1$ (34), where P is the photon flux (photons s^{-1}), ε is the molar extinction coefficient (M^{-1} cm^{-1}), ϕ is the quantum yield for the photochemical or photophysical process(es) that must be considered as discussed below, N_A is Avogadro's number, r is the radius of the focused laser beam (cm), and v is the linear velocity (cm s^{-1}) at which the sample moves through the beam. For pulsed excitation, the corresponding fraction of the molecules in the illuminated volume that are photoconverted during a single pulse is given by

$$F_{pulse} = 2303 \; E \; \varepsilon \; \phi/[\pi \; r^2 \; N_A] \tag{9.2}$$

where E is the pulse energy (photons). Both of these expressions neglect the attenuation of the incident light by the sample, so give upper limits for F.

How one uses these calculated photoalteration values to select the experimental parameters of focused spot size and/or average power depends on the photophysics of the system. Many of the molecules on which resonance Raman spectroscopy is performed undergo rapid excited-state deactivation on a time scale much shorter than the pulse duration or the transit time through a CW beam, forming one or more photoproducts with quantum yield ϕ and repopulating the ground state with quantum yield $(1 - \phi)$. In this case, to avoid significant ground-state depletion it is necessary to keep F small; $F < 0.1$ will insure that, averaged over the pulse duration or transit time through the beam, at least 95% of the molecules will be in their ground states. If on the other hand the ground-state repopulation is slow, the "quantum yield" should be considered as unity for the purpose of estimating saturation effects, even if eventually nearly all photoexcited molecules return to the ground state.

In most experimental configurations the sample is recirculated. If light absorption leads to the formation of photoproducts that are long lived relative to the time between successive illuminations of the same part of the sample, it is also necessary to consider "bulk" photoalteration, the fraction of the entire sample that has been photoconverted. Assuming the sample has a sufficiently high optical density that nearly all of the incident photons are absorbed, the bulk photoalteration for either pulsed or CW excitation is given by

$$F_{\text{bulk}} = P \, \phi \, t/N \tag{9.3}$$

where P is the photon flux (photons s^{-1}), ϕ is the quantum yield, t is the irradiation time, and N is the total number of molecules in the recirculated sample.

Bulk photoalteration can be calculated quite accurately as long as the relevant quantum yields are known, and in liquid or vapor samples it is usually easy to monitor empirically by taking optical absorption spectra of the sample in the reservoir at various time intervals. Single-pass photoalteration tends to be more difficult to calculate accurately. The best way to insure that ground-state depletion during a single pulse is not a problem is to examine the laser power dependence of the Raman intensities of the solute relative to solvent, as the latter, being nonresonant, will experience no saturation.

9.2.5. Intensity Corrections

The relative intensities of Raman lines at different scattered frequencies as directly measured by the detector may not be proportional to the true intensities for several reasons. First, different scattered frequencies generally do not escape the sample with equal efficiencies due to the wavelength dependence of the sample's absorbance (self-absorption). Several approaches exist for handling this problem. Sometimes it is feasible to make the sample concentration low enough, and/or the path length for the emerging light short enough, that reabsorption is negligible. If the excitation is on the red edge of the absorption band, reabsorption of the Stokes-shifted wavelengths may be negligible even if the incident laser is significantly absorbed. Correction for self-absorption is not trivial because even if the sample's absorption spectrum is known accurately, most excitation and collection geometries involve a complicated and often ill-defined range of path lengths for the emitted light (35). A useful empirical approach is to recognize that in the absorbing solution the Raman light scattered by the solvent follows the same optical path and undergoes the same reabsorption as the solute scattering, but in the pure solvent there is generally no self-absorption. Comparison of the relative solvent line intensities in pure solvent and in solution can then be used either to develop a completely empirical function describing the reabsorption correction versus scattered wavelength that can be applied to the solute lines, or to calibrate the effective path length for calculation of the reabsorption correction in an assumed scattering geometry (36). When the optical density must be high, use of a backscattering geometry will simultaneously maximize the signal and minimize the reabsorption correction, since that geometry assures that the incident light intensity is greatest where the path length for the emitted light is shortest. Several equations for calculating the reabsorption correction in backscattering geometries, which make different assumptions about the light penetration depth and the incident beam geometry, have been presented (37–40).

Second, if refractive optics are used to collect and refocus the scattering, different wavelengths may not be transmitted through the spectrograph's entrance slit with equal efficiencies due to chromatic aberrations. This can be prevented by using reflective optics as discussed in Section 9.2.3.

Third, Raman lines having different depolarization ratios may not be detected with equal efficiency due to the polarization dependence of the spectrograph throughput (mostly of the diffraction gratings). This is normally eliminated by passing the scattered light through a polarization scrambler placed before the spectrograph entrance slit (see Section 9.2.8).

Finally, the spectrograph throughput and detector quantum efficiency may have intrinsic variations with wavelength and/or, for multichannel systems, with position on the detector (41). The usual approach to correcting for these effects is to measure the spectrum of a calibrated, broadband standard lamp, set up so as to illuminate the spectrograph entrance slit in as nearly as possible the same geometry as the Raman samples, and then to use the ratio between the known lamp spectrum and the measured apparent one to construct a correction curve that is applied to the Raman spectra measured over the same wavelength region (42). The most commonly used standard lamps are tungsten–halogen from the near-IR to about 250 nm, and high-pressure deuterium from 200–400 nm. The difficult aspect of these measurements is that the sensitivity of a multichannel detector may depend not only on the wavelength but also on the vertical position of the array (i.e., the direction perpendicular to the wavelength dispersion axis). Additionally, the throughput of the spectrograph usually varies across the vertical dimension. Thus the apparent detection sensitivity function can depend strongly on the precise geometry of the light source. It is very difficult to make emission from a lamp, with its comparatively large filament, resemble that from a focused laser spot. To avoid imaging the filament onto the detector it is best to scatter the lamp output from a spectrally flat, diffusely reflecting material, usually a plate coated with $BaSO_4$ placed at the same position as the Raman sample, and when accurate intensities are needed the spectrograph entrance slit height should be made small enough that both the lamp emission and the sample's Raman emission illuminate the slit uniformly.

Normally, Raman spectra are converted to an x axis having units of Raman shift in wavenumbers. This unit conversion must be taken into account when calibrating the spectral sensitivity of the detection system. The theoretical definition of the Raman cross section for a particular vibrational transition involves the intensity integrated over the band profile in units of scattered frequency, so what one really wants is intensity per unit frequency, $I_{Raman,true}(\nu)$. However, grating spectrometers or spectrographs disperse light linearly in wavelength, so what one measures directly is intensity per unit wavelength, $I_{\text{Raman,meas}}(\lambda)$. Since $I(\lambda) = (\partial I/\partial \lambda) = (\partial I/\partial v)(\partial v/\partial \lambda) \propto (\partial I/\partial v)/\lambda^2 = I(v)/\lambda^2$, the desired quantity is related to the measured one by $I_{\text{Raman,true}}(v) = \lambda^2 I_{\text{Raman,meas}}(\lambda)$. The output of calibrated lamps is usually given in units of power or intensity per unit wavelength, $I_{\text{Lamp,true}}(\lambda)$. Therefore, the desired quantity is related to the directly measured Raman spectrum and the measured and calibrated lamp spectra by

$$I_{\text{Raman,true}}(v) = \lambda^2 I_{\text{Raman,meas}}(\lambda)\{I_{\text{Lamp,true}}(\lambda)/I_{\text{Lamp,meas}}(\lambda)\} \tag{9.4}$$

The λ^2 correction to the intensities may be made at the time the x axis is converted from wavelength to wavenumber, or else it may be defined as part of the detector sensitivity correction; in any case, it must be made if accurate intensities are desired.

9.2.6. Integrated Band Intensities

Once the resonance Raman spectra have been converted to a wavenumber axis and the intensities corrected as described above, it is necessary to determine the integrated intensities of the Raman bands of both the sample of interest and the internal or external standard (see Section 9.2.7). If the bands are well separated from each other, it is adequate to subtract whatever background underlies the Raman peaks and then numerically integrate the area under the band. If the bands do overlap significantly, some type of curve fitting must be employed to separate the contributions of different transitions. A wide variety of algorithms have been described in the literature, and most commercial software packages that accompany Raman detectors contain curve-fitting routines. When integrated areas are the only quantities of interest, one can simply choose the band shape (usually Lorentzian, Gaussian, or a Gaussian/Lorentzian convolution, the Voigt lineshape) that gives a qualitative best fit, and fit the intensity corrected data directly. When the band shapes and/or widths are also important, it is necessary to account for the finite resolution of the detection system, either by deconvolving an experimentally determined instrument function from the raw data or by convolving the instrument function with the fitting function before comparing with the data (43,44).

The usual approach to measuring resonance Raman excitation profiles is to record complete Raman spectra at a number of fixed excitation frequencies, integrate all of the Raman bands, and plot the intensity of each band as a function of excitation frequency. It is also possible to directly measure the profile for one Raman band at a time by synchronously scanning both the laser excitation frequency and the detection frequency, using a monochromator with a slit width chosen to transmit the entire Raman band of interest while excluding all other Raman bands (45). This is feasible only if the Raman lines are well separated in frequency and there is negligible fluorescence or other background, such that the detected signal is a sum of the Raman intensity and a constant, subtractable dark signal. There is always some danger of interference from spurious light sources when all the emission within some preselected bandwidth is attributed to a particular Raman transition, but for well-behaved systems this approach is sound and greatly streamlines the measurement of high-resolution excitation profiles.

9.2.7. Absolute Cross Sections

Determination of the relative intensities of different bands in a resonance Raman spectrum as described above is, while not trivial, often fairly straightforward. In contrast, direct measurement of *absolute* intensities, particularly in absorbing samples, is an extremely difficult task. The total absolute cross section for a particular Raman transition at excitation frequency ω_L is defined as

$$\int d\omega_s \, d\Omega [\partial P_{scatt}(\omega_s, \Omega)/\partial\Omega \, \partial\omega_s] = N \, I_{laser}(\omega_L) \, \sigma_R(\omega_L) \tag{9.5}$$

where $[\partial P_{scatt}(\omega_s, \Omega)/\partial\Omega \, \partial\omega_s]$ is the differential Raman scattered power (photons s^{-1} per steradian per unit frequency) at emitted frequency ω_s, the integration extends over all emitted light directions and all frequencies spanning the Raman band, $I_{laser}(\omega_L)$ is the incident intensity (photons area^{-1} s^{-1}), N is the number of molecules in the illuminated volume, and $\sigma_R(\omega_L)$ is the total cross section (area molecule^{-1}) for the Raman transition of interest. The difficulty in making this measurement stems both from the enormous difference between the laser and scattered powers, requiring linearity in power measurement over a huge dynamic range, and from the need to determine the collection and detection efficiency accurately for light that is emitted into all directions, but usually not isotropically. Absorption of the incident and/or scattered light by the sample further complicates the situation. For these reasons, measurements of absolute *resonance* Raman cross sections are usually carried out by ratioing the Raman intensities of the resonant molecules to lines of either the solvent or an added, nonresonant scatterer, usually an ''inert'' gas such as methane or nitrogen in gas-phase samples or a simple ion such as sulfate or nitrate in aqueous samples. The absolute cross section of the nonresonant standard is then either measured directly or determined relative to some other primary cross-section standard as an internal or external reference.

Absolute cross sections for molecules suitable as intensity standards have been measured through a variety of methods (46–56). While the early determinations of absolute cross sections carried rather large uncertainties and different methods often produced widely different values, more recently determined values tend to be more reliable, although there are still few cross sections that can realistically be considered accurate to better than perhaps ±15%. Some of the most recently determined, and presumably most accurate, absolute cross sections reported are those for liquid cyclohexane, acetonitrile, and water from 266 to 200 nm (56), for liquid benzene from 514.5 to 325 nm (53), for liquid benzene from 647 to 351 nm and for liquid cyclohexane and aqueous cacodylate from 647 to 240 nm (52), for aqueous solutions of acetonitrile, SO_4^{2-}, ClO_4^-, and NO_3^- from 640 to 220 nm (51), for aqueous cacodylic acid and sodium selenate from 514.5 to 218 nm (57), and for H_2, D_2, CH_4, N_2, and O_2 gases from 600 to 200 nm (50).

The use of a directly determined internal intensity standard to obtain absolute cross sections for a resonant molecule of interest is fairly straightforward once the appropriate intensity corrections have been applied to insure that the relative line intensities of the solute and the internal standard have been measured properly. An additional consideration is that the absolute cross section may be reported as either a total cross section, σ_R as defined in Eq. (9.5), or a differential cross section $(d\sigma_R/d\Omega)$, the cross section per steradian of solid angle, which also depends on the direction in which the scattering is detected relative to the polarization direction of the incident light. If we define $(d\sigma_R/d\Omega)_{90}$ as the cross section per unit steradian detected at right angles to the laser polarization [Fig. 9.2(a) or (b)], then the two quantities are related by

$$\sigma_R = \int d\Omega \ (d\sigma_R/d\Omega) = \frac{8\pi}{3} \frac{1 + 2\rho}{1 + \rho} \left(\frac{d\sigma_R}{d\Omega}\right)_{90} \tag{9.6}$$

where ρ is the Raman polarization ratio, defined as

$$\rho = I_\perp/I_\parallel \tag{9.7}$$

with $I_\perp$ and $I_\parallel$ being the intensities of Raman scattering having polarizations perpendicular and parallel, respectively, to that of the laser. In most experimental geometries the scattered light is collected over a small range of solid angles around 90°, so the measured intensities are proportional to $(d\sigma_R/d\Omega)_{90}$. The absolute differential cross section for a Raman band of the resonant molecule is then related to the differential cross section for the intensity standard by

$$\left(\frac{d\sigma_R}{d\Omega}\right)_{90,\text{solute}} = \frac{I_\text{solute}}{I_\text{standard}} \frac{[\text{standard}]}{[\text{solute}]} \left(\frac{d\sigma_R}{d\omega}\right)_{90,\text{standard}} \tag{9.8}$$

where I_solute and I_standard are the integrated band intensities and [standard] and [solute] are the molar concentrations. The total cross section may then be obtained from Eq. (9.6) assuming the solute depolarization ratio is known. If, on the other hand, the standard's absolute Raman intensity is reported as a total cross section (the quantity that is naturally measured when an integrating cavity technique is used) (52,56), the depolarization ratios of both solute and standard must be taken into account in converting from measured relative intensities to absolute cross section:

$$\sigma_{R,\text{solute}} = \left(\frac{1 + 2\rho}{1 + \rho}\right)_\text{solute} \left(\frac{1 + \rho}{1 + 2\rho}\right)_\text{standard} \frac{[\text{standard}]}{[\text{solute}]} \left(\frac{I_\text{solute}}{I_\text{standard}}\right) \sigma_{R,\text{standard}} \tag{9.9}$$

Equations (9.6), (9.8), and (9.9) assume that there is no polarization selection in the detection step and that both polarizations of scattered light are detected with equal efficiencies (except in the experiment that measures the depolarization ratio).

9.2.8. Depolarization Ratios

The depolarization ratio ρ in Raman scattering is defined in Eq. (9.7). (Note that in fluorescence spectroscopy the standard convention uses a different definition of polarization, the anisotropy r defined as $r = [(I_\parallel - I_\perp)/(I_\parallel + 2I_\perp).]$ The depolarization ratio is needed in order to convert from differential to total cross sections as described above, and for nonresonant scattering the value of ρ can be quite useful in assigning the symmetries of vibrations. On resonance, the main reason for measuring ρ is to provide information on the nature of the electronic state(s) responsible for the resonance enhancement (7,58–62). If the intensity derives solely from a single, nondegenerate electronic state (or from more than one state having the same transition moment directions in the molecular frame), the sample is initially isotropic, and the electronic dephasing is fast compared to molecular rotation (nearly always true in liquids near room temperature), then the depolarization ratio should be $\frac{1}{3}$ for all bands. If a single, doubly degenerate electronic state contributes to the resonant enhancement, ρ should be $\frac{1}{8}$ (63). Values other than these demonstrate that more than one electronic transition contributes to the intensity. These values do not necessarily apply to gases, where the electronic dephasing is often not fast compared with molecular rotation; in fact, resonance Raman depolarization ratios in gases can be used to determine electronic dephasing times by taking advantage of the molecular rotational time as an internal "clock" (64–66).

Depolarization ratios are most commonly measured by placing a rotatable polarization analyzer in the path of the scattered light and alternately detecting parallel and perpendicular polarizations as shown in Fig. 9.3(a). For the visible and near-UV spectral regions, dichroic sheet polarizers, which preferentially absorb one po-

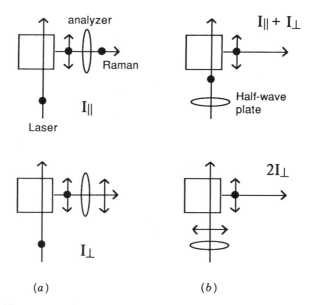

Figure 9.3. (a) Measurement of Raman depolarization ratios in the usual way, with a polarization analyzer in the scattered beam. (b) Measurement of depolarization ratios by rotating the laser polarization. Dots and double-headed arrows represent laser polarization perpendicular and parallel, respectively, to the plane of the paper.

larization while transmitting the other, work reasonably well. Sheet polarizers have the advantage of providing a relatively large acceptance angle (i.e., the scattered light need not be collimated) and are fairly inexpensive and easy to align. However, such polarizers have the disadvantage of absorbing a fairly large fraction (typically one-half or more) of even the desired polarization, and both the transmittance and the extinction ratio (ratio of transmittances of the "wrong" and "right" polarizations) become very poor in the deep UV. Prism-type laser polarizers (e.g., calcite Glan or crystal quartz Rochon) (67) may be used farther into the UV, but these have a comparatively small aperture and very small acceptance angle. Stacked-plate analyzers are probably the most useful alternative for deep-UV work (68). Depolarization ratios in the far-UV are often measured by the alternative method of eliminating the polarization analyzer altogether and instead detecting all polarizations while switching the polarization of the laser, as shown in Fig. 9.3(b) (65,69). The switching may be achieved by using either a standard half-wave plate, if only one or a few excitation wavelengths are to be used, or a Soleil–Babinet compensator, which is expensive but can be adjusted to act as a half-wave plate at any wavelength. The total intensities detected with the two different laser polarizations may then be related to the depolarization ratio defined in Eq. (9.7) as described in Ref. 65. Whichever method is used, it is important to eliminate the natural polarization bias of the spectrograph throughput such that all polarizations are detected with equal efficiencies. This is usually accomplished by placing a polarization scrambler before the spectrograph slit. The scrambler is simply a crystal quartz wedge cut so that rays passing through different physical positions have their polarization rotated by different amounts, resulting in effectively depolarized transmission.

Highly accurate depolarization ratio measurements are fairly difficult and require a number of geometric corrections (70). It should also be noted that the bandshapes of the parallel and perpendicularly polarized components are often quite different, so comparison of integrated intensities, not peak heights, is essential for depolarization ratio determinations (68).

9.2.9. Fluorescence

Interference due to fluorescence is one of the most serious limitations of resonance Raman as a spectroscopic technique, and much of the recent interest in far-red and near-IR excited Raman spectroscopy stems from the desire to avoid electronic excitation of fluorescent chromophores. That is not a useful approach, however, if one is after the excited-state information that can be obtained only from experiments on resonance. As discussed in more detail below, the ratio of the total resonance Raman intensity to the fluorescence intensity for rigorous resonance excitation is approximately the ratio of the excited-state lifetime to the electronic dephasing time. Since electronic dephasing times for molecules in liquids near room temperature are subpicosecond while excited-state lifetimes are more often in the nanosecond range, "typical" molecules have fluorescence emission that is many orders of magnitude stronger than the Raman. Another way of stating the same thing is that resonance Raman quantum yields rarely exceed 10^{-6}, so even "weak" fluorescence with a quantum yield of 10^{-3} or 10^{-4} is very strong compared with the Raman. While the fluorescence is usually broad enough that it can easily be distinguished from the Raman, if the Raman and fluorescence emissions occur in the same spectral region, the weak Raman peaks can be swamped by the photon shot noise from the strong underlying fluorescence background. Even if the S/N ratio is good, Raman peaks appearing as modulation of a few percent or less on top of a broad fluorescence background are very difficult to extract reliably from the channel-to-channel sensitivity fluctuations of most multichannel detectors.

If the excited state of interest is sufficiently short lived, whether because of fast photochemistry, intersystem crossing, or internal conversion, its fluorescence will be fairly weak. For example, higher excited singlet states usually relax quickly to the lowest excited singlet (Kasha's rule), and even if S_1 is fluorescent, its emission is usually Stokes-shifted enough not to interfere severely with the Raman spectrum on resonance with higher singlet states. This is the basis of recent interest in far-UV excited Raman as a purely analytical technique (71). It is also sometimes possible to obtain reasonably good Raman spectra on resonance with fluorescent transitions by exciting on the high-energy side of a broad absorption band such that most of the Raman spectrum emerges at shorter wavelengths than the Stokes-shifted emission (72). A wide variety of approaches based on the different temporal characteristics of the two types of emission have been proposed to allow detection of Raman signals in the presence of strong fluorescence backgrounds (73–75). While many such methods have demonstrated substantial discrimination against fluorescence, the S/N ratio of the remaining Raman spectrum is rarely impressive, and none of these methods have gained widespread use. A technique based on dithering the excitation frequency has been shown to be quite effective in cancelling out the channel-to-channel sensitivity fluctuations that hamper multichannel detection of Raman peaks atop large backgrounds (76), but can do nothing to eliminate the photon shot noise.

At present, resonance Raman spectroscopy remains a technique that is really useful only for electronic states that are not strongly fluorescent. This requirement is not as limiting as it might seem, since the types of excited-state information that resonance Raman can deliver are most interesting for molecules that undergo some fast dynamic process in competition with fluorescence. For longer lived excited states the desired structural information can often be obtained by other methods such as low-temperature, high-resolution absorption and/or fluorescence spectroscopy or direct time-resolved vibrational spectroscopy of the excited state.

9.3.   THEORY AND INTERPRETIVE METHODS

9.3.1.   The Raman versus Fluorescence Question

Every discussion of resonance Raman scattering must, at some point, come to grips with the question of how resonance Raman differs from fluorescence. The answer is not as straightforward as it might seem, for reasons that are partly physical and partly semantic. The same terms are used in different contexts to describe different physical processes, and the experimental distinction between Raman and fluorescence varies depending on the conditions of the experiment. From a theoretical point of view, "Raman scattering" usually refers to a true two-photon process in which the molecular wave function evolves with no change of energy or phase during the time interval between destruction of the incident photon and creation of the scattered photon, with the consequence that paths proceeding through different intermediate states interfere with each other at the level of the quantum mechanical amplitude (Raman polarizability), not the probability (Raman cross section). A process that can be decomposed into separate absorption and emission steps, with no phase memory connecting the two, is usually referred to as fluorescence. Experimentally, Raman scattering is always characterized as emission that appears at a fixed Raman shift ($\omega_L - \omega_S$) as the laser frequency ω_L is scanned, but its ease of separation from the fluorescence component depends strongly on the phase and temperature of the sample. In condensed phases at elevated temperatures, where the absorption spectra are fairly broad, the fluorescence is almost always much broader than the Raman scattering and its bandshape and intensity vary only slowly as the laser frequency is tuned, so there is rarely any difficulty in distinguishing clearly between them; any fluorescence appears as a broad background on which the sharp Raman lines are superimposed (38,77–79), as shown in Fig. 9.4. The distinction is less obvious for molecules in the gas phase or at very low temperatures, where the Raman and fluorescence linewidths become comparable (80). The Raman lines can even be broader than the fluorescence if the laser bandwidth is significant (81).

The traditional starting point for discussions of resonance Raman intensities is the Kramers–Heisenberg–Dirac (KHD) dispersion expression for the Raman polarizability. On resonance with a single nondegenerate electronic state, if the Born–Oppenheimer and Condon approximations hold [negligible coordinate dependence of the electronic transition moment (see below)], the total cross section for a vibrational Raman transition from state $|i>$ to state $|f>$ can be written as (82)

$$\sigma_R(\omega_L) = \frac{8\pi|\mu_{ge}|^4\omega_L\omega_S^3}{9\hbar^2c^4} F_{if}(\omega_L) \tag{9.10}$$

where μ_{ge} is the electronic transition dipole moment, and $F_{if}(\omega_L)$ is given by the KHD formula,

$$F_{if}(\omega_L) = \left| \sum_w \frac{<f|w><w|i>}{i(\omega_L - \omega_{wi}) - \gamma_{wi}} \right|^2 \tag{9.11}$$

where $|w>$ represents a vibrational level of the excited electronic state and γ_{wi} is the lifetime-limited line width of the $|i> \rightarrow |w>$ transition. This expression, derived from second-order time-dependent perturbation theory for a molecule initially in eigenstate $|i>$ interacting with monochromatic radiation of frequency ω_L, describes all emission from an isolated molecule; that is, a monochromatically excited, iso-

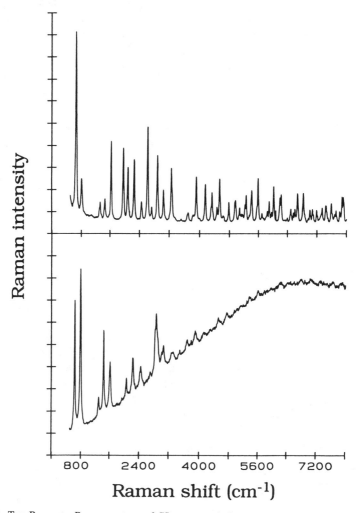

Figure 9.4. Top: Resonance Raman spectrum of CS_2 vapor, excitation wavelength 199.7 nm. Bottom: Resonance emission (Raman and fluorescence) spectrum of CS_2 in hexadecane, excitation wavelength 220.3 nm.

lated chromophore exhibits only Raman scattering. "Fluorescence" can arise in real experiments for two reasons: either the chromophore is not isolated and undergoes interactions with a bath (material pure dephasing), or the incident radiation is not monochromatic (light-induced dephasing).

If the chromophore that undergoes the Raman transition is interacting with some other material system (e.g., the solvent in a liquid or a collision partner in a vapor) whose eigenstates are not part of the $\{|i>, |f>, |w>\}$ basis set of Eq. (9.11), then the states of the chromophore cannot properly be described by a wave function; a density matrix approach must be used to allow averaging over the "uninteresting" bath degrees of freedom. In principle one could simply expand the basis set to include the bath coordinates as well, but usually that is neither practical nor useful, since the bath typically consists of $10^{10}-10^{20}$ molecules weakly coupled to the chromophore. For condensed phase problems the coupling to the environment is often treated as a stochastic fluctuation in the energy levels of the chromophore (83). Usually it is assumed that the coupling to the electronic degrees of freedom is far stronger than to the vibrational ones (solvent shifts of electronic transitions are typically at least an order of magnitude larger than solvent shifts of purely vibrational transitions) so all vibrational levels of the excited electronic state fluctuate together relative to those of the ground electronic state. The energy fluctuations are considered to have a Gaussian distribution and to be described fully by the width of the distribution D and the inverse time scale of the fluctuations Λ. This time scale determines the partitioning of the emission between Raman and fluorescence components (84–88). As the time scale becomes infinitely slow, the limit of "inhomogeneous" broadening is reached. Each molecule's energy levels become static on the time scale of the Raman process (properly determined by the dephasing time for the final vibrational level in the Raman process), and the emission is the same as it is for isolated molecules (Raman scattering only), except that each molecule has a different detuning frequency $\omega_L - \omega_{wi}$. In the opposite limit, as the time scale becomes very fast ($\Lambda >> D$), the "fast modulation" or "motionally narrowed" regime is reached. The absorption lineshape again becomes Lorentzian, but the line width of $\Gamma = D^2/\Lambda$ is dominated by the solvent-induced frequency fluctuations rather than the excited-state lifetime. In addition, the emission spectrum now consists of two distinguishable components: Raman lines centered at $\omega_L - \omega_S = \omega_{fi}$ having line widths determined by the ground-state vibrational dephasing, and fluorescence lines centered at $\omega_S = \omega_{wf}$ having line widths of Γ. When the time scale is intermediate between these two limits, the mathematics becomes considerably more difficult, but the result is that there are now not just two but three distinguishable kinds of emission: Raman, fluorescence, and a component often called "broad Raman," which is centered at Raman resonances but has a lineshape more like that of the fluorescence. Mukamel and co-workers (87) showed that this component should experimentally be associated with the fluorescence (i.e., much broader than the Raman) as long as Λ is large compared to the inverse line width of the Raman lines, and argued that this criterion should usually be met for molecules in solution phases. [These comments apply equally to the physically more realistic Brownian oscillator model for electronic dephasing (89).]

The second issue concerns the line width of the incident radiation. If it is not monochromatic, one can show, through a similar density matrix treatment that considers stochastic fluctuations in the frequency of the driving radiation field, that even in the absence of any interactions of the material system with a bath the emission should again consist of both Raman-like and fluorescence-like components (90). Fluctuations of molecular energy levels relative to a fixed laser frequency are physically similar, though not identical, to fluctuations in the laser frequency relative to fixed energy levels of the chromophore. If the bandwidth of the radiation is large compared with the intrinsic line widths of the molecular transitions, nearly all of the emission will have the spectral characteristics of fluorescence (81), explaining why high-resolution emission spectra of long-lived gas-phase molecules are usually considered as fluorescence rather than Raman spectra. If, on the other hand, the electronic absorption spectra are broad due to either a short excited-state lifetime or rapid pure dephasing, a modest (few wavenumbers) laser bandwidth should be effectively monochromatic, although there has been some recent controversy over this point (91).

9.3.2. Determination of Excited-State Potential Energy Surfaces and Dynamics: General Considerations

The incorporation of experimental resonance Raman intensity data in studies of excited-state potential energy surfaces may be approached from two different directions. One is to attempt to calculate, usually through ab initio or semiempirical molecular orbital methods, the excited-state potential energy surface, generate the resulting resonance Raman spectra, and compare them with experiment. The accuracy with which the experimental spectra are predicted may then be used as a basis for comparing different theoretical potential surfaces. The other is to assume no a priori knowledge of the excited-state potential, and treat it as an unknown to be determined by the data. Unfortunately, there is no generally practical method for directly ''inverting'' the experimental data to give the potential [although progress in this direction is being made by Kinsey, Levine, and co-workers (92,93)], so this process involves parameterizing a model for the surface, calculating the resulting spectra, and then iteratively adjusting the parameters and recalculating the spectra until a best fit is found. While data of high quality and adequate quantity often allow the parameters of a particular model to be defined fairly well, distinguishing between different models is much more difficult (94,95), and this is a problem that future work must address more thoroughly.

Equations (9.5), (9.10), and (9.11) are somewhat oversimplified. The relationship between the Raman scattered power and the incident intensity is expressed more completely as

$$[\partial P(\omega_S)/\partial \omega_S] = I(\omega_L)N \sum_i \mathbf{B}_i \sum_f \sigma_{R,if}(\omega_L)L_{if}(\omega_L - \omega_S) \qquad (9.12)$$

where P is the Raman-scattered power at frequency ω_S, I is the incident intensity at frequency ω_L, N is the number of molecules in the illuminated volume, B_i is the Boltzmann probability that the system is found in initial state $|i>$, and L_{if} is the normalized lineshape of the $|i> \rightarrow |f>$ transition. The signal detected at a particular frequency may have contributions from many different Raman transitions, which may not be experimentally separable due to nearly complete spectral overlap. For example, if anharmonicities are small all $|n> \rightarrow |n + 1>$ transitions in a given mode will be nearly degenerate, and if the temperature is high enough and/or the mode's frequency low enough that several initial states $|n>$ have significant Boltzmann factors, they will all contribute to the band intensity. Thermal population in a low-frequency mode must in general be considered even if the transitions to be modeled do not involve changes in that mode's quantum number. That is, the cross section for the $|0_1 0_2> \rightarrow |1_1 0_2>$ transition, where modes 1 and 2 are a high-frequency mode of interest and some other low-frequency mode, respectively, will generally not be the same as for the $|0_1 1_2> \rightarrow |1_1 1_2>$ transition, but usually the two will fall at almost the same frequency and both sources of intensity must be considered if mode 2 has significant thermal population.

Equation (9.10) assumes the Condon approximation (negligible dependence of the electronic transition moment on nuclear coordinate), and it does not consider the possibility of static inhomogeneous broadening. When these two constraints are relaxed, Eq. (9.10) becomes

$$\sigma_{R,if}(\omega_L) = \frac{8\pi|\mu_0|^4 \omega_L \omega_S^3}{9\hbar^2 c^4} \int_{-\infty}^{\infty} d\delta \, G(\delta) F_{if}(\omega_L, \delta) \tag{9.13}$$

where μ_0 is the electronic transition moment evaluated at the equilibrium nuclear geometry, and any explicit dependence of μ on vibrational coordinates is incorporated into $F_{if}(\omega_L, \delta)$. The function $G(\delta)$ is some normalized inhomogeneous distribution of electronic zero–zero frequency shifts centered at $\delta = 0$, usually but not necessarily taken to be Gaussian. Since different members of the ensemble are assumed to be distinct on the time scale of the Raman process, one must calculate the intensity for each molecule, $F_{if}(\omega_L, \delta)$, prior to taking the average over the inhomogeneous distribution. The explicit form of $F_{if}(\omega_L, \delta)$ is given below.

Resonance Raman scattering and ordinary linear absorption are closely related processes, depending on the same potential energy surfaces and line broadening parameters in slightly different ways. Since the absorption spectrum is usually easy to measure accurately, it provides a strong constraint, and calculation of the absorption spectrum should always accompany modeling of resonance Raman spectra and/or excitation profiles. The absorption cross section at the same level of approximation as Eq. (9.13) is given by

$$\sigma_A(\omega) = \frac{4\pi^2|\mu_0|^2\omega}{3n\hbar c} \sum_i B_i \int_{-\infty}^{\infty} d\delta \, G(\delta) F_{i,A}(\omega, \delta) \tag{9.14}$$

where n is the solvent refractive index. The absorption cross section, in units of square centimeters per molecule (cm^2 molecule^{-1}), is related to the molar extinction coefficient $\varepsilon(\omega)$, in liters per mole per reciprocal centimeter (L mol^{-1} cm^{-1}), by

$$\sigma_A(\omega) = 1000 \ln 10 \, \varepsilon(\omega)/N_A \qquad (9.15)$$

where N_A is Avogadro's number. Again, explicit expressions for $F_{i,A}(\omega,\delta)$ are given below.

Finally, the thermally equilibrated fluorescence from the excited state also depends on the same potential surface and broadening parameters and is given by an expression similar to Eq. (9.14) (96). The relaxed fluorescence spectrum, if available, can thus be used as another constraint in the modeling process (4,97). The difficulty in doing so lies with the fact that if the fluorescence is sufficiently weak to permit high-quality resonance Raman spectra to be obtained, usually the lifetime is sufficiently short that the spectrum has a significant component of "hot" fluorescence from molecules that have not yet thermally equilibrated with the environment. Calculation of the spectrum then requires detailed knowledge of the vibrational relaxation pathways and kinetics on the excited-state surface (98). Also, the absorption and resonance Raman spectra depend mainly on the potential energy surfaces in the vicinity of the ground-state geometry, while the relaxed fluorescence depends on the surfaces near the excited-state equilibrium geometry. If the change in molecular geometry and/or equilibrium solvation structure upon excitation is large, it is perhaps unreasonable to expect simple models for the potential surfaces and solvation dynamics to fit both regions.

In general, Raman intensities can be connected unambiguously to the properties of a particular excited electronic state only when the excitation frequency is rigorously on resonance (i.e., in a region where there is significant absorption to the excited state of interest). Preresonant excitation may seem to be an attractive option since the Raman intensity typically decreases more slowly than the absorption cross section as the excitation frequency is tuned away from resonance, making it easier to obtain high quality Raman spectra without significant reabsorption or photoalteration. Preresonant excitation also affords considerable simplifications in the interpretation of the intensities if a single electronic state dominates the Raman enhancement (1,3,4,99,100). However, single-state enhancement is rarely a safe assumption in the preresonant regime. Recent experimental and theoretical results demonstrate that even when the excitation is rigorously on resonance with a well-isolated electronic transition, contributions from other states may be important if the resonant transition does not have a large oscillator strength (101–104). Since vibrational overtones and combination bands usually have very little nonresonant intensity compared with fundamentals, the observation of unexpectedly weak overtones and combination bands, and/or anomalously shaped excitation profiles for fundamentals (101–103,105–108), may signal significant nonresonant contributions to the fundamental intensities. It is dangerous to draw conclusions about excited-state structure from Raman intensities unless the excitation frequency is rigorously on resonance with an intense and well-isolated absorption band, particularly if overtone and combination band intensities are not analyzed.

9.3.3. Modeling by Sum-Over-States Formalism

The "traditional" KHD expression for the resonance Raman amplitude, introduced in Eq. (9.11), treats the scattering process as involving a virtual absorption from level $|i>$ of the ground electronic state to the set of levels $\{|w>\}$ of the excited electronic state, followed by virtual emission from $\{|w>\}$ to $|f>$. All possible $|i> \rightarrow |w> \rightarrow |f>$ paths, each weighted by its energy denominator and its product of vibrational overlap integrals, must be added at the amplitude level before being modulus squared. A more complete form for Eq. (9.11), consistent with Eqs. (9.12) and (9.13), is

$$F_{if}(\omega_L,\delta) = \left| \sum_w \frac{<f|(\mu(q)/\mu_0)|w><w|(\mu(q)/\mu_0)i>}{i(\omega_L - \omega_0 - \delta - \omega_{wi}) - \Gamma} \right|^2 \qquad (9.16)$$

where ω_0 is the electronic frequency gap between ground and excited electronic states, ω_{wi} represents the purely vibrational frequency difference (see Fig. 9.5), and Γ is the homogeneous line width, assumed here to be the same for all vibrational levels of the excited electronic state. The line width may have contributions from both lifetime decay and pure dephasing as discussed in Section 9.3.1. The quantity $\mu(q)$ is the vibrational coordinate-dependent electronic transition moment, which is usually expanded as a Taylor series about the equilibrium nuclear geometry:

$$\mu(q) = \mu_0 + \sum_j (\partial\mu/\partial q_j)_0 q_j + \cdots \qquad (9.17)$$

with the sum extending over all vibrational modes. If the coordinate dependence of μ is neglected (the Condon approximation), then the numerator becomes a product of purely vibrational overlaps, $<f|w><w|i>$. The corresponding expression for

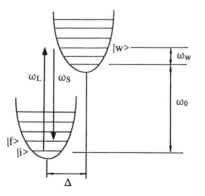

Figure 9.5. Definition of electronic and vibrational frequencies ω and displacement parameter Δ.

optical absorption is

$$F_{i,A}(\omega,\delta) = \sum_w \frac{|<w|(\mu(q)/\mu_0)|i>|^2}{(\omega - \omega_0 - \delta - \omega_{wi})^2 + \Gamma^2} \tag{9.18}$$

where each excited-state vibrational level contributes to the absorption as a nor-malized Lorentzian weighted by a factor that in the Condon approximation is sim-ply the Franck–Condon factor, $|<w|i>|^2$.

Evaluation of Eqs. (9.16) and (9.18) thus requires calculation of vibrational overlap integrals such as $<w|i>$ or the corresponding matrix elements with powers of q, for example, $<w|q_j|i>$. In the most general case, the wave functions are an-harmonic and nonseparable functions of all $3N - 6$ vibrational coordinates, and even for small molecules the number of states $|w>$ that make an important contri-bution to the sum can be enormous. For diatomics it is feasible to calculate the wave functions for arbitrary anharmonic potential surfaces and numerically evaluate the resulting vibrational overlap integrals, although even here approximations are often made (e.g., treating at least the ground state as a Morse potential) (60). In triatomics, rigorous evaluation of Eqs. (9.16) and (9.18) for arbitrary potential sur-faces is currently on the edge of feasibility, and for larger molecules numerous approximations are usually made. The simplest approximation is that both ground- and excited-state potentials are harmonic with no "Duschinsky effect" or normal coordinate rotation (i.e., the normal modes of the excited state are the same as those of the ground state, rather than being linear combinations of them). In this case, the $(3N - 6)$-dimensional vibrational overlaps can be factored into products of one-dimensional overlaps, for example,

$$<w|i> = <w_1 w_2 w_3 \cdots |i_1 i_2 i_3 \cdots> = <w_1|i_1><w_2|i_2><w_3|i_3> \cdots \tag{9.19}$$

where 1, 2, 3, . . . label the vibrational coordinates. The one-dimensional overlaps have simple analytic forms for harmonic oscillators (1). In this case, while the number of different intermediate states (the number of different combinations of quantum numbers w_1, w_2, w_3, etc.) that contribute significantly to the sum may still be enormous, at least the calculation of each overlap is simple. If Duschinsky rotation is to be included, the $(3N - 6)$-dimensional overlaps are no longer products of one-dimensional overlaps (although they factor into blocks if the molecule has any symmetry), and while analytic forms exist for the Duschinsky-rotated harmonic oscillator overlaps, they are considerably more complicated than for the one-dimensional case (109–114). Finally, if coordinate dependence of the electronic transition moment is to be considered, one can take advantage of the fact that the dimensionless normal coordinate q is just a sum of the quantum mechanical raising and lowering operators; therefore,

$$<w_1 w_2 \cdots w_j \cdots |q_j|i_1 i_2 \cdots i_j \cdots> = \sqrt{(i_j + 1)/2}<w_1 w_2 \cdots w_j \cdots \tag{9.20}$$
$$|i_1 i_2 \cdots (i_j + 1) \cdots> + \sqrt{i_j/2}<w_1 w_2 \cdots w_j \cdots |i_1 i_2 \cdots (i_j - 1) \cdots>$$

The sum-over-states approach is most useful in situations where the electronic absorption spectra are sufficiently well resolved that only a small number of inter-mediate states $|w>$ contribute significantly to the resonance Raman enhancement. When the number of intermediate states that must be summed becomes large, as in most larger molecules and directly dissociative small molecules, the time-dependent approach described in the next section usually becomes computationally more efficient. Also, it is more straightforward to include the effects of solvent-induced spectral broadening with an intermediate time scale (see Section 9.3.1) in the time-dependent picture. For these reasons we now focus on the alternative time-dependent approach to resonance Raman and absorption calculations.

9.3.4. Modeling by Time-Dependent Formalism

Writing the energy denominator of Eq. (9.16) as an integral over a time variable followed by some straightforward mathematical manipulations as described, for example, in Ref. 115 results in a time-domain expression for the resonance Raman intensity. This result may alternatively be derived directly from second-order time-dependent perturbation theory without ever going through the sum over interme-diate eigenstates, as developed in Ref. 2. Even when the resonance enhanced chro-mophore is not isolated from its environment and must be described by a density matrix rather than a wave function, the final expression for the Raman part of the emission spectrum takes on nearly the same form as long as the environmental perturbations that broaden the electronic spectrum are in the "rapid fluctuation" limit (see Section 9.3.1) (87). The general time-domain result for the quantity $F_{R,if}(\omega_L,\delta)$ corresponding to Eq. (9.16) is

$$F_{R,if}(\omega_L,\delta) = \left| \int_0^\infty dt <\chi_f|\chi_i(t)> \exp[i(\omega_L - \omega_0 - \delta + \omega_i)t - g(t)] \right|^2 \quad (9.21)$$

where $<\chi_f| = <f|(\mu(q)/\mu_0)$ and $|\chi_i> = (\mu(q)/\mu_0)|i>$ (the ground-state vibrational wave functions multiplied by the coordinate-dependent electronic transition moment), and $|\chi_i(t)> = \exp(-i\mathcal{H}t/\hbar)|\chi_i>$ where $\mathcal{H}$ is the excited-state vibrational Hamiltonian. Equation (9.21) can be interpreted as the half-Fourier transform of a time-dependent overlap of a moving wave packet that represents the initial vibrational wave func-tion propagating on the excited-state potential surface, damped by the function $\exp[-g(t)]$, which represents the effect of both population decay and any solvent-induced pure dephasing (see below). The corresponding result for the absorption spectrum, corresponding to Eq. (9.18), is

$$F_{i,A}(\omega,\delta) = \frac{1}{\pi} \text{Re} \int_0^\infty dt <\chi_i|\chi_i(t)> \exp[i(\omega - \omega_0 - \delta + \omega_i)t - g(t)] \quad (9.22)$$

where Re signifies the real part of the integral.

The interpretation of Eqs. (9.21) and (9.22) in terms of wave packets is depicted in Fig. 9.6. Numerous papers have discussed in detail the interpretation of these

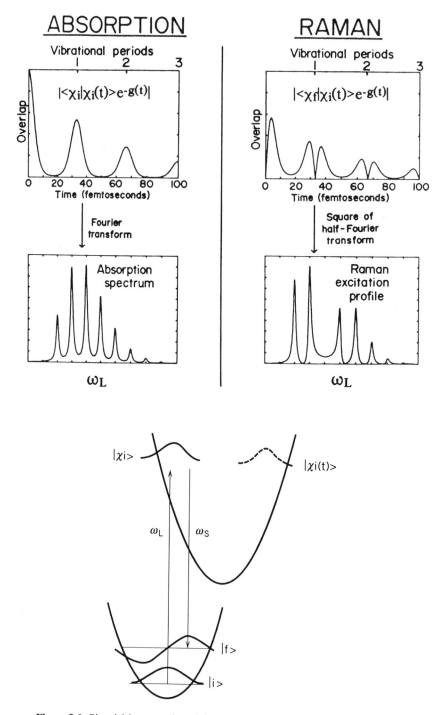

Figure 9.6. Pictorial interpretation of time-domain Raman and absorption expressions.

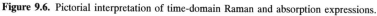

expressions, the manner in which the overlaps depend on the multidimensional potential energy surfaces, and the resulting effects on the absorption spectra and resonance Raman intensities (1,3,4,116). Here we merely summarize a few main points. The time dependence of $|\chi_i(t)\rangle$ represents the way in which the positions of the atoms evolve with time after the potential energy function is instantaneously switched from that of the ground electronic state to that of the excited state. Thus the initial (subvibrational period) time dependence of $|\chi_i(t)\rangle$ reflects the motions of the atoms from their ground-state equilibrium positions toward excited-state equilibrium. If the excited state is bound, the atomic positions will undergo harmonic or quasiharmonic motion on that surface, resulting in a complicated time dependence of the moving wave packet's overlap with itself at time zero, $\langle\chi_i|\chi_i(t)\rangle$. The Fourier transform of $\langle\chi_i|\chi_i(t)\rangle$, which gives the absorption spectrum, should in principle contain peaks at frequencies corresponding to all the bound excited-state vibrations along which geometry changes occur, but in practice these are often not resolvable due to the line broadening induced by a rapidly decaying $e^{-g(t)}$ or to inhomogeneous broadening. This is where the resonance Raman experiment becomes so useful. The intensity of each Raman transition depends on the overlap of $|\chi_i(t)\rangle$ with a particular, different final state, a "projection" of the complex motion of $|\chi_i(t)\rangle$ onto specific ground-state vibrational coordinates (for fundamentals and overtones) or pairs of coordinates (for combination band intensities). In general, those modes that undergo the largest excited-state geometry changes will exhibit the highest fundamental intensities and the longest overtone progressions. However, overtone progressions tend to be fairly weak in large molecules, because the overlaps $\langle\chi_f|\chi_i(t)\rangle$ involve all vibrational coordinates; the overlap for a first overtone ($i \rightarrow i + 2$), for example, will be large only at those times when $|\chi_i(t)\rangle$ has evolved enough from $|\chi_i(0)\rangle$ to gain a good overlap with the ground-state wave function excited by two quanta in the mode of interest, yet still has good overlaps with the initial state in all of the other modes. This is hard to satisfy if many modes undergo large geometry changes, and it becomes progressively harder to satisfy for higher overtones, which take progressively longer to develop good overlaps along the mode of interest. The overall effect of motion of $|\chi_i(t)\rangle$ along many vibrational modes is that the wave packet, once it leaves the neighborhood of the ground-state geometry, virtually never returns (strictly never returns for truly dissociative potentials), and both the absorption spectrum and the resonance Raman intensities become determined almost entirely by the dynamics occurring in the first tens of femtoseconds. One consequence is that the spectra of dissociative molecules, or those with many Franck–Condon active modes, depend mainly on just the first and/or second derivatives of the excited-state potential along the ground-state normal modes; anharmonicities neither influence the intensities strongly nor are readily determined from the data. Molecules with bound excited states and only a few displaced modes may have significant overlaps at much longer times, requiring better models for the excited-state potential surfaces to properly simulate the data.

The time-dependent overlaps, $\langle\chi_i|\chi_i(t)\rangle$ and $\langle\chi_f|\chi_i(t)\rangle$, are 3N-6 dimensional just as the sum-over-states vibrational overlaps of Eqs. (9.16) and (9.18). However, there are often tremendous computational advantages to the time-domain form. If

Duschinsky rotation is neglected, then the time-dependent overlaps separate, in analogy to Eq. (9.19), as

$$<\chi_f|\chi_i(t)> = <\chi_{f,1}|\chi_{i,1}(t)> <\chi_{f,2}|\chi_{i,2}(t)> <\chi_{f,3}|\chi_{i,3}(t)> \cdots \quad (9.23)$$

This presents an enormous advantage, because while evaluation of the sum-over-states expression requires summing over contributions from all possible combinations of excited-state quantum numbers in all modes, the time-domain expression requires a simple multiplication of the overlaps in each of the modes. Another way of stating this is that the absorption spectra and Raman profiles involve convolutions of all modes in the frequency domain, while the Fourier transform of a convolution becomes a product in the time domain. Thus, while the computational time required for the sum-over-states calculation scales as some large power of the number of significantly displaced modes, the time-domain calculation requires nearly constant computational effort regardless of the number of modes, since most of the time is taken by the Fourier transform step, which need be done only once. If Duschinsky rotation and/or anharmonicities are considered, the time-domain form loses some of its advantages because calculation of the overlaps themselves becomes the rate-limiting step. Nevertheless, if only ''short-time'' dynamics are important as discussed above, it is often still much more efficient to compute the full $(3N-6)$ dimensional $<\chi_f|\chi_i(t)>$ for the small number of time steps required than to try to calculate an enormous number of $(3N-6)$-dimensional eigenstates and their overlaps $<w|i>$ and $<f|w>$.

As in the sum-over-states approach, any coordinate dependence of the transition moment is usually expressed as a Taylor series expansion truncated after the linear term. Thus,

$$
\begin{aligned}
<\chi_f|\chi_i(t)> &= <f|(\mu(q)/\mu_0)e^{-i\mathcal{H}t/\hbar}(\mu(q)/\mu_0)|i> \\
&\approx <f|\left[1 + \sum_j (\partial\mu/\partial q_j)q_j/\mu_0\right]e^{-i\mathcal{H}t/\hbar}\left[1 + \sum_j (\partial\mu/\partial q_j)q_j/\mu_0\right]|i> \\
&\approx <f|e^{-i\mathcal{H}t/\hbar}|i> + \sum_j [(\partial\mu/\partial q_j)/\mu_0] \\
&\quad \{<f|q_j e^{-i\mathcal{H}t/\hbar}|i> + <f|e^{-i\mathcal{H}t/\hbar}q_j|i>\} \\
&= <f|i(t)> + \sum_j [(\partial\mu/\partial q_j)/\mu_0] \{\sqrt{f_j/2}<(f-1)_j|i(t)> \\
&\quad + \sqrt{(f+1)_j/2}<(f+1)_j|i(t)> + \sqrt{i_j/2}<f|(i-1)_j(t)> \\
&\quad + \sqrt{(i+1)_j/2}<f|(i+1)_j(t)>\}
\end{aligned}
\quad (9.24)
$$

The Condon approximation retains only the first term, while the second term is the first non-Condon correction.

Explicit forms for $<f|i(t)>$ when $|i> = |0>$, the vibrational ground state, have been given in numerous places in the literature, including the cases of unequal ground- and excited-state vibrational frequencies as well as Duschinsky rotation (1,4,115,117,118). The corresponding analytic expressions for $i \neq 0$ have also been derived (119) but are not widely known, so the one-dimensional expressions will be reproduced here. Reference 119 contains the multidimensional forms needed when Duschinsky rotation is included. In one dimension, $<f|n(t)>$ (here we use n rather than i for the quantum number of the initial state to prevent confusion with $\sqrt{-1}$) is given by

$$<f|n(t)> = [f!n!2^{f+n}\Psi(t)]^{-1/2} \exp[\Delta^2 f(t)] [\alpha(t)]^{f+n}$$

$$\sum_{k=0}^{k^*} [(2k)!/k!] \, \eta_{fnk} \, [\gamma(t)]^k H_{f+n-2k}[\lambda f(t)\Delta/\alpha(t)]$$

(9.25)

where k^* is the integer part of $(f + n)/2$, H_p is the pth Hermite polynomial, Δ is the displacement between ground- and excited-state potential minima in excited-state dimensionless normal coordinates (see Fig. 9.5), and

$$\Psi(t) = (\omega_+^2/4\omega_e\omega_g)[1 - (\omega_-/\omega_+)^2\exp(-2i\omega_e t)]$$

(9.26a)

$$f(t) = -\frac{\omega_g[1 - \exp(-i\omega_e t)]}{\omega_+ - \omega_- \exp(-i\omega_e t)}$$

(9.26b)

$$\alpha(t) = \left[\frac{\omega_- - \omega_+ \exp(-i\omega_e t)}{2\omega_+ - 2\omega_- \exp(-i\omega_e t)}\right]^{1/2}$$

(9.26c)

$$\gamma(t) = \frac{2(\omega_e/\omega_g) + i[(\omega_e/\omega_g)^2 - 1]\sin\omega_e t}{2(\omega_e/\omega_g) - i[(\omega_e/\omega_g)^2 - 1]\sin\omega_e t}$$

(9.26d)

$$\eta_{fnk} = \sum_{q=0}^{2k} (-1)^q \binom{f}{2k - q} \binom{n}{q}$$

(9.26e)

$$\binom{m}{k} = \begin{cases} \dfrac{m!}{k!(m - k)!} & \text{when } m \geq k \\ 0 & \text{when } m < k \end{cases}$$

(9.26f)

$$\omega_\pm = \omega_e \pm \omega_g$$

(9.26g)

with ω_e and ω_g being the excited- and ground-state vibrational frequencies, respectively. If these frequencies are the same and the system starts in the ground vibrational state ($|n> = |0>$), the overlaps assume the following simple form:

$$<f|0(t)> = (f!2^f)^{-1/2} \Delta^f \exp[(\Delta^2/2)(e^{-i\omega t} - 1)] (e^{-i\omega t} - 1)^f$$

(9.27)

Calculations of the overlaps for anharmonic surfaces must be performed numerically. A number of schemes for propagating wave functions on arbitrary potential surfaces have been developed (120–131); these are computationally quite tractable in one dimension, not unreasonable in two dimensions, and quite time-consuming in more than two dimensions, although these generalizations may be obsolete by the time this chapter is published due to continuing improvements in both hardware and software. Applications of the split operator algorithm of Feit and Fleck (120,121) for propagating $|i(t)>$ on one- and two-dimensional anharmonic surfaces are described in Refs. 132 and 133.

The damping function, $e^{-g(t)}$, depends on the nature of the electronic dephasing. For isolated molecules, it should simply be $e^{-t/\tau}$ where τ is the excited-state lifetime. For molecules interacting with a bath, $g(t)$ will usually be dominated by the solvent-induced pure dephasing. The most common approach to evaluating $g(t)$ in condensed phase problems is the stochastic theory. This theory, which was originally developed by Kubo (83) for absorption and has since been adapted by other workers to resonance Raman and fluorescence (84–88), was briefly described in Section 9.3.1. The solvent is treated as a random perturbation that causes the energy levels of the solute to fluctuate with time in a manner that can be described by two parameters, a magnitude D and a characteristic frequency Λ. The effect on the absorption and fluorescence spectra is simply to broaden each vibronic transition of the isolated solute without shifting the bands; as the timescale of the fluctuations ranges from slow ($D \gg \Lambda$; "inhomogeneous" broadening) to fast ($\Lambda \gg D$; "homogeneous" pure dephasing), the lineshape varies from Gaussian to Lorentzian. The standard stochastic theory is incomplete, however, in that it considers only the effect of the solvent on the solute's electronic transition, but not vice versa. Therefore it does not account for the solvent reorganization that causes, among other things, the solvent contribution to the Stokes shift between absorption and emission. A more realistic approach is to model the solvent coordinate(s) as one or more displaced harmonic oscillators that are frictionally damped to varying degrees. Mukamel and co-workers (89,134–136) developed this "Brownian oscillator" model and applied it to a variety of nonlinear spectroscopic processes. In this model, the quantity $g(t)$ can be expressed as an explicit sum over an arbitrary number of vibrational modes, each having its own frequency, excited-state geometry change, and friction coefficient. Here we focus on the simplest treatment in which all the solvent degrees of freedom are lumped into one effective mode. When the friction on this mode is high enough (and/or its frequency low enough) that it is strongly overdamped, as is usually thought to be the case for the most strongly coupled solvent motions (89) (but see also Ref. 137), $g(t)$ takes the following simple form:

$$g(t) = (D/\Lambda)^2(\Lambda t - 1 + e^{-\Lambda t}) + i(D^2/2kT\Lambda)(1 - e^{-\Lambda t}) + t/\tau \quad (9.28)$$

where T is the temperature (here we assume the high-temperature limit where the solvent mode's frequency is very small compared with kT). The real part of Eq. (9.28) has exactly the same form as the expression derived from the stochastic theory, while the imaginary part accounts for the contribution of the solvent to shifting the absorption and emission spectra in addition to broadening them. The t/τ pure lifetime decay term in Eq. (9.28) is often negligible compared with the solvent-induced part.

With the constants numerically evaluated, the time-domain expressions for the absorption and resonance Raman cross sections become (1)

$$\sigma_A(\overline{\omega}) = \frac{5.75 \times 10^{-3} M^2 \overline{\omega}}{n} \int_{-\infty}^{\infty} d\delta \ G(\delta) \sum_i B_i \text{Re} \int_0^{\infty} dt$$

$$\langle \chi_i | \chi_i(t) \rangle \ \exp[i(\omega - \omega_0 - \delta + \omega_i)t - g(t)] \tag{9.29}$$

$$\sigma_{R,if}(\overline{\omega}_L) = 2.08 \times 10^{-20} \ M^4 \overline{\omega}_s^3 \overline{\omega}_L \int_{-\infty}^{\infty} d\delta \ G(\delta) F_{if}(\omega_L, \delta) \tag{9.30}$$

where $\overline{\omega}$ is the wavenumber of the absorbed photon in reciprocal centimeters, $\overline{\omega}_L$ and $\overline{\omega}_s$ are the incident and scattered photon wavenumbers in reciprocal centimeters, M is the electronic transition length in angstroms evaluated at the ground-state equilibrium geometry, σ_A and σ_R are in angstroms squared per molecule, and $F_{if}(\omega_L, \delta)$ is given by Eq. (9.21). The implementation of these expressions to model experimental data depends on what is assumed to be known. If starting from computed potential surfaces, the only unknown parameters will generally be the broadening parameters D, Λ, and $G(\delta)$ (and/or τ), perhaps the electronic transition length M, and usually the electronic zero–zero frequency ω_0. The latter should be chosen to properly position the calculated absorption spectrum in frequency, while M is determined by requiring that the calculated and experimental integrated absorption cross-sections agree (the details of the potential surface and the broadening parameters affect the shape of the absorption spectrum but not its integrated intensity). Fitting to the absorption spectrum alone can fairly well define the total line width, but the absorption is not highly sensitive to the partitioning of the breadth between "homogeneous" and "inhomogeneous" components, or to the time scale of the solvent-induced breadth (usually expressed by the parameter $\kappa = \Lambda/D$). Both the relative and the absolute resonance Raman intensities and excitation profile bandshapes, on the other hand, are quite sensitive to these parameters; in particular, while an inhomogeneous distribution of zero–zero frequencies merely broadens the excitation profiles, homogeneous broadening also decreases the integrated intensities of all of the profiles. Absolute resonance Raman intensities are therefore very useful in distinguishing between homogeneous and inhomogeneous broadening (keeping in mind that such an assumed separation is approximate at best, as discussed in Section 9.3.1).

If little or no a priori knowledge of the potential surfaces is assumed, the usual approach is to begin by choosing the simplest possible model for the surfaces. The ground state is taken to be a multidimensional harmonic oscillator whose frequencies are given by the observed Raman frequencies, and the excited state is assumed to differ from the ground state in the simplest way that can give rise to the observed transitions: equal ground- and excited-state frequencies with a displacement Δ between minima for totally symmetric modes, and for nontotally symmetric modes (which cannot have a displacement and can appear in the resonance Raman spectrum only as overtones or combination bands that are overall totally symmetric), harmonic with different ground- and excited-state frequencies. Initial guesses are then made for the Δs and/or frequency changes as well as the other parameters, the absorption and Raman spectra are calculated, and the parameters are refined iteratively. It is usually possible, although often tedious, to arrive at a fairly well-defined set of best-fit parameters within this simplest model, particularly if absolute cross sections are available. What is more difficult is to determine how bad the fits have to be to justify going to a more complex model (e.g., introducing excited-state frequency changes in totally symmetric modes if there are no direct data on the excited-state frequencies, or adding coordinate dependence of the transition moment, or Duschinsky rotation). The number of adjustable parameters then tends to proliferate well beyond the capability of the data to define them.

9.3.5. Transform Methods

Comparison of Eqs. (9.16) and (9.18), or (9.21) and (9.22), emphasizes the close relationship between resonance Raman scattering and optical absorption. The "transform" approach to analyzing resonance Raman data exploits that similarity. Tonks and Page (138) first showed, starting from the time-correlator formalism of Hiznyakov and Tevher (139), that when the simplest "standard assumptions" hold (separable harmonic modes, Condon approximation, zero temperature, and no inhomogeneous broadening), the resonance Raman excitation profile for each mode can be calculated from the Kramers–Kronig transform of the experimental absorption spectrum multiplied by an overall scaling factor determined by the excited-state displacement Δ. Comparison of experimental and calculated excitation profile bandshapes thus provides a test of the standard assumptions, and if they are found to be valid, provides an easy way to extract the excited-state geometry changes for one or more vibrational modes of interest from a relatively simple mathematical procedure that does not require explicit modeling of all the vibrational degrees of freedom. Others demonstrated shortly thereafter that the same transform relationships could be derived starting from the more familiar sum-over-states formalism (140,141), and the method has since gained considerable popularity. Furthermore, modified transform relationships have been derived to allow treatment of systems that depart from the standard assumptions in many ways. Excited-state frequency changes (142,143), non-Condon effects (67,95,143–147), Duschinsky rotation (95,143), finite temperatures (40,147–149), anharmonicities (150), inhomogeneous broadening (144,151), and energy- or state-dependent electronic dephasing (152) have all been treated. An "inverse transform" allowing calculation of the absorption spectrum for the resonant transition that gives rise to an experimental excitation profile has also been described (153–155).

The transform approach is the easiest way to extract excited-state geometry changes from experimental resonance Raman excitation profiles of large molecules if the standard assumptions are assumed to hold. From a single Kramers–Kronig transform of the absorption, one can easily calculate excitation profile bandshapes for all of the observed Raman transitions. The relevant formulas, given explicitly in Ref. 148, are also summarized in Ref. 1. Scaling of the calculated bandshapes to match the experimental intensities then yields the excited-state displacements (relative Δs for different modes from their relative experimental intensities or absolute Δs if absolute cross sections are available) (156) without the need for any trial-and-error adjustment of parameters. However, the transform expressions quickly become much more complicated when deviations from the standard assumptions are considered, and the usual formulas employed to deal with these deviations are often leading terms in a series expansion that must be evaluated carefully for their limits of applicability. For example, the usual transform expressions employed to treat linear non-Condon scattering are accurate only when the non-Condon modes do not simultaneously have large excited-state displacements (146,157). If physically inappropriate assumptions are made (e.g., if a strongly inhomogeneously broadened absorption spectrum is assumed to be homogeneously broadened), the displacement parameters extracted from the transform analysis may be unrealistic and incapable of reproducing the absorption bandshape in a direct modeling calculation.

Perhaps the greatest advantage that the transform method holds over direct modeling is in treating finite-temperature scattering from molecules with a number of low-frequency vibrations. According to Eq. (9.12), in a direct calculation one must separately evaluate the Raman cross section for each transition originating from each thermally occupied initial state, weight it by its Boltzmann factor, and add together the quasi-degenerate contributions (i.e., transitions that start in different initial states but involve the same quantum number changes in the same modes). Although modes for which the potential surface changes negligibly upon excitation will make no contribution to the time-dependent overlaps and can be ignored, the number of initial states that must be considered still can be enormous for even medium sized molecules at room temperature. Explicit summation over initial states can be avoided in absorption calculations (158), but these methods are not directly applicable to resonance Raman where different initial states must be summed at the level of probabilities, not amplitudes. The transform approach to handling $T \neq 0$ scattering, while still approximate, has proved quite successful and efficient at handling systems having a large number of low-frequency vibrations, each rather weakly coupled to the electronic transition (40,147,149). It should also be noted that since Stokes and anti-Stokes transitions have different excitation profiles, the ratio of their intensities cannot be used directly to determine the temperature as it can in nonresonant scattering. Champion and co-workers (40,159) explored this point in detail both experimentally and theoretically and discussed how to analyze Stokes to anti-Stokes intensity ratios on resonance to obtain vibrational temperatures (see Fig. 9.7).

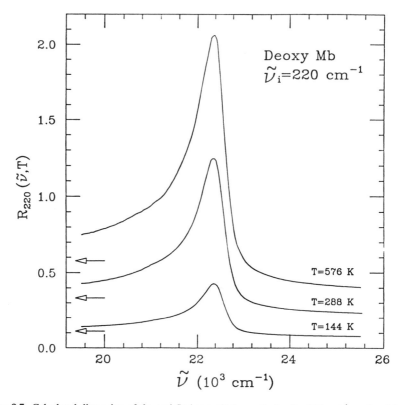

Figure 9.7. Calculated dispersion of the anti-Stokes to Stokes ratio for the 220-cm^{-1} mode of deoxy-myoglobin as the excitation frequency is tuned through the Soret absorption band. The Boltzmann population ratios at the same three temperatures are shown as arrows. [Reproduced from Ref. 40 with permission.]

9.3.6. Conversion from Dimensionless Normal to Internal Coordinates

The ground- to excited-state geometry changes, Δ, obtained as modeling parameters in fitting the experimental spectra, are obtained most naturally in dimensionless normal coordinates. In order to convert them to chemically meaningful bond length and bond angle changes, the normal mode descriptions must be known. In some simple cases (certainly for diatomics!) it may be sufficient to assume that an observed vibration is an isolated stretch or bend, but generally the normal modes are linear combinations of multiple internal coordinates that must be determined separately from a ground-state vibrational analysis.

For a pure stretching coordinate, the dimensionless coordinate q is related to the bond length x by

$$q = (\mu\omega/\hbar)^{1/2}(x - x_0) \tag{9.31}$$

where x and x_0 are the instantaneous and equilibrium bond lengths, μ is the reduced mass of the vibration, and ω is the frequency. More generally, the dimensionless coordinate q_j is related to the normal coordinate by

$$q_j = (\omega_j/\hbar)^{1/2} Q_j \tag{9.32a}$$

$$= 0.17222\ \bar{\omega}_j^{1/2}\ Q_j \tag{9.32b}$$

where Q_j has units of Å-dalton$^{1/2}$ and $\bar{\omega}_j$ is the mode's frequency in reciprocal centimeters. The internal coordinates $\{r_i\}$ (bond lengths, bond angles, and dihedral angles) are related to the normal coordinates $\{Q_j\}$ by

$$r_i = \sum_j A_{ji}\ Q_j \tag{9.33}$$

Therefore, the displacement between ground- and excited-state potential minima along a particular internal coordinate, δ_i, is related to the dimensionless displacements, $\{\Delta_j\}$, through the normal mode coefficients $\{A_{ji}\}$ by

$$\delta_i = 5.8065 \sum_j A_{ji}\ \bar{\omega}_j^{-1/2}\ \Delta_j \tag{9.34}$$

It should be stressed that since the normal modes are usually delocalized over the entire molecule to some degree, the geometry change along a particular internal coordinate will generally have contributions from the dimensionless displacements in a large number of normal modes, each of which may be either positive or negative in sign. Therefore the displacements in all Raman-active modes must generally be included in order to calculate excited-state geometry changes, even if only one or a few internal coordinates are of interest. Also, relatively modest changes in the normal mode coefficients due to changes in the vibrational force field may have rather large effects on the computed excited-state geometry, and in fact the incorporation of intensity data into vibrational force field refinement algorithms has been proposed and implemented (160,161). Finally, fitting of absorption and resonance Raman data to separable harmonic models for the potential surfaces yields as a fitting parameter only the absolute values of the $\{\Delta_j\}$; the signs remain entirely undetermined. For n displaced modes there are 2^n possible sign combinations, corresponding to 2^n different excited-state geometries, that are all equally consistent with the experimental data. This remains a serious problem in the interpretation of resonance Raman intensities of large molecules, although often "chemical intuition" and/or comparison with calculations can be used to eliminate many of the formally possible geometry changes. Intensities of isotopic derivatives, which have the same potential energy surfaces but different normal mode descriptions and sometimes dramatically different intensities (162), can also be very helpful in reducing the ambiguity (1,163,164).

9.3.7. Determination of Electronic Spectral Broadening Parameters

Even if only the excited-state potential surface is of interest, its extraction from experimental absorption and resonance Raman data requires a model for the electronic spectral broadening as well, since the shapes of the excitation profiles and the relative intensities of different Raman lines are sensitive to the broadening function. Even transform analysis requires that assumptions be made about the contribution (if any) of electronic inhomogeneous broadening. The converse, however, is not true; the nature of the broadening may be deduced even in the absence of any model for the potential surfaces, as long as all of the scattering transitions are experimentally observable.

It is well known that the lifetime of an electronically excited state may be determined from the fluorescence quantum yield, as long as the same electronic state is responsible for absorption and emission (165). The fluorescence yield, ϕ_f, is simply the ratio of the radiative rate k_r to the sum of radiative and nonradiative rates $(k_r + k_{nr})$; in other words, $\phi_f = \tau/\tau_r$, where τ is the excited-state lifetime having contributions from both radiative and nonradiative decay routes, and τ_r is the natural radiative lifetime, which depends only on the electronic transition moment and emission wavelength and can be calculated from the integrated absorption strength. A corresponding relationship connects the resonance scattering quantum yield with the electronic *dephasing* time, at least if the dephasing is assumed to be exponential in time (fast modulation limit). In terms of the Raman cross sections, the relationship is (78)

$$\frac{\sum_i B_i \sum_f \sigma_{R,if}(\bar{\omega})/\bar{\omega}_{S,if}^3}{\sum_i B_i \sigma_{i,A}(\bar{\omega})} = \frac{2n^2 A_0}{\bar{\Gamma}} \qquad (9.35)$$

where ω is the incident laser frequency in reciprocal centimeters, $\bar{\omega}_{S,if}$ is the scattered frequency for the $|i\rangle \rightarrow |f\rangle$ Raman transition, Γ is the total homogeneous line width (inverse of total electronic dephasing time) in reciprocal centimeters, n is the solvent refractive index, and

$$A_0 = \int d\bar{\omega}\ \sigma_A(\bar{\omega})/\bar{\omega} \qquad (9.36)$$

is the integrated absorption cross section divided by the incident frequency, a quantity proportional to the square of the electronic transition moment. The left-hand side of Eq. (9.35) is simply the resonance Raman quantum yield, corrected for the dependence of the emission rate on the cube of the emitted frequency. Equation (9.35) is straightforward to define but less straightforward to evaluate, since it requires summation over all scattering transitions including the often numerous and weak ones at large Raman shifts as well as the Rayleigh line, which usually cannot be measured directly. Nonetheless it provides a very useful estimate of the homogeneous line width in the presence of arbitrary inhomogeneous broadening, independent of any knowledge of the ground- or excited-state potential surfaces. The idea of relating the total Raman yield to excited-state relaxation parameters was apparently first exploited by Stockburger and co-workers (166,167), who attributed the nonradiative relaxation to an electronic depopulation process (internal conversion), probably a reasonable interpretation in light of the negligible broad fluorescence observed from their systems.

The implication of Eq. (9.35) that the quantum yield should be independent of excitation wavelength holds only if the electronic dephasing is exponential in time. A nonconstant quantum yield implies a frequency-dependent Γ, that is, nonexponential decay of the electronic coherence. In principle, the full time-dependent electronic dephasing function [$g(t)$ in Eqs. (9.21) and (9.22)] can be extracted from measurements of the total Raman yield as a function of detuning from resonance over a wide frequency range (78,84). This is almost never feasible in practice due to the increasing importance of the Rayleigh scattering and nonresonant contributions to the scattering as the excitation is tuned away from resonance. Several possible approaches to circumventing these problems are discussed in Ref. 78. We also note that if a single excited-state vibronic resonance (e.g., the origin) dominates the resonance enhancement, then the Raman yield may be obtained from the ratio of the sharp Raman scattering to the broad fluorescence for transitions to the same final states, if the fluorescence lifetime has been separately measured (79) (Fig. 9.8). This eliminates the need for absolute Raman cross-section measurements and/ or estimates of the Rayleigh scattering, but is feasible only when the detuning from the resonance is small enough that higher vibronic levels do not contribute significantly to the Raman enhancement.

9.4. RECENT APPLICATIONS

9.4.1. Small Molecules

As described in the Introduction, resonance Raman spectroscopy is generally most useful for molecules that show little vibrational resolution in their electronic absorption spectra due to either environmental perturbations or very fast photochemistry. However, analyses of resonance Raman depolarization dispersion curves and resonance rovibrational excitation profiles have proved valuable for determining lifetimes of bound but predissociative states of small isolated molecules in which the electronic spectra do show considerable vibrational structure. The reader is referred to the original papers for theoretical details (64–66,168–171), but the basic idea is that both the rovibrational intensities and the depolarization ratios are sensitive to the excited-state lifetime when that lifetime is comparable to thermally averaged molecular rotational periods. These methods have been employed by the Myers and Ziegler groups to determine excess energy or vibronic level dependences of the predissociation lifetimes of the A state of ammonia (64,169,171,172), the S_3 state of CS_2 (65,66), the B state of methyl iodide (64,170,171,173), the Schumann–Runge bands of O_2 (171,174), and the $3s$ Rydberg state of the methyl

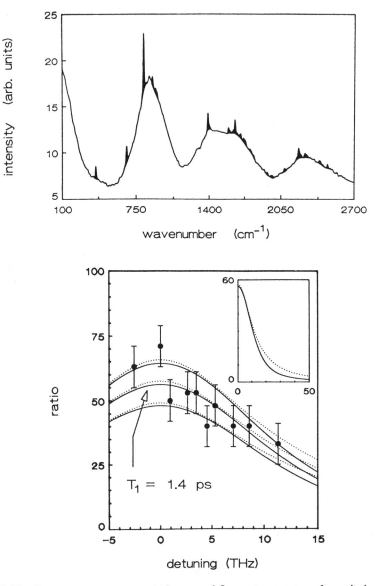

Figure 9.8. Top: Room temperature resonance Raman and fluorescence spectrum for excitation near the $S_1 \leftarrow S_0$ origin transition of azulene in cyclohexane. The sharp Raman lines are darkened. Bottom: Ratio of the fluorescence to Raman yields (filled circles) and calculated ratios using two different master equations for the solvent-induced electronic dephasing (solid and dashed curves) for three values of the lifetime ranging from 1.6 (top) to 1.2 ps (bottom). The insert shows that at larger frequency detunings a clear distinction between the two models can be made. [Reproduced from Ref. 79 with permission.]

radical (171,175) (see Figs. 9.9 and 9.10). All of these experiments employed tunable far-UV excitation in the 215–190-nm region. Measured depolarization ratios have also been used to determine rotational temperatures in jet-cooled methyl iodide (176). Far-UV vibrational resonance Raman scattering has been used in assigning vibronic spectral features and determining aspects of the excited state potential surfaces of ammonia (177), CS_2 (38,178), the methyl radical (179), ketene (180), and acetylene (181).

Absolute resonance Raman cross sections have been employed to study solvent-induced electronic spectral broadening in I_2 (59,60,182,183), CS_2 (38,78), and SO_2 (133). For I_2 in several solvents the solvent-induced broadening was found to be mostly inhomogeneous. In contrast, the solvent-induced spectral breadth in CS_2 and SO_2 was found to be mostly homogeneous, but dominated by electronic dephasing in CS_2 and by population decay in SO_2. The frequency dependence of the resonance Raman quantum yields was used to deduce the nonexponential nature of the electronic dephasing function for CS_2 in several solvents (78) (see Fig. 9.11).

9.4.2. Photodissociation

Since the pioneering studies of Kinsey and co-workers (184–187) on the *A* state of methyl iodide and the Hartley bands of ozone first brought the resonance Raman technique to the attention of the chemical dynamics community, there has been considerable activity in the use of resonance Raman to explore the dynamics of direct or near-direct photodissociation processes. A more quantitative analysis of the absorption and resonance Raman spectra of ozone has since been reported (188). The photodissociation dynamics in both vapor and solution phases, as well as the effects of resonant–nonresonant interferences on the fundamental intensities (see Fig. 9.12), have been mapped out in far greater detail for both methyl iodide (101,102,189,190) and higher alkyl iodides (103,191,192). Solvation effects on the short-time photodissociation dynamics of the alkyl iodides appear to be minor. Depolarization ratios of the *A*-state resonant scattering have been analyzed as a probe of the curve crossing dynamics involved in the photodissociation process (69,193), although more recent studies question this interpretation (68). Continuously scanned excitation profiles of several modes of iodobenzene excited into the *B* state continuum have also been analyzed (45).

The photodissociation of water has been examined in great detail through both vacuum-UV resonance Raman and absorption spectroscopy and high-level theory (104,194–196). Good agreement between experimental and theoretical spectra has been found (see Fig. 9.13), and the effects on the spectra of the coordinate dependence of the transition dipole and the form of the ground-state vibrations for highly excited stretching modes have been explored. The analogous photodissociation reaction of H_2S has also been examined, although in less detail (197,198). Raman depolarization ratios in the vapor (199) and spectra in both vapor and solution phases (200) have been obtained for photodissociating nitromethane and deuterated nitromethane; they exhibit resolved vibrations at excess energies nearly up to the lowest ground-state dissociation limit, but have not been particularly helpful in unraveling the complex photodissociation dynamics.

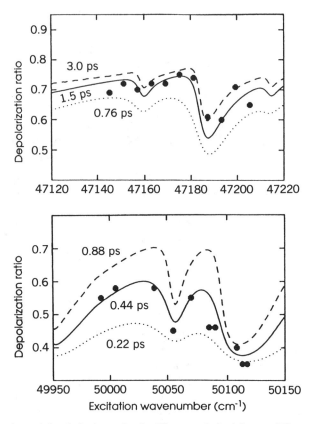

Figure 9.9. Experimental depolarization ratios for CS_2 vapor (points) in two different excitation frequency regions, and calculated curves assuming the indicated excited-state lifetimes. The solid curve is determined to be the best fit in both cases. Parameters are from Ref. 66.

9.4.3. Inorganic Compounds

Transition metal based compounds have long been popular subjects for resonance Raman studies due to their visible absorption bands and generally weak fluorescence. Quantitative evaluations of excited-state geometries have been made for a variety of such compounds based on absorption and preresonance Raman spectra as well as from full resonance Raman excitation profiles (201). A recent review by Zink and Shin (4) summarizes much of that group's work in this area and presents detailed descriptions of how to go about extracting excited-state geometry changes from various types of data. The data of Shin et al. (202) on two complexes have been modeled both by themselves and by Lee and Lee (203), with both groups obtaining similar excited-state displacement parameters and homogeneous line widths. Zink and co-workers (108) observed and modeled resonance Raman deenhancement due to resonance-preresonance interference for several transition metal complexes. Adelman and Gerrity (204) obtained spectra of $Cr(CO)_6$

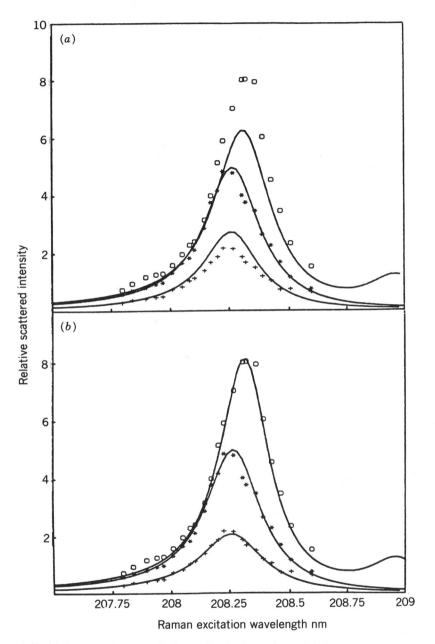

Figure 9.10. (*a*) Resonance Raman excitation profiles for the S3 (open circles), S5 (filled circles), and S7 (crosses) rotational transitions of the $\nu_1 + \nu_2^s$ combination band of ammonia vapor. Solid curves are the calculated excitation profiles calculated with a line width for the resonant vibronic transition of Γ = 68 cm^{-1}. (*b*) Same using the best-fit line widths of 56, 68, and 75 cm^{-1} for S3, S5, and S7, respectively, showing the sensitivity of the fits to line width. [Reproduced from Ref. 172 with permission.]

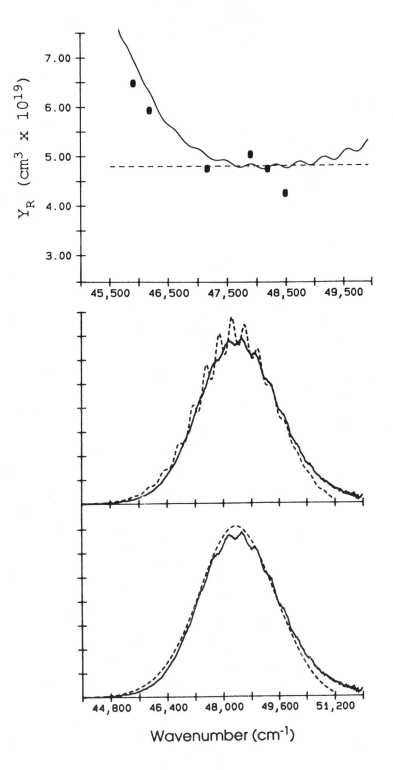

Wavenumber (cm⁻¹)

and W(CO)$_6$ in cyclohexane from 355 to 253 nm, the region of the two lowest allowed charge-transfer absorption bands. They observed significant intensity only in vibrational fundamentals, from which they concluded that the rapid loss of CO from these complexes upon photoexcitation results from rapid coupling to another, photochemically reactive excited state, but these conclusions must be considered preliminary without absolute cross sections and quantitative modeling of the data. Dines (205) analyzed previously published overtone progressions and/or excitation profiles of ClO$_2$, O$_3^-$, S$_3^-$, and TiI$_4$ in solid or solution environments to determine excited-state geometries.

9.4.4. Polyenes and Stilbene

Isoprene (2-methylbutadiene) and 1,3,5-hexatriene were the first molecules for which resonance (actually preresonance) Raman intensities were interpreted via the time-domain picture (115). Since then, rigorous resonance spectra of a number of small polyenes have been analyzed quantitatively. The simple linear dienes and trienes are of interest mainly because their relatively diffuse electronic spectra probably arise from a rapid curve crossing from the lowest allowed excited electronic state to a nearby one-photon forbidden state. Relative and absolute cross sections of 1,3-butadiene in both vapor and solution phases (206–208) and of isoprene in solution (209) have been analyzed to determine the lifetime and potential energy surface of the allowed state, as well as the location of the forbidden state and its influence on the intensities of overtones and combination bands of nontotally symmetric vibrations. Vacuum-UV spectra of butadiene vapor have also been obtained to aid in assigning the electronic states in this region (210). Relative and absolute cross sections of both *cis-* and *trans-*1,3,5-hexatriene in both vapor and solution phases have been analyzed to obtain the excited-state population decay and dephasing times and potential surface parameters for the allowed state (132,211–213). A significant dependence of the potential surface for out-of-plane motion on solvent polarizability was observed (Fig. 9.14) (132,213). Raman spectra of *cis-*hexatriene vapor enhanced by resonance with the forbidden state have been reported and analyzed qualitatively (214). In related work, very detailed preresonance Raman profiles of the symmetric C=C stretch of diphenyldecapentaene have been simulated to characterize the properties of the optically forbidden state (215).

While the small linear polyenes exhibit relatively little photochemistry, the cyclic ones undergo efficient photochemical ring-opening and/or hydrogen migration reactions. Mathies and co-workers studied the femtosecond dynamics of these processes through detailed excitation profile and absolute Raman and fluorescence in-

---

Figure 9.11. Top: Points show the experimental modified Raman quantum yield [the left-hand side of Eq. (9.35)] for CS$_2$ in pentane. Solid curve and dashed line are fits assuming stochastic solvent-induced electronic broadening near the slow modulation limit and purely exponential electronic dephasing, respectively. Middle: Experimental absorption spectrum of CS$_2$ in pentane (solid) and spectrum calculated by shifting the zero–zero and convolving the experimental vapor-phase absorption with the Lorentzian lineshape function used to generate the dashed line in the upper plot. Bottom: Same for the non-Lorentzian lineshape function in the top plot. Parameters are from Ref. 78.

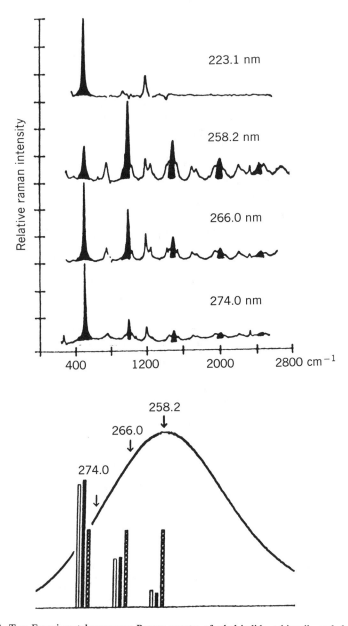

Figure 9.12. Top: Experimental resonance Raman spectra of ethyl iodide, arbitrarily scaled relative to one another, as the excitation frequency is tuned from the red side to the maximum (258 nm) of the directly photodissociative *A* band, and then into the gap between the *A* and *B* bands (223 nm). The C–I stretching progression is shaded, showing the anomalous deenhancement of the fundamental relative to the overtones as the absorption maximum is reached. Bottom: Superimposed on the experimental absorption spectrum are the experimental C–I stretch fundamental intensities (open bars), intensities calculated from a single state resonant assumption (hatched bars), and those calculated from a model considering resonant-preresonant interference (solid bars). Parameters are from Ref. 103.

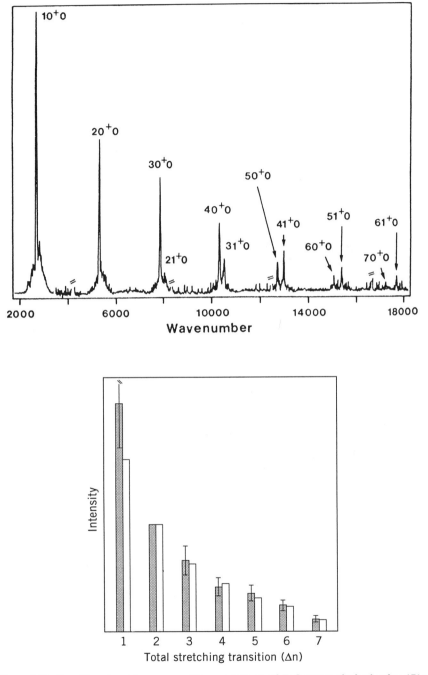

Figure 9.13. Top: Experimental resonance Raman spectrum of D_2O vapor obtained using 171-nm excitation. Bottom: Comparison of experimental (with error bars) and calculated relative intensities of the members of the stretching progression. [Reproduced from Ref. 104 with permission.]

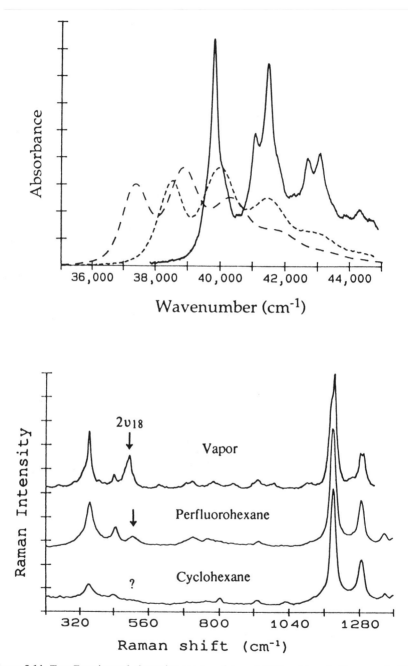

Figure 9.14. Top: Experimental absorption spectra of *trans*-1,3,5-hexatriene in the vapor (solid) and in perfluorohexane (short dashes) and cyclohexane (long dashes). Bottom: Resonance Raman spectra for origin excitation in the vapor and in perfluorohexane and cyclohexane solvents, demonstrating the pronounced damping of intensity in the overtone of the central C=C torsion (ν_{18}) on going to more red-shifting solvents. See Ref. 132 for details.

tensity measurements on 1,3-cyclohexadiene (216,217), 1,3,5-cyclooctatriene (218), 1,3,5-cycloheptatriene (219), and cyclobutene (220) in solution (see. Fig. 9.15). All of these results are consistent with photoreaction occurring from an essentially planar ring system, and with rapid internal conversion from the optically prepared state to a nearby forbidden state on which the actual photochemistry occurs.

Stilbene and other substituted ethylenes, as well as some conformationally perturbed polyenes, undergo rapid and efficient cis–trans photoisomerization processes whose dynamics can be further elucidated through resonance Raman analyses. Ethylene itself has a planar ground state but has a 90° twisted equilibrium geometry in its lowest V excited state, resulting in strong Raman activity in overtones of the C=C torsional mode (221). Early resonance Raman studies of stilbene elucidated the initial dynamics of out-of-plane distortion in the cis isomer (222) and quantitated spectral broadening mechanisms in trans-stilbene (72). More recently, Myers and co-workers obtained resonance Raman spectra for trans-stilbene in the vapor phase (223) and for cis-stilbene in several different solvents (224) in order to explore solvation effects on the excited-state torsional potentials. A slight dependence of the out-of-plane Raman intensities of cis-stilbene on solvent polarity correlates well with solvent polarity effects on the partitioning between torsionally isomerized and ring-closed photoproducts, suggesting that the quantum yields for these competing processes are being dictated at least in part by the dynamics in the Franck–Condon region.

9.4.5. Aromatic Hydrocarbons

Since most aromatic hydrocarbons exhibit fairly strong fluoresence from their lowest singlet states, resonance Raman studies of such molecules have generally involved excitation into higher excited singlet states. One exception is azulene, in which rapid $S_1 \rightarrow S_0$ internal conversion renders the resonance Raman scattering reasonably strong compared with the S_1 relaxed fluorescence. Raman scattering on resonance with the red $S_1 \leftarrow S_0$ transition has been analyzed by several groups. Page and co-workers (95,225,226) employed both transform methods and direct modeling of excitation profiles for a number of transitions to extract excited-state geometry changes, non-Condon couplings, and Duschinsky rotations from the profiles in CS_2 solution. Sue et al. (87) used the stochastic theory of line broadening to analyze the same experimental data, and obtained their best fit with the solvent-induced broadening in the slow modulation regime, but not inhomogeneous. Nibbering et al. (79,227) analyzed the temperature-dependent absorption lineshapes of the $S_1 \leftarrow S_0$ region of azulene in perfluorohexane, isopentane, cyclohexane, and CS_2 solutions and fitted them to a stochastic model, obtaining intermediate time scales for the solvent fluctuations. They further showed that the frequency dependence of the Raman to fluorescence quantum yields for near-origin excitation is reasonably consistent with these fitting parameters.

Cable and Albrecht (67) studied the excitation profiles on resonance with the S_4 state of azulene as a test of transform theory. They concluded that the "standard assumptions" are violated for at least some modes. Hudson and co-workers (228–

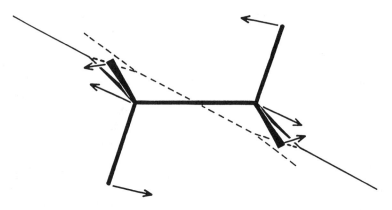

Figure 9.15. Resonance Raman-deduced excited-state geometry changes in 1,3-cyclohexadiene, 10 fs after photoexcitation. Arrows represent the superposition of the motion along the four dominant normal coordinates, multiplied by a factor of 20, with motions of the diene moiety suppressed for clarity. [Reproduced with permission from Ref. 217.]

231) studied the far- and vacuum-UV resonance Raman spectra of benzene and substituted benzenes, mainly with the aim of unraveling the complex vibronic coupling mechanisms and state assignments. Jones and Asher (232) measured Raman spectra on resonance with the S_2, S_3, and S_4 excited states of pyrene in solution and used a transform theory approach to obtain excited-state geometry changes and non-Condon coupling parameters for several modes on resonance with S_4.

9.4.6. Biological Chromophores

Resonance Raman spectroscopy has long been a valuable tool for both ground-state vibrational analysis and excited-state structural and dynamic studies of biopolymers because of its ability to selectively observe vibrations of a particular chromophoric moiety. Proteins containing porphyrin or retinal chromophores, which absorb strongly in the visible, have been the most widely studied. There is an enormous literature on the resonance Raman spectroscopy of porphyrins, both free in solution and protein bound, but most of this work focuses on the extraction of ground-state properties. Some recent studies that address the excited-state information contained in resonance Raman intensities can be found in Ref. 233. Champion and co-workers (40,147,149,234,235) examined the Soret band resonant Raman spectra of heme proteins in great detail, including careful measurements of absorption bandshapes and absolute cross sections as a function of temperature, and absolute excitation profiles for both Stokes and anti-Stokes transitions. These data have been analyzed by a combination of transform methods and direct modeling to determine excited-state displacements and non-Condon coupling strengths for the numerous Raman-active vibrations of the porphyrin, as well as to discriminate among vibronic activity of low-frequency vibrations, environmental pure dephasing, and inhomogeneous broadening contributions to the electronic spectral breadth. In related work, Reinisch and co-workers (236,237) examined the absolute intensity of the quasielastic light scattering (Rayleigh and Brillouin) from cytochrome c and metmyoglobin, and found it to be considerably larger than predicted from the Kramers–Kronig transform of the absorption spectrum. This discrepancy remains to be satisfactorily explained. Recent Raman studies of the retinal-containing chromophores, rhodopsin, bacteriorhodopsin, and their relatives, focused mainly on vibrational analysis of electronic ground states. Loppnow and Mathies (238) did report a partial excitation profile study of rhodopsin at room temperature from which the excited-state potential parameters for 25 normal coordinates, plus the environment-induced broadening, were extracted.

Carotenoids, with their extremely strong, visible absorption bands, weak fluorescence, and relatively high photostability, were among the earliest biologically relevant chromophores studied by resonance Raman techniques (239). More recently, Page and co-workers (151) obtained and modeled excitation profiles of β-carotene at two different temperatures in order to separate thermal and inhomogeneous site broadening contributions to the spectral breadth. They found no evidence for significant inhomogeneous broadening. Saito and Mikami (240), assuming a simple separation of Lorentzian homogeneous and Gaussian inhomogeneous broadening, also concluded that homogeneous broadening dominated their room temperature excitation profiles of lycopene in toluene. Sue and Mukamel (241), in contrast, modeled several groups' published profiles of β-carotene as a function of solvent, temperature, and pressure to extract the line-broadening parameters, and obtained their best fits with some inhomogeneous broadening that increased with increasing temperature. They also modeled profiles for tetradesmethyl-β-carotene using a stochastic model that did not include inhomogeneous broadening, and found the solvation dynamics to be in the slow modulation regime (98). Watanabe et al. (85) measured and modeled resonance Raman and fluorescence excitation profiles of β-carotene in isopentane at 177 K, and also found that stochastic line broadening close to the slow modulation limit fit their data best. None of these experimental data included absolute cross sections, which would have provided a more stringent test of the conclusions.

Resonance Raman spectroscopy of proteins and nucleic acids and their constituents also has a long history. The pioneering studies of Peticolas and co-workers (141,242–244) on nucleic acid components were among the first resonance Raman experiments performed with UV excitation. More recently, Fodor and co-workers (245) reported qualitative studies with excitation deeper into the UV. Much recent work in UV resonance Raman spectroscopy has focused on protein constituents— amino acids and their analogs (162,246–251), peptides (252), and simple models for the peptide bond (94,253,254). In most of these studies the intensities have been interpreted only qualitatively, although detailed excitation profiles have been reported for the aromatic amino acids (248,250), the absorption and resonance Raman spectra of aqueous N-acetylpyrrolidine have been modeled quantitatively (94), and the effects of interferences between electronic states on the excitation profiles of formamide have been modeled (253).

9.4.7. Electron and Proton Transfer

A topic of considerable recent interest is the connection between Raman intensities on resonance with electron-transfer transitions and the rates of the corresponding nonphotochemical return electron-transfer processes (96). Both processes depend on the "reorganization energies" (proportional to the squares of the displacements Δ) for the solvent and each of the Franck–Condon active modes, and determination of these parameters from resonance Raman should be helpful in understanding and/or modeling the return electron-transfer rates. Hupp and co-workers (255–259) have been examining Raman spectra of a variety of photoinduced charge-transfer transitions to extract the vibrational reorganization energies. These analyses, some based on single-frequency pre- or postresonance spectra, utilize simple "short-time" approximations to the Raman and absorption spectra. The intermolecular charge-transfer complex between hexamethylbenzene and tetracyanoethylene, one of the earliest systems for which Raman excitation profiles were reported (260,261), has recently been analyzed in detail by two groups. Myers and co-workers (157) measured absolute excitation profiles in CCl_4 and analyzed them by direct modeling with non-Condon terms included and the solvent treated as a Brownian oscillator, while McHale and co-workers (262) used CH_2Cl_2 as the solvent and obtained the excited-state displacements in each mode from a transform analysis that included non-Condon contributions. While the two groups' analyses are qualitatively similar, the absolute Raman intensities and profile lineshapes differ considerably. Myers et al. (157) also found that perdeuteriation of the hexamethylbenzene donor has only very minor effects on the strongly Raman-active modes, an observation that is difficult to rationalize with the significant isotope effect observed on the nonphotochemical return electron-transfer rate. Further work will be needed to determine the applicability of the Raman-derived vibrational reorganization energies in modeling of the return electron-transfer kinetics (96).

Efforts have been made to probe the dynamics of fast excited-state proton-transfer reactions through resonance Raman intensities. Spectra obtained on resonance with the proton-transfer electronic absorption band of 2-hydroxyacetophenone showed intensity in many vibrations, none of which involved significant proton motion (263). This result implies that the proton transfer occurs subsequent to structural rearrangement along other coordinates.

9.5. CONCLUSIONS AND PROSPECTS

It is clear that resonance Raman intensities, coupled with appropriate quantitative modeling, can be very useful in revealing information about the potential energy surfaces of excited electronic states in the region near the ground-state geometry. While generally only a rather small part of the excited-state surface contributes significantly to the experimental intensities, and a number of possible complications must be overcome in order to extract the desired information from the spectra, resonance Raman remains the most direct way to explore the structure and dynamics of excited states when the pure electronic (absorption or fluorescence) spectra do not yield vibrational resolution.

Future applications of the technique appear certain to involve an increasing role of theory for predicting spectra to which the experimental results will be compared. Improvements in computational speed and in methods for numerical solution of the time-dependent Schrödinger equation now make it possible to calculate spectra for (small) multidimensional anharmonic potential energy surfaces. Comparison of calculated and experimental resonance Raman spectra will become increasingly important as a technique for evaluating the accuracy of calculated potential surfaces for photoreactive excited states. Even with larger molecules, where neither electronic structure calculations nor wave packet propagation methods can yet handle the full multidimensional complexity of the problem, more approximate theoretical methods for predicting spectra are likely to play an increasingly major role in the interpretation of experiments.

Resonance Raman studies should make important future contributions to developing a clear and consistent picture of electronic and vibrational dephasing processes of molecules in condensed phases. While there is abundant evidence, both experimental and theoretical, that resonance Raman and fluorescence intensities and bandshapes are quite sensitive to electronic dephasing processes, it is not yet clear under what conditions resonance Raman cross sections in solution may be properly expressed as the modulus squared of an amplitude, how prevalent or diagnostic the "broad Raman" component of the emission may be, and how the answers to these questions depend on the specific model chosen for the solvent-induced dephasing. Once experimental and theoretical studies in concert can resolve these issues, the origin and time scale of dephasing processes in liquids can be attacked through a combination of resonance Raman and fluorescence spectroscopies together with other optical probes, such as the photon echo.

REFERENCES

1. Myers, A. B. and Mathies, R. A. In *Biological Applications of Raman Spectroscopy*; Spiro, T. G. (Ed.). Wiley, New York, 1987, Vol. 2, p. 1.

2. Lee, S.-Y. and Heller, E. J. *J. Chem. Phys.* **1979**, *71*, 4777.

3. Heller, E. J., Sundberg, R. L., and Tannor, D. *J. Phys. Chem.* **1982**, *86*, 1822.

4. Zink, J. I. and Shin, K.-S. K. *Adv. Photochem.* **1991**, *16*, 119.

5. Clark, R. J. H. and Dines, T. J. *Angew. Chem. Int. Ed. Engl.* **1986**, *25*, 131.

6. Champion, P. M. and Albrecht, A. C. *Annu. Rev. Phys. Chem.* **1982**, *33*, 353.

7. Sonnich Mortensen, O. and Hassing, S. *Adv. Infrared Raman Spectrosc.* **1980**, *6*, 1.

8. Siebrand, W. and Zgierski, M. Z. In *Excited States*; Lim, E. C. (Ed.). Academic, New York, 1979, Vol. 4, p. 1.

9. Warshel, A. *Annu. Rev. Biophys. Bioeng.* **1977**, *6*, 273.

10. Johnson, B. B. and Peticolas, W. L. *Annu. Rev. Phys. Chem.* **1976**, *27*, 465.

11. Porterfield, D. R. and Campion, A. *J. Am. Chem. Soc.* **1988**, *110*, 408.

12. Schulte, A., Lenk, T. J., Hallmark, V. M., and Rabolt, J. F. *Appl. Spectrosc.* **1991**, *45*, 325.

13. Schulte, A. *Appl. Spectrosc.* **1992**, *46*, 891.

14. Buesener, H., Renn, A., Brieger, M., vonMoers, F., and Hese, A. *Appl. Phys. B* **1986**, *39*, 77.

15. Miyazaki, K., Sakai, H., and Sato, T. *Opt. Lett.* **1986**, *11*, 797.

16. Kato, K. *IEEE J. Quantum Electron.* **1986**, *QE-22*, 1013.

17. Asher, S. A., Johnson, C. R., and Murtaugh, J. *Rev. Sci. Instrum.* **1983**, *54*, 1657.

18. Borsutzky, A., Brunger, R., Huang, C., and Wallenstein, R. *Appl. Phys. B* **1991**, *52*, 55.

19. Wilke, V. and Schmidt, W. *Appl. Phys.* **1979**, *18*, 177.

20. Fodor, S. P. A., Rava, R. P., Copeland, R. A., and Spiro, T. G. *J. Raman Spectrosc.* **1986**, *17*, 471.

21. Hudson, B. and Sension, R. J. In *Vibrational spectra and structure*; Durig, J. and Bist, H. D. (Eds.). Amsterdam: Elsevier, Vol. 17.

22. Stoicheff, B. P. *Pure Appl. Chem.* **1987**, *59*, 1237.

23. White, J. C. *Top. Appl. Phys.* **1992**, *59*, 115.

24. Jones, C. M., DeVito, V. L., Harmon, P. A., and Asher, S. A. *Appl. Spectrosc.* **1987**, *41*, 1268.

25. Gustafson, T. L. *Opt. Commun.* **1988**, *67*, 53.

26. Benson, R. L., Iwata, K., Weaver, W. J., and Gustafson, T. L. *Appl. Spectrosc.* **1992**, *46*, 240.

27. Hudson, B., Kelly, P. B., Ziegler, L. D., Desiderio, R. A., Gerrity, D. P., Hess, W., and Bates, R. In *Advances in Laser Spectroscopy*; Garetz, B. A. and Lombardi, J. R. (Eds.). Wiley, New York, 1986, Vol. 3, p. 1.

28. Devlin, G. E., Davis, J. L., Chase, L., and Geschwind, S. *Appl. Phys. Lett.* **1971**, *19*, 138.

29. Schlotter, N. E., Schaertel, S. A., Kelty, S. P., and Howard, R. *Appl. Spectrosc.* **1988**, *42*, 746.

30. Murray, C. A. and Dierker, S. B. *J. Opt. Soc. Am.* **1986**, *3*, 2151.

31. Bilhorn, R. B., Epperson, P. M., Sweedler, J. V., and Denton, M. B. *Appl. Spectrosc.* **1987**, *41*, 1125.

32. Bilhorn, R. B., Sweedler, J. V., Epperson, P. M., and Denton, M. B. *Appl. Spectrosc.* **1987**, *41*, 1114.

33. Chase, B. and Talmi, Y. *Appl. Spectrosc.* **1991**, *45*, 929.

34. Mathies, R. In *Chemical and Biological Applications of Lasers*; Moore, C. B. (Ed.). Academic, New York, 1979, Vol. IV, p. 55.

35. Ludwig, M. and Asher, S. A. *Appl. Spectrosc.* **1988**, *42*, 1458.

36. Strekas, T. C., Adams, D. H., Packer, A., and Spiro, T. G. *Appl. Spectrosc.* **1974**, *28*, 324.

37. Shriver, D. F. and Dunn, J. B. R. *Appl. Spectrosc.* **1974**, *28*, 319.

38. Myers, A. B., Li, B., and Ci, X. *J. Chem. Phys.* **1988**, *89*, 1876.

39. Womack, J. D., Mann, C. K., and Vickers, T. J. *Appl. Spectrosc.* **1989**, *43*, 527.

40. Schomacker, K. T. and Champion, P. M. *J. Chem. Phys.* **1989**, *90*, 5982.

41. Purcell, F. J., Kaminski, R., and Russavage, E. *Appl. Spectrosc.* **1980**, *34*, 323.

42. Ouillon, R. and Adam, S. *J. Raman Spectrosc.* **1982**, *12*, 281.

43. Tanabe, K. and Tachiya, M. *Spectrochim. Acta* **1981**, *37A*, 895.

44. Michaelian, K. H. and Friesen, W. I. *Appl. Spectrosc.* **1988**, *42*, 1538.

45. Kung, C.-Y., Chang, B.-Y., Kittrell, C., Johnson, B. R., and Kinsey, J. L. *J. Phys. Chem.* **1993**, *97*, 2228.

46. Murphy, W. F., Holzer, W., and Bernstein, H. J. *Appl. Spectrosc.* **1969**, *23*, 211.

47. Kato, Y. and Takuma, H. *J. Chem. Phys.* **1971**, *54*, 5398.

48. Kato, Y. and Takuma, H. *J. Opt. Soc. Am.* **1971**, *61*, 347.

49. Penney, C. M., Goldman, L. M., and Lapp, M. *Nature Phys. Sci.* **1972**, *235*, 110.

50. Bischel, W. K. and Black, G. In *Excimer Lasers-1983*; Rhodes, C. K., Egger, H., and Pummer, H. (Eds.). American Institute of Physics, New York, 1983, p. 181.

51. Dudik, J. M., Johnson, C. R., and Asher, S. A. *J. Chem. Phys.* **1985**, *82*, 1732.

52. Trulson, M. O. and Mathies, R. A. *J. Chem. Phys.* **1986**, *84*, 2068.

53. Schomacker, K. T., Delaney, J. K., and Champion, P. M. *J. Chem. Phys.* **1986**, *85*, 4240.

54. Eysel, H. H. and Bertie, J. E. *J. Raman Spectrosc.* **1988**, *19*, 59.

55. Eysel, H. H. *J. Raman Spectrosc.* **1988**, *19*, 223.

56. Li, B. and Myers, A. B. *J. Phys. Chem.* **1990**, *94*, 4051.

57. Song, S. and Asher, S. A. *Biochem.* **1991**, *30*, 1199.

58. Henneker, W. H., Penner, A. P., Siebrand, W., and Zgierski, M. Z. *J. Chem. Phys.* **1978**, *69*, 1704.

59. Sension, R. J., Kobayashi, T., and Strauss, H. L. *J. Chem. Phys.* **1987**, *87*, 6221.

60. Sension, R. J. and Strauss, H. L. *J. Chem. Phys.* **1986**, *85*, 3791.

61. Harris, R. A., Wedlock, M. R., Butler, L. J., and Freed, K. F. *J. Chem. Phys.* **1992**, *96*, 2437.

62. Shang, Q.-Y. and Hudson, B. S. *Chem. Phys. Lett.* **1991**, *183*, 63.

63. Strommen, D. P. *J. Chem. Educ.* **1992**, *69*, 803.

64. Ziegler, L. D., Chung, Y. C., Wang, P., and Zhang, Y. P. *J. Chem. Phys.* **1989**, *90*, 4125.

65. Li, B. and Myers, A. B. *J. Chem. Phys.* **1988**, *89*, 6658.

66. Li, B. and Myers, A. B. *J. Chem. Phys.* **1991**, *94*, 2458.

67. Cable, J. R. and Albrecht, A. C. *J. Chem. Phys.* **1986**, *84*, 1969.

68. Wang, P. G. and Ziegler, L. D. *J. Phys. Chem.* **1993**, *97*, 3139.

69. Lao, K. Q., Person, M. D., Xayariboun, P., and Butler, L. J. *J. Chem. Phys.* **1990**, *92*, 823.

70. Dawson, P. *Spectrochim. Acta* **1972**, *28A*, 715.

71. Asher, S. A. *Anal. Chem.* **1993**, *65*, 201.

72. Myers, A. B., Trulson, M. O., and Mathies, R. A. *J. Chem. Phys.* **1985**, *83*, 5000.

73. Carroll, M. K., Miller, R. M., Keller, R. A., and Hieftje, G. M. *Appl. Spectrosc.* **1992**, *46*, 442.

74. Kamogawa, K., Fujii, T., and Kitagawa, T. *Appl. Spectrosc.* **1988**, *42*, 248.

75. Wirth, M. J. and Chou, S.-H. *Anal. Chem.* **1988**, *60*, 1882.

76. Shreve, A. P., Cherepy, N. J., and Mathies, R. A. *Appl. Spectrosc.* **1992**, *46*, 707.

77. Friedman, J. M. and Rousseau, D. L. *Chem. Phys. Lett.* **1978**, *55*, 488.

78. Myers, A. B. and Li, B. *J. Chem. Phys.* **1990**, *92*, 3310.

79. Nibbering, E. T. J., Duppen, K., and Wiersma, D. A. *J. Chem. Phys.* **1990**, *93*, 5477.

80. Hochstrasser, R. M. and Nyi, C. A. *J. Chem. Phys.* **1979**, *70*, 1112.

81. Zhang, Y. P. and Ziegler, L. D. *J. Chem. Phys.* **1990**, *93*, 8605.

82. Albrecht, A. C. *J. Chem. Phys.* **1961**, *34*, 1476.

83. Kubo, R. In *Fluctuation, relaxation, and resonance in magnetic systems*; ter Haar, D., (Ed.). Oliver and Boyd, Edinburgh, 1962, p. 23.

84. Yan, Y. J. and Mukamel, S. *J. Chem. Phys.* **1987**, *86*, 6085.

85. Watanabe, J., Kinoshita, S., and Kushida, T. *J. Chem. Phys.* **1987**, *87*, 4471.

86. Takagahara, T., Hanamura, E., and Kubo, R. *J. Phys. Soc. Jpn.* **1977**, *43*, 802.

87. Sue, J., Yan, Y. J., and Mukamel, S. *J. Chem. Phys.* **1986**, *85*, 462.

88. Lee, D. and Albrecht, A. C. *Adv. Infrared Raman Spectrosc.* **1985**, *12*, 179.

89. Li, B., Johnson, A. E., Mukamel, S., and Myers, A. B. *J. Am. Chem. Soc.* **1994**, *116*, 11039.

90. Melinger, J. S. and Albrecht, A. C. *J. Phys. Chem.* **1987**, *91*, 2704.

91. Shapiro, M. *J. Chem. Phys.* **1993**, *99*, 2453.

92. Remacle, F., Levine, R. D., and Kinsey, J. L. *Chem. Phys. Lett.* **1993**, *205*, 267.

93. Remacle, F. and Levine, R. D. *J. Chem. Phys.* **1993**, *99*, 4908.

94. Harhay, G. P. and Hudson, B. S. *J. Phys. Chem.* **1993**, *97*, 8158.

95. Lu, H. M. and Page, J. B. *J. Chem. Phys.* **1990**, *92*, 7038.

96. Myers, A. B. *Chem. Phys.* **1994**, *180*, 215.

97. Barqawi, K. R., Murtaza, Z., and Meyer, T. J. *J. Phys. Chem.* **1991**, *95*, 47.

98. Sue, J., Mukamel, S., Okamoto, H., Hamaguchi, H., and Tasumi, M. *Chem. Phys. Lett.* **1987**, *134*, 87.

99. Tang, J. and Albrecht, A. C. In *Raman Spectroscopy*; Szymanski, H. A. (Ed.). Plenum, New York, 1970; Vol. 2, p. 33.

100. Harris, R. A., Mathies, R., and Myers, A. *Chem. Phys. Lett.* **1983**, *94*, 327.

101. Galica, G. E., Johnson, B. R., Kinsey, J. L., and Hale, M. O. *J. Phys. Chem.* **1991**, *95*, 7994.

102. Markel, F. and Myers, A. B. *J. Chem. Phys.* **1993**, *98*, 21.

103. Phillips, D. L. and Myers, A. B. *J. Chem. Phys.* **1991**, *95*, 226.

104. Sension, R. J., Brudzynski, R. J., Hudson, B. S., Zhang, J., and Imre, D. G. *Chem. Phys.* **1990**, *141*, 393.

105. Stein, P., Miskowski, V., Woodruff, W. H., Griffin, J. P., Werner, K. G., Gaber, B. P., and Spiro, T. G. *J. Chem. Phys.* **1976**, *64*, 2159.

106. Schick, G. A. and Bocian, D. F. *J. Raman Spectrosc.* **1981**, *11*, 27.

107. Reber, C. and Zink, J. I. *J. Phys. Chem.* **1992**, *96*, 571.

108. Shin, K.-S. K. and Zink, J. I. *J. Am. Chem. Soc.* **1990**, *112*, 7148.

109. Warshel, A. and Dauber, P. *J. Chem. Phys.* **1977**, *66*, 5477.

110. Warshel, A. *Isr. J. Chem.* **1973**, *11*, 709.

111. Siebrand, W. and Zgierski, M. Z. *Chem. Phys. Lett.* **1979**, *62*, 3.

112. Hemley, R. J., Dawson, J. I., and Vaida, V. *J. Chem. Phys.* **1983**, *78*, 2915.

113. Morris, D. E. and Woodruff, W. H. *J. Phys. Chem.* **1985**, *89*, 5795.

114. Kupka, H. and Cribb, P. H. *J. Chem. Phys.* **1986**, *85*, 1303.

115. Myers, A. B., Mathies, R. A., Tannor, D. J., and Heller, E. J. *J. Chem. Phys.* **1982**, *77*, 3857.

116. Heller, E. J. *Acc. Chem. Res.* **1981**, *14*, 368.

117. Tannor, D. J. and Heller, E. J. *J. Chem. Phys.* **1982**, *77*, 202.

118. Myers, A. B., Harris, R. A., and Mathies, R. A. *J. Chem. Phys.* **1983**, *79*, 603.

119. Yan, Y. J. and Mukamel, S. *J. Chem. Phys.* **1986**, *85*, 5908.

120. Feit, M. D., Fleck, J. A., Jr., and Steiger, A. *J. Comp. Phys.* **1982**, *47*, 412.

121. Feit, M. D. and Fleck, J. A., Jr. *J. Chem. Phys.* **1983**, *78*, 301.

122. Kosloff, R. *J. Phys. Chem.* **1988**, *92*, 2087.

123. Untch, A., Weide, K., and Schinke, R. *Chem. Phys. Lett.* **1991**, *180*, 265.

124. Heather, R. and Metiu, H. *J. Chem. Phys.* **1989**, *90*, 6903.

125. von Dirke, M. and Schinke, R. *Chem. Phys. Lett.* **1992**, *196*, 51.

126. Guo, H. *Chem. Phys. Lett.* **1991**, *187*, 360.

127. Bandrauk, A. D. and Shen, H. *J. Chem. Phys.* **1993**, *99*, 1185.

128. Manthe, U. and Hammerich, A. D. *Chem. Phys. Lett.* **1993**, *211*, 7.

129. Messina, M. and Coalson, R. D. *J. Chem. Phys.* **1989**, *90*, 4015.

130. Founargiotakis, M. and Light, J. C. *J. Chem. Phys.* **1990**, *93*, 633.

131. Someda, K., Kondow, T., and Kuchitsu, K. *J. Phys. Chem.* **1991**, *95*, 2156.

132. Ci, X., Pereira, M. A., and Myers, A. B. *J. Chem. Phys.* **1990**, *92*, 4708.

133. Yang, T.-S. and Myers, A. B. *J. Chem. Phys.* **1991**, *95*, 6207.

134. Mukamel, S. and Yan, Y. J. In *Recent trends in Raman spectroscopy*; Banerjee, S. B. and Jha, S. S. (Eds.). World Scientific, Singapore, 1989, p. 160.

135. Bosma, W. B., Yan, Y. J., and Mukamel, S. *Phys. Rev. A* **1990**, *42*, 6920.

136. Mukamel, S. *Annu. Rev. Phys. Chem.* **1990**, *41*, 647.

137. Rosenthal, S. J., Xie, X., Du, M., and Fleming, G. R. *J. Chem. Phys.* **1991**, *95*, 4715.

138. Tonks, D. L. and Page, J. B. *Chem. Phys. Lett.* **1979**, *66*, 449.

139. Hiznyakov, V. and Tehver, I. *Phys. Stat. Sol.* **1967**, *21*, 755.

140. Champion, P. M. and Albrecht, A. C. *Chem. Phys. Lett.* **1981**, *82*, 410.

141. Chinsky, L., Laigle, A., Peticolas, W. L., and Turpin, P.-Y. *J. Chem. Phys.* **1982**, *76*, 1.

142. Tonks, D. L. and Page, J. B. *J. Chem. Phys.* **1982**, *76*, 5820.

143. Lu, H. M. and Page, J. B. *J. Chem. Phys.* **1988**, *88*, 3508.

144. Stallard, B. R., Champion, P. M., Callis, P. R., and Albrecht, A. C. *J. Chem. Phys.* **1983**, *78*, 712.

145. Stallard, B. R., Callis, P. R., Champion, P. M., and Albrecht, A. C. *J. Chem. Phys.* **1984**, *80*, 70.

146. Chan, C. K. *J. Chem. Phys.* **1984**, *81*, 1614.

147. Morikis, D., Li, P., Bangcharoenpaurpong, O., Sage, J. T., and Champion, P. M. *J. Phys. Chem.* **1991**, *95*, 3391.

148. Chan, C. K. and Page, J. B. *J. Chem. Phys.* **1983**, *79*, 5234.

149. Schomacker, K. T., Bangcharoenpaurpong, O., and Champion, P. M. *J. Chem. Phys.* **1984**, *80*, 4701.

150. Patapoff, T. W., Turpin, P.-Y., and Peticolas, W. L. *J. Phys. Chem.* **1986**, *90*, 2347.

151. Lee, S. A., Chan, C. K., Page, J. B., and Walker, C. T. *J. Chem. Phys.* **1986**, *84*, 2497.

152. Li, P. and Champion, P. M. *J. Chem. Phys.* **1987**, *88*, 761.

153. Cable, J. R. and Albrecht, A. C. *J. Chem. Phys.* **1986**, *84*, 4745.

154. Lee, D., Stallard, B. R., Champion, P. M., and Albrecht, A. C. *J. Phys. Chem.* **1984**, *88*, 6693.

155. Rava, R. P. *J. Chem. Phys.* **1987**, *87*, 3758.

156. Gu, Y. and Champion, P. M. *Chem. Phys. Lett.* **1990**, *171*, 254.

157. Markel, F., Ferris, N. S., Gould, I. R., and Myers, A. B. *J. Am. Chem. Soc.* **1992**, *114*, 6208.

158. Yan, Y. J. and Mukamel, S. *J. Chem. Phys.* **1988**, *89*, 5160.

159. Li, P., Sage, J. T., and Champion, P. M. *J. Chem. Phys.* **1992**, *97*, 3214.

160. Peticolas, W. L., Strommen, D. P., and Lakshminarayanan, V. *J. Chem. Phys.* **1980**, *73*, 4185.

161. Lagant, P., Derreumaux, P., Vergoten, G., and Peticolas, W. *J. Comput. Chem.* **1991**, *12*, 731.

162. Markham, L. M., Mayne, L. C., Hudson, B. S., and Zgierski, M. Z. *J. Phys. Chem.* **1993**, *97*, 10319.

163. Wright, P. G., Stein, P., Burke, J. M., and Spiro, T. G. *J. Am. Chem. Soc.* **1979**, *101*, 3531.

164. Schick, G. A. and Bocian, D. F. *J. Am. Chem. Soc.* **1984**, *106*, 1682.

165. Strickler, S. J. and Berg, R. A. *J. Chem. Phys.* **1962**, *37*, 814.

166. Ranade, A. and Stockburger, M. *Chem. Phys. Lett.* **1973**, *22*, 257.

167. Grundherr, C. V. and Stockburger, M. *Chem. Phys. Lett.* **1973**, *22*, 253.

168. Myers, A. B. and Hochstrasser, R. M. *J. Chem. Phys.* **1987**, *87*, 2116.

169. Ziegler, L. D. *J. Chem. Phys.* **1986**, *84*, 6013.

170. Wang, P. G. and Ziegler, L. D. *J. Chem. Phys.* **1989**, *90*, 4115.

171. Ziegler, L. D., Chung, Y. C., Wang, P. G., and Zhang, Y. P. *J. Phys. Chem.* **1990**, *94*, 3394.

172. Ziegler, L. D. *J. Chem. Phys.* **1987**, *86*, 1703.

173. Wang, P. G. and Ziegler, L. D. *J. Chem. Phys.* **1991**, *95*, 288.

174. Zhang, Y. P. and Ziegler, L. D. *J. Phys. Chem.* **1989**, *93*, 6665.

175. Westre, S. G., Kelly, P. B., Zhang, Y. P., and Ziegler, L. D. *J. Chem. Phys.* **1991**, *94*, 270.

176. Wang, P. G., Zhang, Y. P., Ruggles, C. J., and Ziegler, L. D. *J. Chem. Phys.* **1990**, *92*, 2806.

177. Ziegler, L. D. and Hudson, B. *J. Phys. Chem.* **1984**, *88*, 1110.

178. Desiderio, R. A., Gerrity, D. P., and Hudson, B. S. *Chem. Phys. Lett.* **1985**, *115*, 29.

179. Westre, S. G., Gansberg, T. E., Kelly, P. B., and Ziegler, L. D. *J. Phys. Chem.* **1992**, *96*, 3610.

180. Liu, X., Westre, S. G., Getty, J. D., and Kelly, P. B. *Chem. Phys. Lett.* **1992**, *188*, 42.

181. Berryhill, J., Pramanick, S., Zgierski, M. Z., Zerbetto, F., and Hudson, B. S. *Chem. Phys. Lett.* **1993**, *205*, 39.

182. Sension, R. J., Kobayashi, T., and Strauss, H. L. *J. Chem. Phys.* **1987**, *87*, 6233.

183. Sension, R. J. and Strauss, H. L. *J. Chem. Phys.* **1988**, *88*, 2289.

184. Imre, D. G., Kinsey, J. L., Field, R. W., and Katayama, D. H. *J. Phys. Chem.* **1982**, *86*, 2564.

185. Imre, D., Kinsey, J. L., Sinha, A., and Krenos, J. *J. Phys. Chem.* **1984**, *88*, 3956.

186. Sundberg, R. L., Imre, D., Hale, M. O., Kinsey, J. L., and Coalson, R. D. *J. Phys. Chem.* **1986**, *90*, 5001.

187. Hale, M. O., Galica, G. E., Glogover, S. G., and Kinsey, J. L. *J. Phys. Chem.* **1986**, *90*, 4997.

188. Johnson, B. R. and Kinsey, J. L. *J. Chem. Phys.* **1987**, *87*, 1525.

189. Myers, A. B. and Markel, F. *Chem. Phys.* **1990**, *149*, 21.

190. Markel, F. and Myers, A. B. *Chem. Phys. Lett.* **1990**, *167*, 175.

191. Phillips, D. L., Lawrence, B. A., and Valentini, J. J. *J. Phys. Chem.* **1991**, *95*, 9085.

192. Phillips, D. L., Myers, A. B., and Valentini, J. J. *J. Phys. Chem.* **1992**, *96*, 2039.

193. Wedlock, M. R., Jensen, E., Butler, L. J., and Freed, K. F. *J. Phys. Chem.* **1991**, *95*, 8096.

194. Engel, V., Staemmler, V., Van der Wal, R. L., Crim, F. F., Sension, R. J., Hudson, B., Andresen, P., Hennig, S., Weide, K., and Schinke, R. *J. Phys. Chem.* **1992**, *96*, 3201.

195. Sension, R. J., Brudzynski, R. J., and Hudson, B. S. *Phys. Rev. Lett.* **1988**, *61*, 694.

196. von Dirke, M., Heumann, B., Schinke, R., Sension, R. J., and Hudson, B. S. *J. Chem. Phys.* **1993**, *99*, 1050.

197. Person, M. D., Lao, K. Q., Eckholm, B. J., and Butler, L. J. *J. Chem. Phys.* **1989**, *91*, 812.

198. Brudzynski, R. J., Sension, R. J., and Hudson, B. *Chem. Phys. Lett.* **1990**, *165*, 487.

199. Lao, K. Q., Jensen, E., Kash, P. W., and Butler, L. J. *J. Chem. Phys.* **1990**, *93*, 3958.

200. Phillips, D. L. and Myers, A. B. *J. Phys. Chem.* **1991**, *95*, 7164.

201. Clark, R. J. H. and Stewart, B. *J. Am. Chem. Soc.* **1981**, *103*, 6593.

202. Shin, K. S., Clark, R. J. H., and Zink, J. I. *J. Am. Chem. Soc.* **1990**, *112*, 3754.

203. Lee, S.-Y. and Lee, S. C. *J. Chem. Phys.* **1992**, *96*, 5734.

204. Adelman, D. and Gerrity, D. P. *J. Phys. Chem.* **1990**, *94*, 4055.

205. Dines, T. J. *Spectrochim. Acta* **1988**, *44A*, 1087.

206. Chadwick, R. R., Gerrity, D. P., and Hudson, B. S. *Chem. Phys. Lett.* **1985**, *115*, 24.

207. Chadwick, R. R., Zgierski, M. Z., and Hudson, B. S. *J. Chem. Phys.* **1991**, *95*, 7204.

208. Phillips, D. L., Zgierski, M. Z., and Myers, A. B. *J. Phys. Chem.* **1993**, *97*, 1800.

209. Trulson, M. O. and Mathies, R. A. *J. Phys. Chem.* **1990**, *94*, 5741.

210. Strahan, G. D. and Hudson, B. S. *J. Chem. Phys.* **1993**, *99*, 5780.

211. Myers, A. B. and Pranata, K. S. *J. Phys. Chem.* **1989**, *93*, 5079.

212. Amstrup, B., Langkilde, F. W., Bajdor, K., and Wilbrandt, R. *J. Phys. Chem.* **1992**, *96*, 4794.

213. Ci, X. and Myers, A. B. *J. Chem. Phys.* **1992**, *96*, 6433.

214. Westerfield, C. and Myers, A. B. *Chem. Phys. Lett.* **1993**, *202*, 409.

215. Sztainbuch, I. W. and Leroi, G. E. *J. Chem. Phys.* **1990**, *93*, 4642.

216. Trulson, M. O., Dollinger, G. D., and Mathies, R. A. *J. Am. Chem. Soc.* **1987**, *109*, 586.

217. Trulson, M. O., Dollinger, G. D., and Mathies, R. A. *J. Chem. Phys.* **1989**, *90*, 4274.

218. Lawless, M. K. and Mathies, R. A. *J. Chem. Phys.* **1994**, *100*, 2492.

219. Reid, P. J., Shreve, A. P., and Mathies, R. A. *J. Phys. Chem.* **1993**, *97*, 12691.

220. Lawless, M. K., Wickham, S. D., and Mathies, R. A. *J. Am. Chem. Soc.* **1994**, *116*, 1593.

221. Sension, R. J. and Hudson, B. S. *J. Chem. Phys.* **1989**, *90*, 1377.

222. Myers, A. B. and Mathies, R. A. *J. Chem. Phys.* **1984**, *81*, 1552.

223. Ci, X. and Myers, A. B. *Chem. Phys. Lett.* **1989**, *158*, 263.

224. Rodier, J.-M. and Myers, A. B. *J. Am. Chem. Soc.* **1993**, *115*, 10791.

225. Brafman, O., Chan, C. K., Khodadoost, B., Page, J. B., and Walker, C. T. *J. Chem. Phys.* **1984**, *80*, 5406.

226. Chan, C. K., Page, J. B., Tonks, D. L., Brafman, O., Khodadoost, B., and Walker, C. T. *J. Chem. Phys.* **1985**, *82*, 4813.

227. Nibbering, E. T. J., Duppen, K., and Wiersma, D. A. *J. Photochem. Photobiol. A* **1992**, *62*, 347.

228. Gerrity, D. P., Ziegler, L. D., Kelly, P. B., Desiderio, R. A., and Hudson, B. *J. Chem. Phys.* **1985**, *83*, 3209.

229. Sension, R. J., Brudzynski, R. J., and Hudson, B. *J. Chem. Phys.* **1991**, *94*, 873.

230. Sension, R. J., Brudzynski, R. J., Li, S., Hudson, B. S., Zerbetto, F., and Zgierski, M. Z. *J. Chem. Phys.* **1992**, *96*, 2617.

231. Li, S. and Hudson, B. *Chem. Phys. Lett.* **1988**, *148*, 581.

232. Jones, C. M. and Asher, S. A. *J. Chem. Phys.* **1988**, *89*, 2649.

233. Spiro, T. G., Czernuszewicz, R. S., and Li, X.-Y. *Coord. Chem. Rev.* **1990**, *100*, 541.

234. Bangcharoenpaurpong, O., Schomacker, K. T., and Champion, P. M. *J. Am. Chem. Soc.* **1984**, *106*, 5688.

235. Schomacker, K. T. and Champion, P. M. *J. Chem. Phys.* **1986**, *84*, 5314.

236. Reinisch, L., Schomacker, K. T., and Champion, P. M. *J. Chem. Phys.* **1987**, *87*, 150.

237. Chiarello, R. and Reinisch, L. *J. Chem. Phys.* **1988**, *88*, 1253.

238. Loppnow, G. R. and Mathies, R. A. *Biophys. J.* **1988**, *54*, 35.

239. Rimai, L., Heyde, M. E., and Gill, D. *J. Am. Chem. Soc.* **1973**, *95*, 4493.

240. Saito, M. and Mikami, Y. *Bull Chem. Soc. Jpn.* **1986**, *59*, 969.

241. Sue, J. and Mukamel, S. *J. Opt. Soc. Am. B* **1988**, *5*, 1462.

242. Blazej, D. C. and Peticolas, W. L. *Proc. Natl. Acad. Sci. USA* **1977**, *74*, 2639.

243. Peticolas, W. L. and Blazej, D. C. *Chem. Phys. Lett.* **1979**, *63*, 604.

244. Blazej, D. C. and Peticolas, W. L. *J. Chem. Phys.* **1980**, *72*, 3134.

245. Fodor, S. P. A., Rava, R. P., Hays, T. R., and Spiro, T. G. *J. Am. Chem. Soc.* **1985**, *107*, 1520.

246. Rava, R. P. and Spiro, T. G. *J. Am. Chem. Soc.* **1984**, *106*, 4062.

247. Rava, R. P. and Spiro, T. G. *J. Phys. Chem.* **1985**, *89*, 1856.

248. Asher, S. A. and Johnson, C. R. *J. Phys. Chem.* **1985**, *89*, 1375.

249. Caswell, D. S. and Spiro, T. G. *J. Am. Chem. Soc.* **1986**, *108*, 6470.

250. Fodor, S. P. A., Copeland, R. A., Grygon, C. A., and Spiro, T. G. *J. Am. Chem. Soc.* **1989**, *111*, 5509.

251. Sension, R. J., Hudson, B., and Callis, P. R. *J. Phys. Chem.* **1990**, *94*, 4015.

252. Copeland, R. A. and Spiro, T. G. *Biochemistry* **1987**, *26*, 2134.

253. Hildebrandt, P., Tsuboi, M., and Spiro, T. G. *J. Phys. Chem.* **1990**, *94*, 2274.

254. Mayne, L. C. and Hudson, B. *J. Phys. Chem.* **1991**, *95*, 2962.

255. Doorn, S. K. and Hupp, J. T. *J. Am. Chem. Soc.* **1989**, *111*, 1142.

256. Doorn, S. K. and Hupp, J. T. *J. Am. Chem. Soc.* **1989**, *111*, 4704.

257. Doorn, S. K., Blackbourn, R. L., Johnson, C. S., and Hupp, J. T. *Electrochim. Acta* **1991**, *36*, 1775.

258. Blackbourn, R. L., Johnson, C. S., and Hupp, J. T. *J. Am. Chem. Soc.* **1991**, *113*, 1060.

259. Blackbourn, R. L., Johnson, C. S., Hupp, J. T., Bryant, M. A., Sobocinski, R. L., and Pemberton, J. E. *J. Phys. Chem.* **1991**, *95*, 10535.

260. Michaelian, K. H., Rieckhoff, K. E., and Voigt, E.-M. *Proc. Nat. Acad. Sci. USA* **1975**, *72*, 4196.

261. Jensen, P. W. *Chem. Phys. Lett.* **1977**, *45*, 415.

262. Britt, B. M., Lueck, H. B., and McHale, J. L. *Chem. Phys. Lett.* **1992**, *190*, 528.

263. Peteanu, L. A. and Mathies, R. A. *J. Phys. Chem.* **1992**, *96*, 6910.

Chapter **X**

ULTRAFAST VIBRATIONAL SPECTROSCOPY: METHODS, THEORY, AND APPLICATIONS

Gilbert C. Walker* and Robin M. Hochstrasser
Department of Chemistry,
University of Pennsylvania, Philadelphia, Pennsylvania

*Present address: Department of Chemistry, University of Pittsburgh

Laser Techniques In Chemistry, Edited by Anne B. Myers and Thomas R. Rizzo.
Techniques of Chemistry Series, Vol. XXIII.
ISBN 0-471-59769-4   © 1995 John Wiley & Sons, Inc.

10.1. INTRODUCTION

Describing the evolution of molecular coordinates of solutions following excitation by short light pulses is a major challenge to chemists. Both the solvent and solute must be considered; formerly common treatments of the solvent as an unstructured bath are being replaced by microscopic descriptions of local solute environments and the pathways of solute molecular dynamics. Central to developing a detailed understanding of the molecular processes are improvements in the materials and computational technology of ultrafast spectroscopies that can provide bond-by-bond detail, for both ensemble averaged and/or isolated chromophores. These methods are promising not only for solute–solvent systems with modest local order but also in biological systems with significant short- and long-range order.

Ultrafast methods used to study chemical processes yield transient electronic and vibrational spectra. Condensed phase electronic spectra are diffuse due to fast dephasing processes, but can yield vibrational structure if coherent dynamical responses to impulsive excitations due to Franck–Condon allowed Raman processes are measured. However, the restriction to Franck–Condon active modes does not apply to transient infrared (IR) spectroscopy. Typical condensed phase vibrational spectra exhibit lines whose inverse widths yield dephasing times on the order of 100 fs to 10 ps. Therefore, spectra are sufficiently resolved for convenient structural analysis on these time scales. On even shorter time scales, with correct treatment of the observed dynamical response, structural detail can be obtained.

An unexpected bonus to applying IR spectroscopy to biological systems has been the time-resolved exposure of IR electronic transitions. These results provide an important challenge to the electronic structure theory of symmetric systems, and demonstrate access to spectral regions where there is no interference from ground-state electronic transitions.

10.2. METHODS

Experimental approaches for ultrafast time-resolved IR spectroscopy have usually involved the following principles:

1. Conventional pump–probe methods: A delayed IR pulse probes the sample, and its loss or gain is measured by IR detectors that integrate the whole transmitted field.

2. Spectrally analyzed pump–probe methods: As in (1) except that the different frequency components of the probe field are detected separately by means of a monochromator, filters, or array detection of the dispersed probe light.

3. Gated detection of quasi-continuous wave probe radiation: Here a slowly varying, tunable IR probe pulse envelope is recorded with high time resolution. The gating is accomplished by sum frequency mixing of the probe envelope with an optical pulse in a nonlinear medium.

For studies of kinetics on time scales much slower than the inverse vibrational bandwidth, $1/\Delta\nu$, all these approaches yield the same signals and kinetic information. In the coherent regime, defined here as times less than $1/\Delta\nu$, each method gives a different type of signal. A detailed understanding of the signal generation is required to interpret the results, and a summary of the principles needed to interpret transient IR spectra in the coherent regime is provided in the next two sections.

Infrared absorption cross sections, $\sigma(cm^2)$, fall in the range 10^{-20} for weak transitions, through about 5×10^{-19} for many common fundamentals, to 10^{-17} for very strong transitions such as those found in metal carbonyls. Thus, sensitive methods are needed to detect common IR transitions. Expressing the detection limit in optical density units is convenient, in which case the current achievement in picosecond experiments is an optical density change of a few times 10^{-5} in several seconds of collection at a fixed delay and frequency. Still greater sensitivity will be required to detect the light induced changes of the weaker IR lines in solution.

10.2.1. Pulsed Infrared

Most commonly used methods for IR pulse generation follow a three-wave mixing process, beginning with either one or two input pulses and a nonlinear material. The first method, which generates a useful pulse of light born of amplified noise (optical parametric generation) is usually followed by amplification to produce pulses of useful energy per unit frequency. The second method generates the third field by difference frequency mixing of two input fields in a nonlinear material. Both methods have enjoyed popularity.

Parametric generation is a simple method for generating tunable light. In a single pass, one crystal serves as both a generating and tuning element. The generation of light is coupled to the driving field through the second-order polarizability of the nonlinear material, subject to conservation of energy and momentum (wavevector). Amplification can start either from quantum noise or from a weak input at the signal or idler frequencies. Materials limitations have led to the development of multipass devices known as optical parametric oscillators (OPOs). In this configuration, the signal or idler is resonated to lower the threshold. The recent development of high peak power and low average power pulses makes possible efficient parametric generation without crystal damage.

Traditional materials used for generating IR have been $LiIO_4$, KTP, and adenosine dephosphate (ADP). However, recent developments have provided a number of other promising options for efficient IR generation. We will now summarize some of the nonlinear materials available for IR generation. Particularly useful, as shown by early work by Elsasser et al. (1), is $AgGaS_2$, which can be phase matched from $1.5-10$ μm, with transparency over the whole region. Photon conversion efficiencies of $10^{-1}-10^{-3}$ were found by using two $AgGaS_2$ crystals. Vanherzeele (2) reports high-efficiency tunable IR generation from $1.2-4.5$ μm using KTP by synchronously mixing the output of a mode locked dye laser and Nd:YLF laser. Moreover, the authors provide a phase-matching scheme for generation of IR pulses by using crystal birefrigence to compensate for group velocity mismatch, by cleavage along the natural (100) KTP plane. In this case, the group velocity mismatch contribution to broadening can be limited to less than 100 fs mm^{-1}. The KTP shows advantages over LiNbO$_3$ and LiIO$_3$ due to its high-damage threshold. Kato and Masutani (3) report a widely tunable 90° phase-matched KTP parametric oscillator. This solution was attempted to overcome limitations of angle tuned parametric oscillators that raise oscillator threshold and lead to spectral broadening as a result of aperture walk-off.

Clark and co-workers (4) report a particularly useful solution to the problems of beam walk-off and displacement found in parametric generation. In their method, a single pass optical parametric source (OPS) uses a pair of nonlinear crystals as both generating and tuning elements. The OPS was aligned by rotating the first crystal to cause the pump to be at normal incidence. The second crystal was rotated to make the residual pump and parametric outputs colinear. Counterrotation of the two crystals over the entire wavelength range (450–650 and 800–1600 nm) caused no significant deflection in the direction of the output or the pump. In this configuration, the effects of dispersion and birefringence approximately cancel.

Phase matching controls the spectral and spatial profiles of the OPS beam. Parametric fluorescence emerges from the first crystal at any phase-matched frequency and direction. Thus, the gap between the crystals selects the frequency distribution of the amplified radiation in the second crystal. This eventually leads to the limit of colinear amplification at large crystal separations. Clark and co-workers (4) suggest some solutions to the problems of less-than-diffraction-limited beam and the relatively broad spectral width.

A recent report (5) from our laboratory considered the development of a high repetition rate, IR-pump, IR-probe spectrometer. In this work, Iannone et al. (5) generated two independently tunable 25-ps IR pulses at 500 Hz using two dye lasers, a regenerative amplifier, and two lithium iodate OPAs. The IR pump pulse was generated by mixing the output of a dye laser pulse, amplified by one-half of the frequency doubled output of a regenerative amplifier, with the remaining one-half in a 2-cm crystal. The second IR pulse was obtained by mixing the output of a second, unamplified dye laser and was used for a probe beam. Though this method produced time-resolved IR data at somewhat worse detectivity (2 × 10^{-3}OD^{-1}) than available with upconverting CW IR, subsequent improvements by Raferty et al. (6) using shot-by-shot normalization improved detectivity by nearly two orders of magnitude!

One of the most exciting recent technical developments in our laboratories, performed by Wynne and co-workers (7), has been the construction of a laser system delivering amplified (40 μJ) short (35 fs) pulses at 800 nm and moderate (5 kHz) repetition rate and the subsequent generation of short (40–120 fs) IR pulses. These experiments are fulfilling the promise of earlier expectations, and represent the current state-of-the-art, following the lead of Mourou and co-workers (8), who developed chirped pulse amplification using grating pairs for solid state amplifiers in the mid-1980s. Beginning with the very short pulses available from a self-mode-locked Ti:Al$_2$O$_3$ oscillator design by Asaki et al. (9) Wynne and co-workers (7) developed an improvement to the basic Ti:Al$_2$O$_3$ regenerative amplifier design by minimizing the dispersion in all parts of the amplifier by adding intra-cavity prisms. This approach is distinct from and simpler than methods that use delicate off-axis positioning of the grating pairs to correct for dispersions due to the stretcher as well as the amplifier up to third order all at once (8). The resulting short, powerful pulses were then directly exploited to generate short pulse IR.

The primary challenge to generating femtosecond short pulses using the OPG/OPA method is the difference in group delay between pump, signal, and idler fields. In β-barium borate (BBO) the group delay is zero for downconversion of 735 nm to 1.47 μm + 1.47 μm, see Fig. 10.1. Wynne and Reid performed downconversion at 800 nm, and produced an OPO tunable from 1.2 to 2.4μm, with 40 fs pulse widths at 1.2 μm and 120 fs at 1.6 μm. Careful optimization of the OPG/OPA geometry has led to less than 5% root-mean-square (GW) (rms) noise in the generated pulse intensity.

As shown in a recent paper by Zinth and co-workers (10), AgGaS$_2$ may not be as easily mated with modern Ti:Sapphire technology as BBO for ultrashort pulse generation. With difference frequency generation using 120-fs Ti:sapphire pulses at 815 nm (near its maximum power wavelength), pulses from a Ti:sapphire pumped, traveling wave dye laser tunable from 870 to 1000 nm, and crystals 1mm long, they produced pulses broadened to 400 fs in the IR, with 10-nJ energy. The dispersion by AgGaS$_2$ would permit parametric generation of shorter (100 fs) pulses with careful choice of input fields frequency and crystal lengths (e.g., there are differences in group velocity of 3.3 ps cm^{-1} between 9398 and 2080 cm^{-1}), but only by sacrificing power, unless a scheme to prechirp the input pulses or deeper IR OPO pump sources are employed.

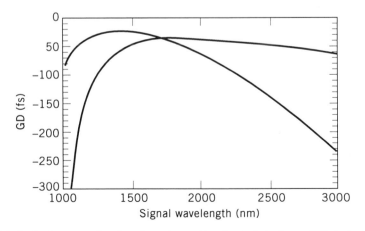

Figure 10.1. In the femtosecond regime the main consideration in designing an OPO is the difference in group delay between the signal, idler, and pump waves. In BBO this difference is zero for downconversion from 735 nm → 1.47 μm + 1.47 μm.

10.2.2. Signal Generation Theory in Transient Infrared Spectroscopy

Infrared spectra generally consist of separated spectral lines. While this is precisely the strength of IR as a structural tool it raises important issues for time-resolved IR methods. Both the equilibrium widths and the line separations of the IR transitions influence the transient response. We will now consider the three different approaches described above to studying IR transients:

1. Conventional pump–probe.
2. Pump–probe with a frequency selective detector.
3. Gated IR detection of a single-frequency IR source.

To summarize the situation we indicate here, as functions of the generated IR fields $\varepsilon_{gen}(t, \tau)$, the incident IR fields $\varepsilon_0(t)$, and the gating fields $\varepsilon_{vis}(t, \tau)$, the quantities that must be calculated to obtain the signal currents in arbitrary units for the three above cases:

$$\int_{-\infty}^{\infty} dt \varepsilon_{gen}(t,\tau)\varepsilon_0(t) \tag{10.1}$$

$$\varepsilon_{gen}(\omega,\tau)\varepsilon_0(\omega) \tag{10.2}$$

$$\int_{-\infty}^{\infty} dt \varepsilon_{gen}(t)\varepsilon_0(t)\varepsilon_{vis}^2(t - \tau) \tag{10.3}$$

where $\varepsilon(\omega)$ is the Fourier transform of the field, τ is the pump–probe delay or gating time, and ω is the filter or CW laser frequency. In each case the signal is due to the interference of the generated and incident fields or their unconverted replicas at a square-law detector. We now proceed with a more detailed account of the coherent transient effects that these signals will expose in experiments done on time scales that are shorter than the phase-loss processes.

10.2.3. Transient IR Spectral Measurements in the Coherent Regime

The probe field radiating from the sample provides all the required transient spectral information. This field consists of the incident probe field and the field generated by the interaction of the probe radiation with the sample at some time delay from the pump. The propagation of these fields, their generation from the macroscopic polarization in the sample, their effect on detectors, and the effect on them of filters or monochromators is dealt with by textbook methods. The polarization on the other hand arises from the interaction of the probe pulse with a quantum mechanical sample that is also coupling to the pump. The usual way to begin calculating the polarization is to obtain the density matrix $\rho(t)$ up to the required order in the fields, assuming that perturbation theory can be used, then to find the dipole induced in the sample from $\mathrm{Tr}\{\rho\mu\}$, where μ is the dipole operator. This procedure (11–13) leads to results that can have many independent contributions that are important when the time scale of the experiment is shorter than the dephasing times of the pairs of levels involved. Fortunately there are a number of diagrammatic methods that allow one to reach these results quickly (13–15) but the systematic, predictable nature of the perturbation expansion provides a straight-forward prescription for any particular situation (13b,16,17). With reference to transient IR spectroscopy there are some special cases that will be of interest.

There are two common situations that arise when a set of levels is pumped: (1) If the pulses are shorter than the inverse line width (we will refer to this as T_2 and it incorporates *all* types of line broadening) the experiment is in the coherent transient regime that is fully discussed by Allen and Eberley (18) for two-level systems. There are no essential differences between the gas and condensed-phase descriptions except that solution T_2 values are many orders of magnitude shorter than in low-pressure gases; (2) If the pulses are longer than T_2, such that there is no phase memory in the system, the signal is proportional to the convolution of the pump intensity with the population decay function of the level pair.

10.2.3A. Optical Pump/IR Probe Experiments.

Density matrix expansions have been previously employed (19) to describe some aspects of this experiment and a full treatment, showing the development of all the expressions in this article, was published recently (19a). Generally, the dephasing times for electronic transitions are very short in comparison with those for vibrational transitions. For example, vibrational spectral lines in solutions seldom display line widths in excess of 50 cm^{-1} and 10–20 cm^{-1} is common. Thus vibrational dephasing times are not expected to be shorter than 200 fs. Perhaps in extremely inhomogeneous systems, for example, certain glasses, associated solvents, or proteins, some dephasing times may be shorter than this, but they will be the exceptions rather than the rule. The characteristic times for electronic dephasing are in the range 6–20 fs. Given this inequality between electronic and vibrational dephasing times the expressions for the polarization can be greatly simplified. The simplification arises because in the limit that the electronic dephasing time approaches zero, both optical fields that are needed in order to generate the pumped states must couple to the system at the same instant. This means that only two types of time-ordered diagram need be considered: One where the IR field couples before and another with it coupling after the sample is optically pumped. Without this separation of time scales it would have been necessary to include the IR field coupling while the system contained electronic-phase coherence. Of course this third time ordering can be included if necessary. The pump and probe conditions are most easily demonstrated by time lines, as shown in Fig. 10.2. In Fig. 10.2(a) the system first couples to an IR field, then evolves coherently for some time before radiating the generated IR field. The coupling to the optical field E_0 occurs only at some later time and has no effect on the IR spectrum. This configuration gives the unpumped signal. Figure 10.2(b) diagrams the situation where the pump fields couple while the system is undergoing coherent evolution in the vibrational states, and the generated field is detected later. Finally, in Fig. 10.2(d) the pump fields first cause the change in the sample that is later probed by the IR fields. The common theme in these diagrams is that the

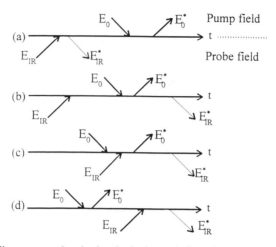

Figure 10.2. Time lines representing the four basic time orderings of the fields involved in a spectroscopic experiment. The optical and IR fields are E_0 (above the time line) and E_{IR} (below the time line) or their complex conjugates.

system responds only to the intensity $[E_0(t)E_0^*(t)]$ of the optical fields. Figure 10.2(c) corresponds to the omitted sequence. These qualitative considerations frame the discussion of the various types of transient IR spectroscopy presented in this chapter. In what follows the basic features expected in coherent transient IR spectroscopy are outlined for pairs of states. There are a few frequently occurring situations for which the results can be readily obtained. One example arises when a photoreaction induced by the pump pulse causes removal of the initially populated states and does not replace them with populations that have transitions in a similar spectral region. The generated field for these conditions apart from constants that determine its amplitude and polarization is given by

$$\varepsilon_{gen}(t,\tau) = \sigma_{gg'} n_g(t) \cdot \mathrm{Re}\left\{ \int_{-\infty}^{t} dt' \; \varepsilon_0(t' - \tau) e^{\Omega_{gg'}(t'-t)} \right\} \tag{10.4}$$

where $\Omega_{gg'} = (i\Delta_{gg'} + \gamma_{gg'}$ is the complex detuning frequency with $\Delta_{gg'} = \omega_{gg'} - \omega_{IR}$ and $\omega_{gg'}$ being the spacing between the vibrational levels g and g' of the ground-state surface. The IR cross section is $\sigma_{gg'}$. The population of the initial state $n_g(t)$, which is driven by the pump intensity determines the amplitudes of a field that is the convolution of the probe field envelope $\varepsilon_0(t)$ with the complex frequency response of the system. In the pump–probe method the pump (centered at $t = 0$) is delayed from the probe by τ. The current in the detector circuit arises from the intensity of the probe pulse and from the time integrated product of the probe and generated fields. The former part, which is the unpumped signal, is usually subtracted off to leave the difference signal displayed. The integration is needed because the detector is slow compared with the responses of the system. In the gated detection method the time t in Eq. (10.4) is the delay between the pump and the gate pulse, and the probe envelope is constant allowing the integral to be obtained analytically. Again the difference signal originates with the product of the generated and incident probe fields [see Eq. (10.3)] and in this case assuming that the gate width is very small it should have the form:

$$\varepsilon_{gen}(t) = \frac{\varepsilon_0 \gamma_{gg'} n_g(t) \sigma_{gg'}}{\Delta^2_{gg'} + \gamma^2_{gg'}} \tag{10.5}$$

Thus only the spectral amplitude changes with time. The spectral shape is the same before, during, and after the application of the pump pulse.

The generated field is different if the new population created by the pump also contributes to the signal. An additional field having the following form is then created:

$$\varepsilon_{gen}(t,\tau) = \sigma_{ee'} \text{Re} \left\{ \int_{-\infty}^{t} dt' n_e(t') \varepsilon_0(t' - \tau) e^{\Omega_{ee'}(t' - t)} \right\} \tag{10.6}$$

where the spectrum is now time dependent, consisting of the convolution of the product of the probe field and the new population that, again would be obtained from a kinetic analysis, with the complex response of the new (excited state) transition between vibrational levels e and e'. If the probe frequency were chosen at the peak of the initial absorption the effect of Eq. (10.6) would be to cause the bleaching to be slower than the decay of $n_g(t)$ and to exhibit oscillations with an amplitude and frequency dependent on the pump induced shift of the line. When the probe frequency is chosen at the peak of the new absorption the new signal grows in with the dephasing time of the new absorption band and decays with the lifetime of the new state. Also, in this situation there will be an oscillatory component whose amplitude and period depends on the shift. In general there will be a number of possible transitions and the generated field will consist of Eqs. (10.5) and (10.6) summed over the various complex frequencies involved. When the gating method is used a simple form can again be found for the low pump intensity signal, since in that case Eq. (10.6) can be integrated by parts. To lowest order the result is:

$$\varepsilon_{gen}(t) = \text{Re} \left\{ \left(\frac{\sigma_{ee'} n_g(0) \varepsilon_0}{\Omega_{ee'} - \gamma} \right) \int_{-\infty}^{t} dt' \ I(t') \left\{ e^{-\gamma(t' - t)} - e^{\Omega_{ee'}(t' - t)} \right. \right.$$
$$\left. \left. \left[1 - \left(\frac{\sigma_{gg'}}{\sigma_{ee'}} \right) \left(\frac{\Omega_{ee'} - \gamma}{\Omega_{gg'}} \right) \right] \right\} \right\} \tag{10.7}$$

where $I(t)$ is the intensity of the pump field, γ is the inverse lifetime of the new state, and σ is the specified optical absorption cross section. It should be noted that the observed new spectrum may have its width determined by only the pure dephasing of the transition. Not only can the spectrum be narrower than the equilibrium width but it is broad with sidebands at the earliest times. If the new population does not decay significantly the spectrum again starts out very broad and eventually scales down to its equilibrium shape. These characteristics imply that kinetic results will be different at each probe frequency, and therefore care is needed in analyzing data. Frequently, the coherence transferred from the initial into the final state through an IR–pump–pump process will be important. This appears in Eq. (10.7), which is specifically for the CW probe signal, through the term $(\Omega_{ee'} - \gamma)/\Omega_{gg'}$, which is required to be added to the pump–pump–IR term of Eq. (10.6) for a complete description.

When the new state is an excited electronic state created directly by the pumping process, there is depletion of the initially populated states concomitant with the creation of a decaying population that will usually have quite a similar IR spectrum. For example, this arises in our experiments involving the excitation of the special pair in the Reaction Center (20). In such cases, which were dealt with in our earlier papers (19,19a), there can occur a complex interference between all the different transitions associated with the two electronic potential surfaces. Analysis of this pattern in time and frequency naturally provides the parameters of the two-potential energy surfaces. Special conditions lead to particularly useful, transparent analytical forms for the field even in this case, for example, if the two surfaces are very similar such that optical transitions do not exhibit Franck–Condon progressions. This situation is approximately upheld for the electronic transitions of a few large molecules. If, as might frequently arise in practice, there is little or no shift of the vibrational frequencies on excitation, the bleaching of the initial state is not observed until the excited-state transition is dephased. An example of this effect, which is readily verified by adding Eqs. (10.5) and (10.7) for a system of four levels with $g = e$ and $g' = r'$, has been observed experimentally (20).

The discussion up to now has treated the material as having a set of fixed transition frequencies and widths that determine the coherence dynamics and an associated set of populations whose relationships are determined by kinetics induced by the pump field. In real chemical transformations the frequencies and widths change continuously, so we have assumed that the time intervals during which these changes occur are too small to measure. That will not always be the case, especially as the time resolution of IR spectroscopy gets closer to that of electronic dephasing. However, it is necessary even on slower time scales to consider how to treat such situations. The next example involves the changes in the IR spectrum of a cofactor embedded in a protein resulting from structural transitions of the protein. The probed population is not changing but its spectrum is. It turns out to be relatively easy to handle these situations as outlined below.

When the bath to which the dynamical system is coupled undergoes changes without inducing significant changes of the probed states' population distribution, the spectral changes can be calculated by propagating the complex frequency over a continuous range of times. In other words we introduce widths and frequencies having a time dependence determined by the bath dynamics and the system-bath coupling. In this case the generated field takes the form:

$$\varepsilon_{gen}(t) = \sigma_{gg'} n_g(0) \ \text{Re}\left\{ \int_{-\infty}^{t} dt' \varepsilon_0(t') e^{\int_{t'}^{t} d\tau \, \Omega(\tau)} \right\} \tag{10.8}$$

where $n_g(0)$ is the fixed population of absorbing oscillators. In principle Eq. (10.8) incorporates the (a), (b), and (d) contributions diagrammed in Fig. (10.2). As an example of Eq. (10.8), the solution of which would normally require a dynamical model for the propagator followed by numerical analysis, we will consider the signal resulting from an impulsive perturbation that brings about a sudden change in complex frequency from $\Omega_0 = i\Delta_0 + \gamma_0$ to $\Omega_1 = i\Delta_0 + \gamma_1$. With the condition of δ-function pulses to initiate the changes and also to gate the IR field the difference signal takes the following form:

$$\text{Re}\left\{ \left(\frac{1}{\Omega_1} - \frac{1}{\Omega_0} \right)(1 - e^{-\Omega_1 t}) \right\} \tag{10.9}$$

where t is the gating or delay time, which is considered to be greater than or equal to zero. When t is less than zero the umpumped signal is observed. Equation (10.9) shows that difference signal grows in on the time scale of the dephasing of the *new* transition and it is oscillatory when the probe field is detuned from the new resonance. If this shift is large enough the oscillations become so rapid that the gating pulse averages them out. The corresponding pump–probe difference signal with a filter between the sample and the detector or with an array detector is found to be

$$\text{Re}\left\{ \left(\frac{1}{\Omega_1} - \frac{1}{\Omega_0} \right)e^{\Omega_0 \tau} + \left(\frac{1}{\Omega_1} - \frac{1}{\Omega_0} \right) \right\} \tag{10.10}$$

where the first term is for negative delays and the second is exclusively for the case where the probe arrives after the pump [diagram (d) in Fig. 10.2]. Note that the first term always grows in with the time constant determined by the *unperturbed* resonance and the peak signal occurs at zero delay. In the gated experiment the integral in Eq. (10.8) only extends up to the gating time or the pumping time, whichever comes latest. On the other hand, in the pump–probe experiment the detector senses the free induction decay of the initial transition starting from the arrival time of the probe. Figure 10.3 shows the time variations of Eq. (10.9) and (10.10) for model parameters that dramatizes the difference in signal expected from the IR gate and pump–probe methods.

10.2.3B.   Modified Bloch Equations for Coherent Transient IR Spectroscopy of Optically Pumped States.   A simple set of equations consisting of terms having obvious physical origin and whose solutions give the results of the previous

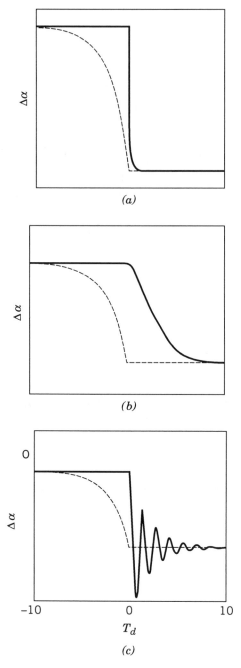

Figure 10.3. Simulation of the difference between signals expected for pump–probe (dashed line) and quasi-CW gated method (solid ine). (*a*) Signals for instantaneous line broadening, in this example γ_0 = 5 cm^{-1}, $\gamma_1 = 5\gamma_0$, and $\Omega_0 = \gamma_0$. (*b*) Signals for an instantaneous shift by 2 cm^{-1} with $\Omega_0 = \gamma_0$, $\Omega_0 = \gamma_0$, $\gamma_1 = 5\gamma_0$, γ_0 = 5 cm^{-1}. (*c*) Instantaneous shift of 50 cm^{-1}; $\Omega_0 = \gamma_0$; $\gamma_0 = \gamma_1$ = 5 cm^{-1}. The presence of the large oscillation is due to the δ-pulse gating. In practice an average over the gate width as in Eq. (10.3) is needed.

section, can be set up. These equations are solidly based on the explicit use of perturbation theory of the Liouville equation to obtain the coherences. All the time-ordered diagrams needed for the model are automatically incorporated. These orderings (see Fig. 10.2) include IR–pump–pump and pump–pump–IR but omit the ordering pump–IR–pump on the grounds that the electronic dephasing γ_{el} is faster than all the relaxation processes involving the vibrations. We present here a limited model in which a set of ground-state levels $\{g\}$, electronically excited-state levels $\{e\}$, and photoproduct state levels $\{p\}$ are considered. The $\{e\}$ manifold is reached by an optical pumping process in our experiments, the $\{p\}$ manifold results from the spontaneous decay of the levels of $\{e\}$, and the $\{p\}$ manifold itself may decay away by some other processes that need not be specified. The set of equations describing the evolution of the populations and coherences in such a system will be referred to as the modified Bloch equations for the model system. It is convenient to write these equations in the interaction representation of the zero-order Hamiltonian that is diagonal in the eigenstates $\{g\}$, $\{e\}$, and $\{p\}$. The equations of motion for the populations of the vibrational levels of the three states of the model are as follows:

$$\dot{n}_g = -[\gamma_g + B_{gg}(t)]n_g + \sum_{g'} \gamma_{g'g}n_{g'} \tag{10.11}$$

$$\dot{n}_e = -\gamma_e n_e + \sum_{g} A_{gg\,ee}(t)n_g + \sum_{e'} \gamma_{e'e}n_{e'} \tag{10.12}$$

$$\dot{n}_p = \gamma_p n_p + \sum_{e} \gamma_{ep}n_e + \sum_{p'} \gamma_{p'p}n_{p'} \tag{10.13}$$

where the optical pumping parameters are defined as:

$$B_{gg}(t) = \sum_{e} A_{gg\,ee}(t) \tag{10.14}$$

and

$$A_{gg'\,ee'}(t) = \frac{2V_{ge}(t)V_{e'g'}(t)}{\hbar^2 \gamma_{el}} \tag{10.15}$$

When $i = j = g$, $B_{gg}(t)$ is just $\sigma I(t)$, where σ is the absorption cross section and $I(t)$ is the effective pump intensity. In these and subsequent expressions, and neglecting spatial (wavevector) effects for the moment, $V_{ij}(t)$ is the matrix element of the system–radiation interaction in the interaction representation and incorporating the rotating wave approximation, so that:

$$V_{ij}(t) = \mu_{ij} \cdot \varepsilon(t)e^{i(\omega_{ij}+s\omega)t} \tag{10.16}$$

where $\varepsilon(t)$ is the field envelope and s is chosen as ± 1 to make $(\omega_{ij} + s\omega)$ smallest where ω is the frequency of the field. We use γ_i to designate the total rate coefficient for decay of the population in level i and γ_{ij} as the specific rate constant for transitions between levels i and j. A full calculation of the B and A quantities will require knowledge of all the Franck–Condon factors involved (i.e., the surfaces) in the pumping process.

The coherences ρ_{ij} are generated by the IR field coupling the level pairs in each of the manifolds but the pump field and the rate processes may transfer coherence from one manifold to another. The equations of motion are as follows:

$$\dot{\rho}_{gg'} = (i/\hbar)V_{gg'}^{(IR)}(t)(n_g - n_{g'}) - \{\Gamma_{gg'} + \tfrac{1}{2}[B_{gg}(t) + B_{g'g'}(t)]\}\rho_{gg'}$$

$$+ (i/\hbar) \sum_{g''} (\rho_{gg''}V_{g''g'}^{(IR)}(t) - V_{gg''}^{(IR)}(t)\rho_{g''g'}) \qquad (10.17)$$

$$- \tfrac{1}{2} \sum_{g'' \neq g,g'} [B_{gg''}(t)\rho_{g''g'} + \rho_{gg''} B_{g'g'}(t)]$$

where $V_{ij}^{(IR)}(t)$ is given by (Eq. 10.16) for the IR field resonances. The first term is the IR field driving the coherence in the gg' level pair and the second is the loss of coherence due to equilibrium dephasing and to the pumping away of the populations of the g and g' levels. The last two terms, calculated up to the third order, correspond respectively to the effect of Raman pumping, where coherences in the initial manifold are generated by the pump, and to the losses caused by pumping away populations of states other than gg' that may be involved in the initial coherent interaction with the IR field. The first and third terms correspond to the field time ordering pump–pump–probe [Fig. (10.2(d)] while the second and fourth correspond to probe–pump–pump [Fig. 10.2(b)]. The equations of motion for the e-manifold are as follows:

$$\dot{\rho}_{ee'} = (i/\hbar)V_{ee'}^{(IR)}(t)(n_e - n_{e'}) - \Gamma_{ee'}\rho_{ee'} + \sum_{g,g'} A_{gg'ee'}(t)\rho_{gg'}$$

$$+ (i/\hbar) \sum_{e'' \neq e,e'} (\rho_{ee''}V_{e''e'}^{(IR)}(t) - V_{ee''}^{(IR)}(t)\rho_{e''e'}) \qquad (10.18)$$

Again each term is assumed third order in the applied fields. The third term in Eq. (10.18) [time ordering probe–pump–pump as in Fig. 10.2(b)] corresponds to the pump induced transfer of coherence from g to the e manifold. The last term again involves the Raman induced coherences coupling with the IR field (in a thin sample) with $\rho_{ee'}$ calculated to second order. The p-manifold spectrum is obtained from the density matrix $\rho_{pp'}$ given by

$$\dot{\rho}_{pp'} = (i/\hbar)V_{pp'}^{(IR)}(t)(n_p - n_{p'}) - \Gamma_{pp'}\rho_{pp'} + \tfrac{1}{2} \sum_{e,e'} (\gamma_{ep} + \gamma_{e'p'})\rho_{ee'} \quad (10.19)$$

where the third term is due to coherence transfer from the e to the p manifold.

These equations lead to different physical responses depending on which model siutations are considered. A common spectroscopic configuration that occurs with biological cofactors such as hemes or chlorophylls is that Franck–Condon progressions do not occur to any significant extent. Thus, the excitation process is dominated by transitions $g \rightarrow e$, $g' \rightarrow e'$, and so on of roughly equal cross section, involving no change in vibrational quantum number. This model is also suitable for other situations. Additionally, in this chapter we will consider experiments involving only high-frequency modes with only $v = 0$ populated initially and which cannot be Raman pumped. The pump light intensity will be kept low enough that none of the feeding terms in the population equations for $\{g\}$ and $\{e\}$ need consideration. In that case these equations become:

$$\dot{n}_g = - A_{gg\,ee}(t) n_g \tag{10.20a}$$

$$\dot{n}_g = A_{gg\,ee}(t) n_g - \gamma_e n_e \tag{10.20b}$$

$$\dot{\rho}_{gg'} = i\, V_{gg'}^{(IR)}(t) n_g - \{\Gamma_{gg'} + \tfrac{1}{2}\, [A_{gg\,ee}(t) + A_{g'g'e'e'}(t)]\} \rho_{gg'} \tag{10.20c}$$

$$\dot{\rho}_{ee'} = i\, V_{ee'}^{(IR)}(t) n_e - \Gamma_{ee'} \rho_{ee'} + A_{gg'ee'}(t)\, \rho_{gg'} \tag{10.20d}$$

where all the A factors are basically $\sigma I(t)$ times their frequency factors [Eqs. (10.15) and (10.16)]. These equations are easily integrated and yield the results given in the previous section that assume just one ground state g is populated. It is straightforwad to apply these equations to any two combining surfaces.

10.3. BIOLOGICALLY RELEVANT SYSTEMS

The search for the various factors that determine the dynamical aspects of protein structure is extremely promising. The goal is to obtain a microscopic interpretation, in terms of quantum and statistical mechanics, of the dynamics of biological structures. These structures can be well defined in terms of average configurations that are often known from X-ray diffraction or NMR. The methods of ultrafast spectroscopy have already proved to be very useful in this field particularly for these proteins whose functions can be triggered by light pulses. As with the other examples in this chapter IR methods complement the ongoing optical spectroscopic studies of biological chromophores (cofactors). However, in the case of protein dynamics the features available from transient IR allow one to address in addition a whole new class of questions that relate directly to the protein rather than to its cofactors. This section reviews some recent results from IR studies of the vibrational responses of three canonical protein types whose operation can be triggered by optical pulses (23). These are the hemoproteins (21), the Reaction Centers of photosynthetic bacteria (20,22,24), and bacteriorhodopsins (23). In each case there is a vast literature of important work that forms the basis for the questions that we are asking and these references will provide some background with particular emphasis on ultrafast spectroscopy as well as an entry into the more biological literature.

10.3.1. Hemoproteins

Hemoglobin (Hb) and myoglobin (Mb) bind oxygen and other ligands at an iron porphyrin (heme). Carbon monoxide is the ligand most often incorporated in IR spectroscopic studies. Bound to the heme, CO absorbs near 1950 cm^{-1}, while unbound CO absorbs near 2100 cm^{-1}. The IR transparency of aqueous solutions in this region permits the measurement of bound and free CO spectra within the protein. Studies in the University of Pennsylvania labs have explored three themes for MbCO and HbCO: the geometry of CO bound to the heme, the flow of energy within the protein, and relaxation of bound CO vibrational excitation.

Two kinds of experiments have been performed to obtain geometric information, both of which depend on the polarization spectroscopy method previously described (19). The first type (21b) has employed absorption of a visible photon by the heme, leading to the dissociation of bound CO. The resultant change in transmission for IR radiation polarized parallel and perpendicular to the photolysis pulse depends on the angle between the CO and heme plane absorption dipoles. The second type of experiment has employed IR excitation and IR probing. These measurements yield T_1, or the population relaxation time of the excited vibrational mode, and the orientational relaxation times.

10.3.1A. Orientational Features and Rigidity.

In visible pump/IR probe experiments using 300 fs photolysis pulses Locke et al. determined that the 1951 cm^{-1} band of HbCO has the CO at an angle of 10° ± 2.5° from the normal to the plane containing the heme absorption dipoles. To explore the extent of ultrafast heme motion, the anisotropy in HbCO was measured with 300-fs time resolution (25). The anisotropy from 30 fs to 1 ns is shown in Fig. 10.4. We wish to note that the anisotropies discussed here have been corrected according to Locke et al. (25b). The solid line is the expected decay if the heme is held rigid and only protein reorientation contributes. The dashed lines assume the heme can wobble within 10° and 20° of its equilibrium position and that the porphyrin undergoes damped harmonic motions about in-plane axes along and perpendicular to the projection of the CO onto the heme plane. The average measured anisotropy for delay times of 30 fs to 1 ns was nearly identical indicating that the heme is rigidly held within the protein on a picosecond time scale.

Recently, ultrafast IR spectroscopy has been used to study the geometry of CO bound to HbBoston (26), a mutant in which the distal histidine E7 is replaced by a tryosine. The bound CO of the alpha subunits in HbBoston shows an IR absorption at 1972 cm^{-1}, while the normal beta subunit has the IR absorption at 1952 cm^{-1}. The measured anisotropy was nearly identical for both subunits and corresponded to values for the angle between CO and the normal to the heme plane of 17° ± 2.5° for the beta subunit and 16.6° ± 2.5° for the alpha subunit. The results for HbCOBoston support the contention that polar interactions affect the bound CO absorption frequency.

10.3.1B. Energy Flow Experiments: Energy Transfer through the Protein.

The photodissociation of carboxyhemoglobin (HbCO) into Hb and CO oc-

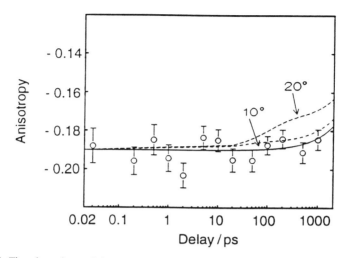

Figure 10.4. Time dependence of the anisotropy data monitored at the peak of the bound CO absorption of HbCO, taken from Locke et al. (25). The solid line shows the expected anistropy decay due to overall protein rotation with the heme rigidly in place. The curves marked 10° and 20° are the expected decays assuming the heme has a 90% probability of being within 10° or 20° of its equilibrium position calculated from a model in which the heme undergoes small damped harmonic oscillators about its equilibrium position. The anistropy-axis has been transferred from that found in figure 1b of reference 25a by the equation $r_a = 1.076r + 0.038r^2 - 0.011$, as described in reference 25b. [After Ref. 25 with permission.]

curs in less than 100 fs. As a result of severing the Fe—CO bond there occurs an immediate relaxation of the heme pocket structure that inhibits the rebinding of CO. The excess energy is released into low-frequency modes of the protein by means of translational and rotational cooling of the photogenerated CO and vibrational cooling of the heme product of reaction. The released energy ultimately warms not only the protein but also the surrounding solvent. This picture of the effects of photodissociation was used to explain results obtained using subpicosecond IR spectroscopy (21). The excitation of deoxyHb generates vibrationally excited ground states extremely rapidly, so that this system may serve as a model for the process occurring in the early stages of the HbCO photodissociation in which vibrationally excited-ground-state deoxyHb is also used. In order to understand the energy-transfer mechanism in heme proteins, we had to understand the effect of rapid heating of water on its IR spectrum. A model system for studying the heating of water should contain only water and a well-understood heat source. The system we decided to use as a model is malachite green in D_2O. Malachite green was chosen because its excited-state dynamics and vibrational relaxation in the ground state are well studied (27–32).

After absorbing a visible photon, a deoxy heme is excited to an excited electronic state (Q band of the metal porphyrin), which then relaxes to the ground electronic state within 200 fs. During this relaxation process, the energy of the absorbed photon is converted into vibrational energy of the heme. The internal conversion process, at least for the low-frequency modes, is expected to involve a large enough number of modes that the vibrational energy distribution would be difficult to distinguish from a Boltzmann distribution and therefore can be defined by a temperature (33–35). The vibrationally hot heme will cool down via collisions with the surrounding protein. There are approximately 100 van der Waals contacts between the protein and the heme, which is a substantial thermal contact. The cooling of the heme will cause the temperature of the protein to rise. The heated protein meanwhile will be transferring its excess energy to the surrounding water, which is at a lower temperature. In summary, the energy of the absorbed photon heats the heme and is eventually transferred through the protein to the surrounding water until the whole system, defined by the active volume excited by the pump pulse, reaches a new thermal equilibrium at a higher temperature. The observed IR signal in deoxyHb solution (shown in Fig. 10.5) had a similar magnitude to that observed in malachite green in D_2O using the same pump energy and sample volume. The change in transmission is caused by the temperature induced shift of the water IR spectrum.

The observed bleaching signal in the 1800-cm^{-1} region was attributed to the

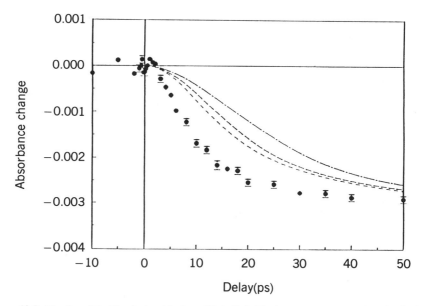

Figure 10.5. Kinetics of the bleach signal in deoxyHb in D_2O. The instrument response function limited spike at $t = 0$ has been subtracted from the kinetics and is represented by the full circles. Also shown in the figure are the calculated heating signals from D_2O for deoxyHb in D_2O using classical diffusion theory. The broken curves represent the calculated results for three heme cooling times of 3, 5, and 10 ps with the slower heme cooling time corresponding to slower curves in the figure. This comparison clearly illustrates that these measured kinetics cannot be described by the classical diffusion model. [After Ref. 21c with permission.]

shift of the water IR spectrum as a result of heating and the response time of this spectral shift to heating was found to be 4 ± 3 ps. A similar bleach signal was observed for heme protein and D_2O solutions and attributed to the heating of surrounding water. The signal in Hb and also Mb (shown in Fig. 10.5) was fitted to a model that consisted of a fast and a slow component leading to the explanation that the heat transfer from heme to the surrounding water in heme proteins proceeds through at least two different processes or possibly a range of processes. The slow component, about 40% of the total signal, was attributed to the diffusion of the heat through the protein. There was good agreement between the measured slow kinetics and the kinetics calculated using a classical heat diffusion theory with a 5-ps heme cooling time. The fast component in the observed signal, with amplitudes and time constants of 60%, 7.5 ± 2.0 ps and 63%, 8.5 ± 1.5 ps for Mb and Hb, is faster than the time scale for a thermal diffusion process. It was proposed that the energy transfer from the heme to the surrounding water in heme protein solutions proceeds through both a diffusive process in the protein and a nondiffusive process that involves the collective motions of the protein. These collective motions must be heavily damped by the surrounding water, thereby causing water molecules to heat faster than the diffusional process. The time scales for the energy propagation from the heme to the protein surface through these collective motions are determined by their group velocities in the protein. These experiments constitute a direct observation of the rate and magnitude of the temperature rise of solvents surrounding a protein.

10.3.1C.   Vibrational Relaxation of Liganded Proteins.   The desire to understand the nature of cooperative ligand binding and dissociation of heme proteins has inspired many experimental and theoretical studies. Many factors, including those arising from primary up to quaternary structural features, may be important in the ligand-binding dynamics. It is clearly useful to examine various environmental factors that are involved, including the ligand-binding strength and geometry, as well as more subtle influences, such as those from the solvent and protein host. The vibrational population relaxation time (T_1) of the ligand should be sensitive to its environment, so that T_1 measurement for various heme proteins might reveal evidence about the ligand bond and perhaps about the heme-ligand dynamics, where the latter obviously is of crucial importance in ligand-binding processes.

The CO ligand vibrational frequencies (~ 1950 cm^{-1}) in heme$-$CO molecules are similar to those for terminal CO ($\sim 1950-2030$ cm^{-1}) in metal carbonyls. The IR absorption coefficient for the CO stretch vibration is greatly increased by the metal$-$CO bonding, but the vibrational frequencies are not as strongly affected. The heme CO and terminal metal carbonyl (and top-site surface adsorbate CO) frequencies are only somewhat lower than that of the free carbon monoxide (2143 cm^{-1}) while organic carbonyls (1600$-$1700) and bridged metallocarbonyl and bridge surface groups ($\sim 1700-1800$ cm^{-1}) are considerably lower. The question of interest in our work (36$-$39) is the mechanism of coupling vibrational energy in the CO bond into other degrees of freedom of the metal carbonyl compound (IVR) and the competing process in which energy is transferred directly to surrounding solvent motions (e.g., by external vibrational redistribution, EVR) in CO bound to heme and hemeproteins.

We carried out two-color IR pump–probe studies of the vibrational relaxation times of CO stretch vibrations for HbCO, MbCO, and photoporphyrin–CO. The IR–IR anisotropy was also measured from the polarization dependent signals, and this revealed information concerning the rigidity of the CO orientation in the heme pocket while in the ground electronic state. These studies represent unique ways of exploring important questions regarding the nature of the forces exerted by protein motions on the vibrational coordinate as well as on nearby closely coupled modes. A study of the temperature dependence of the vibrational relaxation of HbCO and MbCO obtained by pumping and probing with the same frequency recently came to our attention (40).

The IR difference spectra in Fig. 10.6 for HbCO and MbCO illustrate a number of important points. The intensity of the bleach signal is nearly equal to that of the new absorption caused by the population in $v = 1$. This is in accord with a harmonic model for the IR cross sections for which the $v = 1$ to $v = 2$ absorption cross section is twice as large as that of the $v = 0$ to $v = 1$ transition since it depends on q_{12}^2 rather than q_{01}^2. Frequency shifts between the bleach and new absorption peaks provide a value of $2x_{ij}\omega$ for the CO oscillator. Finally, the data indicate that the line widths of the $v = 0 \rightarrow v = 1$ and $v = 2$ transitions are both approximately 8 cm^{-1}. Information of this nature is scarce in condensed phases. The result is consistent with the width of the transition arising from anharmonic coupling of the CO stretch to low-frequency modes as described by an exchange model.

The relaxation times for the hemoprotein carbonyls ($T_1 = 18 \pm 2$ ps) are considerably faster than most of these measured for most other metallocarbonyls (41). We considered whether it makes sense to attribute this increase in rate to the large number of states of nuclear motion of the porphyrin structure and the effective coupling of the CO stretching mode to this set of states. The CO stretch vibration must be directly coupled to the bending and stretching motions of the FeCO group as in any "triatomic" molecule. An important question is whether the coupling is strong enough to account for such a fast relaxation time.

The Fe–CO modes involve motions of the iron atom that are both in and out of the plane of the porphyrin ring. A pathway for relaxation of the CO stretch involving the anharmonic coupling of the ligand and ring modes is opened by these ligand induced motions of the iron atom but it has two features that seem unusual. One is that the small displacements of the heavy iron atom could be expected to be relatively inefficient in mediating the transfer of energy from the ligand to the porphyrin ring. The other is that the relatively low frequencies of the other FeCO modes (580 and 496 cm^{-1} for the FeCO bend and FeC stretch, respectively) implies that at least three quanta would be needed in order to come within a few hundred reciprocal centimeters of the energy of the CO stretching mode. The known features of the electronic structure of the heme may provide a justification for such an interaction. The FeC stretch and the CO stretch modes are both known to have frequencies that are very sensitive to the amount of charge shift (backbonding) from the iron to the carbon. Both these frequencies are very strongly correlated

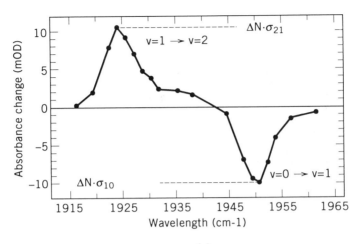

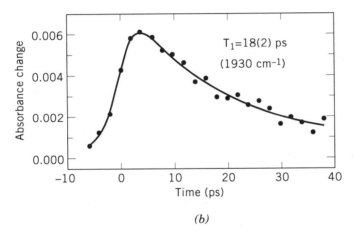

Figure 10.6. (*a*) The IR pump–probe experiment on HbCO. The figure shows the CO bleach at 1950 cm^{-1} and the generated $v = 1 \rightarrow v = 2$ transition at 1925 cm^{-1}. (*b*) Transient absorption of the carbonyl vibration in HbCO [After Ref. 39].

with structure such that a charge shift decreases the CO stretch frequency and proportionately increases the FeC frequency (42). Such a charge shift also alters the potential governing the ring modes as a result of the decrease of the partial charge on the iron. Thus the CO stretching motion is coupled to the ring modes by this mechanism, which has as its key feature not only the small displacements of the iron atom but the alteration of its effective charge during the vibrational motion. Put another way, the relaxation of the CO stretching mode could arise from the forces exerted on it by the FeCO stretching and bending motions responding to the fluctuating charges on the iron atom from the motions of the ring modes.

The IR anisotropies of the vibrationally excited transients provide interesting information. The vibrational transition dipole is directed along the CO bond axis so the pump–probe anisotropy yields the orientational correlation function for that bond. In the case of the heme protein the value of 0.36(0.05) for the anisotrophy at 10 ps indicates that the CO direction is fixed inside the protein quite rigidly. The protein undergoes overall rotation with a time constant of 7 ns so this motion does not contribute to the loss of anisotropy at the few picosecond time scale. The results show that the order parameter for the CO bond axis is greater than 0.92 at 10 ps. A systematic study of the decay of such orientational correlation functions for different chemical bonds could provide a unique dynamical picture of a protein.

10.3.2. Bacteriorhodopsin

The application of picosecond transient IR methods to the study of intermediates in the bacteriorhodopsin photocycle has recently been accomplished (43). These results will be summarized here without the backdrop of a detailed review of the subject. Bacteriorhodopsin (BR) serves as a light driven proton pump in the purple membrane of *Halobacterium halobium*. The BR has been studied by a wide range of methods. Its structure has been determined by Henderson et al. (44) to an accuracy of 3.5 Å in the plane of the membrane. The kinetics of the chromophore have been monitored by optical spectroscopic methods throughout the whole time range of the photophysical events of excitation and isomerization on the femtosecond time scale (45). Various IR spectroscopic methods have been used to study the dynamics of protein and the chromophore down to the nanosecond time regime (46). Low-temperature Fourier transform infrared (FTIR) techniques have been employed to study the early response to the isomerization under steady-state conditions, where the motion of the backbone and of the side chains is strongly hindered (46,47). Picosecond time resolved IR studies on BR discussed here concern the BR–K differences. The K structure is formed in a few picoseconds and lasts for microseconds. The BR–K difference spectra of a hydrated purple membrane film display the protein responses in the amide I and II regions and cofactor transitions (43). The kinetics of a few of the bands are shown in Fig. 10.7. We have observed three types of responses: (a) an immediate response of a peptide carbonyl group to a change in the nearby chromophore (appearance of the amide I band at 1660 cm^{-1}); (b) a response involving both the chromophore and protein relaxation (the band at 1556 cm^{-1}, which must relax rapidly at ambient temperatures because it is only seen in the picosecond spectra); (c) a slow nanosecond response, most likely involving a protein relaxation (slow rise of the signal in the carboxylic region). Type (a) direct responses to nearby chromophore changes may be dominant ones. Type (b) responses illustrate the significance of time-resolved studies at ambient temperatures with respect to low-temperature studies. The protonation pathway and consequently the function of BR are critically dependent on the temperature and on very specific dynamic properties of the protein. Both type (b) and type (c) responses may permit detailed structural studies of the origins of the non-Arrhenius-like characteristics of the BR dynamics.

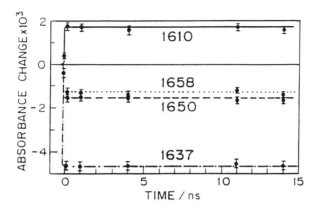

Figure 10.7. Infrared absorbance kinetics following the excitation of BR 570 at 550 nm. Probe as indicated at 1658, 1650, 1637, and 1610 cm^{-1}. [Reproduced with permission from R. Diller, M. Dannone, B. R. Cowen, S. Maiti, R. A. Bogomolni, and R. M. Hochstrasser, *Biochemistry*, **1992**, *31*, 5567. Copyright © (1992) American Chemical Society.]

At present, detailed structural conclusions from either time-resolved or stationary IR studies are still limited by the lack of assignments, especially from amide I and amide II bands that concern the protein backbone. By the use of structural information and isotopic labeling genetic techniques, it should be possible to gain more insight into the proton pump path as the current research demonstrates, and also on the concomitant dynamics of other residues and the backbone responding to the primary process.

Very recent results have been obtained in collaboration with Diller et al. (48) using a Ti:sapphire based system. The new Ti:sapphire laser technology enables us to generate light pulses of few hundred femtoseconds width and a spectral energy density of about 10 nJ nm^{-1} (@540 nm). This is enough to efficiently pump chromoproteins. We have used this technique together with a CW IR upconversion scheme to extend our picosecond IR studies on BR to the femtosecond regime. We have already observed the very fast bleach of the IR band at 1637 cm^{-1}, the C=NH stretch vibration of BR$_{570}$. We monitored the protein response, probed by amide I contributions at around 1660 cm^{-1} and the dynamics of the J and K intermediate, respectively, as displayed by the absorption features between 1630 and 1580 cm^{-1}.

10.3.3.   IR Studies of Photosynthetic Reaction Centers

In bacterial photosynthesis the primary event is a photoinduced charge separation that takes place in a transmembrane pigment–protein complex known as the reaction center (RC), see Fig. 10.8. The central part consists of two protein subunits L and M, related by approximate C_2 symmetry, interfaced at a bacteriochlorophyll dimer (P) (49). Absorption of a near-IR photon (~ 870 nm) leads to the formation of the singlet excited state P*. Approximately 3 ps later the electron arrives at a

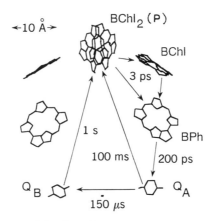

Figure 10.8. Schematic diagram of the disposition of cofactors and electron-transfer rates in the reaction centers. The process initiates with excitation of P, culminating in reduction of quinone B (Q$_B$).

bacteriopheophytin (BPh$_L$) on the L side to form the positively charged dimer state P$^+$, with a quantum efficiency of greater than 0.98 (50). Despite the near C_2 symmetry, there is negligible electron transfer to BPh$_M$. The role of an accessory bacteriochlorophyll that lies between the special pair and the bacteriopheophytin has been the subject of debate and is thought to serve as either a superexchange or sequential intermediate, or both (23b). The structure and dynamics of the cofactors and protein units may be critical to understand the rate and efficiency of the electron-transfer process.

10.3.3A. Vibrational Spectroscopy. These time-resolved data rest against a large backdrop of static FTIR difference spectra (24). We observed many vibrational bands in the region between 1650 and 1800 cm^{-1} at delays corresponding to many tens of picoseconds, corresponding to times at which there are no subtleties with regard to vibrational coherence. Infrared difference spectra were first recorded at 70 and 600 ps (20). The differences between these spectra were small and revealed that most of the spectral change occurred earlier, and primarily in the cofactors. No major structural change in the peptide backbone was observed. Samples prepared as dry films exhibited different responses than solution samples, especially in the amide I region, suggesting the importance of intraprotein waters.

On the other hand, interpretation of the vibrational dynamics on the subpicosecond time scales requires consideration of the coherence as discussed above. One very interesting example occurs when probing at 1682 cm^{-1}, which from Fig. 10.9 is seen to be at the location of the main bleached transition. The 1682-cm^{-1} band is due to a depleted P (ground-state) transition, specifically from the C=O stretch of the 9-ketocarbonyl group. Figure 10.10 clearly shows that the bleaching signal does not appear immediately folowing the light absorption (the pulse width used in this experiment was ~ 300 fs). Instead the bleach takes about 3 ps to appear, which is the electron-transfer time. How can this be explained? The explanation

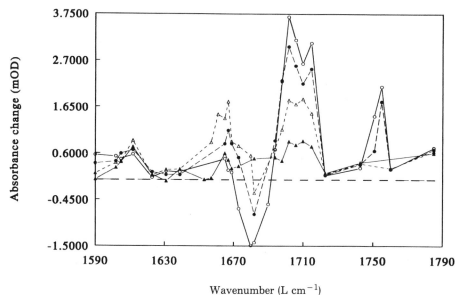

Figure 10.9. Room temperature IR difference spectra of Rhodobacter sphaeroides RCs, after photoexcitation at 870 nm at four representative time delays of lps (filled triangles), 2.5 ps (open triangles), 5 ps (filled circles), and 35 ps (open circles). The dashed line represents zero absorbance change. Uncertainty is less than or equal to 0.25 mOD. [After Ref. 20 with permission.]

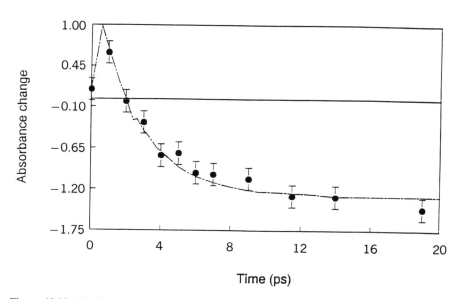

Figure 10.10. The kinetics of the primary bleach of the RC at 1682 cm^{-1} corresponding to the 9-ketocarbonyl group of the special pair [After Ref. 20 with permission.]

involves the transfer of vibrational coherence in the photolysis step. Although the naive expectation is a population response according to Eq. (10.5), in fact, the pump–pulse creates vibrational coherence in the P* state, which implies that a polarization at the IR frequency still exists after the ground-state P population has gone. In this case, the signal is the sum of Eq. (10.5) and (10.7) with equal frequencies and absorption cross sections in the ground and excited states, so that $\Omega_{ee'}-\gamma_e \approx \Omega_{gg'}$ yielding a difference IR spectrum proportional to

$$-\text{Re}\left\{\left(\frac{\sigma}{\Omega_{gg'}}\right) \int_{-\infty}^{t} dt' \, I(t') \left[1 - e^{-\gamma_e(t-t')}\right]\right\}$$

where γ_e is the rate at which the electronically excited state undergoes electron transfer [γ_e is the γ of Eq. (10.7)]. Thus the spectrum has the same shape at all times but its time evolution does not follow that of the bleached population. This result is also obtained directly by adding $\rho_{ee'}$ and $\rho_{gg'}$ from Eq. (10.20c) and (10.20d), assuming fairly narrow band excitation. When excited off-resonance or when the ground- and excited-state frequencies are not close enough, the signal shows oscillations at the appropriate detuning frequency. These and other experiments we have carried out clearly indicate that the vibrational absorbances of P* for the IR active modes investigated are very similar to those of P. The major changes in these modes occur only after the formation of the ions.

The 1702-cm^{-1} region, corresponding to the 9-ketocarbonyl, is displayed in the three-dimensional spectrum on Fig. 10.11, which shows the emerging IR absorption at a changed frequency. Coherent effects are evident in the data. In particular, the peak (1702 cm^{-1}) signal grows in apparently more slowly than do the wings on the new transition. This may be explained by Eq. (10.7). A fit yields a single homogeneously broadened band with 9-cm^{-1} hwhm shifting from 1682 to 1702 cm^{-1} with a time constant of 2.5 ps. *No* parameter is separately varied to fit the kinetics at the different frequencies. Although different kinetics are observed at each wavelength this does not imply inhomogeneity but instead signifies coherent coupling.

In reaction centers these analytic forms are also of critical value in determining the existence of species that may be short lived, for example, the accessory bacteriochorophyll monomer anion. Analysis of that cofactor's IR spectrum near 1665 cm^{-1} during the first few picoseconds following preparation of P*, specifically vibrations of 9-ketocarbonyl, in conjunction with anisotropy studies, are consistent with the presence of the anion, although definitive IR pump IR probe experiments in this spectral region remain to be made in order to settle this issue (20). This situation presents an example of the complementarity of the various approaches to coherent transient IR that were discussed in 10.2.

10.3.3B. Electronic Spectroscopy in the IR. The IR methods now have time resolution and sensitivity comparable to the optical measurements and permit not only studies of transient vibrational states but also of low-energy electronic transitions (51). The nature of electronic structure of the cofactors in RCs, and in

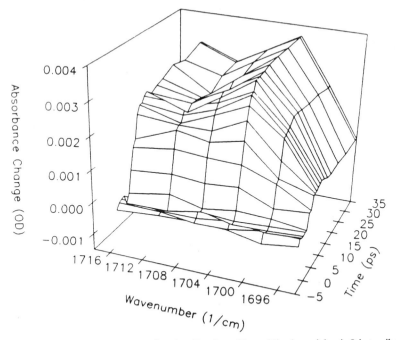

Figure 10.11. Kinetics in the 1702-cm^{-1} region (dominated by oxidized special pair 9-keto vibrational band) shown in a three dimensional (3D) plot. Error bars are less than or equal to 0.2 mOD. Points are joined by straight lines for clarity. [After Ref. 20 with permission.]

particular of the excited state P* and the ionic state P$^+$, must be understood to fully explain the remarkable features of the fast, efficient, and unidirectional electron transfer. The dominant contribution to the P* state undoubtedly comes from the lower (Q_{y-}) of the excitonic states (Q_{y-} and Q_{y+}) that arise from the interaction of the individual Q_y states of the two monomer constituents (52). The P* state may also contain a significant admixture of intradimer charge-transfer states (53–60). Wavefunctions of the electronic states of P obtained from the most recent calculations (61) do show that the P* state can be reasonably well represented by a linear combination of the excitonic (Q_{y-} and Q_{y+}) and charge-transfer ($P_L^+P_M^-$ and $P_L^-P_M^+$) states.

Verification of the extent to which the zero-order excitonic and charge-transfer states are mixed requires direct probing of the low-energy transitions of P*. Unfortunately, there are many overlapping electronic states associated with the multitude of cofactors in the RC making it difficult to obtain unambiguous information about these eigenstates from the ground-state visible–near-infrared (vis–near-IR) spectrum at ambient temperatures. Time-resolved IR methods serve to alleviate this complexity, as shown below. We observed transitions *between* the excited states of P in the 5–6-μm wavelength region. The energies and transition dipole moments of excited-state transitions, which should contain a wealth of information regarding the specific nature of the individual excited states, have been only recently explored.

The transient IR spectra following sample excitation were obtained by upconverting a CW CO laser with an optical pulse from a Ti:sapphire regenerative amplified system, as described previously (51,52). Figure 10.12(a) shows the absorption spectrum in the 1700–2000-cm^{-1} region of reaction centers at 30 ps. Two features are apparent. A relatively narrow band at lower frequency is assigned to the 9-ketocarbonyls of the special pair cation, P$^+$ (62,63). The feature to higher energy is much broader and does not exhibit the typical line structure of vibrational transitions. The width is more typical of an electronic transition. Figure 10.12(b) and (c) shows the absorption kinetics monitored at the extremes of the displayed, broad spectral feature, at 1743 and 1960 cm^{-1}, respectively. The 1743-cm^{-1} kinetics are fit to a sum of two contributions—an instantaneous rise with an exponential decay (67%) and an exponential rise with no decay (33%). The first process is the signature of a P* species that appears immediately on absorption of the pump pulse. The second feature is associated with the first step of electron transfer, taking about 3 ps. The 1960-cm^{-1} kinetics are fit to the same model but the amplitudes of 34 ± 10% and 66 ± 10%, for the instantaneous and slower processes. Figure 10.12(d) shows the plot of the two component amplitudes for the different probe frequencies. Systematically, the instantaneous component is smaller and appears to be somewhat shifted to lower frequency. We assigned these two components to distinct electronic transitions of the reaction center's special pair.

Absorbance changes measured parallel ($I_\parallel$) and perpendicular ($I_\perp$) to the pump polarization were transformed into anisotropies, and then to a mean-square cosine of an angle between pumped and probed transition dipoles, as described earlier. At 1960 cm^{-1}, r(0.5 ps) = 0.185, or $<\theta>$ = 37.0 ± 5°, and r(30 ps) = 0.17 or $<\theta>$ = 38.0 ± 5°. At 1723 cm^{-1} r(0.5 ps) = 0.19 ± 0.02 or $<\theta>$ = 36.5 ± 5°, and r(30 ps) = 0.17 or $<\theta>$ = 38.3 ± 5°.

The broad transitions observed in the region 2000–1700 cm^{-1} are assigned as electronic transitions. Vibrational transitions would be expected to be narrow (~ 10 cm^{-1}). Furthermore, there are no vibrational fundamentals expected in this spectral region (64). On the other hand, identical chromophores that are spatially nearby to one another, as found in the special pair of reaction centers, should exhibit states of an excitonic nature. All previous electronic structure calculations predict the existence of such states (58,61,65) and they are inferred from observations of nearby transitions in optical spectra (66). We have employed a model based on charge-transfer exciton theory (67,68) that can explain the existence of these low-frequency electronic transitions and also account for the observed anisotropies. The spectrum at many picoseconds is consistent with the recent results of Breton et al. (69) on the static IR spectra of P$^+$ (see below).

Infrared spectra are expected to reveal critical properties of the electronic structure of the excited states of reaction centers. In order to provide a simple theoretical format for describing our results we have employed the charge-transfer exciton model with parameters that yield wave functions and energies quite close to those obtained from more sophisticated theoretical calculations on the dimer (61). The

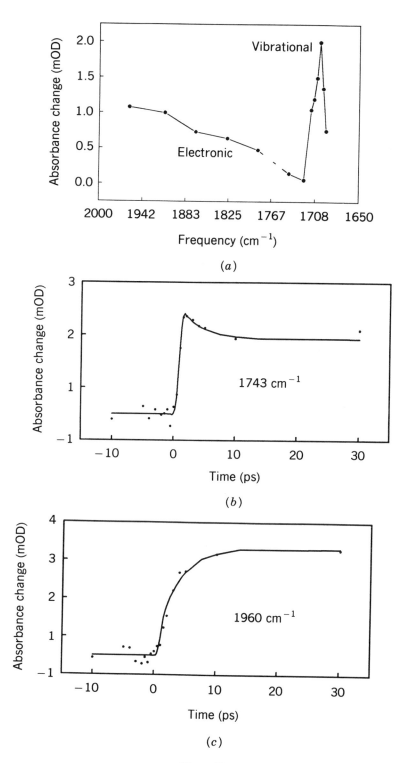

Figure 10.12a–c

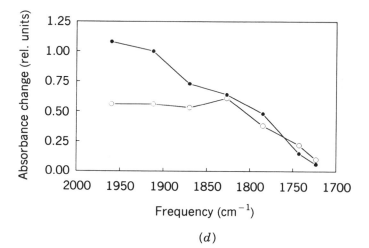

(d)

Figure 10.12. Observations of electronic transitions of the reaction center in the IR. (*a*) Absorption spectrum of the RC between 1700 and 2000 cm^{-1} 30 ps after photoexcitation. The dashed portion is known to contain another vibrational band arising from the 10a ester of P. (*b*) Absorption kinetics of the RC at 1743 cm^{-1}. (*c*) Absorption kinetics at 1960 cm^{-1}. (*d*) Relative contribution of the P* (open circles) and P$^+$ (closed circles) components to the electronic absorption band in the region 1720–1960 cm^{-1}. [After Ref. 52.]

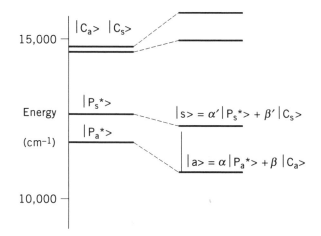

Figure 10.13. Energy level diagram for RC excited states yielded by charge-transfer exciton theory. Left: zero-order states; Right: eigenstates. The arrow indicates the proposed nature of the transition. [After. Reference 52.]

zero order and fully mixed states are shown in Fig. 10.13. The exchange of electronic excitations between the components of P will give rise to exciton states $|P_s^*\rangle$ and $|P_a^*\rangle$, which are symmetric (s) or antisymmetric (a) combinations of $|P_L^*P_M\rangle$ and $|P_LP_M^*\rangle$. If the monomer undergoes a change of dipole moment $\Delta\mu_L$ or $\Delta\mu_M$ on electronic excitation, this transition is allowed with transition dipole $\Delta\mu = 1/2$ $(\Delta F\mu_L - \Delta\mu_M)$ and is polarized perpendicular to the assumed C_2 symmetry axis that interchanges the monomers. Charge-transfer exciton theory for a dimer considers the states to be mixtures of $|P_s^*\rangle$, $|P_a^*\rangle$, $|C_a\rangle$, and $|C_s\rangle$, where $|C_s\rangle$ and $|C_a\rangle$ are symmetric and antisymmetric combinations of $|P_L^+P_M^-\rangle$ and $|P_L^-P_M^+\rangle$. *A robust feature of the charge-transfer exciton model is that* **strong** *interexciton transitions can now occur as a result of the large transition dipole* $\langle C_a|\mu|C_s\rangle$. The transitions between $|C_a\rangle$ and $|C_s\rangle$ are again polarized perpendicular to the assumed C_2 axis and have magnitude μ_{LM}, which is the electronic charge times the vector joining the charge centers of P_L and P_M. When $|P_a^*\rangle$ is expected to be the lowest energy exciton state, the lowest charge-transfer exciton state is a mixture of $|P_a^*\rangle$ and $|C_a\rangle$, as indicated in Fig. 10.13. Transitions from this state to $|C_s\rangle$ become strongly allowed when the mixing becomes significant and are present even when the monomer dipole change is small or zero.

Two additional features are important in the reaction centers: one involves the coupling to the other cofactors, particularly the accessory bacteriochlorophyll (BChl), and the other concerns the intrinsic asymmetry of the L and M branches. The delocalization of the P* excitation onto BChl* has been considered by a number of investigators. For example, Parsons and Warshel (58,59) and Zerner and coworkers (70) both predict five states within 3000 cm^{-1} of the lowest P* excitation, and considerable BChl* character is present in some of these states. The inequivalence of the P_L and P_M sites should give rise to an asymmetric site splitting in the spectra. The asymmetry of the two branches of the RC is predicted by the structural work on reaction centers (49). Calculations based on this structural asymmetry predict nonidentical electron dynamics along the two branches (70–74). With inclusion of asymmetry, the excitation and electron exchange described above then take place between inequivalent bacteriochlorophylls in P* so that the probability to find the excitation or the electron (or hole) should not be the same on each half of the dimer in any of the states of P: Within the framework of the charge-transfer exciton model there is now an intrinsic charge asymmetry associated with each state. The existence of this charge asymmetry is made evident by Stark effect measurements on reaction centers (53–57).

After the electron is transferred to BPh, a positive hole will be left on the dimer. The hole can exchange between halves of the dimer giving rise to the hole-transfer states $|h_a\rangle$ and $|h_s\rangle$. These will also be modified by the asymmetry.

10.3.3C. Structural Basis for the Anisotropy.

A significant result of these studies was that the values of $r(0)$ obtained when probing P* or P$^+$ transitions are the same within experimental error. If the excited states of the dimer with ground state $|P_LP_M\rangle$ are those of a charge-transfer exciton, it can be shown that there should in fact be a relationship between the two measurements. For simplicity, the dimer states that are symmetric (s) and antisymmetric (a) to interchange of the monomers by virtue of the approximate C_2 axis are written as:

$$|s\rangle = \alpha' \, |P_s^*\rangle + \beta' \, |\, C_s\rangle \qquad |a\rangle = \alpha \, |P_a^*\rangle + \beta \, |C_a\rangle \qquad (10.21)$$

These are the two lowest energy states shown in Fig. 10.13. Another robust feature of the CT–exciton model is that $|\alpha| \approx |\alpha'|$ and $|\beta| \approx |\beta'|$. From published calculations values of β are found in the range $0.06–0.4$ $(60,62,75)$. The ground-state (870-nm) absorption then involves the transition dipole moment $\langle P_L P_M | \mu | a \rangle \equiv \mu_{oa}$. The probed transition dipole of the P* absorption is proposed to be $\mu_{as} = \langle a | \mu | s \rangle$. If the P^+ absorption is assumed to correspond to the hole-transfer transition (54) from a state that incorporates an antisymmetric to one that contains a symmetric combination of $|P_L^+ P_M\rangle$ and $|P_L P_M^+\rangle$, the probed transition dipole would be $\gamma \mu_{LM} = \gamma \langle P_L^+ P_M^- | \mu | P_L^+ P_M^- \rangle$, where γ is a parameter determined by the asymmetric site splitting of the hole-transfer states and by mixing of these states with other cofactor excitations. In this model the values of the measured projection for the excitonic, $\cos \theta_{ex}$, and hole transfer, $\cos \theta_h$, probes are given by

$$\cos \theta_{ex} = \hat{\mu}_{oa} \cdot \hat{\mu}_{as} \qquad (10.22a)$$

$$\cos \theta_h = \hat{\mu}_{oa} \cdot \hat{\mu}_{LM} \qquad (10.22b)$$

where the transition dipole moments are written as unit vectors ($\hat{\mu}$). The dipole μ_{LM} is given by $e\mathbf{R}_{LM}$, where $\mathbf{R}_{LM}$ is the vector separation of the centroids of charge in the monomers. From the X-ray coordinates it is found that $|\mathbf{R}_{LM}| = 7.66$ Å so $|\mu_{LM}| = e|\mathbf{R}_{LM}| = 36.7$ D. We require this for evaluating the moment μ_{as} since from Eq. (10.4), assuming $\alpha = \alpha'$ and $\beta = \beta'$:

$$\mu_{as} = \alpha^2 \Delta\mu + \beta^2 \, \mu_{LM} \qquad (10.23)$$

The magnitude $\Delta\mu$, which is obtained from one-half the vector difference between the permanent dipole changes on P_L and P_M on excitation, was found to be about 1.8 D from Stark effect measurements on the monomeric bacteriochlorophyll (57). When β^2 is sufficiently large (i.e., greater than ~ 0.1), the first term in Eq. (10.23) can be neglected in comparison with the other. Then both $\cos \theta_{ex}$ and $\cos \theta_h$ reduce to the same form, namely,

$$\cos \theta = \hat{\mu}_{oa} \cdot \hat{\mathbf{R}}_{LM} \qquad (10.24)$$

Both measurements are mainly determined by the projection of the Q_y^- transition dipole onto the line joining the centers of P_L and P_M monomers. This angle can be calculated from the X-ray coordinates (49) and the known direction of the y axis in a monomer frame, and employing the Q_y transition polarization according to Parson and Warshel (58), the angle is then found to be $\langle \theta \rangle = 33°$. By using the observed anisotropies, we find that the ensemble average angles are $38° \pm 5°$ for the P* probe and $39° \pm 5°$ for the hole-transfer probe transition. The agreement with the prediction is quite good.

Incorporating an asymmetric P* state leads to the difference between the dipole moment of the Q_y^- and the ground states exhibiting a component perpendicular ot the pseudo C_2-axis. This difference vector is easily seen to lie nearly parallel to $\hat{R}_{LM}$, with modest asymmetry. *Thus, the results of anisotropy experiments in which P is pumped at 870 nm and that probe the P* exciton transition* **or** *the hole-transfer transition* **or** *the change in dipole moment via an oriented field (43) can all be seen to be mainly determined by the same angle—that between a monomer y axis and the line joining the centers of the monomers.*

10.3.3D. Assignments. There is theoretical as well as experimental evidence for low-frequency transitions based on electronically excited states of the reaction center. Although none of the theoretical papers describing the excited states of P has discussed the properties of the spectrum arising from a population in the lowest energy exciton-like state, which is presumably Q_y^-, one can utilize the published wave functions for the first few calculated excited states of P and *estimate* the properties of transitions *between* them. The results of such estimates is shown in Table 10.1. We note here that these calculations are based on the RC from *Rhodopseudomonas viridis*, while our experimental results were obtained from the RC of *Rhodobacter sphaeroides*.

Table 1. Comparison of Theoretical Results with Experimental Observations

Author	Transition	Parentage of Transition	Energy (cm^{-1})	Dipole Moment (De)	$<\cos^2\theta>$
Parson and Warshel (60)	1st exc → 2nd exc	$\|Pa>$ + $\|Ca>$ → $\|Bs>$ + $\|Ps>$	1345	2.2	0.86
	1st exc → 4th exc	$\|Pa>$ + $\|Ca>$ → $\|Ps>$ + $\|Cs>$	1878	2.7	0.86
Zerner and co-workers (62)	Qyl → Qy2	$\| Pa>$ + $\| Ca>$ → $\| Ps>$ + $\| Cs>$	1559	2.7	0.79
	Qyl → Qy3	$\| Pa>$ + $\| Ca>$ → $\| Cs>$ + $\| Ps>$	3209	15.4	0.69
Scherer and Fischer (75)	1st exc → 2nd exc	$\| Pa>$ + $\| Ca>$ → $\| Bm^*>$ + $\| Ps>$	1774	5.0	0.82
	1st exc → 3rd exc	$\| Pa>$ + $\| Ca>$ → $\| Bm^*>$ + $\| Ps>$ + $\| Cs>$	1875	7.1	0.68 ←
	1st exc → 4th exc	$\| Pa>$ + $\| Ca>$ → $\| Bl^*>$ + $\| Ps>$ + $\| Cs>$	2082	8.2	0.68 ←
	1st exc → 6th exc	$\| Pa>$ + $\| Ca >$ → $\| Cs>$ + $\| Pl^*>$	2485	18.3	0.73
Experiment (52)		$\| a>$ → $\| s>$	1700	≥ 1.0	0.63

Experimentally, we observe a transition that onsets around 1700 cm^{-1}, reaches a peak extinction coefficient of $\varepsilon \sim 500$ L/(mol^{-1}cm^{-1}), is polarized in the $\hat{\mathbf{R}}_{LM}$ direction, and exhibits $\mu^2 > 1$ D. The theoretical energy predictions presumably refer to Franck–Condon maxima and should be thus somewhat larger than given by the zero–zero frequency. Also shown in Table 10.1 is the dominant parentage of the calculated level in terms of the basis states of the CT exciton model used to guide the discussion of this chapter. The calculated states include not only the Q_y and CT states of P, but also contributions from BChl and BPh. So we have calculated the dipole directions of those transitions that are effectively polarized perpendicular to the approximate C_2 axis. If the P* absorption we observe is the lowest energy transition of P*, then our results would confirm those calculations that place a state that is a mixture of $|P_s\rangle$ and $|C_s\rangle$ in the region 1720 cm^{-1} (onset). The transitions that are in reasonable agreement with the observations are marked by an arrow. It is clear that the spectrum of P*, in the range 5–>1 μm, should contain many other transitions of great interest to testing and refining theoretical calculations. It should be noted that holeburning experiments locate another transition at ca.1250 cm^{-1} above P* (75) at 4.2 K. Could this be the location of the time O–O transition of P* in ambient temperature solutions also?

10.4. FUTURE DIRECTIONS

Coherent IR spectroscopy has great promise, born of bond-by-bond detail available from vibrational IR, to observe local events that happen to take place on time scales shorter than vibrational dephasing. In addition to the examples previously mentioned, a number of bright research areas are presenting themselves. For example, energy harvesting systems, mimicking in part natural systems, transfer energy or electrons primarily via electronic excitation, along free energy gradients between states. These energy–electron transfers can be very rapid and are sensitive functions of the relative orientations of the chromophores yet the overall energy conversion efficiencies have been thought to be less than might be hoped for. New methods that additionally exploit energy harvesting along vibrational degrees of freedom, perhaps in largely symmetric systems, are desirable. For enhanced efficiency these energy transfers must occur on time scales $<T_1$ for high-frequency modes in the condensed phase.

Ultrafast intramolecular electron-transfer systems are predicted to have rates that are vibrational state dependent. Detailed understanding of these processes, especially regarding energy redistribution among non-Franck Condon active modes or when multiple electronic states are involved, will require coherent IR transient spectroscopy. However, such state dependence is unlikely to be relevant only for electron transfer in small molecules but also in much longer hydrogen-bonded networks. If as one might expect the spacing between constituents in the hydrogen-bond are important, possible empirical approaches include inducing coherent non-equilibrium distributions of hydrogen-bond vibrational states. Additional quantum dynamical effects, such as the role of increased positional certainty of deuterium bonds may be expected.

Another promising area is nanotechnology, where site specific reaction monitoring and control is a critical need. For example, a memory or logic device's function can depend on molecular conformational state change, triggered by visible or IR pulses, which is then read by IR pulses. Just as NMR methods now routinely employ sophisticated pulse sequences to isolate individual nuclear spin resonances in highly congested systems, so coherent IR spectroscopy holds promise for interrogating highly integrated molecular circuits. A somewhat different approach that may prove fruitful in the IR is the study of individual or small numbers of IR chromophores using near-field microscopy. This method can dramatically narrow spectral lines by imaging objects at much less than ($<\lambda/100$) the diffraction limit ($\lambda/2$). The applications are numerous, such as characterizing complex polymeric samples or sites on an individual macromolecule and coupling to IR biosensors at zeptomolar detectivity.

ACKNOWLEDGMENTS

This research was supported by grants from the NSF and NIH. G.C.W. thanks NIH for a postdoctoral fellowship. R.M.H. is indebted to H.P. Trommsdorff and the University of Grenoble where some of this writing was completed.

REFERENCES

1. Elsasser, T., Seilmeir, A., Kaiser, W., Koidl, P., and Brandt, G. *Appl. Phys. Lett.*, **1984**, *44*, 383.
2. Vanherzeele, H. *Opt. Lett.*, **1989**, *44*, 728.
3. Kato, K. *IEEE, J. Quant. Elect.*, **1991**, 27, 1137. Kato, K. and Masutani, M. *Opt. Lett.*, **1992**, *17*, 176.
4. Anton, D. W., Nathel, H., Guthals, D. M., and Clark, J. H. *Rev. Sci. Inst.*, **1987**, *58*, 2064.
5. Iannone, M., Cowen, B. R., Diller, R., Maiti, S., and Hochstrasser, R. M. *Appl. Opt.*, **1991**, *30*, 5247.
6. Raferty, D., Gooding, E., and Hochstrasser, R. M. *Proc. Biomed, Opt. SPIE*, in press.
7. Wynne, K., Reid, G., and Hochstrasser, R. M. *Opt. Lett.* **1994**, *19*, 895.
8. Strickland, D. and Mourou, G. *Opt. Commun.*, **1985**, *56*, 219. Rudd, J. V., Korn, G., Kane, S., Aquier, J., Mourou, G., and Bado, P. *Opt. Lett.* (in press).
9. Asaki, M. T., Huang, C. P., Garvey, D., Zhou, J., Kapeyn, H. C., and Murnane, M. M. *Opt. Lett.*, **1993**, *18*, 977.
10. Hamm, P., Lauterwasser, C., and Zinth, W. *Opt. Lett.*, **1993**, *18*, 1925.
11. Bloembergen, N., "Nonlinear Optics," W. A. Benjamin, New York, 1964.
12. Butcher, P. N. and Cotter, D. "The Elements of Nonlinear Optics," Cambridge Study of Modern Optics, Cambridge University Press, 1991.
13. (a) Shen, Y. R., "The Principles of Nonlinear Optics," Wiley, New York, 1984; (b) Mukamel, S. "Principles of Nonlinear Optical Spectroscopy," Oxford University Press, New York, 1995.
14. Yee, T. K. and Gustaffson, T. K. *Phys. Rev. A*, **1978**, *18*, 1597.
15. Druet, S. A. J., Taran, J. P. E., and Borde, C. J. *Physique*, **1979**, *40*, 819.
16. Dick, B., Hochstrasser, R. M., and Trommsdorff, "Resonant Molecular Optics" in Nonlinear Optical Properties of Organic Molecules and Crystals, D. Chemla (Ed.). Academic, New York, 1986.

17. Dick, B. and Hochstrasser, R. M. *J. Chem. Phys.*, **1983**, *78*, 3398.

18. Allen, L. D. and Eberley, J. H., "Optical Resonance and Two Level Atoms," Wiley, New York, 1975.

19. Anfinrud, P., Han, C., Lian, T., and Hochstrasser, R. M. *J. Phys. Chem.*, **1990**, *94*, 1180.

19(a). K. Wynne and R. M. Hochstrasser, *Chem. Phys.*, **1995**, *192*.

20. Maiti, S., Walker, G. C., Cowen, B., Dutton, P. L., and Hochstrasser, "Ultrafast Infrared Spectroscopy of the Reaction Center Dynamics," *Proc. Natl. Acad. Sci. USA 1994, 91*, 10360.

21. (a) Anfinrud, P. A. and Hochstrasser, R. M. *Proc. Natl. Acad. Sci. USA*, **1989**, *86*, 8387; see also Ultrafast Phenomena VIII, Harris, C. B., Ippen, E. P., Mourou, G., Zewail, A. (Eds.). **1991**. (b) Moore, J. N., Hansen, P. A., and Hochstrasser, R. M., *Chem. Phys. Lett.* **1987**, *138*, 110. (c) Lian, T., Locke, R. B., Kholodenko, Y., and Hochstrasser, R. M., in "Energy Flow from Solute to Solvent Probed by IR Spectroscopy" *J. Phys. Chem.* **1994**, *98*, 11648.

22. Maiti, S., Cowen, B., Dutton, P. L., and Hochstrasser, R. M. *Proc. Natl. Acad. Sci. USA*, **1993**, 5247.

23. (a) Locke, R. B., Diller, R., and Hochstrasser, R. M., "Ultrafast Infrared Spectroscopy and Protein Dynamics," in Biomolecular Spectroscopy, Part B., Vol. 20, Clark, R. H. J. and Hester, R. E., Wiley, 1993. (b) Hochstrasser, R. M. and Johnson, C. K., "Biological Processes Studied by Ultrafast Laser Techniques," in Ultrashort Laser Pulses and Applications, Kaiser, W. (Ed.) Springer-Verlag, New York, 1988.

24. Mäntele, W., Nabedryk, E., Tavitian, B. A., Kreutz, W., and Breton, J. *Proc. Natl. Acad. Sci. USA*, **1988**, *85*, 8468. Mäntele, W., in "The Photosynthetic Reaction Center," Deisenhofer, J. and Norris, J. (Eds.) Academic, New York, Vol. 2, 239, 1993. Buchanan, S., Michel, H., and Gerwert, K., in "Reaction Center of Photosynthetic Bacteria," (Ed.) Michel-Beyerle, Springer, Berlin-Heidelberg, 75, 1990. Morita, E. H., Hayashi, H., and Tasumi, M. *Chem. Lett.*, **1991**, *10*, 1853.

25. (a) Locke, B., Lian, T., and Hochstrasser, R. M. *Chem. Phys.*, **1991**, *158*, 409. (b) Locke, B., Lian, T., and Hochstrasser, R. M. *Chem. Phys.*, **1995**, *190*, 155.

26. Lian, T., Locke, R. B., Kitagawa, T., Nagai, M. and Hochstrasser, R. M. *Biochemistry* **1993**, *32*, 5809.

27. Ippen, E. P., Shank, C. V., and Bergman, A., *Chem. Phys. Lett.* **1976**, *38*, 611–614.

28. Mokhtari, A., Chebira, A., and Chesnoy, J., *J. Opt. Soc. Am. B*, **1990**, *7*, 1551–1557.

29. Mokhtari, A., Fini, L., and Chesnoy, J., *J. Chem. Phys.* **1987**, *87*, 3429–3435.

30. Migus, A., Antonetti, A., Etchepare, J., Hulin, D., and Orszag, A., *J. Opt. Soc. Am.* **1985**, B2. 584–694.

31. Robl, T. and Seilmeier, A., *Chem. Phys. Lett.* **1988**, *147*, 544–550.

32. Ben-Amotz, D. and Harris, C. B., *J. Chem. Phys.*, **1987**, *86*, 4856–4870.

33. Gottfried, N. H., Seilmeier, A., and Kaiser, W., *Chem. Phys. Lett.*, **1984**, *111*, 326.

34. Wondrazek, F., Seilmeier, A., and Kaiser, W., *Chem. Phys. Lett.* **1984**, *104*, 121.

35. Henry, E. R., Eaton, W. R., and Hochstrasser, R. M., *Proc. Natl. Acad. Sci. USA* **1986**, *83*, 8922–8986.

36. Owrutsky, J., Diller, R., Iannone, M., Cowen, B. R., Maiti, S., Li, M., Sarisky, M., Kim, Y.-R., Locke, B. R., Lian, T., and Hochstrasser, R. M., *Proc. SPIE-Int. Soc. Opt. Eng.* **1991**, *1599*, 52.

37. Owrutsky, J., Li, M., Culver, J. P., Sarisky, M., Yodh, A., and Hochstrasser, R. M., in "Time Resolved Vibrational Spectroscopy VI, A. Lau, F. Siebert and W. Werncke, eds. (Springer-Verlag, 1993) p. 63.

38. Cowen, B. R., Lian, T., Walker, G. C., Maiti, S., Moser, C. C., Locke, B. R., Dutton, P. L., and Hochstrasser, R. M., *Proc. Int. Sym. Photochem. Photobiol., Photomed.*, **1993**.

39. Owrutsky, J. C., Li, M., Culver, J., Yodh, A., and Hochstrasser, R. M., "Vibrational Relaxation of Bound Carbonyls in Myoglobin and Hemoglobin," *J. Phys. Chem.*, in press.

40. Dlott, D. and Fayer, M. D. (private communication).

41. Heilweil, E. J., Cassassa, M. P., Cavanaugh, R. R., and Stephenson, J. C., *Ann. Rev. Phys. Chem.*, **1983**, *40*, 143.

42. Li, X. Y. and Spiro, T. G., *J. Am. Chem. Soc.*, **1988**, *110*, 6024.

43. Diller, R., Iannone, M., Cowen, B. R., Maiti, S., Bogomolni, R. A., and Hochstrasser, R. M. *Biochemistry*, **1992**, *31*, 5567.

44. Henderson, R., Baldwin, J. M., Ceska, T. A., Zernlin, F., Beckman, E., and Downing, K. H. *J. Mol. Biol.*, **1990**, *213*, 899.

45. Mathies, R. A., Brito Cruz, C. H., Pollard, W. T., and Shank, C. V. *Science*, **1988**, *240*, 777. Zinth, W., Dobler, J., Dressler, K., and Kaiser, W. in "Ultrafast Phenomena VI" Yajima, T., Yoshihara, K., Harris, C. B., and Shionoya, S. (Eds.). Springer, Berlin, **1988**, pp. 581–583.

46. Siebert, F., Mäntele, W., and Kreutz, W. *Can. J. Spectroscopy*, **1981**, *26*, 119. Gerwert, K. and Hess, B. *Microchem. Acta II A/6.04*, **1987**.

47. Braiman, M. S., Bousché, O., and Rothschild, K. J. *Proc. Natl. Acad. Sci. USA*, **1991**, *82*, 388. Bagley, K., Dollinger, G., Eisenstein, L., Singh, A. K., and Zimanyi, L., *Proc. Natl. Acad. Sci. USA*, **1982**, *79*, 4972. Gerwert, K. and Siebert, F. *EMBO*, **1986**, *5*, 805.

48. Diller, R., Walker, G. C., Maiti, S., Cowen, B. R., Pippenger, R. S., Bogomolni, R., and Hochstrasser, R. M. *Biophys. J.* **1994**, *66*, A113.

49. Deisenhofer, J., Epp, O., Miki, K., Huber, R., and Michel, H. *J. Mol. Biol.*, **1984**, *180*, 385. Deisenhofer, J., Epp, O., Miki, K., Huber, R., and Michel, H. *Nature (London)* **1985**, *318*, 618. Michel, H., Epp, O., and Deisenhofer, J., *EMBO J.*, **1986**, *5*, 2445. Deisenhofer, J. and Michel, H., "The Photosynthetic Bacterial Reaction Center; Structure of Dynamics," Breton, J. and Vermiglio, A. (Eds.). NATO ASI Series 149, Plenum Press, New York, 1988, pp. 1–3. Chang, C. H., Tiede, D., Tang, J., Smiyh, U., Noris, J. R., and Shigger, M. *FEBS Lett.*, **1986**, *205*, 82. Yates, T. O., Komiya, H., Chirino, A., Rees, D. C., Allen, J. P., and Feher, G. *Proc. Natl. Acad. Sci. USA*, **1988**, *85*(21), 7993.

50. Breton, J., Martin, J.-L., Petrich, J., Migus, A., and Antonetti, A. *FEBS Lett.*, **1986**, *209*. Martin, J.-L., Breton, J., Hoff, A. J., Migus, A., and Antonetti, A. *Proc. Natl. Acad. Sci. USA*, **1986**, *83*, 957–961. Fleming, G. R., Martin, J.-L., and Breton, J. *Nature (London)*, **1988**, *12*, 190. Kirmaier, C. and Holten, D. *Proc. Natl. Acad. Sci. USA*, **1990**, *87*, 3552. Breton, J., Martin, J.-L., Migus, A., Antonetti, A., and Orszag, A. *Proc. Natl. Acad. Sci. USA*, **1986**, *83*, 5121. Woodbury, N. W., Becker, M., Middendorf, D., and Parson, W. W. *Biochemistry*, **1985**, *24*, 7516. Kirmaier, C. and Holten, D. *FEBS Lett.*, **1988**, *239*, 211. Breton, J., Martin, J.-L., Fleming, G. R., and Lambry, J. C. *Biochemistry*, **1988**, *27*, 8276. Wraight, C. A. and Clayton, R. K. *Biochim. Biophys. Acts*, **1973**, *333*, 246–260.

51. Walker, G., Maiti, S., Cowen, B. R., Moser, C. C., Dutton, P. L., and Hochstrasser, R. M. *J. Phys. Chem.*, **1994**, *98*, 5778.

52. Vermeglio, A. and Paillotin, G. *Biochim. Biophys. Acta.*, **1982**, *681*, 32.

53. DeLeeuv, D., Malley, M., Butterman, G., Okamura, M. Y., and Feher, G. *Biophys. Soc. Abstr.*, **1982**, *31*, 111a. Lösche, M., Feher, G., and Okamura, M. Y., "The Photosynthetic Bacterial Reaction Center—Structure and Dynamics," Breton, J., Vermeglio, T. (Eds.). Plenum Press, New York, 1988, p. 221.

54. Lockhart, D. J. and Boxer, S. G. *Biochemistry*, **1987**, *26*, 664, 2958.

55. Lockhart, D. J. and Boxer, S. G. *Proc. Natl. Acad. Sci. USA*, **1988**, *85*, 107.

56. Lösche, M., Feher, M., and Okamura, M. Y. *Proc. Natl. Acad. Sci. USA*, **1987**, *84*, 7537.

57. Boxer, S. G., Goldstein, R. A., Lockhart, D. J., Middendorf, T. R., and Takiff, L. *J. Phys. Chem.*, **1989**, *93*, 8280.

58. Warshel, A. and Parson, W. W. *J. Am. Chem. Soc.*, **1987**, *109*, 6143.

59. Parson, W. W. and Warshel, A. *J. Am. Chem. Soc.*, **1986**, *131*, 153.

60. Scherer, P. O. J. and Fischer, S. F. *Chem. Phys. Lett.*, **1986**, *131*, 153.

61. Thompson, M. A., Zerner, M. C., and Fajer, J. *J. Phys. Chem.*, **1991**, *95*, 5693.

62. Mäntele, W., Wollenweber, A., Nabedryk, E., and Breton, J. *J. Proc. Natl. Acad. Sci. USA.*, **1988**, *85*, 8468. Maiti, S., Cowen, B. R., Diller, R., Iannone, M., Moser, C. C., Dutton, P. L., and Hochstrasser, R. M. *Proc. Natl. Acad. Sci. USA*, **1993**, *90*, 5247.

63. Maiti, S., Cowen, B. R., Diller, R., Iannone, M., Moser, C. C., Dutton, P. L., and Hochstrasser, R. M. *Proc. Natl. Acad. Sci. USA*, **1993**, *90*, 5247.

64. Lim-Vien, D., Colthup, N. B., Fately, W. G., and Grasselli, J. G. "The Handbook of Infrared and Raman Characteristic Frequencies of Organic Molecules," Academic, New York, 1991.

65. Eccles, J., Honig, B., and Schulten, K. *Biophys. J.*, **1983**, *53*, 137.

66. Breton, J. *Biochim. Biophys. Acta.*, **1985**, *810*, 235. Lyle, P. A., Kolaczkowski, S. V., and Small, C. J. *J. Phys. Chem.*, **1993**, *97*(26), 6924.

67. Michl, J. and Bonačić-Koutecký, V., "Electronic Aspects of Organic Photochemistry," Wiley, New York, 1990.

68. Philpott, S. R., Rice, M. J., Bishop, A. R., and Campbell, D. K. *Phys. Rev. B.*, **1987**, *36*, 1735.

69. Breton, J., Nabedryk, E., and Parson, W. W. *Biochemistry*, **1992**, *31*, 7503.

70. Thompson, M. A. and Zerner, M. C. *J. Am. Chem. Soc.*, **1991**, *113*, 8210.

71. Plato, M., Möbius, K., Michel-Beyerle, M. E., Dixon, M., and Jortner, J. M. *J. Am. Chem. Soc.*, **1988**, *110*, 7279. Gehlen, J. N., Chandler, D., and Marchi, M. *J. Am. Chem. Soc.*, **1993**, *115*, 4178 and references cited therein.

72. Warshel, A., Creighton, S., and Parson, W. W. *J. Phys. Chem.*, **1988**, *92*, 2692.

73. (a) Freisner, R. A. and Won, Y. *Biochim. Biophys. Acta.*, **1989**, *977*, 99. (b) Won, Y. and Freisner, R. A. *Biochim. Biophys. Acta.*, **1989**, *935*, 32.

74. Scherer, P. O. J. and Fischer, S. F. *J. Phys. Chem.*, **1989**, *93*, 1633. Scherer, P. O. J. and Fischer, S. F. *Chem. Phys.*, **1989**, *131*, 115.

75. Tang, D., Kankowoik, R., Hayes, J. M., Piedi, D., and Small, G. J. 22nd Jerusalem Symposium on Quantum Chemistry and Biochemistry: Perspectives in Photosynthesis. J. Jortner and B. Pullman eds., (Kluver Academic Press, Dordrecht, 1990), p. 99.

INDEX